GOLD FORUM on Technology and Practices – 'World Gold '89'

Proceedings of the First Joint International Meeting Between SME and AusIMM, Nov. 5-8, 1989, Reno, Nevada

Editors

Roshan B. Bhappu
Rod J. Harden

Sponsored by

Society for Mining, Metallurgy, and Exploration, Inc.
The Australasian Institute of Mining and Metallurgy

Cooperating Sponsors

Canadian Mineral Processors Division of CIM
The Institute of Metals
The Institution of Mining and Metallurgy
The Mining, Geological & Metallurgical Institute of India
The South African Institute of Mining and Metallurgy

Published by
Society for Mining, Metallurgy, and Exploration, Inc.
Littleton, Colorado • 1989

Printed in the United States of America
by Cushing-Malloy, Inc., Ann Arbor, Michigan

Library of Congress Catalog Card Number 89-62338
ISBN 0-87335-085-5

Preface

In the last two decades, gold mining has become a very important part of the mining industry worldwide. This is especially true for the United States and Australia where gold mining is the major activity in all phases of the industry, including exploration, mining, processing, metallurgical extraction, environmental aspects, and economics. Thousands of professionals in both these countries are currently engaged in exploration, construction, operation, research and development, technology transfer, management, and financing.

It is primarily this creative environment in both Australia and the United States that provided the impetus in the summer of 1987 for organizing the first Society for Mining, Metallurgy, and Exploration (SME) and The Australasian Institute of Mining and Metallurgy (AusIMM) joint conference – Gold Forum on Technology and Practices - 'World Gold '89.' The SME International Advisory Committee, headed by Alfred Weiss, provided early support to this conference to broaden SME's involvement in the international mining arena. Additional worldwide participation and interest was provided by the five Cooperating Sponsors – the Canadian Mineral Processors Division of CIM, The Institute of Metals, The Institution of Mining and Metallurgy, The Mining, Geological & Metallurgical Institute of India, and The South African Institute of Mining and Metallurgy.

The primary aims of this joint SME-AusIMM conference are:

- To share the developing technology of gold exploration, mining, and metallurgy in both the US and Australia.
- To discuss the latest operating practices in the gold industry of those countries.
- To brief participants on the latest research supporting the effective mining and extraction of gold.

For this important and timely conference, Reno, Nevada, was selected as the meeting place and the conference scheduled for November 5-8, 1989. Reno is the center of gold mining activity in the United States and it allows a lively location for technical meetings, social activities, and field trips.

The organization for the conference consisted of appointing respective Organizing Committees, headed by Rod Harden in Australia and Roshan Bhappu in the United States and soliciting papers for technical sessions including Keynote, Geology/Exploration, Mining, Processing, and Finance/Economics. Each session was developed to include papers from the US and Australia as well as contributions from the Cooperating Sponsors as available. Most of the presentations at the conference have been submitted for publication in the conference proceedings volume to be published by SME and available to the participants at the conference. Moreover, field trips, organized by SME and emphasizing the geology, mining, and metallurgy of the Carlin Trend, are scheduled following the conference.

As can be appreciated, the preparation for this joint SME/AusIMM conference was a Herculean task and the credit goes to both the Society and the Institute, their Organizing Committees, Editorial Committees, and their staffs. Especially, the diligent efforts of Darline Daley, Meetings Manager of SME, and Judy Webber, Conventions and Marketing Coordinator of AusIMM, are greatly appreciated.

It is hoped that these initial efforts in organizing this first joint SME/AusIMM conference will be the basis of many other conferences to follow and to promote a continuous dialogue between the mining communities of both countries.

July 18, 1989

Roshan B. Bhappu
Rod J. Harden

Organizing Committees

Society for Mining, Metallurgy, and Exploration

Roshan B. Bhappu, *Chairman*
Mountain States R&D International
Vail, Arizona

Leonard Harris
Newmont Mining Corp.
Denver, Colorado

Laurence D. Hartzog
Coeur d'Alene Mines, Inc.
Coeur d'Alene, Idaho

J. Brent Hiskey
University of Arizona
Tuscon, Arizona

Bruce A. Kennedy
Pelsart Resources N.L.
Jakarta, Indonesia

Donald L. Simpson
Echo Bay Mines, Ltd.
Reno, Nevada

David L. Stevens
Freeport-McMoRan Gold Co.
Reno, Nevada

Milton H. Ward
Freeport-McMoRan Inc.
New Orleans, Louisiana

The Australasian Institute of Mining and Metallurgy

Rod J. Harden, *Chairman*
Fluor Daniel Australia Ltd.
Melbourne, Victoria

David S. Tyrwhitt
Ashton Mining Ltd.
Melbourne, Victoria

Colin E. Palethorpe
BHP Gold Mines
Melbourne, Victoria

John O. Reynolds
Victorian Chamber of Mines
Melbourne, Victoria

K. John Burns
The Australasian Institute of Mining and Metallurgy
Melbourne, Victoria

Alban J. Lynch
University of Queensland
Brisbane, Queensland

Angus M. Robinson
Geological and Mining Museum
Sydney, New South Wales

Don M. Morley
Western Mining Corp. Holdings Ltd.
Melbourne, Victoria

Paul A.R. Odd
Big Bell Mines Pty. Ltd.
Geraldton, Western Australia

Judy M. Webber
The Australasian Institute of Mining and Metallurgy
Melbourne, Victoria

Contents

1

Geology/Exploration I
Geochemistry and Sampling

D. Giles
Cimarron Gold Corp.
Santa Fe, New Mexico

D.S. Tyrwhitt
Ashton Mining Ltd.
Melbourne, Vic., Australia

Chapter 1

ROCK GEOCHEMISTRY IN EPITHERMAL GOLD EXPLORATION: AN EXAMPLE FROM MISIMA ISLAND, PAPUA NEW GUINEA

Donald S. Clarke

Epithermal Mineralisation Research Unit, Geology Department, University of Auckland, New Zealand

Abstract

The Umuna epithermal precious metal deposit has provided the target for a simulated rock geochemistry exploration program, directed at bedrock exposed in the drainage system of eastern Misima Island. A geochemical halo to the deposit is readily recognised in the mafic meta-volcanic country rock. This halo is defined by anomalous results for both major (K, Na, Ca, Mg) and minor (Au, As, Sb, Cu, Pb, Zn, Ba, Rb, Sr, Cs, Bi, Ga, S) element geochemistry, extending further than 1km from the deposit. Such halos are sufficiently extensive to provide adequate exploration targets in the search for epithermal mineralisation, complementing mineralogic and secondary media geochemical investigations.

INTRODUCTION

The Umuna gold-silver deposit is an epithermal precious metal resource presently under development. Study of the deposit has been conducted as the basis of a doctoral research program (Clarke, 1988). The geologic and trace element (neutron activation analysis - NAA) results of this study have been reported elsewhere (Clarke et al., 1989); major and minor geochemical data (X-ray fluorescence - XRF) are combined here with the previous results to evaluate the bedrock response (from mafic meta-volcanic host-rock) at Umuna to a simulated exploration program.

Misima Island lies 625km east of Port Moresby (Fig. 1). It is 40km long and has a maximum width of 9km. Mount Sisa forms the highest point in eastern Misima Island, being 413m above sea level. The Umuna deposit lies west and south of this central watershed.

Lode gold was exploited at Umuna from 1904; total recorded production up to 1951 was 6,804 kg Au and 20,412 kg Ag (Macnab, 1976). Following the Second World War, a number of efforts were made to re-evaluate the Umuna Au mineralisation, and associated weak Cu mineralisation to the northeast of Mount Sisa (the Mararoa porphyry-Cu prospect; Patterson and Small, 1971). Exploration interest turned in 1975 to the open pit potential of the Umuna deposit (Macnab, 1976); in 1986 Placer (PNG) Pty Ltd announced an in situ proved and probable reserve of 62.1 million tonnes of ore containing 1.35 grams per tonne Au and 20.0 grams per tonne Ag at a 0.7 grams per tonne Au equivalent cut-off grade (Wilson, 1987).

Keyser (1961) presented a detailed description of the geologic setting of the Umuna deposit. The regional geology of Misima Island was subsequently remapped (Williamson and Rogerson, 1983) to complement the increased exploration activity.

GEOLOGY OF EASTERN MISIMA

Stratigraphy: The Tertiary basement rocks of eastern Misima Island consist of the Sisa Association, and the Boiou intrusive suite; these rocks host the Umuna deposit (Fig. 2). The Sisa Association is metamorphosed to

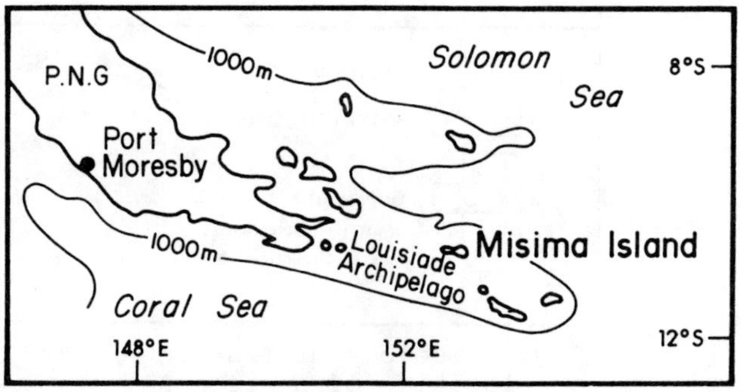

Fig.1: Locality map, Misima Island.

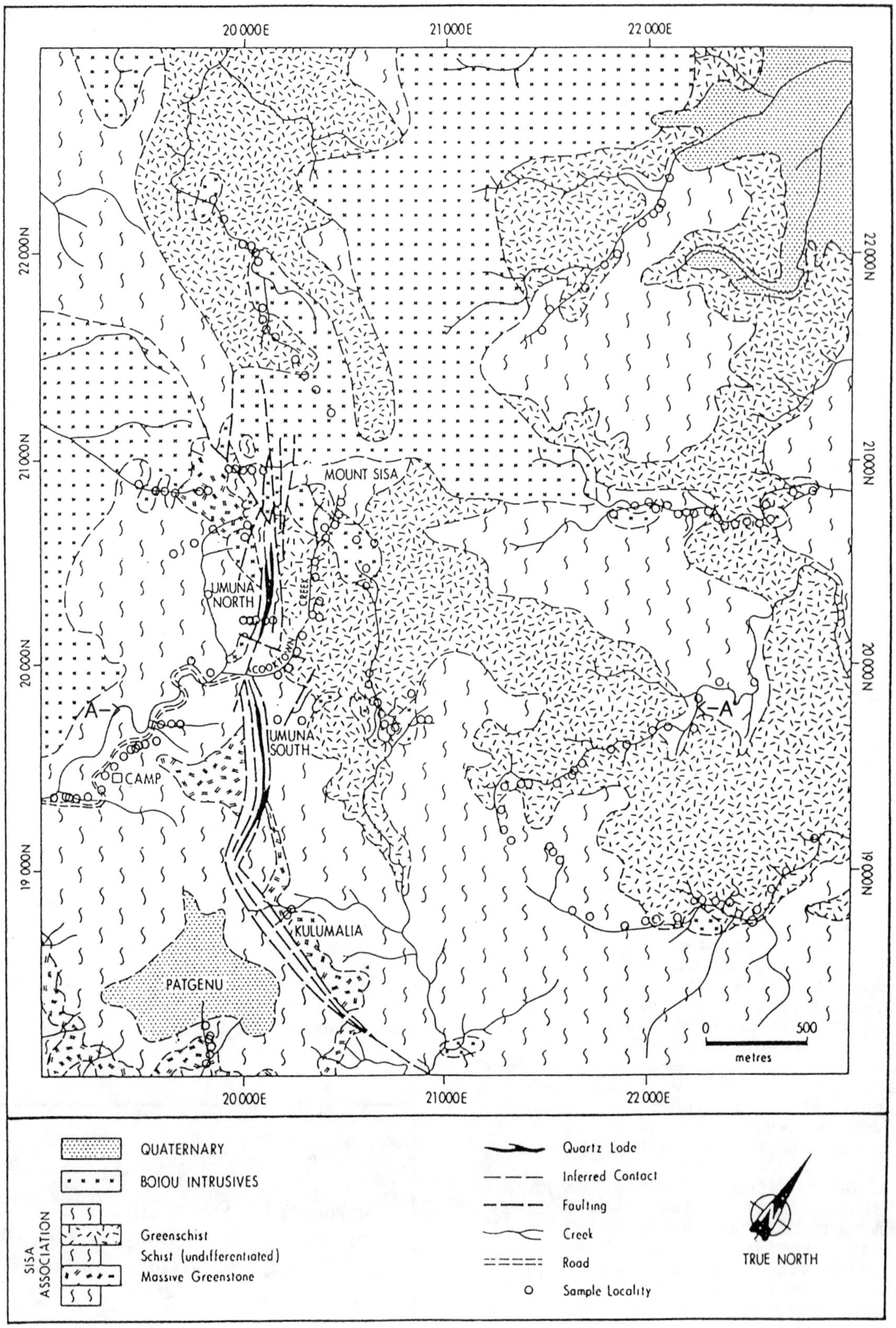

Fig.2: Geology of eastern Misima Island, with bedrock sample localities and location of section AA' (Fig.3). Modified from Williamson and Rogerson (1983) and Clarke (1988).

greenschist facies. It is an apparently conformable sequence of meta-basic rocks (greenstone and greenschist) within dominantly psammopelitic meta-sedimentary rocks (schists).

Bedrock exposure of basement rocks in the study area is mostly confined to creek exposures. Generally these outcrops are partially oxidized but range from very weathered to very fresh. This basement is only partly obscured by a unconformable cover sequence.

Regional Structure: Penetrative deformation accompanying regional metamorphism is preserved within the Sisa Association as lineation, foliation, and crenulation fabrics. Isoclinal to chevron folding is developed locally, but in general regional east-west elongate domal folding gives only shallow dips to strata (Fig. 3). The Boiou intrusive rocks postdate this deformation. High angle faulting occurred during and after emplacement of these porphyries.

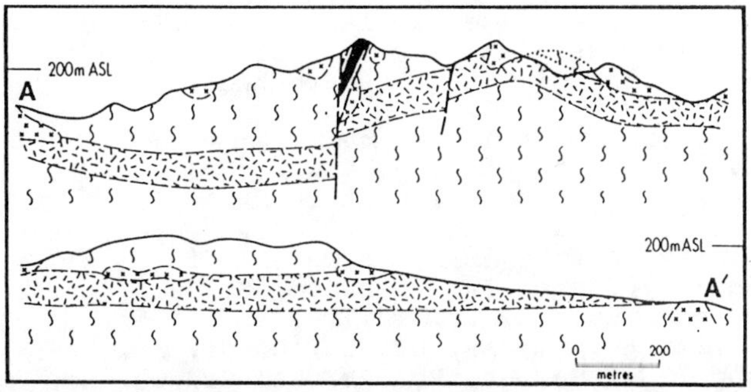

Fig.3: Simplified geological cross section of eastern Misima Island. Location of section is shown on Fig.2; legend as for Fig.2 (from Clarke, 1988).

Rock Types: The schists exhibit well developed slaty parting, commonly with a crenulation cleavage, subparallel to layering. Dark grey carbonaceous and light grey micaceous quartz-plagioclase-sericite-chlorite schists are dominant in the sequence.

Massive greenstone is the most common mafic rock type on eastern Misima island, and appears to form one continuous unit. It is generally fine-grained and felted to lineated to weakly foliated, and can be well fractured; definite volcanic structures (e.g. pillow forms, flow top breccias) have not been recognized. The characteristic mineral assemblage in greenstone is amphibole (hornblende), quartz, plagioclase, and chlorite, with variable prominence locally of clinozoisite, epidote, sericite, biotite, magnetite, and carbonate. Plagioclase and sparse magnetite commonly form porphyroblastic phases.

Foliated to schistose mafic rocks, greenschist, are found intercalated with schist; very thin units appear gradational to the more chloritic of the schist, and may represent mafic tuffs. In practice there is little to differentiate between strongly sheared greenstone and apparently bedded greenschist.

A discontinuous thin horizon of carbonate rocks at the upper contact to the greenstone consists of massive, fine to medium grained, pale grey marble and pale green calc-silicate rocks. Banded to brecciated quartz - ferromanganese oxide lenses are also present at the greenstone upper contact to the north and east of the Umuna deposit. These lenses are very poorly exposed at surface due to their incompetent nature.

The Boiou intrusive rocks appear relatively unmetamorphosed. The dioritic to granodioritic porphyries form small to large irregular stocks and sills, and intrude all units of the Sisa Association. Minor mafic dykes of uncertain affinity also occur. The characteristic mineral assemblage for the porphyries is quartz, plagioclase, hornblende, and biotite.

UMUNA DEPOSIT

Structure: The Umuna Fault Zone is a 100-300m wide north-west trending high angle fracture system recognized over a 3km strike length. The internal structure is complex, and many phases of faulting in a variety of orientations are present (Fig. 2). The dominant west dipping, north-north-west striking orientation of the fault zone is the major control to mineralized veining and brecciation forming the Umuna deposit, with the contacts to porphyry bodies providing a secondary influence. The zone of faulting narrows and changes orientation to the south (Kulumalia area, Fig. 2). To the north (the Mount Sisa area) the width of the fault zone is maintained, but the intensity of mineralized fracturing and brecciation gradually decreases.

Mineralisation: The Umuna Lode is the most distinctive feature of mineralisation within the Umuna deposit. It forms a thick spine of complex quartz veining in the upper 120m section of the deposit. The lode is commonly enveloped by a stockwork of laminated quartz veins. The lode structure at depth is marked by a zone of schist and porphyry breccia with angular fragments in a sandy quartz - rock flour matrix. An extensive stockwork of quartz-limonite-clay veinlets is exposed at surface immediately adjacent to the Umuna Fault Zone in the central - eastern portion of the deposit.

In the Mount Sisa area, the main lode and branch lodes exhibit a gradation from quartz to calcite. Calcite in the lodes forms a coarse-grained matrix (cement) to breccia clasts. A broad zone of anastomosing breccia veins and laminated fractures, filled with a massive black, earthy matrix of amorphous ferromanganese oxide, gritty quartz and brown clay, lies to the east of the main lode.

In the most easterly part of the Mount Sisa area, light brown siliceous veins with minor pyrite are prominent, and fractures are stained with limonite and rare malachite. This style of mineralisation is apparently continuous with that described by Patterson and Small (1971) as forming the weak Cu mineralisation to the northeast.

Alteration: The Umuna deposit is generally deeply oxidized to below sea level from a surface elevation of 150-300m. The depth of oxidation throughout the deposit is apparently controlled by the intensity of deformation. Wilson (1987) has attributed the oxidation to weathering although it may, at least in part, result from the mineralizing process (Williamson and Rogerson, 1983). Epithermal alteration is only partially obscured by this oxidation.

A zone of argillic alteration (smectite, sericite) with local silicification envelops the Umuna Lode in the central part of the Umuna deposit. To this assemblage is added chlorite in the remainder of the Umuna Fault Zone. Phyllic alteration (sericite, chlorite) peripheral to the fault structure grades outwards to propylitic alteration (more chlorite and epidote). The lateral extent of the latter alteration zone is not clear, largely due to surficial weathering and the poor resolution of propylitic alteration and greenschist metamorphic assemblages in the host-rocks. Zonation is not evident for alteration in the Mount Sisa portion of the deposit; here phyllic to propylitic alteration dominates.

Peripheral Gold Occurrences: Past gold mining to the northeast of the Mount Sisa area has been directed at quartz vein networks in massive magnetite - garnet lodes, and massive silica lodes. Both of these two lode types appear to represent skarn replacement of carbonate rock. Associated with, and east of, these workings greenstone and porphyry commonly contains fine irregular limonitic fracturing over a wide area with weak Cu mineralisation.

A number of minor occurrences of low grade gold mineralisation lie within one kilometre east of the Umuna deposit. The gold is associated with small lenses of quartz-ferromanganese oxide within and on the contacts of porphyry intrusions, and with fracture-fill of similar material in porphyry, schist and greenstone. These occurrences have not been exposed below the base of surficial weathering; the best grades appear to be associated with wire gold developed very near surface.

GEOCHEMISTRY

Sampling and Analyses

Surface sampling of host-rock exposed up to 3.5 km from the Umuna deposit was performed as a series of creek and road-cutting traverses. The location of samples and their relationship to the general geology is shown in Fig. 2. Drillhole samples of altered wall-rocks and lode material from one section across each of the Umuna North and Mount Sisa areas were also analysed for comparison. These samples were generated as part of the economic evaluation of the deposit, and are 2 metre bulked whole core composites. The lode geochemistry is included here as a reference for the host-rock data.

Major and minor elements were analysed by XRF at the Univ. of NSW, precious metals and trace element by NAA at Becquerel Laboratories, Sydney. Detection limits for the analyses were as follows: Au - 0.005ppm; Ir - 0.02ppm; Sb, Sm, Sc - 0.2ppm; Eu, La, Lu, Ta, Th, Yb - 0.5ppm; Cs, Hf - 1ppm; As, Br, Ce, U, Ni, Sr, Zr - 2ppm; Cu, Ga, Nb, Pb, Rb, Zn, Cr, Co, Mo, Se, Ag, W, Y - 4ppm; Sn - 15ppm; Ba - 20ppm; major oxides and S - 0.02% (Clarke, 1988). Precision was generally better than +/-10%.

Lode geochemistry

Analytical results for representative drillhole samples of lode material from the Umuna deposit are included in Table 1. Similar results have been presented by Williamson and Rogerson (1983), including "significant" Bi contents.

The elements considered to characterize the Umuna mineralisation are listed in Table 2. Such a list is very subjective, depending on what is successfully analysed for and what is considered a high concentration. The reported geochemical signatures of a number of other Southwest Pacific epithermal deposits, largely based on similar subjective assessments (including reported vein mineralogy), are summarized in Table 2. Of the elements ascribed to the other deposits, only Te and Se do not appear to be "significant" components in the Umuna veining (Tl and Hg have not been assessed).

XRF DATA:

No	SiO_2	TiO_2	Al_2O_3	Fe_2O_3	MnO	MgO	CaO	Na_2O	K_2O	P_2O_5	S
145	96.30	0.02	1.07	0.77	0.01	0.10	0.23	0.7	0.12	0.06	0.18
303	69.98	0.50	14.16	4.54	0.05	1.29	0.04	<0.1	0.04	0.28	0.03
374	11.38	0.02	0.34	0.75	0.59	0.21	46.68	<0.1	<0.02	0.02	0.06
373	66.64	0.02	3.38	8.76	1.66	3.87	0.40	0.6	<0.02	0.83	0.02

No	Cu	Zn	Ni	Ba	Nb	Ga	Zr	Sb	Mo	Pb	Sr	Rb	Y	Sn
145	444	1475	10	4273	<4	27	9	<15	<8	1164	20	15	15	28
303	407	2669	35	3333	4	93	135	<15	18	4569	124	45	24	27
374	195	5326	2	<20	<4	39	24	<15	<8	2132	356	<4	7	<15
373	2558	21550	49	529	<4	781	85	31	11	39888	20	<4	63	<15

NAA DATA:

No	Sb	As	Ba	Br	Ce	Cs	Cr	Co	Eu	Au	HFf	Ir	%Fe	La
145	6.1	16	2400	<2	<2	<1	280	2	<0.5	6.3	<1	<0.02	0.52	1
303	7.7	35	2500	<2	39	<1	120	10	1.3	15.4	3	<0.02	3.00	23
374	0.8	4	<100	<2	16	<1	11	<1	2.6	1.2	<1	<0.02	0.41	11
373	6.3	120	<100	<2	15	<1	100	6	3.8	58.5	<1	<0.02	5.84	18

No	Lu	Mo	Rb	Sm	Sc	Se	Ag	Ta	Th	W	U	Yb
145	<0.5	<5	<20	0.3	0.3	<5	300	<0.5	<0.5	7	<2	<0.5
303	<0.5	19	55	3.8	9.3	<5	150	<0.5	5.6	6	<2	0.8
374	<0.5	<5	<20	2.2	0.7	<5	15	<0.5	<0.5	<5	<2	<0.5
373	<0.5	<5	<20	3.2	3.1	<5	48	<0.5	<0.5	<5	<2	1.1

Table 1: Analyses of lode from the Umuna deposit, for four representative drillhole samples. Oxides, S and Fe as percent, minor elements as ppm. No.145 - massive quartz, Umuna Lode; No.303 - porphyry sandy quartz breccia, Umuna Lode; No.374 - massive carbonate lode, Mount Sisa; No.373 - ferromanganese oxide vein, Mount Sisa.

Bedrock Geochemistry

In this study a contrast is made between the geochemical signature of *wallrock* samples from the Mount Sisa (sampling area 1) and Umuna North (area 2) mineralisation, and samples from the predicted *halo* to the northeast (area 5) and east (area 8) of the deposit, with that of *regional* samples taken some distance, greater than 1200m, from the deposit (area 9).

Only data for sampling within greenstone are presented here. These results are displayed in Fig. 4 and 5 as a series of multiple box-and-whisker plots (Tukey, 1977) for the elements analysed. Such plots are a simple method of presenting the distribution information required to compare two or more population categories (in this case sampling areas). For each element, each area is nominated a position on the X axis with element concentration scaled on the Y axis. A central box is drawn above each area category with the upper and lower sides of the box incorporating the middle 50% of the data values, between the upper and lower quartiles, for that area. Lines (whiskers) extend out from the box to the minimum and maximum of the data, except when these points are within 1.5 times the inter-quartile range. Such outliers are plotted as single points. Central tendency of the data is indicated by a line across the box at the median value.

Elements showing the most consistent enrichment in both the deposit wallrock (area 1 and 2; Fig. 4 and 5) and halo samples (area 8) relative to regional greenstone samples (area 9) are K, Rb, Ba, As, Ga and the "ore" elements: Au, Sb, Zn, Cu, Pb and S. To this list can be added Ag (at economic grades in all wallrock samples), Mo (a generally inadequate analytical sensitivity preventing any examination of regional levels) and Bi (based on a small number of unreported AAS analyses). Cs and Cr are similarly enriched in most halo samples (areas 5 and 8); these samples also contain the highest concentrations of Cu, Co, K and Rb. The concentrations of other elements in halo samples are similar to, or slightly lower than, regional values, except for Mg, Ca, Na, and Sr which tend to be depleted.

Si and Th are enriched in the Umuna area wallrock samples (area 2), and Rb, K,

DEPOSIT	LODE GEOCHEMISTRY
PHILIPPINES	Au,Ag,As,Sb,Cu,Pb,Zn, Se,Hg,Te,Bi,Mo,W,Cd,Tl
BABELTHUAP ISLAND Palau	Au,Ag,As,Cu,Pb,Zn,Te, Hg,Sb
EMPEROR GOLDMINE Fiji	Au,Ag,As,Sb,Cu,Pb,Zn, Ni,Hg
GUNUNG LIMBUNG W. Java, Indonesia	Ba,As,Pb,Zn
HAURAKI GOLDFIELD North Is., N.Z.	Au,Ag,As,Sb,Cu,Pb,Zn, Se,Hg,Te,Bi,Mo,Mn,Ba
KABANG PROSPECT Ambitle Is, P.N.G.	Au,As,Sb
LADOLAM DEPOSIT Lihir Is., P.N.G.	Au,Ag,As,Sb,Cu,Pb,Zn, Te,Mo
MANGANI MINE W. Sumatra, Indonesia	Au,Ag,As,Sb,Cu,Pb,Zn, Mn,Se,Sn,Bi,B
MOROBE GOLDFIELD Morobe Prov, P.N.G.	Au,Ag,As,Cu,Pb,Zn,Cd, Co,Mn,Ni,Cr
UMUNA DEPOSIT Misima Is, P.N.G.	Au,Ag,As,Sb,Cu,Pb,Zn, Ba,Bi,Sn,Ga,Mo,Mn,Cr

Table 2: Geochemistry of lode material for Southwest Pacific epithermal gold deposits, as summarized by Clarke and Govett (1988) and determined for the Umuna deposit in this study.

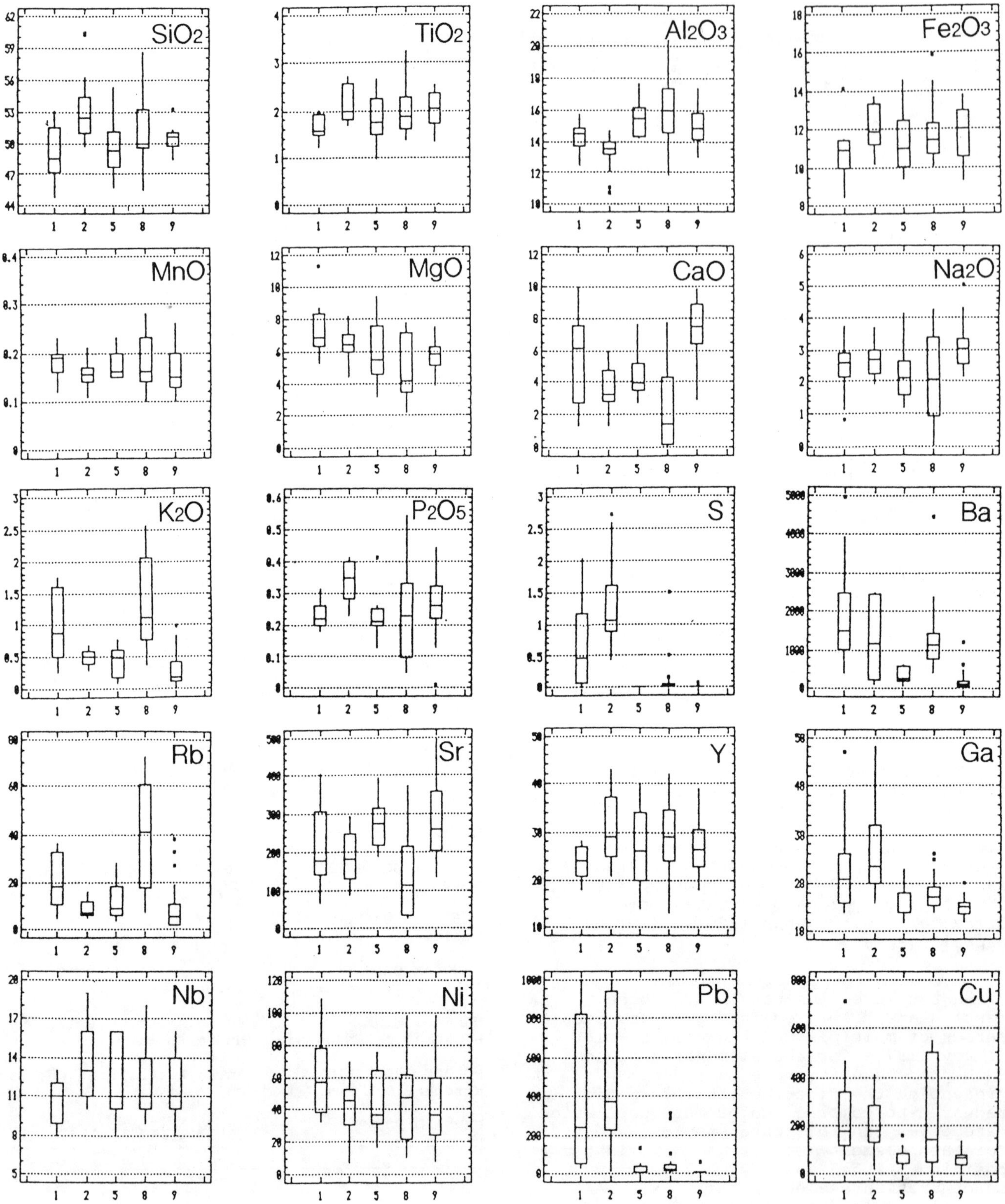

Fig.4: Multiple box-and-whisker plots for analyses of greenstone bedrock samples; from Mount Sisa wallrock (1), Umuna North wallrock (2), northeastern halo (5), eastern halo (8) and regional area (9). Analyses by XRF; results in percent for oxides and S, ppm for others.

Mg, Mn, Cs, Cr and Ni are more elevated at Mount Sisa (area 1). Sc, Co, Al, Ca, Na, and Sr are depleted in both areas of the deposit.

In a routine exploration rock geochemistry survey (as simulated by the sampling program in this study), the position of the target is, of course, not known. A common method of evaluating such data is to plot the results on plan, and use symbols denoting subdivisions of the distribution of results (content plans) to indicate anomalous areas. The content of Ba and Ca in the greenstone samples has been plotted in plan using this technique (Fig. 6). The plots show depletion for Ca and enrichment for Ba for most samples within the "halo" areas east and northeast of Umuna.

DISCUSSION

The anomalous geochemistry of the few Southwest Pacific epithermal deposits for which a geochemical halo has been reported

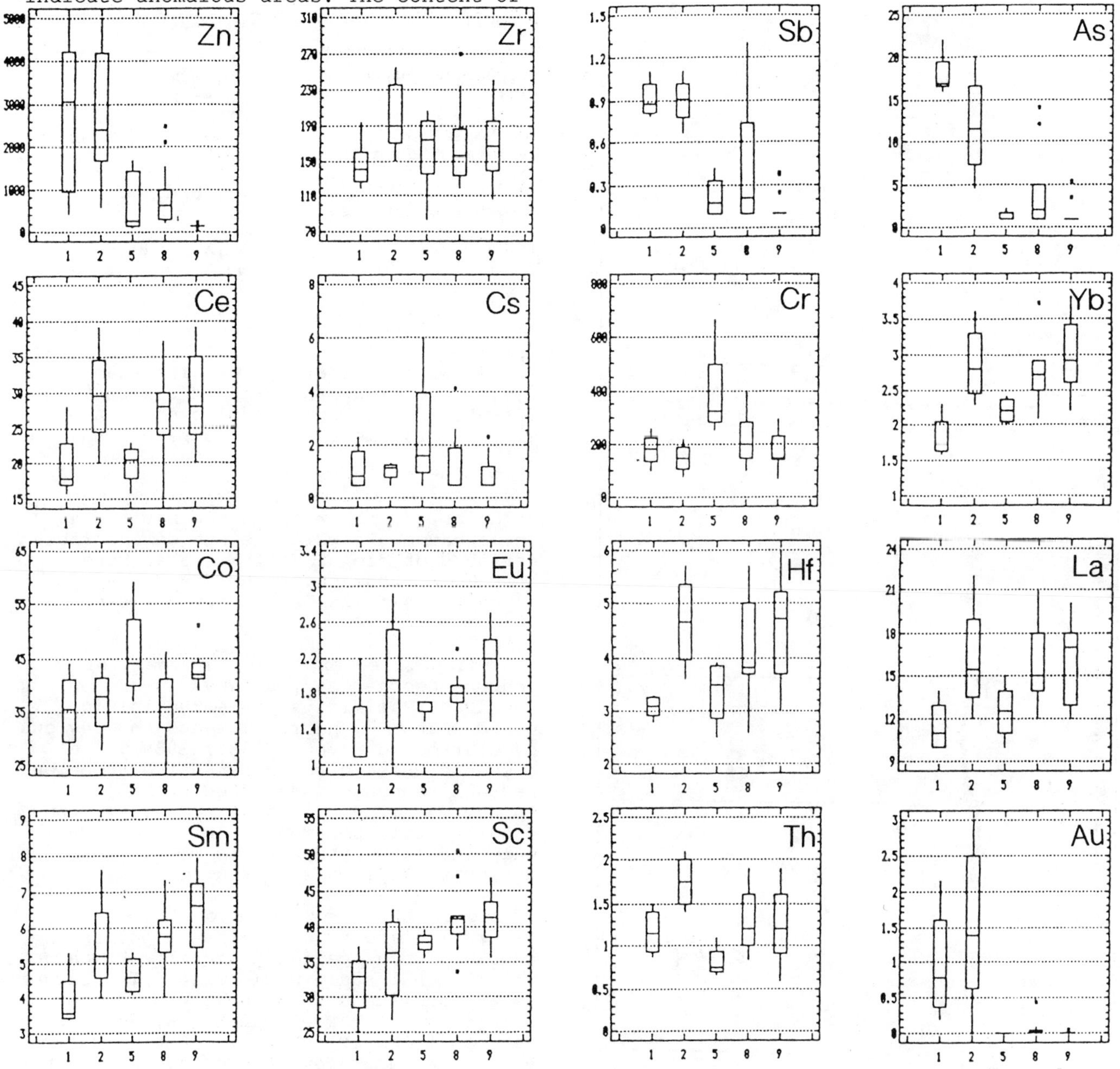

Fig.5: Multiple box-and-whisker plots for analyses of greenstone bedrock samples; from Mount Sisa wallrock (1), Umuna North wallrock (2), northeastern halo (5), eastern halo (8) and regional area (9). Analyses by XRF for Zn and Zr, NAA for others; results in ppm.

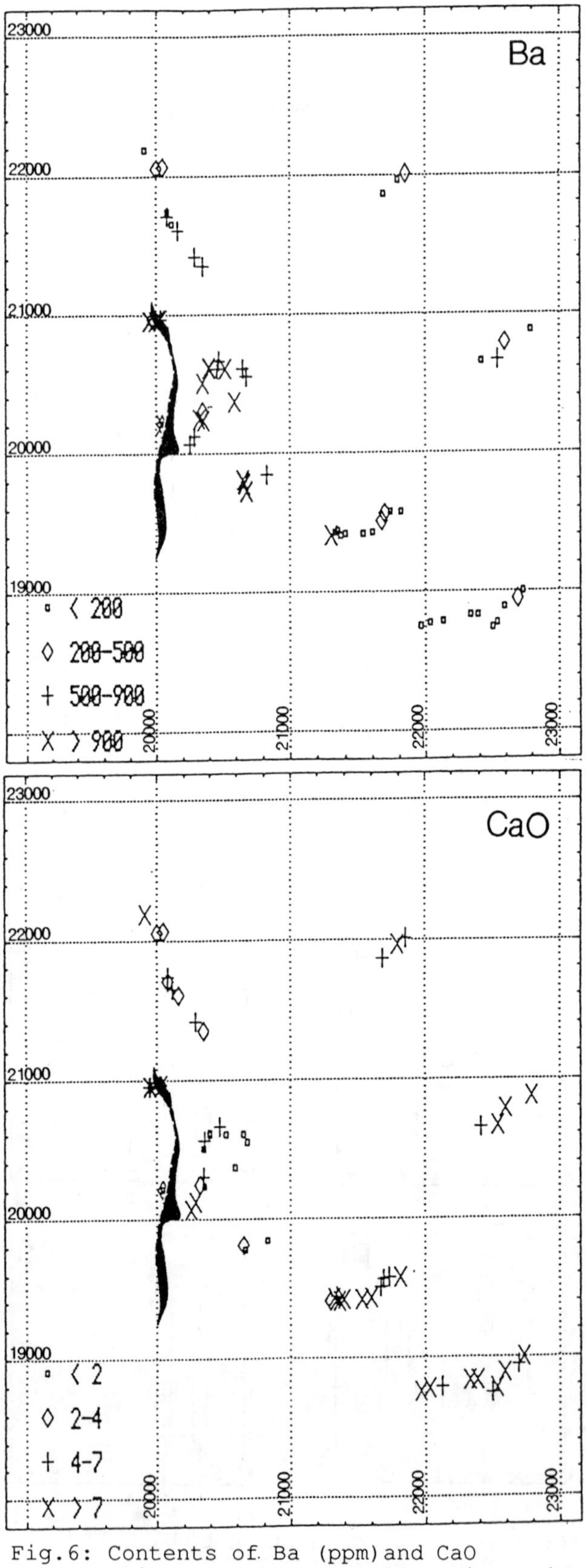

Fig.6: Contents of Ba (ppm) and CaO (percent) in greenstone samples. Mine grid as in Fig.2. Umuna deposit shown in black.

are listed in Table 3 (as reviewed by Clarke and Govett, 1989), together with the generalized situation for the Umuna deposit. The number of elements shown as forming halos to the other deposits is relatively restricted. It is probable that many more elements would be reported as forming halos, if more elements were examined at adequate detection limits; as at Umuna, these probably would include most of the elements "enriched" in the vein material of these deposits (cf Table 2) plus a number of the major elements.

DEPOSIT	PERIPHERAL HALO
EMPEROR GOLDMINE Fiji	+ Au,Hg,K - Ca,Mg,Fe
GUNUNG LIMBUNG W. Java, Indonesia	+ K,Ba,Fe,Cu,Pb,Zn - Ca,Na
MOROBE GOLDFIELD Morobe Prov, P.N.G.	+ Ag,Pb,Zn,Mn
UMUNA DEPOSIT Misima Is, P.N.G.	+ Au,As,Sb,Mo,Cu,Pb,Zn + Ba,Rb,Cs,K,Ga,Mn,Bi - Na,Ca,Mg

Table 3: Anomalous bedrock geochemistry as halos to Southwest Pacific epithermal deposits, as summarized by Clarke and Govett (1988) and determined for the Umuna deposit in this study. Halos shown are either enrichments (+) or depletions (-).

Of the elements that are reported to form halos to the other Southwest Pacific deposits, all show similar behaviour at Umuna except Fe which is not demonstrably enriched or depleted. It is also noted that Mn enrichment around the Umuna deposit (as might be predicted) is not a general feature.

A simplified exploration model for bedrock geochemistry surveys in the Southwest Pacific is presented in Fig. 7, based on the Umuna results and the reviews by Clarke and Govett (1988; 1989). The elements which give an extensive anomalous bedrock response associated with epithermal mineralisation will vary from area to area; the model presented here only indicates those which might be sought during an exploration program.

Regionally pervasive propylitic alteration around epithermal deposits should form a spatially significant target for regional-scale exploration. The target is a significant zone anomalous in the more mobile major elements - K, Na, Ca, Mg (and to a lesser extent Si, Fe and Mn). Associated anomalous concentrations of the characteristic pathfinder elements (Ag, Sb, As, Hg, Mn), as well as the base metals, may equally define spatially extensive anomalies, and indicate that the hydrothermal system was potentially mineralizing. Geochemically anomalous rocks at Umuna occur at least 1.2km to the east of the Umuna Fault Zone. The size of the geochemical halo to Umuna is certainly of the order expected from analogy with active geothermal systems (Henley and Ellis, 1983; Clarke and Govett, 1989).

The geochemical signatures for Ba, Sr and Rb at Umuna (predictably) mimics that of major elements defining the extensive propylitic alteration around the deposit. Commonly, the major elements are not available in routine analytical packages; in any exploration programme where it is suspected that "alteration geochemistry" anomalies may be more extensive than "ore element geochemistry", lithophile minor elements such as Ba, Sr and Rb may readily take the part of the major elements in anomaly definition.

For prospect-scale exploration for epithermal mineralisation within an extensive zone of alteration, the most prospective site for mineralisation will be associated with the focus of the geothermal activity (Silberman and Berger, 1985). The target for this focus should be

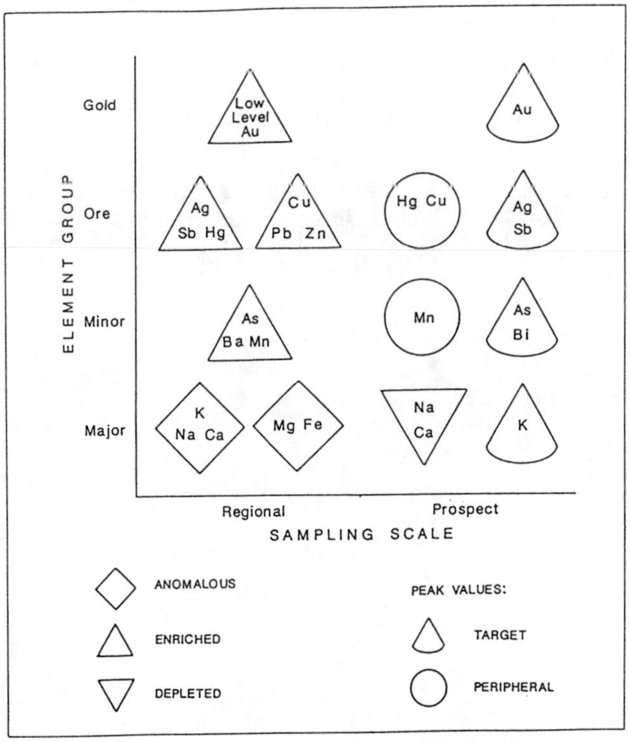

Fig.7: Characteristic rock geochemistry responses to be sought for regional or prospect scale epithermal exploration. "Anomalous" - either enriched or depleted relative to background; "Target" - peak analytical values centred over mineralisation; "Peripheral" - peak values offset from centre of mineralisation.

provided by an internal zonation of the alteration, indicating the most intense, highest temperature alteration. Peak levels of As, Sb, Bi, and the precious metals also may define a target. Peripheral alteration zones may have anomalous Hg and Mn and, locally, base metals (Clarke and Govett, 1989). It is likely that at this scale Au will be the best indicator of Au mineralisation.

The highest values of Cu in bedrock at Umuna occur in greenstone to the east of the Mount Sisa area, rather than coincident with the main mineralisation. Williamson and Rogerson (1983) considered this weak Cu mineralisation and the associated "skarns" to predate the Umuna precious metal deposition. As a preferred alternative to the superposition of different hydrothermal systems, it seems likely that the different styles of mineralisation represent different stages in an evolving system. At Umuna it may be that major Cu deposition occurred prior to focussing of ore deposition within the Umuna fault zone structure.

The other style of peripheral mineralisation on eastern Misima Island is the stratiform type that forms minor gold occurrences to the east of Umuna. That these occurrences lie within the recognized geochemical halo to the Umuna deposit supports the interpretation that they were all formed by the same mineralizing event (Clarke et al., 1989).

This study has emphasised the potential of rock geochemistry in epithermal Au exploration. The results of this study are also pertinent to other exploration techniques. Investigation of the secondary media geochemical response of the Umuna deposit did not form a part of this study; however, the value of understanding primary dispersion patterns to the design of appropriate drainage or soil surveys is obvious, although it appears to be given little consideration in the literature. Also, rock geochemistry is an obvious complement to mineralogical studies, the latter not always being amenable to routine quantitative documentation.

ACKNOWLEDGEMENTS

This study was supported by Placer Pacific Pty Ltd and is presented with their permission. Staff at the Dept. of Applied Geology, Univ. of NSW, and at Becquerel Laboratories, Sydney, assisted with analyses.

REFERENCES

Clarke, D.S., 1988, "Geology and Rock Geochemistry of Epithermal Gold-Silver Mineralisation, Misima Island, Papua New Guinea," PhD Thesis, University of New South Wales, Sydney, 358 pp.

Clarke, D.S. and Govett, G.J.S., 1988, "Rock Geochemistry in the Exploration for Southwest Pacific Epithermal Gold Deposits (Abs.)," Proceedings of the Bicentennial Gold 88 Conference, Geol. Soc. Aust. Abstracts, No. 23, Vol. 1, pp 369.

Clarke, D.S. and Govett, G.J.S., 1989, "Southwest Pacific Epithermal Gold: a Rock Geochemistry Perspective," Epithermal Gold Mineralisation of the Circum-Pacific: Geology, Geochemistry, Origin and Exploration, J.W. Hedenquist, N.C. White, and G. Siddeley, eds., J. Geochem. Explor. Special Volume (in press).

Clarke, D.S., Lewis, R. and Waldron, H.M., 1989, "Geology and Trace Element Geochemistry of the Umuna Gold-Silver Deposit, Misima Island, Papua New Guinea," Epithermal Gold Mineralisation of the Circum-Pacific: Geology, Geochemistry, Origin and Exploration, J.W. Hedenquist, N.C. White, and G. Siddeley, eds., J. Geochem. Explor. Special Volume (in press).

Henley, R.W. and Ellis, A.J., 1983, "Geothermal Systems, Ancient and Modern: A Geochemical Review," Earth-Sci. Review., Vol. 19, pp 1-50.

Keyser, F. de, 1961, "Misima Island - Geology and Gold Mineralisation," Bur. Miner. Resour. Aust. Rept., No. 57, 17 pp.

Macnab, R.P., 1976, "The Open-pit Lode Gold Mining Potential of Misima Island, Papua New Guinea," Meneses Explorations Pty Ltd, 47 pp.

Patterson, K.R., and Small, A.K., 1971, "Final Report, Misima Island, Territory of Papua and New Guinea," Noranda Aust. Ltd Rep., 169 pp.

Silberman, M.L., and Berger, B.R., 1985, "Relationships of Trace Element Patterns to Alteration and Morphology in Epithermal Precious-Metal Deposits," Geology and Geochemistry of Epithermal Systems, B.R. Berger and P.M. Bethke, eds., Soc. Econ. Geol. Reviews in Economic Geology, Vol. 2, pp. 203-232.

Tukey, J.W., 1977, Exploratory Data Analysis, Reading, Addison-Wesley, Mass., 506 pp.

Williamson, A., and Rogerson, R., 1983, "Geology and Mineralisation of Misima Island," Geol. Surv. PNG. Rep., 83/2, 171 pp.

Wilson, G.I., 1987, "Misima Project Update," Papua New Guinea Mineral Development Symposium, Australas. Inst. Min. Metall., Madang, p.85-88.

Chapter 2

GOLD EXPLORATION IN THE TERTIARY PALAEODRAINAGE SYSTEMS OF WESTERN AUSTRALIA.

E.L.SMYTH [1] & A.BUTTON [2]

[1] BHP-Utah Minerals International, P.O. Box 619, Hawthorn. Australia. 3122
[2] Ashton Mining, 441 St Kilda Rd, Melbourne, Australia. 3001

ABSTRACT

The apparent absence of significant Tertiary auriferous palaeoplacers in the Yilgarn Block of Western Australia can be explained by the dissolution of the placer gold into the highly saline, acid, oxidising ground waters which currently occur in upper portions of the palaeochannels proximal to the gold sources. The dissolved gold has then been redeposited in redox fronts further downstream. The placer gold is preserved in more distal portions where the ground waters are reducing.

Exploration for these placers can be effective using the gravity contrasts that exist between the Tertiary sediments and the Archaean bedrock and SIROTEM which can detect the highest salinity zones which occur in the deepest parts of the channels. Final assessment of the gold content of the sediments must be determined by drilling.

INTRODUCTION

Up until 1977 alluvial gold deposits in Tertiary and Recent river gravels in Australia had yielded a third of Australia's 6,000 tonnes (193 million oz.) of gold production (Woodall,1979). The majority of this alluvial gold was produced in south eastern Australia with the Ballarat field alone having produced 330 tonnes (10.6 million oz.). Woodall (1979) also pointed out that alluvial gold deposits are virtually absent over the Archaean terrain (dominantly the Yilgarn Block of Western Australia),and despite 2,200 tonnes (70 million oz.) of lode gold production, recorded production from related alluvials has only been 10 tonnes (300,000 oz.).

Most lode gold provinces elsewhere in the world,where erosional and preservation regimes are favourable, have some associated placer gold (Boyle 1979).Tertiary placers have produced significant amounts of gold from the Sierra Nevada placers in California (55 million ounces), the Klondike placers in The Yukon (30 million ounces) and the Siberian placers (at least 40 million ounces). The average grade of these placers is extremely variable.

Bedrock lode deposits in the Yilgarn Block have undergone considerable erosion and alluvial channel deposition in the Tertiary, but no significant placer deposits have been located. From 1981 to 1987 BHP Minerals Ltd and BHP Gold Mines Ltd examined this enigma and conducted a major exploration programme to investigate the extensive Tertiary palaeochannels in the Yilgarn Block for secondary and placer gold mineralisation. This paper presents some of the results of this work.

THE TERTIARY PALAEOCHANNELS

The Tertiary palaeodrainage systems currently evident over the Yilgarn Block of Western Australia are characterised by a series of saline lakes which are either partially connected or isolated along current topographic lows. These lakes rarely fill with surface water and generally have no present day surface outlet to the sea. The current surface expression of these channels is poorly developed compared with the subsurface erosional profile. These drainages have been partially described by Bunting et al (1973) and Van de Graaff et al(1977), and some of the climatic implications have been discussed by Kemp (1978) and Benbow (1982).

The palaeodrainages can be subdivided into three major basins(Figure 1), each containing a number of channels:- the Eastern Yilgarn system which drained to the east and outlet to the Southern Ocean via numerous, no longer active, rivers; the Southern Cross system which drained to the west and outlet to the Indian Ocean via numerous rivers which are no longer active and; and the Murchison system which drained to the north and/or west and outlet to the sea via river systems most of which are coincident with current day drainages (principally the Murchison, Chapman, Irwin and Moore Rivers). These modern drainages were probably reactivated after uplift along the northern portion of the Darling Scarp.

During the Late Cretaceous and Early Tertiary, the channels within these basins deeply incised the topography and were actively carrying bedloads of coarse sand and gravel. As the climate became dryer and runoff decreased, the rivers gradually choked up with sand and later with clay (presumably from the eroding lateritised terrain). Throughout this period, regional tilt of the Yilgarn Block.caused the most active portion of the channels to migrate southward. The current expression of the drainages lows therefore rarely coincide with the deepest parts of the palaeochannels. In the Eastern Yilgarn system for instance, the deepest part of the

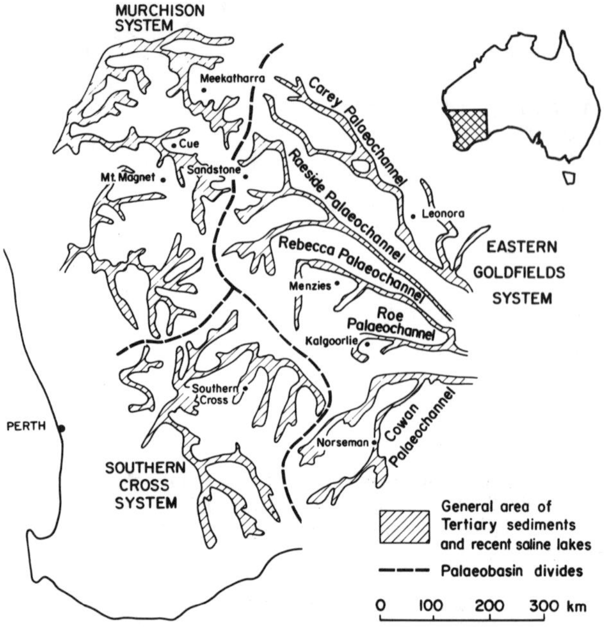

FIGURE 1 Palaeodrainage systems and channels of the Yilgarn Block, Western Australia.

palaeochannels are commonly 2 to 3 kilometers north of the current salt lakes.

Most of the work described below involved exploration in the Eastern Yilgarn system however limited work was also completed in the other two palaeochannel systems.

MODEL DEVELOPMENT

Major advances were made in the 1960s and 70s in the development of geochemical models aimed at explaining the concentration of certain metals at geochemical fronts. The most successful of these is the 'roll front' model, developed for Mesozoic and Cainozoic sandstone-hosted uranium deposits of the United States. The model was widely applied and resulted in the discovery of numerous deposits, both in the U.S. and elsewhere (Harshman and Adams, 1981).

During 1982 a conceptual model, based on the uranium roll-front model, was developed by Button (1982) that explained the absence of the expected rich alluvial gold deposits within the Tertiary palaeochannels of the Eastern Yilgarn

The general requirements for the formation of a significant deposit at a geochemical front are straight forward, and include the following conditions:-

i) An adequate amount of the metal is present in a state such that it is accessable to the transporting medium (usually ground water).

ii) An adequate transfer of the metal from source material to the ground water occurs.

iii) The chemistry of the ground water is appropriate to the transport of the metal for a significant distance.

iv) An appropriate geochemical barrier is present in a suitable location.

v) The volume and duration of the ground water flow is sufficient to produce an economic deposit of the metal.

Under normal circumstances the very low concentrations of gold in groundwater (average 0.002 ppb; Rose et al, 1979) make gold an unfavourable candidate for this type of economic deposit. However, in some circumstances, gold does move in solution, as illustrated by the extensive supergene enrichment of some gold deposits (Boyle, 1979, p. 435-445).There is also extensive literature on the solution chemistry of gold (e.g. Larkin et al, 1974 and Boyle, 1979), but of particular interest for the saline groundwater conditions of Western Australia, is the solution of gold in the chloride system.

Cloke and Kelly (1964) present an Eh-pH- chloride diagram that allows prediction of the solubility of gold as the $AuCl_4^-$ complex. Their work shows that significant concentrations of gold can be sustained in solution, as the $AuCl_4^-$ complex ion, where groundwaters are acid, chloride-rich and oxidising. Their work also helps in predicting conditions under which gold will precipitate from similar chloride-rich waters. Any one of (or a combination of) increasing pH, decreasing Eh or decreasing salinity will serve to reduce gold solubilities to very low concentrations. It appears from the study of the available evidence that reduction fronts are probably the most effective geochemical barriers for gold in solution but other effective barriers include sharp changes in pH, salinity and the availability of iron oxides.

In applying these principles to the Tertiary palaeochannel systems in Western Australia it has become clear that as the major channels filled with sediment the subsurface waters became salt-rich and acid, and started to transport precious and base metals in solution as metal- chloride complexes. This transport mechanism is still active. These solutions move downslope, eventually ending up in the sand-filled channels. Here, sulfates in the brine are reduced to sulfides, particularly around organic particles in the sand such as fossilised wood. Iron in the environment is precipitated as pyrite or marcasite. Other metals in solution are also precipitated, sometimes as crusts around pyrite grains (Ni and Co) and sometimes as base metal sulfides (Pb,Zn,Cu). Silver is precipitated in solid solution with lead and copper sulfides. Some gold remains free but some appears to be in solid solution with iron sulfides (Figure 2).

Some of the exploration holes drilled to test for secondary gold intersected gold-bearing gravels in a reducing. environment. Application of the same geochemical principles shows how placer gold is modified and redistributed above the water table, destroyed in the oxidising zone below the water table and preserved in the reducing zone which occurs further downstream (Figure3). The deeper, reduced gravel portions of the channels therefore still contain much of the original placer gold deposited during the Early Tertiary.

TERTIARY PALAEOCHANNEL STRATIGRAPHY

The stratigraphy of a typical Eastern Goldfields palaeochannel consists of a layer of unconsolidated Recent sediments, usually sandy clays, overlying a 30 to 60 meter thickness of stiff Tertiary clays.These clays are underlain by clean quartz sands varying in grain size from fine to pebbly and in thickness from 20 to 40

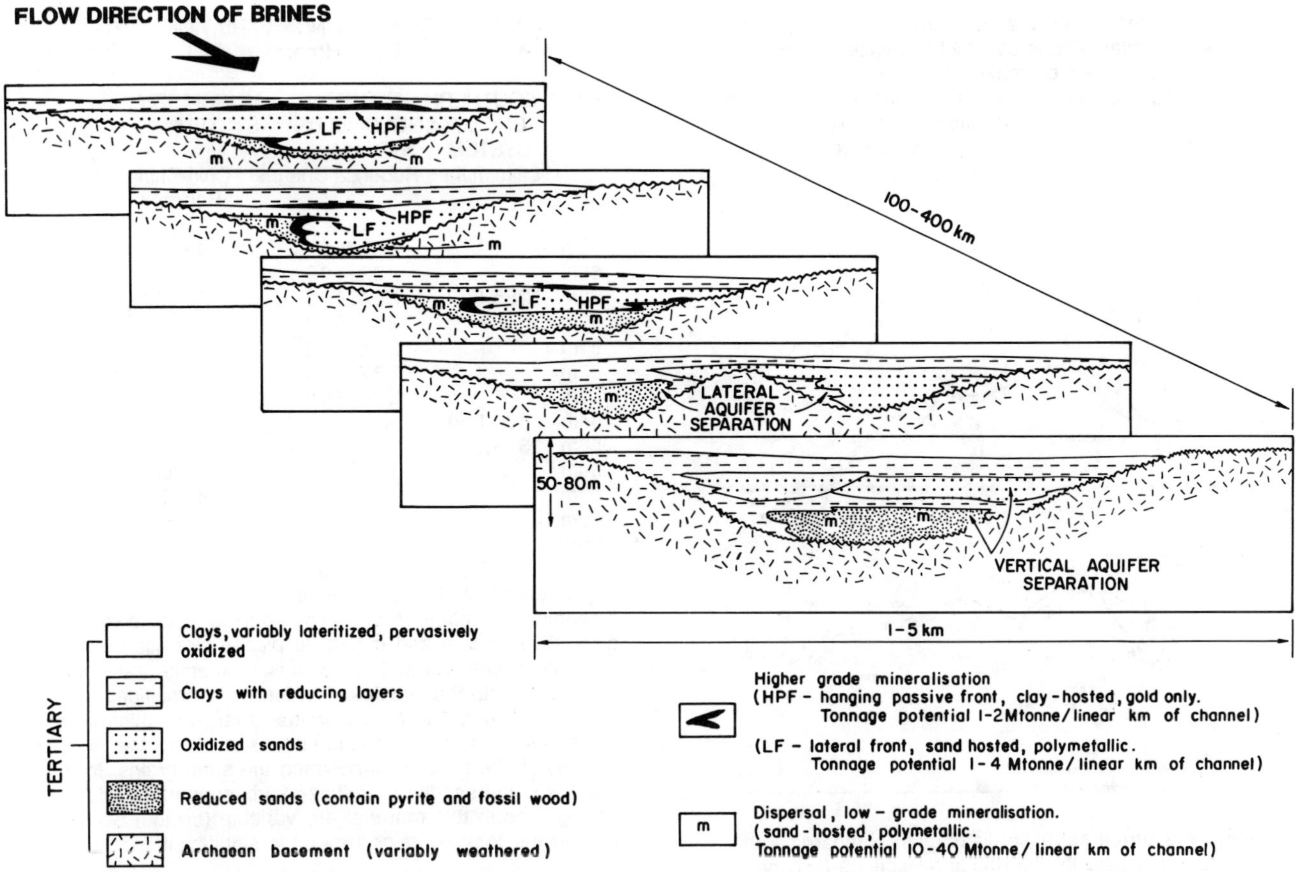

FIGURE 2. Styles of redox-controlled secondary mineralisation in Tertiary palaeochannel sediments, Yilgarn Block, Western Australia.

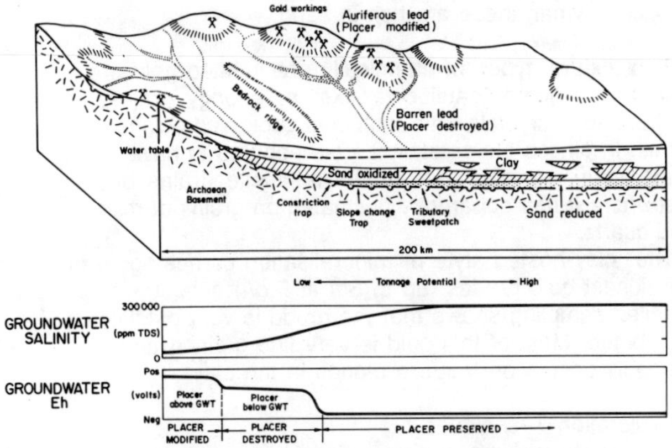

FIGURE 3. Idealised block diagram showing the typical distribution of modified, destroyed and preserved auriferous Tertiary palaeoplacers in the Yilgarn Block, Western Australia.

meters.The sands are generally well sorted with numerous fining upward sequences. Layers of carbonaceous material consisting of wood clasts and other vegetable matter are common, especially in the lower parts of the sequence.

The sands are unconsolidated and saturated with extremely saline brines. The Tertiary sands overlay Archaean basement material that is generally saprolitic. The saprolite is thickest on the channel margins and is often almost non-existent beneath the deepest part of the channel.

Palynological studies carried out by Monash University and Analabs of Perth on palynomorphs extracted from drill samples of the carbonaceous lower sands confirm that the channels represent a Tertiary fluviatile environment. There are no marine indicators in any of the material and the assemblages are consistent with an alluvial depositional environment. The age of the sands is considered Late Eocene to Early Oligocene. The miospore assemblage appears consistent with that described by Balme & Churchill (1959), from 'deep leads' near Coolgardie, and Cockbain (1968), from the widespread Plantagenet Group to the south.

Various principles can be used to help predict the location of the deepest parts of the palaeochannels and the channel morphology:-

i) For any one palaeochannel system there is a reasonably constant relationship between the present day salt lake distribution and the palaeochannels. In the Eastern Yilgarn system the palaeochannels lie to the east or north of the salt lakes (Figure 4).

ii) In the Eastern Yilgarn system the maximum erosional depth decreases from north to south. The Carey palaeochannel, for instance, commonly has maximum depths of 80 to 110m whereas most of the Roe palaeochannel is in the order of 50 to 65m. This is apparently related to regional tilt during the Tertiary.

iii)Traces of the palaeochannels are generally straight except where palaeoridges constrict the valleys or the drainage is influenced by major structural features.

iv) The proportion of sand to clay is partially dependent on the local bedrock types. Granitic bedrock areas produce significantly more sand than the greenstone belts.

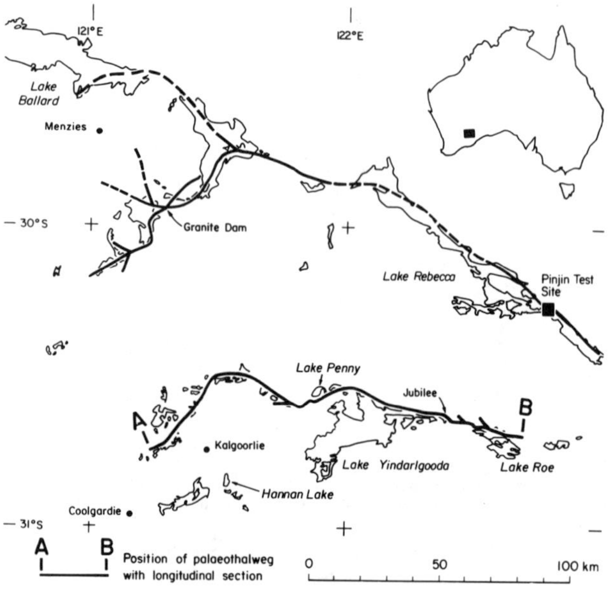

FIGURE 4 Current salt lake distribution compared with thalweg trace for the Roe and Rebecca Tertiary palaeochannels, Western Australia.

The ground waters in the channels vary in salinity with depth, distance down the channel and the particular palaeosystem. In general the Southern Cross system is the most saline (up to 319,000ppm TDS) with the Eastern Goldfields system producing TDS values of up to 287,000ppm and the Murchison system producing a maximum of only 59,200ppm TDS. The variation of TDS with depth in the different systems is illustrated in Table1. The salinity levels are much higher than open sea water (35,000ppm TDS) and 'closed seas" with high evaporation rates such as the Red Sea and the Persian Gulf where TDS values reach 40,000ppm (Mason,1966).

Superficially the Southern Cross and Eastern Goldfields systems appear to have similar salinities but the composition of the dissolved solids is quite different. Magnesium levels, for instance, at 78m in the samples shown in Table 1 are 15,400 ppm and 7,900 ppm respectively.

EXPLORATION FOR SECONDARY GOLD MINERALISATION

Within the Tertiary sand and clays there exists a variety of forms of secondary mineralisation (Figure 2):

i) Polymetallic deposits entirely enclosed within the reduced sands.

ii) Clay-hosted gold mineralisation where reduced clays are in contact with oxidised sands (hanging passive redox front)

iii) Gold mineralisation at or near the contact between oxidised and reduced sands (lateral redox front).

TABLE 1. TYPICAL SALINITY VARIATION WITH DEPTH(ppm).

DEPTH	SOUTHERN CROSS SYSTEM Lake Julia	EASTERN GOLDFIELDS SYSTEM Rebecca Channel	MURCHISON SYSTEM Chin Chin Rocks
45m		213,400	4,900
60m		262,400	
78m	255,000	254,400	28,700
80m		263,000	
81m		244,600	
82m		227,000	
83m		271,600	
84m	300,000		
88m	319,000		
90m			31,200
97m			35,200
99m			34,200
105m			35,700

The sand-hosted polymetallic deposits are predominantly made up of sulfide nodules which contain gold, silver, nickel and cobalt. In places significant amounts of uranium and leucoxenised ilmenite grains are associated with the sulfides. The form of the sulfides varies from very fine single grains to large nodules(up to 15cm long) of sulfide cemented sands in which the sulfides fill the interstices between the sand grains. In a few cases the sands are sulfide-matrix supported. The gold grades in this material are variable (up to 1.5 g/t)and the total contained metal is significant.

On any one section several different types of redox boundaries can occur (Figure 2). The depth to the main redox boundary in the sand is reasonably constant but it may vary where major tributaries join the main channel, especially when these are rich in sand.

Two textural types of chemically precipitated gold are evident; delicate "cauliflower-like" or spongy grains, and euhedral crystals with combined cubic and octahedra forms. In places the crystalline gold appears to have preferentially encrusted detrital andalusite grains but elsewhere it has apparently nucleated on grains of mica and quartz.

The clay-hosted style of mineralisation carries some of the richest gold grades (up to 5m at 5 g/t) although detailed sampling shows that the grade is very patchily distributed. Most of this gold is very fine grained and probably occurs on fracture planes in the clays.

Exploration Techniques.

The main problems in exploration for these secondary gold deposits are:-

i) location of the main Tertiary channel

ii)location of redox fronts in both sand and clay.

Although gravity techniques can be used to locate the main channel position it has limited application in further reducing the target area. Drilling is the only effective exploration technique that can be used to locate these types of gold deposits.

Trials using self potential were conducted over the Roe palaeochannel, however the data is difficult to interpret due to the variable influence of noise on the profiles. This is thought to be due to the high resistances between the base of the porous pots and the dry, sandy soils.

Assessment of the value of contained gold and other heavy minerals in the different geological environments is difficult. Considerable emphasis has been placed in obtaining representative samples and then determining the heavy mineral content. At Jubilee this was achieved by:-

i)Collecting all the sample returned for each 2m drill interval.

ii)Treating all of the sample at an in-house heavy minerals laboratory by dry screening and then crushing the +2mm material.

iii)Producing a clean heavy mineral concentrate using a Wilfley table, a heavy liquid (TBE) centrifuge and, finally, submitting a split of each concentrate for analysis.

Most of these concentrates were also observed by experienced mineral observers and many grains were microprobed for more definitive identification. General observations made in relation to these concentrates are:-

i)Silver values can be up to 297ppm and these are related to high copper (up to 0.69%)and lead (up to 5.7%).

ii)Nickel and cobalt are present in violarite and bravoite with values of up to 0.859% Ni and 0.430% Co.

iii) Gold values up to 3.54g/t .

iv)Industrial minerals such as leucoxene (which average 65% TiO_2), andalusite, pyrite and marcasite occur in significant quantities.

In order to economically assess this assortment of heavy minerals, appropriate, for the time, metal prices were allocated and then a contained metal value was calculated for the concentrates. These values were then recalculated to give a sand value to each heavy mineral concentrate.

Economic Testwork.

Economic modelling of the deposits, using contained metal values, and based on the results of two years exploration for these secondary styles of mineralisation, showed that mining would not be profitable. Typically the deposits were at depths of 30 to 55m below surface and had in situ gold grades averaging 0.1g/t at Jubilee and 0.18g/t at Granite Dam. The total heavy mineral contents averaged 0.88% at Jubilee and 0.64% at Granite Dam.

Four mining methods were investigated and rejected;

i) Open Cut Mining: The thickness of stiff clay would be expensive to move and the underlying saline water-saturated sands made ground stability a major problem. The low average value of contained metal made this option unattractive.

ii) Dredging: The depth of overburden, most of which was stiff plastic clays, downgraded this as an option.

iii) Borehole Mining: The available technology was found to be unsuitable for these types of orebodies due to dilution caused from sand collapse from above or the strength of the containing clays. It was concluded that the amount of high risk investment required to upgrade this technology was not warranted.

iv)In-situ Mining: Due to the lack of a fluid-confining base to the mineralisation and the low value of total contained metal this method appeared to be uneconomic.

EXPLORATION FOR PRIMARY PLACER GOLD MINERALISATION

Placer gold commonly occurs in the reduced gravels of all of the major palaeochannels as particles ranging in size from less than 0.1mm to about 1mm in diameter.The gold is confined to reduced, coarse grained, conglomeratic sands within 4 meters of the contact between the Tertiary and the Archaean. The gold bearing zone is commonly less than 1m thick but can be up to 3m thick. Although the grade of this mineralisation is difficult to determine without expensive cable tool drilling, reverse circulation (R.C.) drill samples, over 1 to 2 meter intervals, commonly average1g/tonne in gravels in the deepest parts of the palaeochannel.

In late 1985, BHP and Uranerz Australia Limited, in a Joint Venture, explored part of the Rebecca Palaeochannel on Pinjin Station where some of the best placer gold results had been obtained (Figure 4). Cost modelling indicated that with reasonable gold dissolution rates solution mining might be economically viable. The Pinjin Project then developed into a major testing ground for a number of exploration and mining techniques and ideas (Whincup, McDonald & Duke, 1989; this volume). Many of the techniques described below were tested or used to explore this test site.

The fast and economical location of the deepest part of the palaeochannel became the most important issue in the search for the Tertiary placer gold deposits. Due to the high cost of drilling 60 to 100m deep R.C. holes, surface techniques were needed to define the target zones prior to drilling

Several techniques proved to be very useful in these environments:-

Topography and Satellite Imagery.

Although the approximate position of a Tertiary palaeochannel can be deduced from regional topographical and geological maps the trace of the deepest portion of the palaeochannel can not. NOAA imagery is useful but is limited by its poor resolution.

Geophysics

Gravity. There is a significant density contrast between the Tertiary channel sediments (1.8 to 2.0 tonnes per cubic meter)and the underlying Archaean basement (2.4 to 3.0 tonnes per cubic meter).The gravity profile, in constant bedrock conditions, will therefore reflect the mass deficit resulting from the displacement of the higher density Archaean bedrock by the Tertiary sediments (Winer, 1986a). The use of gravity profiles can have considerable cost savings by reducing the number of R.C. holes needed to locate the channel. In practice a number of geological conditions complicate the gravity response and can make the interpretation of the profile difficult:

i) In some areas the anomalies are caused by bedrock variations.

ii) Highly weathered saprolitic bedrock has a similar density to the Tertiary sediments and hence the gravity contrast is poor. Since the thickness of the weathered zone is greatest on the flanks of the channel, this can result in the lateral stretching of the gravity low.

iii)The density contrast between the Archaean bedrock and the Tertiary sediments can be quite different on lines only 10 km apart. This can cause significant differences in the magnitude of the gravity response and make it difficult to trace the channel position from section to section.

iv) Strong regional gravity gradients can distort the gravity profile if they are not removed from the data before interpretation.

The use of gravity to locate the Tertiary drainages has been successful on about 50% of the survey lines where drilling has then been used to define the main channel. However if the channel is greater than 500m wide the cost effectiveness of the method is greatly reduced. A typical example of a drill-confirmed anomaly is shown in Figure 5. The success rate of gravity in locating the deepest part of the channel is greatly improved when used in combination with SIROTEM.

EM and SIROTEM. Model development by in-house geophysicist, M. Asten, shows that electromagnetic techniques can be used to measure the ground conductivity as a function of depth. Since the Tertiary sands are saturated with extremely saline groundwater, the salinity of the groundwater increases with depth and the volume of saline water in the sediments is greater than in the bedrock, this technique can be used to define the deepest part of the channel and to interpret the spurious gravity anomalies.

SIROTEM is even more successful after the data are inverted.The computer inversion package GRENDLB uses a starting model with estimated thicknesses and resistivity (inverse of conductivity) of each of the normally intersected layers. The programme then adjusts the thicknesses and resistivities to a best fit model with the field data. This is an expensive computer exercise, but the resultant profile is very useful in defining the Tertiary-Archaean unconformity. Figure 5 shows the value of retesting the deepest part of the channel(holes J24-J27) eventhough the original test holes (J1-J4) produced only moderately encouraging results.

A typical inversion starting model is shown below:-

	---------- 0 m	
Surface clays		5 ohm .m.
	---------- 5 m	
Tertiary clays		0.5 ohm.m.
	------------- 4 0 m	
Tertiary sand		20/(D-30) ohm.m.
	unconformity-----------------	D(>40m)
Archaean Bedrock		100 ohm.m.

where D= Depth below surface.

A semi-quantative estimate of total conductance can be obtained in the field by overlaying a nomogram (showing apparent resistivity asymptotes for values of total conductance) onto the field plot of apparent resistivities (Asten,1987).This can be used as a field guide to test whether thick channel sediments have been located.

The most cost effective use of SIROTEM is to use the nomogram on single station readings over every gravity anomaly on a profile to select which one reflects the true channel position. Once the channel is located a detailed line of 100m spaced stations is run and the data points are inverted and final drill hole positions selected.

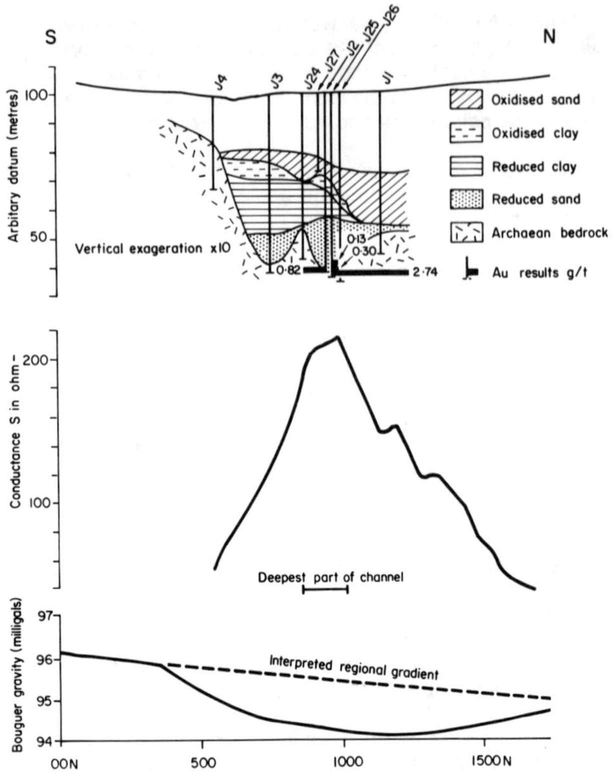

FIGURE 5 Geological cross section, conductance curve and gravity profile for drill line 1, Jubilee, Roe palaeochannel,Western Australia.

Seismic Surveys. Trial surveys of both seismic reflection and refraction have been carried out at the Pinjin test site. Both these methods were successful in defining the position of the centre of the channel but they could not clearly define the saprolite/sediment contact.In general the high cost of seismic techniques make them unsuitable for general exploration. Figure 6 gives an example of a seismic refraction section from the Pinjin test site.

Down-hole Gamma Logs. The use of down-hole gamma logs proved very useful in accurately locating the Tertiary/ Archaean unconformity. The R.C. drilling technique often produces enough sand contamination to obscure the true unconformity position, and therefore all holes are gamma logged. The location of the unconformity is important for determining the true thickness of the gold-bearing zone.

Figures 6 and 7 show comparitive geophysical methods and drill interpreted geology for Line 4 at the Pinjin test site (Figure 4).

The most successful procedure for defining the target drill sites is to use a combination of geological and geophysical methods. The most prospective length of channel is selected using general principles (such as channel narrowing, bedrock highs etc)and then to define the target using gravity and SIROTEM.The best target is then drilled.

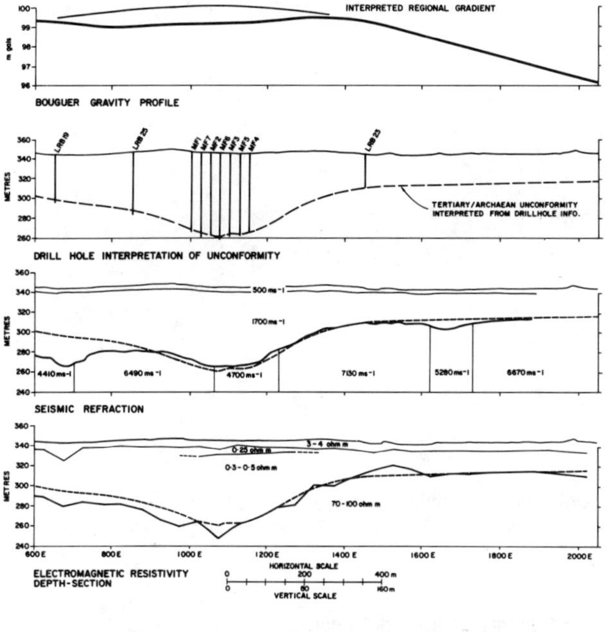

FIGURE 6. Comparison of bouger gravity, seismic refraction and resistivity and the interpreted position of the Tertiary/Archaean unconformity derived from drill hole data. Pinjin test site. (After Winer 1986b).

Drilling

The difficulties of sampling gold placers by the use of small diameter or uncased drill holes is well documented (Wells, 1970; Boyle, 1979) but the expense of drilling large diameter exploration holes through up to 100m of overburden, as in the Yilgarn, is prohibitive. Since the main aim of the exploration phase of this drilling programme is to locate the trace of the deepest part of the palaeochannel (thalweg), and to test whether it is auriferous, it was decided to get the greatest coverage per dollar by using a compromise method of small diameter (100mm) reverse circulation drilling. It was not until this exploration had located a favourable test area that more conventional cable tool drilling techniques were used.

<u>R.C. Drilling.</u> It is considered, after numerous trials, that for the majority of the placer exploration programme the reverse circulation (R.C.) drilling techniques developed by Wallace Drilling of Kalgoorlie provides the best potential to obtain relatively uncontaminated, samples at low cost.

Once 6 to 8 drill sections have been established for a particular palaeochannel, a down-channel longitudinal section is used to check whether the deepest part of the palaeochannel has been located on each section. Figure 8 illustrates the use of this method.It shows that holes PKG 242, WP 52 and R 30 appear to be too shallow and

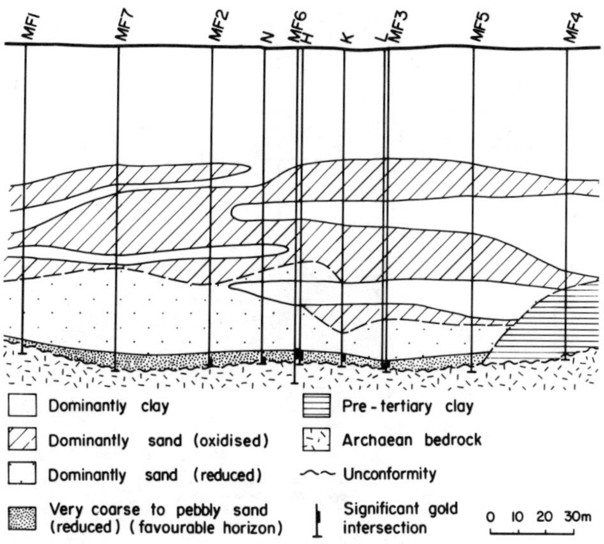

FIGURE 7. Geological cross section of Line 4 at the Pinjin test site, Rebecca palaeochannel, Western Australia.

therefore do not test the thalweg position for gold mineralisation. In some sections of the Roe palaeochannel this technique was used successfully to select old section lines where the thalweg had obviously not been tested by the original drilling and then to use SIROTEM to position new drill sites (Figure 5).

Most drill lines are about 10 to 15 km apart, with drill holes spaced at 100 to 200m and later infilled to 50 or 25m. Although the channels apear to be relatively straight when drilled at 10 to 15 km spacing, in the few areas where lines are spaced at 1 km the channels prove to be more meandering.

<u>Cable Tool Drilling</u> It became obvious after a lot of test drilling that the relationship between the absolute in situ gold grade and the R.C. drill indicated grade was very important. It was therefore decided to drill some cable tool holes as close as practicable to existing R.C. holes (1.5m).

The drilling was conducted at the Pinjin test site. This work showed that the R.C. drilling technique had been giving elevated gold grades due to the preferred recovery of the finer grained material up the 100mm diameter internal rod annulus. Since the largest diameter pebbles (150mm) produced in the cable tool drilling were conciderably less than than the 200mm internal diameter of the cable tool drill rods, it is assumed that this problem did not extend into the cable tool drilling results.

Examination of the sediment sizing and gold distribution produced from this test drilling programme, indicates a wide variation in gold values between adjacent holes. This variation is probably partly due to the normal variability of gold in an alluvial environment. However, the bulk weights of the various size fractions shows a 130% increase in the -2mm size fraction in the R.C. samples compared with the cable tool drill samples. This confirms that the excess sample being produced during R.C. drilling is preferentially finer grained. Since the fine material

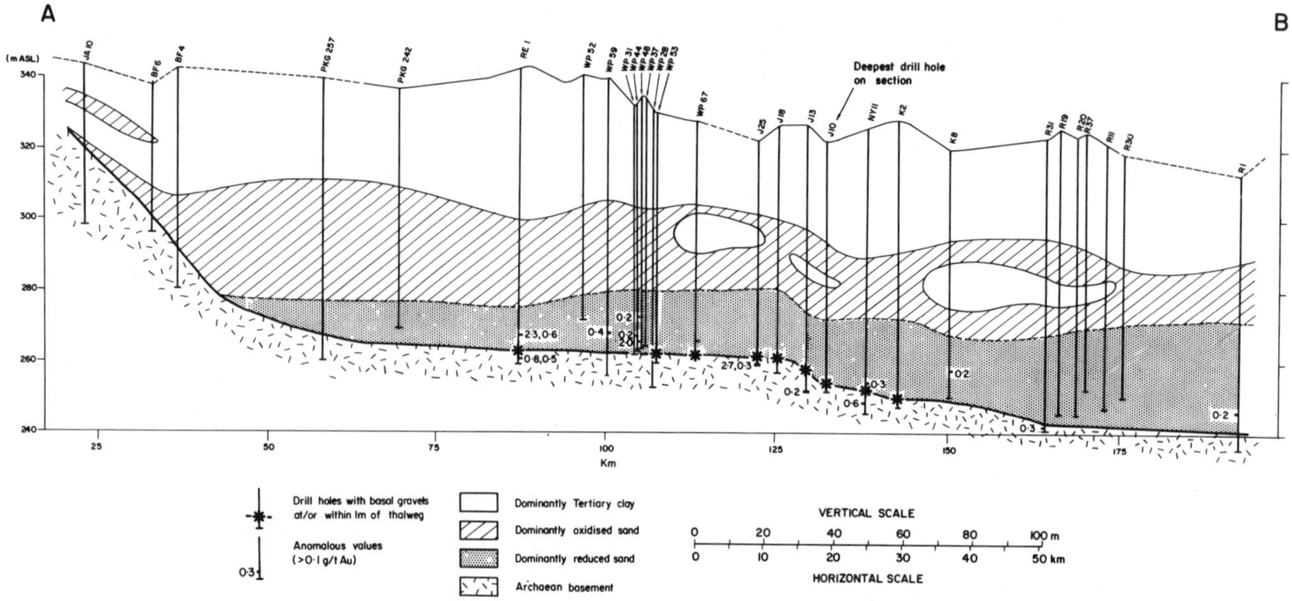

FIGURE 8 Longitudinal section down the Roe palaeochannel through deepest drillhole intersections on each drill section. The best fit thalweg position shows several drill sections (PKG 242, WP 52 and R30) that have not tested the deepest part of the palaeochannel.

contains far more gold than the coarse material it is concidered that the R.C. method it is high grading results by about 30%.

Geochemistry

Obtaining repeatable gold values from placer deposit drill cuttings is a notorious problem. It was important in this programme to establish both a reliable drill technique that allowed collection of a consistent sample type and an
assay technique that was reasonably cheap and yet repeatable. Due to the nugget effect of the coarse placer gold, the fire assay technique using a 50g (FA 50) sample charge did not prove very satisfactory although it did, in combination with panning, provide a method for determining some order of magnitude for the gold grade. The more expensive screen fire assay (SFA) was more repeatable because it uses a larger sample size thereby reducing the nugget effect.

Screen Fire Assay The screen fire assay technique requires screening of the initial 1kg -2mm material into 7 size fractions (+3/16inch; -3/16inch+10mesh; -10mesh+20mesh; -20mesh +60mesh; -60mesh +80mesh; -80mesh+200 mesh; -200mesh) plus the calico bag (to ensure any fine gold in the bag fabric is included). Each fraction is then pulverised and weighed and all of the material is fire assayed. The calico bag is weighed, burnt and the ash assayed. For the -200mesh material two 50g fire assays are completed and averaged. All of the results are recalculated according to weight proportion. Corrections are made for the +2mm material and the salt content.

This screen fire analysis method showed that 43% to 57% of the gold in the sample occured in the -60mesh +200mesh size fraction and a further 17% to 21% occurs in the -3/16inch +10mesh size fraction. Since the majority of the gold occured in the -60mesh +200mesh size fraction the FA50 technique could be used to give some indication of in situ grade during the exploration drilling phase.

Cyanide Leach Test work was also carried out on duplicate samples using cyanide (CN) leach as the analytical method. The majority of analyses involved 24 hour agitated (bottle-roll) leaching of -2mm splits from samples obtained from two of the cable tool holes H and K at the Pinjin test site.

Results Table 2 gives comparative results for 1kg screen fire assay, 2kg cyanide leach and 1kg panned grain counts for -2mm material from two of the cable tool drill holes at the test site. It illustrates the effect of the high quantity of low grade oversized material. Testwork shows that the percentage of the total gold that occurs in the +2mm size fraction varies from 0.73 to 4.28%.

There is still concilderable variation in the SFA/CN leach assays and this is thought to be due to sampling problems, the proportion of oversized and the nugget effect. Some of the differences are probably also related to the short period of time that the samples were exposed to the leachant.

Table 3 shows a comparison of screen fire assay results and the proportion of the oversized +2mm material for the R.C. hole, MF6, and the adjacent cable tool drill hole, H. It illustrates that the variation in the results cannot be totally explained by the proportion of coarse sediments alone.

The coarse grained interval at the test site gave consistent average values based on screen fire assay and CN leach assay of 2.25 m at 0.67 and 0.62g/t respectively for hole H and 2.5m at 0.62 and 0.71 g/t respectively for hole K. Although individual samples

TABLE 2. COMPARATIVE GOLD RESULTS FROM CABLE TOOL DRILL HOLES AT PINJIN TEST SITE.

INTERVAL m	-2mm SFA oversize excluded (g/t)	-2mm SFA * oversize corrected (g/t)	CN LEACH* head grade (g/t)	GOLD grains panned (No)
HOLE H				
77.50-78.00	.040	.040	-	-
78.00-78.50	.128	.128	.634	-
78.50-79.00	.081	.066	.155	4
79.00-79.25	1.360	.824	.733	100
79.25-79.50	.398	.240	.199	50
79.50-79.75	.500	.327	.314	80
79.75-80.00	.507	.309	.555	120
80.00-80.25	1.347	.753	.458	100
80.25-80.50	2.373	1.373	1.239	200
80.50-80.75	.734	.426	.481	80
80.75-81.00	.208	.175	.261	80
81.00-81.25	.167	.135	1.310	60
81.25-81.50	.105	.100	.555	20
81.50-81.75	.042	.040	.144	10
81.75-81.94	.062	.054	.087	6
81.94-81.97	.804	.703	1.049	50
81.97-82.20	1.917	2.023	.834	40
HOLE K				
78.10-78.50	.028	.025	<.01	6
78.50-79.00	.366	.259	.756*	60
79.00-79.25	2.183	1.232	.841	120
79.25-79.50	1.270	.624	.949	60
79.50-79.75	1.727	.781	.841	130
79.75-80.00	2.857	1.358	1.217	140
80.00-80.25	1.014	.678	.942	140
80.25-80.50	.673	.454	.331	90
80.50-80.75	.448	.238	.260	80
80.75-81.00	.418	.295	.162	60
81.00-81.25	.118	.107	.092	50
81.25-81.43	.062	.057	.101	20
81.43-81.75	.011	.011	.021	10

* Assumes no gold in +2mm fraction
** 1kg grab sample normalised to 2kg

TABLE 3. COMPARITIVE SCREEN FIRE ASSAY GOLD RESULTS AND +2mm SIZE FRACTION FOR R.C. HOLE MF 6 AND CABLE TOOL DRILL HOLE H

HOLE	SCREEN FIRE ASSAY		+2 mm SIZE FRACTION		MF6 Au H Au
	MF6	H	MF6	H	
m	g/t	g/t	%	%	%
78.0-79.0	.033**	.098	7.1	15.0	33.7%
79.0-80.0	.257*	.425	19.8	40.2	60.5%
80.0-80.5	.946*	1.063	30.6	44.9	90.0%
80.5-81.0	1.037*	.514	25.6	31.6	201.8%
81.0-81.5	.005**	.118	10.4	15.0	4.2%

* Single screen fire assay
**Average FA 50 fire assay.

show significant variation between their SFA and CN results, the overall grades and thicknesses of the coarse sediment intervals are remarkably consistent both between holes and between assay techniques, suggesting that these grades and thicknesses are realistic.

CONCLUSIONS

This programme of work has shown that extensive placer gold deposits associated with the Tertiary erosion of the Archaean lode deposits of Western Australia do exist. The reason placer deposits from below the water table have not been located up until recently is that in the highly saline, acid, oxidising groundwater the precious and base metals in the sediments are dissolved and transported in solution away from the down-channel position near the source lode deposits.

This secondary gold was eventually redeposited (often in conjuction with polymetallic sulfides) at various types of redox boundaries in reduced sands and clays. These deposits are commonly under, or close to, widespread salt pans which are difficult to cross with normal drill rigs and equipment. Traditional thinking gave little encouragement to drill for basement bedrock gold under conditions so hostile to exploration and mining.These secondary gold deposits have therefore not been intentionally or accidentally investigated in the past.

The placer gold in these channels invariably occurs in reduced sands in the distal parts of the palaeochannels. Since these do not occur much above 50m below the present ground surface the chances that they were intersected, or recognised as being gold bearing, by earlier explorers were small. Once the placer target was identified, traditional geophysical methods had to be modified to enable the application of gravity and EM techniques in selecting the most favourable drill sites.

The presence of significant palaeoplacer gold deposits in the Tertiary drainage channels of the Yilgarn of Western Australia has been confirmed.

ACKNOWLEDGEMENTS.

This is the result of 6 years work by BHP Minerals Ltd and BHP Gold Mines Ltd. It has been a team effort with major contributions from a large number of Company geoscientists. In particular the authors would like to thank Geoff Woad, Nick Winer, Mike Asten, John Surman, Roger Cooper, Margaret Whitehead, Marlene Kelly and David Gilbert for their technical contributions and the budget and management support recieved from D.A.C. Williams, C.E. Palethorpe and R.J. Carter. David Moore helped by reviewing the paper before submission. Much of the work carried out in the period 1985 to 1987 was in Joint Venture with Uranerz Australia Limited and their contributions and permission to publish have been appreciated .

REFERENCES.

ASTEN, M. W., 1987, "Re-interpretation of Tertiary Channel SIROTEM." BHP Company Memo CM 2250 Unpublished.

BALME, B.E. and CHURCHILL, D.M., 1959,"Tertiary Sediments at Coolgardie, Western Australia." Jrn.Royal.

Soc. West. Aust., 42, pp37-43.

BENBOW, M.C., 1982,"Tertiary Climate of Australia." Report BK No 81/105. Dept of Mines and Energy South Australia.

BOYLE, R.W., 1979,"The Geochemistry of Gold and its Deposits." Geol. Surv. Canada, Bull 280. 584 pp.

BUNTING, J.A., VAN DE GRAAFF, W.J.E. and JACKSON, M.J. 1973," Paleodrainages and Cainozoic Palaeogeography of the Eastern Goldfields, Gibson Desert and Great Victoria Desert." Ann.Rept. 1973, Geol. Surv. West. Aust., pp. 45-50.

BUTTON, A., 1982,"Theoretical Basis for the Development of Gold Deposits at Geochemical Fronts in Parts of Australia." BHP Company Report, CR 3324. Unpublished.

CLOKE, P.L. and KELLY, W.C., 1964," Solubility of Gold Under Inorganic Supergene Conditions." Econ. Geol., v59, pp. 259- 270.

COCKBAIN, A.F., 1968," The Stratigraphy of the Plantagenet Group, Western Australia." Ann.Rept. 1967, Geol. Surv. West. Aust. pp61-63.

HARSHMAN, E.N. and ADAMS, S.S., 1981,"Geology and Recognition Criteria for Roll-type Uranium Deposits in Continental Sandstones." Report GJBX-1(81), U.S. Dept. Energy, 185 pp.

KEMP, E.M., 1978," Tertiary Climatic Evolution and Vegetation History in the Southeast Indian Ocean." Palaeogeography, Palaeoclimatology, Palaeoecology, 24 1978, pp. 169-208.

LARKIN, H.W., CURTIN, G.C. and HURBERT, A.E., 1974, "Geochemistry of gold in the Weathering Cycle." Bull. 1330, U.S. Geol. Surv., 80 pp.

MASON, B., 1966; Principals of Geochemistry. John Wiley & Sons, London. 3rd Edition.

ROSE, A.W.,HAWKES, H.E. and WEBB, J.S., 1979: Geochemistry in Mineral Exploration. Academic Press.

VAN DE GRAAFF W.J.E., CROWE, R.W.A., BUNTING, J.A. and JACKSON, M.J., 1977," Relict Early Cainozoic Drainages in Arid Western Australia. Zertschrift fur Geomorpologie N.F. 21(4) pp. 379-400.

WELLS, J.H., 1970," Placer Examination Principles and Practice." Technical Bull 4.U.S. Dept Interior & Bureau Land Management. 207 pp.

WHINCUP, P., McDONALD, D. & DUKE, H., 1989," In Situ Leaching of Western Australian Palaeochannel Gold. "Gold Forum on Technology and Practices- World Gold '89. Reno, Nevada, Nov 1989. This Volume.

WINER, N., 1986a," Discussion of Exploration Procedures Used by the Tertiary Channels Group." BHP Company Report CR 5063. Unpublished.

WINER, N., 1986b,"Trial Gravity, Electromagnetics, Seismic Refraction and Seismic Reflection Surveys. Magpies Find. W.A." BHP Company Report, CR 5064. Unpublished.

WOODALL, R.W., 1979," Gold - Australia and the World." Gold Mineralization Geol. Dept. & Extension Service, Univ. West. Aust., No.3 pp. 1-34.

Chapter 3

Application of Non-parametric Geostatistics to the Manhattan Gold Deposit

Bruce H. Van Brunt
Dr. Danny L. Taylor

Mackay School of Mines
University of Nevada-Reno
Reno, Nevada 89557

Abstract. Estimation of local recovered tonnage and its grade above a cutoff is of paramount importance during the exploration and development stages of any mineral deposit. The spatial distribution of grades as well as the cumulative distribution must be defined.

In this paper, Probability Kriging, an estimator based upon rank order statistics and the estimation of the distribution of point grades within a panel, is applied to the Manhattan gold deposit. Mapping the conditional probabilities that estimated panel grades exceed a cutoff is useful to the understanding of the distribution of mineralization in a structurally controlled gold deposit.

INTRODUCTION

Conventional estimation techniques such as inverse distance weighting and ordinary kriging, estimate the average grade of a block dependent upon a series of weights assigned to neighboring drill holes. During the development phase of the project large blocks must be included or rejected in total from the mineral inventory on the basis of this estimated average grade. Without detailed geologic information to suggest that portions of a block might not be mineralized and should, therefore, be excluded from the reserve, the use of an estimated average grade will result in the overestimation of mineable reserves.

The development of non-parametric or distribution-free estimation techniques at Stanford University (Journel, 1983), (Sullivan, 1985), offers an interesting alternative to conventional reserve estimation techniques. Probability Kriging is a data dependent estimator that uses rank order statistics to estimate the distribution of grades within a block (Sullivan, 1985). Data within the user defined neighborhood is used to estimate a conditional probability distribution for a given block. The knowledge of such a distribution allows the estimation of the proportion of a block above a particular cutoff grade and the probabitity that a particular cutoff grade is exceeded.

SCOPE OF THE PAPER

The purpose of this paper is three-fold: 1) to compare the performance of probability kriging, ordinary kriging, and the inverse distance weighting method in estimating reserves involving exploration data which has a skewed data distribution; 2) to demonstrate that the practice of clipping high-grade samples from a distribution may lead to over valuation of the deposit; and 3) to evalute the mapping of conditional probabilities as a guide to the spatial continuity of grade in a mineral deposit.

Kriging and inverse distance weighting is conducted in 3-D, although only one bench in the West Pit is used to determine the accuracy of each estimator. Quantity of metal and tons above a cutoff grade is calculated using each method. A comparison is made with data provided by Echo Bay Minerals Co. and the Smoky Valley Common Operation that was the result of a detailed reserve estimate of the West Pit. A discriminator kriging approach is used for the estimate, utilizing exploration data, and blastholes on mined out benches.

The usefulness of mapping conditional probabilities calculated from exploration data is evaluated by a comparison with a contoured map of blasthole data from the test bench.

The mathematical development of each of the methods has been well documented. Therefore, in the interest of brevity and to allow for the discussion of applications in greater detail, the development of these methods is not addressed in this paper.

HISTORY AND GEOLOGY OF THE MANHATTAN GOLD DEPOSIT

The Manhattan mining district is located approximately 80 km (50 mi) north of Tonopah, Nevada in the Toquima Mountain Range (Figure 1). The discovery of gold in 1905 by John C. Humphrey led to the development of the Manhattan district. The Manhattan Mine, currently operated by the Smoky Valley Gold Corp., is located on the western flank of the Toquimas at an elevation of 2150 m (7000 ft).

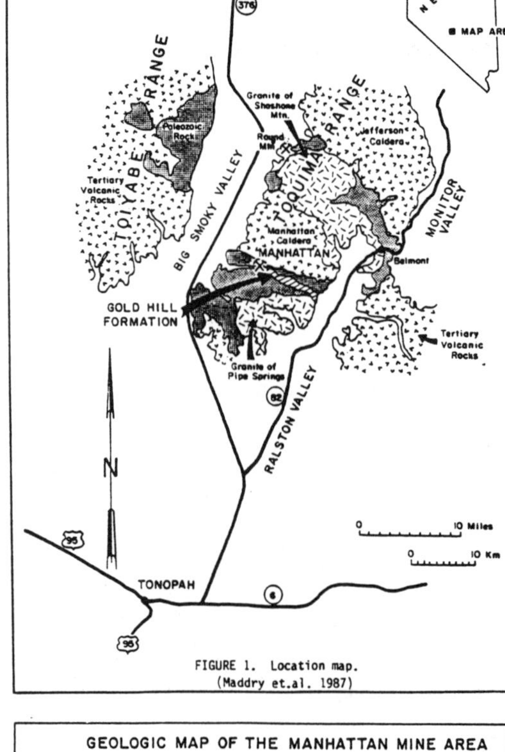

FIGURE 1. Location map.
(Maddry et.al. 1987)

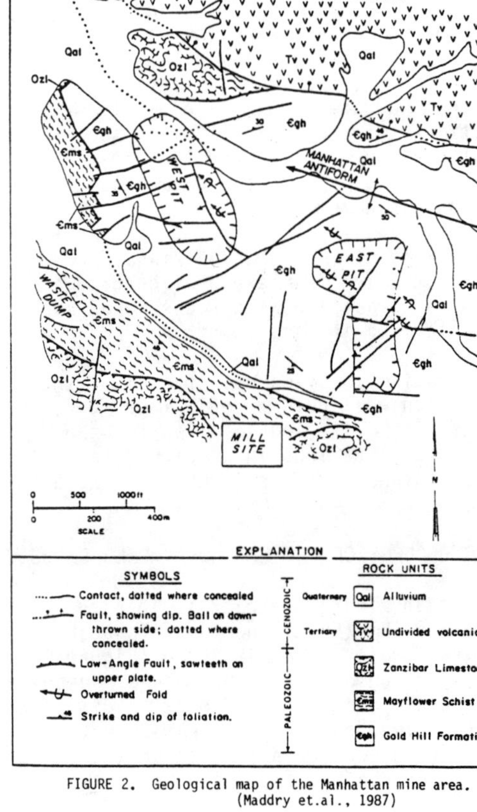

FIGURE 2. Geological map of the Manhattan mine area.
(Maddry et.al., 1987)

The host rock for the Manhattan gold deposit is the Cambrian Gold Hill Formation. The Gold Hill Formation consists of phyllite, quartzite, marble, and sandy phyllite. Immediately to the north of the deposit, Paleozoic rocks are overlain by Tertiary volcanic rocks of the Manhattan Caldera (Maddry, et.al., 1987).

Data for this paper is from the West Pit orebody of the Manhattan Mine. Historically, several underground operations preceeded the open pit on the Little Grey Fault. Limited production records indicate that prior to the 1940's, the total production from mines operating along the Little Grey Fault was 68,000 tonnes (75,000 st) yielding 1,200 kg (38,000 tr oz) (Kral, 1951). The coarse gold in the West Pit is associated with drusy quartz and adularia lining and filling fractures along and adjacent to the Little Grey Fault. Gold particles range in size from 0.04 mm to 10.0 mm (400 mesh - 0.4 in.) and average 700 fine.

Exploration drilling in the Little Grey Mine area was sporadic from 1974 through 1983. By 1985 reserves were considered sufficient to begin production from what is now the West Pit, at a cutoff grade of 0.62 g/t (0.018 oz/st) and a stripping ratio of 3.9:1 (Maddry, 1986). A breccia zone approximately 6 m (20 ft) in width lies at the base of the Little Grey Fault zone. Sympathetic fractures extend mineralization up to 60 m (200 ft) into the hanging wall (Maddry et.al., 1987). The upper portion of the orebody is oxidized. Supergene oxidation extends downward via permeable zones. To date, no wall rock alteration has been identified. The vein mineralogy includes quartz, adularia, Fe-oxides, and calcite in addition to native gold.

DATA ANALYSIS

Statistical analysis of the composited exploration data is a prerequisite to performing a spatial analysis such as kriging. The successful application of parametric geostatistical methods is directly related to the data distribution. For non-parametric estimation techniques such as probability kriging the analysis of the distribution of data is crucial in the selection of indicator cutoffs.

A total of 1,386 exploration composites exist in the West Pit database. Of these, 790 are within the Little Grey Fault zone. Exploration drilling in the West Pit is oriented along the major axis of the Little Grey Fault zone. Drill grid spacing is approximately 21 m (70 ft). Drill holes were sampled every 1.5 m (5 ft) and fire assayed for gold. Assays are then composited into 6 m (20 ft) benches. The spatial distribution of composited drill hole assays is examined by posting the composite location for each bench. Several of the drill holes were angled to intercept the fault zone in its true thickness. The effects of spatial clustering of data were negated by declustering according to the cell method (Journel, 1983) using a 15 m x 15 m x 6 m (50 ft x 50 ft x 20 ft) block system.

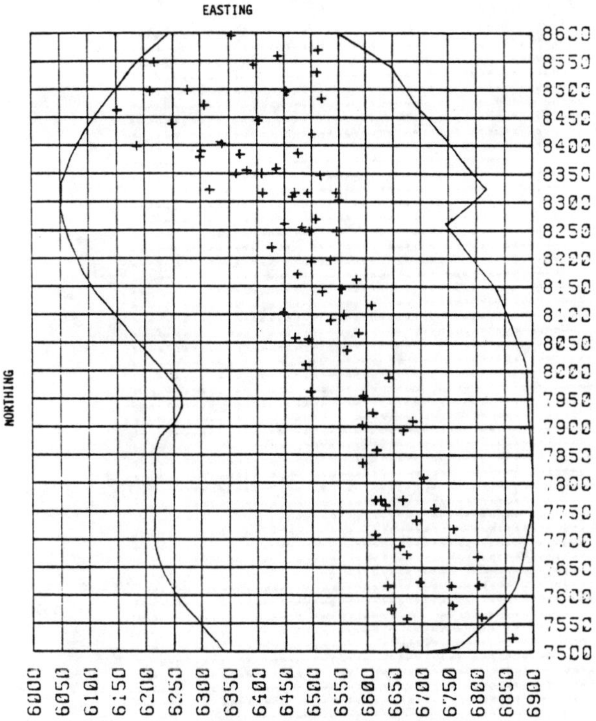

FIGURE 3. 6820 Bench map showing pit outline and composite locations within the Little Grey Fault zone.

Figure 3 shows the outline of the pit walls on the 6820 Bench and the location of exploration composites.

A cumulative distribution function plot of the exhaustive set of composites (Figure 4) indicates that a normal distribution is not present. A plot of the histogram of the composited data (Figure 5) shows a positively skewed distribution with a mean grade of 0.62 g/t (0.018 tr oz/st), a variance of 4.78 g^2/t^2 (0.004 tr oz^2/st^2), and a spike of values at zero containing 35% of the data. The highest value in this distribution is 45.27 g/t (1.313 oz/st), and the coefficient of variation of the data is 3.24.

Subdivision of the database into geologically distinct zones on the basis of structural

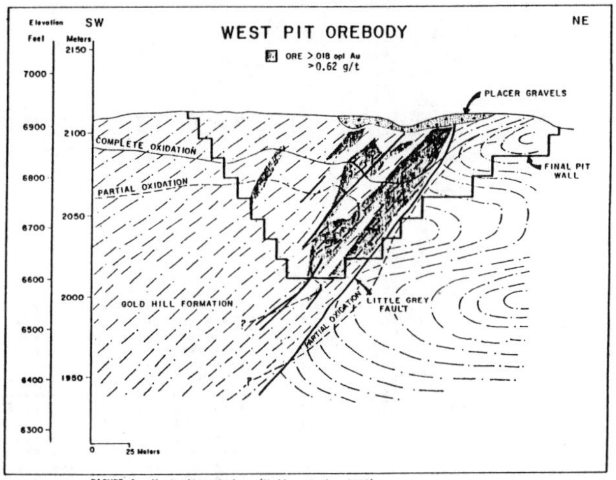

FIGURE 6. West pit orebody. (Maddry et.al., 1987)

position and the presence of drusy quartz-adularia mineralization incorporates into the estimation process the natural physical boundaries of the deposit. Defining populations helps satisfy the assumption of second order stationarity of the random function which describes the grade distribution. A cross section perpendicular to the long axis of the Little Grey Fault zone (Figure 6) indicates the relative position of the three zones. Most of the mineralization in the deposit is contained within the Little Grey Fault zone which includes both the breccia along the Little Grey Fault and the sympathetic fractures extending into the adjacent Cambrian sandy phyllite of the Gold Hill Formation. The upper boundary of the Little Grey Fault zone is a parallel structure which limits the extent of fracturing, and controlled the flow of mineralizing fluids. Above the Little Grey Fault zone is a relatively barren zone designated the Hanging Wall. The lower zone which does contain significant mineralization is designated the Footwall. Statistics of the exhaustive data set and the individual zones are given in Table 1.

For this study estimation is restricted to the Little Grey Fault zone. A cumulative distribution function (Figure 7) plot of the 790 composite data within this zone indicates that the distribution tends to lognormality, not unlike the original data set. A histogram (Figure 8) of

TABLE 1. STATISTICS OF EXPLORATION AND BLASTHOLE DRILLING DATA

	Exhaustive Exploration Data	Hanging Wall	Little Grey Fault Zone	Footwall	Blastholes
Number of Data	1386	431	790	165	2853
Mean (g/t):(oz/st)	0.62:0.018	0.21:0.006	0.90:0.026	0.41:0.012	0.78:0.023
Variance $(g/t)^2$:$(oz/st)^2$	3.57:0.003	0.20:$1.7x10^{-4}$	7.13:0.006	2.38:0.002	32.10:0.027
Coefficient of Variation	3.24	2.06	2.85	3.54	

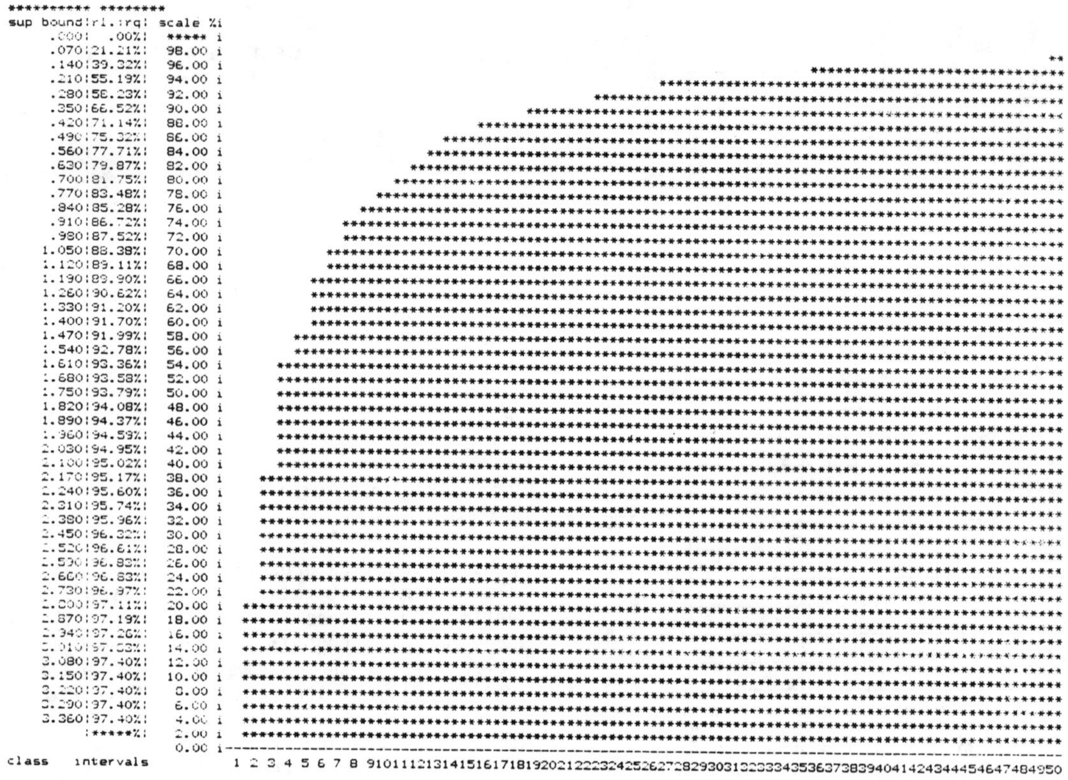

FIGURE 4. Cumulative distribution function plot of the exhaustive exploration composite data.

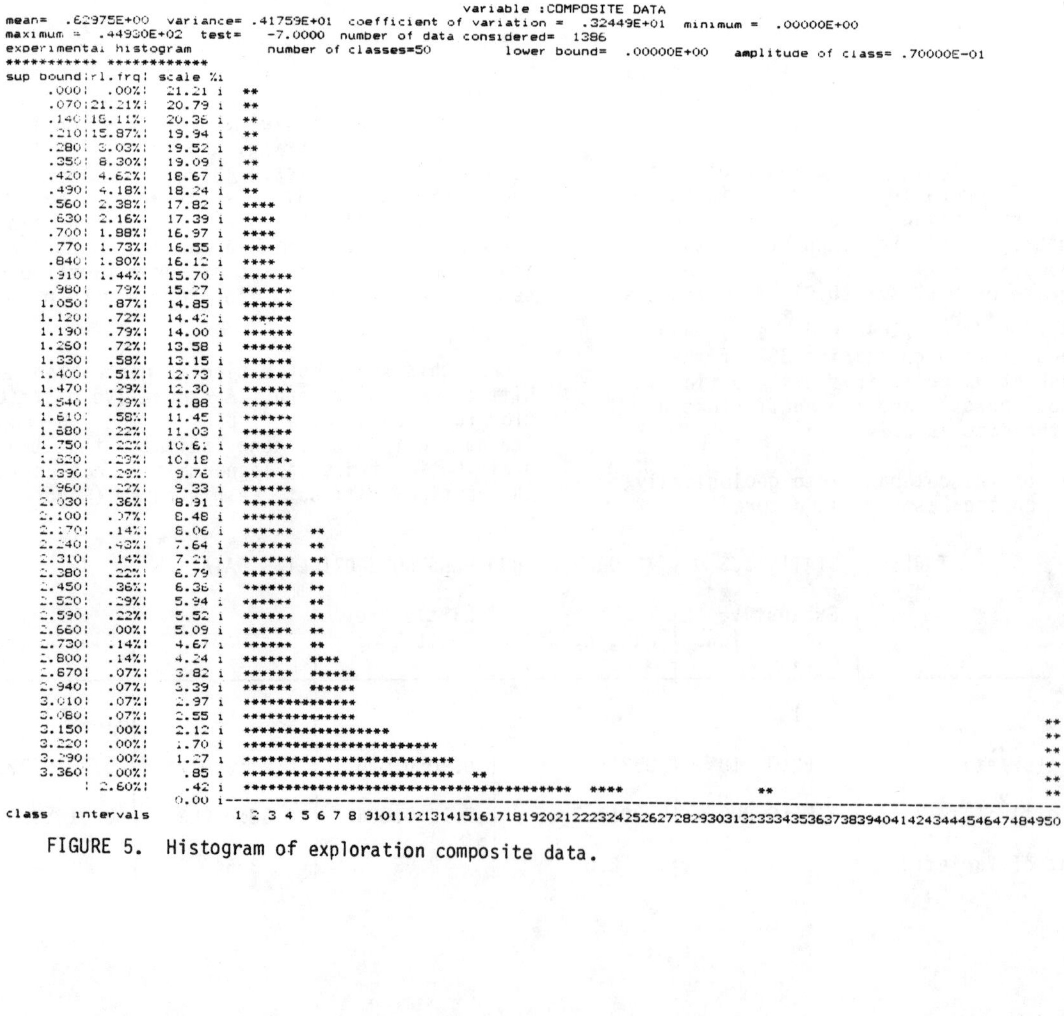

FIGURE 5. Histogram of exploration composite data.

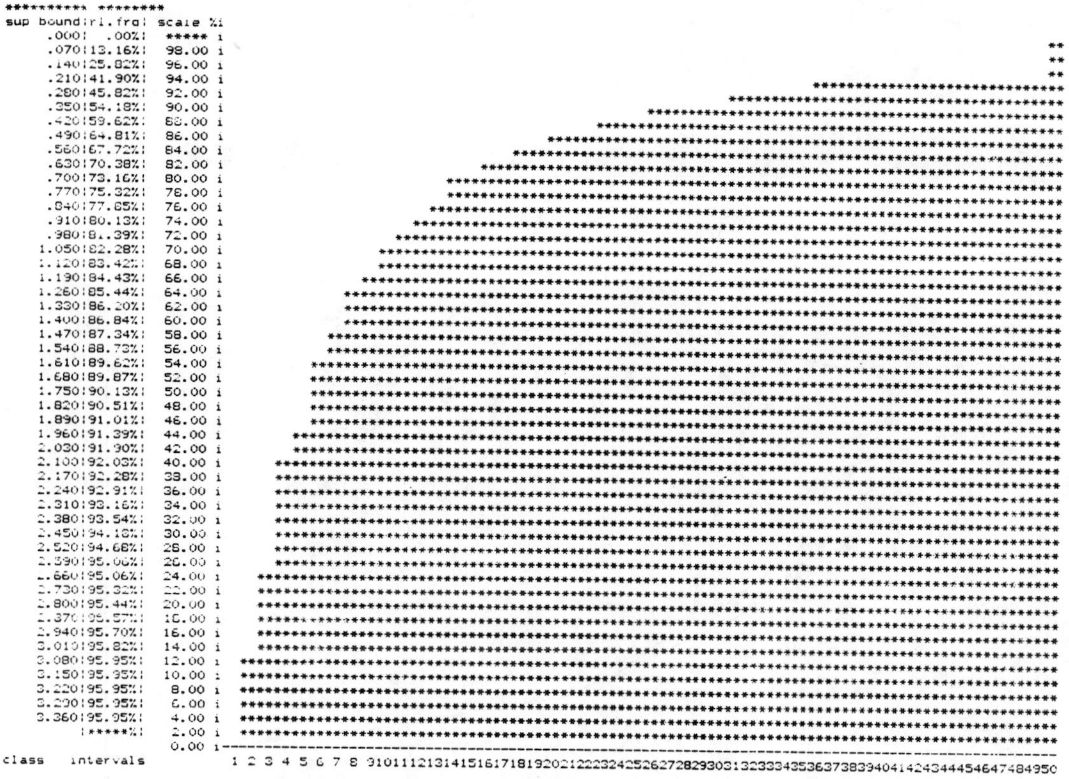

FIGURE 7. Cumulative distribution function of Little Grey Fault zone composites.

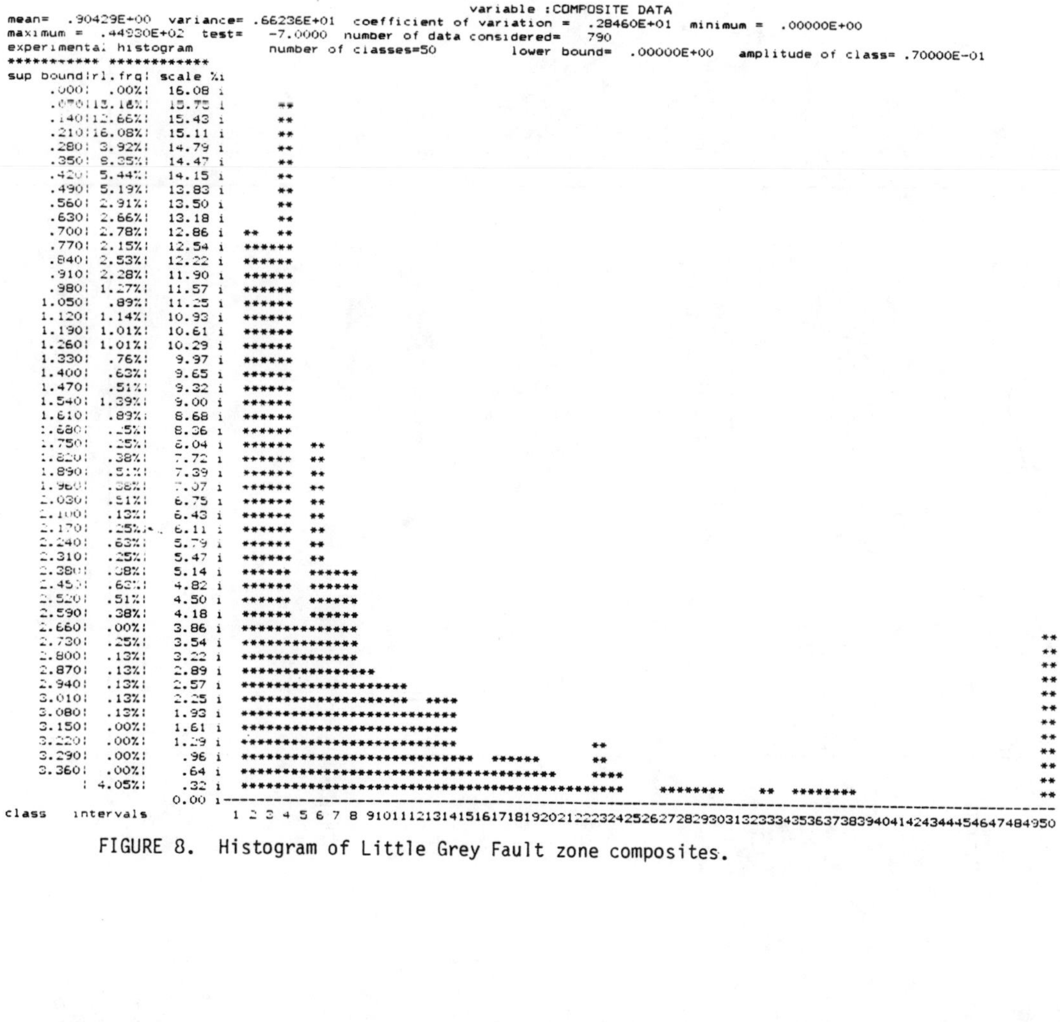

FIGURE 8. Histogram of Little Grey Fault zone composites.

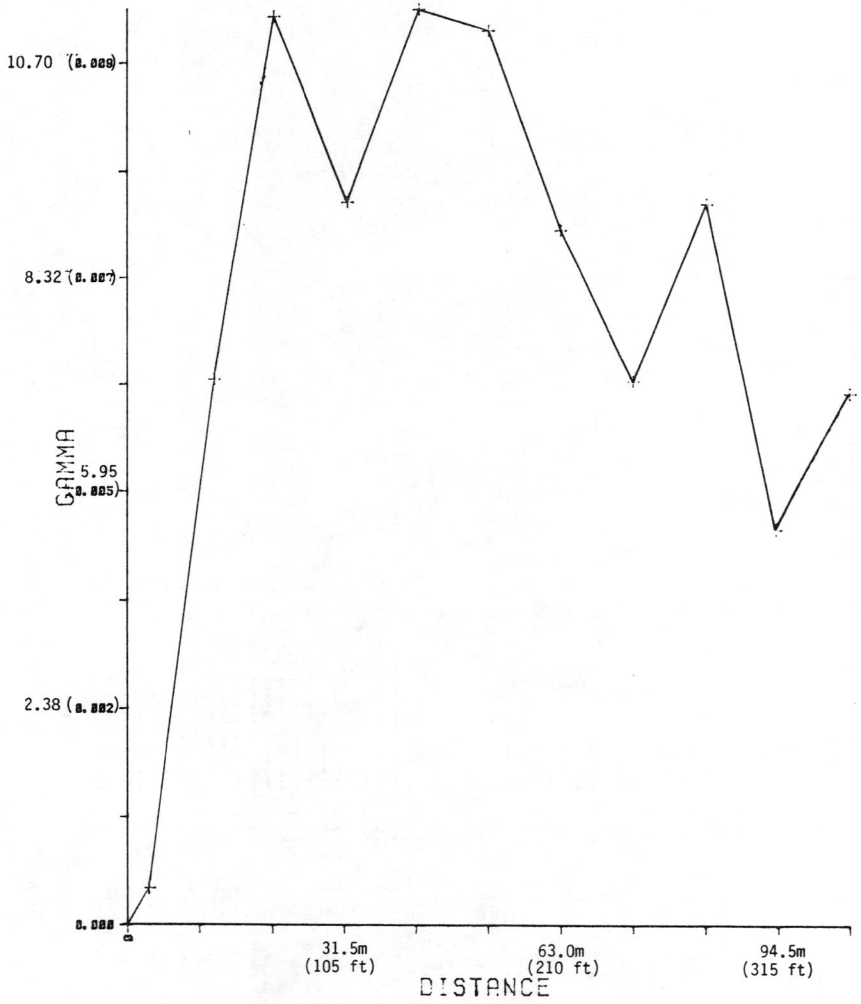

FIGURE 9. Variogram of raw exlporation data.

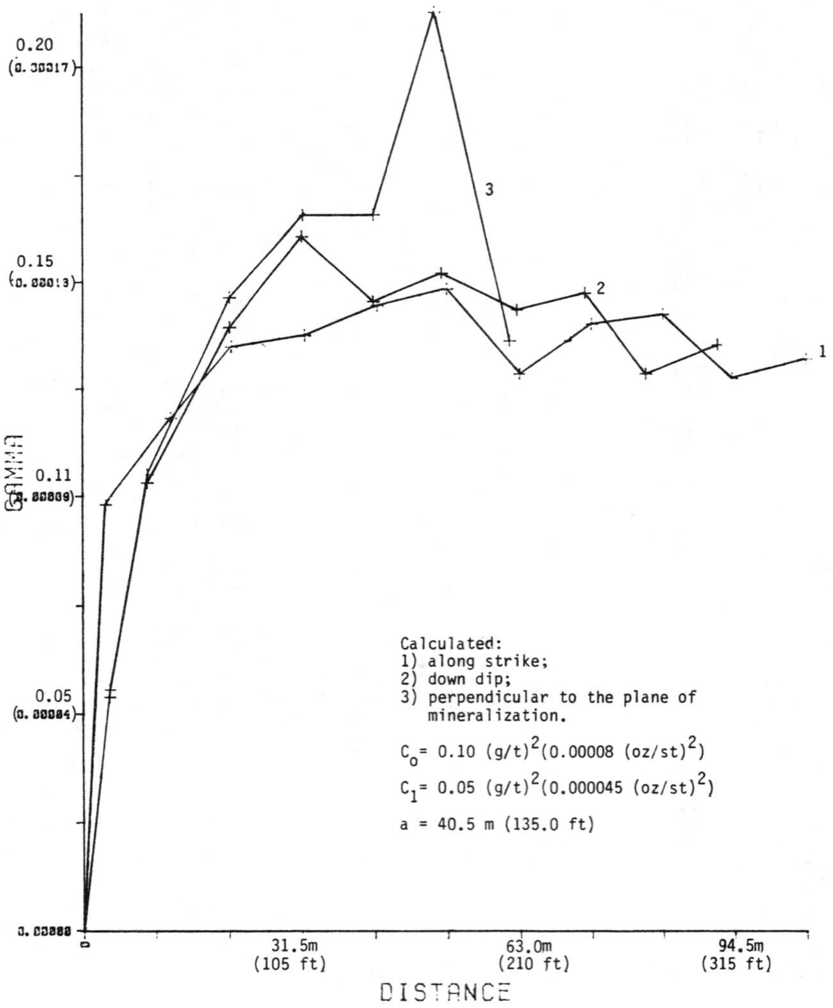

FIGURE 10. Variogram of trimmed exploration data.

the data and the related statistics show that the distribution is positively skewed. Nine indicator cutoffs were chosen in such a way as to divide the data into classes each containing approximately 10% of the data. The indicator cutoff grades are: 0.07 g/t (0.002 oz/st), 0.14 g/t (0.004 oz/st), 0.21 g/t (0.006 oz/st), 0.35 g/t (0.010 oz/st), 0.41 g/t (0.012 oz/st), 0.55 g/t (0.016 oz/st), 0.76 g/t (0.022 oz/st), 1.03 g/t (0.030 oz/st), and 1.72 g/t (0.050 oz/st). In practice, consecutive cutoffs that produce indicator variograms that can be fitted with the same model represent the same spatial phenomena and therefore are unnecessary. The number of cutoffs in such a case would be reduced, and new variograms would be estimated.

Statistics for the production blastholes used for comparison in the conditional probability analysis are also listed in Table 1. Blastholes are drilled on 4.6 m (15 ft) centers, to a depth of 6.6 m (22 ft).

VARIOGRAPHY

The purpose of the variogram is to quantify the structural features of the mineralization into an operational form. The variogram of the raw data (Figure 9) shows no discernible structure. Inclusion of so called outlier data, those data residing in the tail of the distribution, in the variogram calculation, frequently causes the experimental variogram to be highly erratic. The practice of trimming high grade data, i.e. removing the data from the calculation of the variogram and reintroducing the data into the kriging calculations, has been tested as a method to reduce the effects of high grades and allow the underlying structure of the data to be modelled.

For this study 10% of the high grades were removed resulting in a variogram (Figure 10) that can easily be fitted with a simple spherical model. It is evident that the deposit is anisotropic, with the range of the experimental variogram along the strike of the Little Grey Fault being greater than the ranges calculated down dip and perpendicular to the plane of the fault. It must be noted that the mean of the data less the upper 10% has been reduced to 0.38 g/t (0.011 oz/st). The resultant variogram will be representative of grades near this mean value. However, the mining cutoff grade is 0.62 g/t (0.018 oz/st), and the mill cutoff is substantially higher at 1.72 g/t (0.050 oz/st). The underlying assumption when cutting high grades is that the parameters describing the resultant experimental variogram will also describe the spatial structural of the high grades.

For the probability kriging portion of the study indicator and cross variograms were calculated at each of the cutoffs listed above. Also, a variogram of the uniform transform of the data was calculated. Comparison of the indicator variograms for the 0.41 g/t (0.012 oz/st) cutoff (Figure 11) and the 0.76 g/t (0.022 oz/st) cutoff grades (Figure 12) oriented along the strike of the Little Grey Fault zone shows that the range

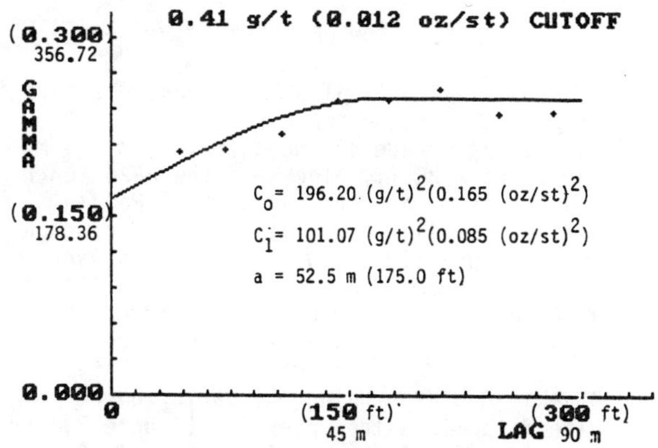

FIGURE 11. Indicator variogram calculated along the strike of the Little Grey Fault zone.

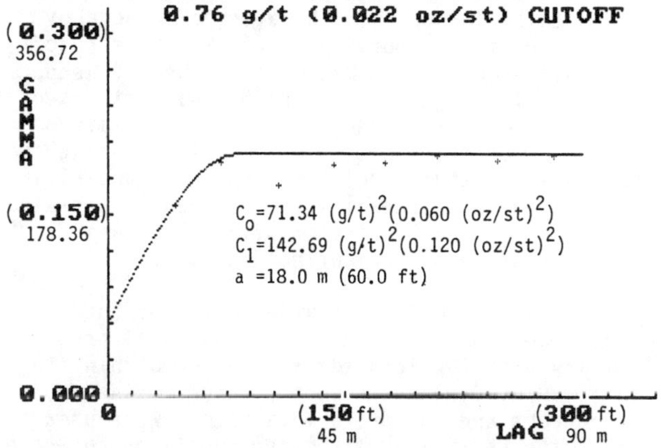

FIGURE 12. Indicator variogram calculated along the strike of the Little Grey Fault zone.

of influence of the higher grades is substantially less than at the lower cutoff. The 0.41 g/t (0.012 oz/st) cutoff indicator variogram shows a range of influence that is approximately equal to the variogram of the raw data less the upper 10%, which has a mean grade of 0.38 g/t (0.011 oz/st). The indicator variograms showed a distinct decrease in range beginning with the 0.55 g/t (0.016 oz/st) cutoff grade, and continued with the higher cutoffs. The nugget variance as a percentage of the total variance was about 50% for the lower cutoffs, increasing to 70% for the intermediate cutoffs, and decreasing to 40% or less for the upper cutoff grades. The sill values were greatest for those cutoffs near the mean of the deposit. Indicator variograms calculated in a direction perpendicular to the plane of the mineralization were completely random above the 0.21 g/t (0.006 oz/st) cutoff. At the 1.72 g/t (0.050 oz/st) cutoff the indicator variogram calculated parallel to the dip of the Little Grey Fault zone was also random, no structure was observed. The variography suggests that the high grade mineralization is occuring in isolated pockets. The long range observed at lower cutoffs suggests that the low grade material occurs throughout the deposit, and, in fact, the ore and waste material is mixed together throughout the fault zone.

ESTIMATION OF RESERVES

INVERSE DISTANCE WEIGHTING - Inverse distance weighting to the second power was used to estimate the average grade of the 15 m x 15 m x 6 m (50 ft x 50 ft x 20 ft) blocks on the 6820 Bench of the West Pit within the Little Grey Fault zone. A search ellipse defined by the variogram of trimmed exlporation composite data was used to restrict input from adjacent composites. Blocks were accepted into the reserve or rejected in total based on the estimated average grade.

ORDINARY KRIGING - Using the variogram of composite data minus the upper 10% (Figure 7) the blocks on the 6820 Bench within the Little Grey Fault zone were kriged using ordinary kriging. The distribution of estimated block grades were found to be lognormal. The average grade above cutoff and the proportion of blocks above cutoff was estimated using a method described by Rendu (1981). The logarithmic variance was estimated from the volume-variance relationship obtained from the logarithm of the block values, using the variance additivity relationship. The underlying assumption in this estimation procedure as stated above is that the variogram of trimmed data is representative of the entire data set.

PROBABILITY KRIGING - Nine indicator cutoff grades were used to estimate conditional probability distributions for each block within the Little Grey Fault zone on the bench. Nine cross variograms and one uniform variogram were used to solve the PK system of kriging equations for each block. Determination of recoverable tonnage and ore grade is based on the estimated conditional probability distributions. Tonnage is estimaed by integrating the area beneath the portion of the conditional probability distribution function that exceeds the specified cutoff grade. Sullivan (1985) has defined two methods of calculating the average grade above a specified cutoff, the direct method, and the indirect method. Both were used in this study.

The direct method of estimating average grade above a cutoff utilizes the following equation:

$$Q^*(zc_k) = \sum_{\alpha=k}^{nc-1} [\emptyset^*(A,zc_{\alpha+1})-\emptyset^*(Azc_\alpha)]*c_{\alpha+1} /1-\emptyset^*(A,zc_k)$$

where: nc = number of cutoffs

zc_k = cutoff of interst

$c_{\alpha+1}$ is a measure of the central tendency for the material between cutoffs $zc_{\alpha+1}$ and zc_α

$\emptyset^*$ (A,zc) is the proportion of the estimated conditional probability distribution function below the cutoff.

The direct method is based on the tonnage estimates and the values used as measures of the central tendency of a class of data.

The indirect method of estimating average grade above a cutoff utilizes the equation:

$$Q^*(zc_k) = \left(m^* - \sum_{\alpha=1}^{k-1} [\emptyset^*(A,zc_{\alpha+1})-\emptyset^*(A,zc_\alpha)]*c_{\alpha+1} - \emptyset^*(A,zc_1)*c_1\right) /1-\emptyset^*(A,zc_k)$$

where: m^* is an estimate of the mean grade.

PERFORMANCE COMPARISON

A comparison of prediction performace was restricted to the 6820 Bench of the West Pit of the Manhattan Mine, at a cutoff grade of 0.55 g/t (0.016 oz/st). The results are summarized in Table 2. Based on these results one can conclude that in the presence of a skewed data distribution with a high coefficient of variation, the PK method using an indirect approach to estimate grade is superior to ordinary kriging and the inverse distance weighting method.

TABLE 2. ACTUAL AND ESTIMATED RESERVES

Method	Avg. Grade Above Cutoff		Tonnage Above Cutoff		Total Recovered Metal	
	g/t	oz/st	t	st	kg	oz
Actual	1.17	0.034	147,000	162,000	172	5,508
IDW	1.48	0.043	242,000	267,000	358	11,481
OK	1.79	0.052	194,000	214,000	347	11,128
PK (direct)	1.45	0.042	137,000	151,000	199	6,343
PK (indirect)	1.31	0.038	137,000	151,000	179	5,739

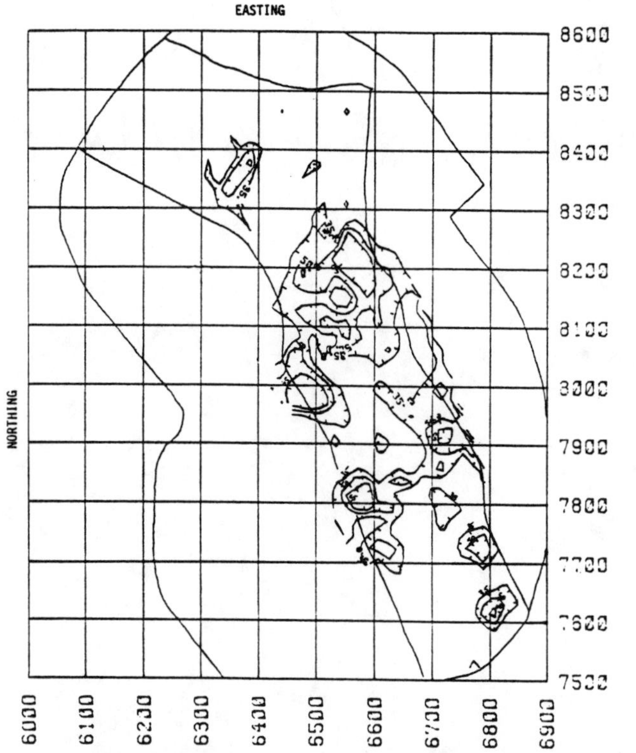

FIGURE 13. 6820 Bench showing contoured conditional probabilities within the Little Grey Fault zone> 0.55 g/t (0.016 oz/st).

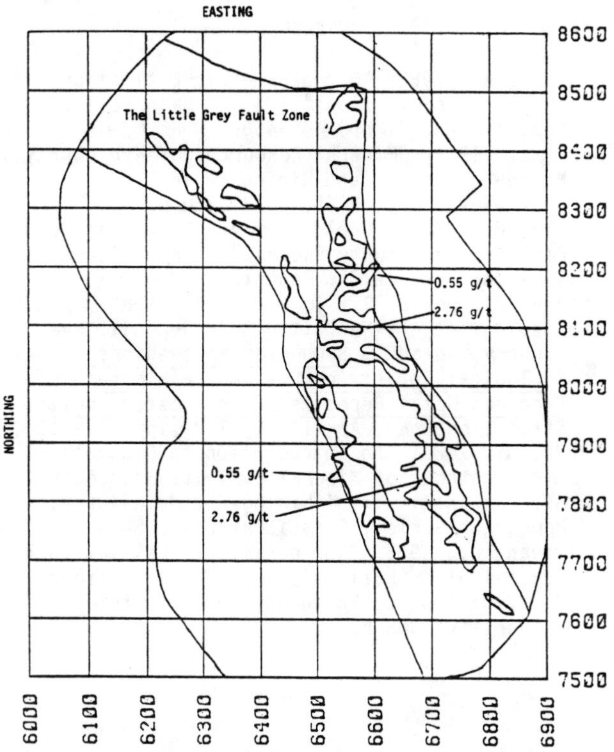

FIGURE 14. 6820 Bench contoured blasthole map with 0.55 g/t(0.016 oz/st) and 2.76 g/t(0.080 oz/st) contours. Outline of Little Grey Fault zone.

MAPPING CONDITIONAL PROBABILITIES

An indication of the spatial distribution of mineralization during the exlporation stage would be useful in understanding the geology of a deposit as well as in the initial mine planning and feasibility work. Conditional probabilities of a block exceeding the 0.55 g/t (0.016 oz/st) cutoff grade have been mapped (Figure 13). Probabilities below 35% have been excluded so that the map is not cluttered. For comparison, the blasthole map for the 6820 Bench has been contoured within the Little Grey Fault zone (Figure 14).

The 35% conditional probability contour line in Figure 13 is a good indicator of the limits of mineralization defined by the 0.55 g/t (0.016 oz/st) contour line in Figure 14. The 50% and 65% conditional probability contours tend to outline the 2.76 g/t contour more so than the lower grade. Of particular interest is the inclusion of a zone of <35% probability, likely to be waste, in the block centered at 8150N 6550E in Figure 13. This zone is immediately between two zones of >60% probability. The presence of waste in this portion of the Little Grey Fault zone between pods of mineralization exceeding 2.76 g/t (0.080 oz/st) is verified by Figure 14.

CONCLUSIONS

The results of this study lead to the following conclusions:

1. The difference in the results (Table 2) using the direct versus the indirect method of determining average grade is due to the inclusion of outliers in the class of data defined by the cutoff 1.72 g/t (0.050 oz/st) and the limit of the data. This class of data is incorporated into all estimates with the direct method. The indirect method will only use this class when estimating the average grade above the 1.72 g/t (0.050 oz/st) cutoff. Therefore, when data sets are positively skewed and the integrity of high grade data may be suspect, the indirect approach should be used to estimate average grade above cutoff.
2. Trimming data to generate a smooth variogram prior to ordinary kriging will introduce serious error when the resultant variogram is not representative of the spatial structure of the high grades.
3. Mapping conditional probabilities for the mining cutoff grade successfully identified the boundary of mineable mineralization as compared to a contoured blasthole map. The map of conditional probabilities also gave an indication of the degree to which internal waste is present in the ore zone.

ACKNOWLEDGMENTS

This paper represents part of a Master's thesis in progress at the Mackay School of Mines. The generous financial support from Echo Bay Minerals Co. and Round Mountain Gold Corp. has made this project possible. A special thanks is due to John Witner, Hal Elson, Bruce Veek, and Kevin Francis who have assisted me throughout the project.

REFERENCES

Journel, A., 1983, "Non-parametric Estimation of Spatial Distributions", J. of Math Geol., Vol. 15, no. 3, pp. 445-468.

Kral, V., 1951, "Mineral Resources of Nye County, Nevada", Geology and Mining Series, No. 50, Univ. Nev. Bull. v.45, no.3, 220 p.

Maddry, J., 1986, "Geology of the Manhattan Gold Deposit, Nye County, Nevada", unpublished Tenneco Minerals Co. Report.

Maddry, J., Coyle, R., & Pettit, P., 1987, "Geology of the Manhattan Gold Deposit, Nye Country, Nevada", Symposium Proceedings, Geological Society of Nevada, Bulk Mineable Precious Metal Deposits of the Western United States, Reno, NV, April 6-8, 1987, pp. 119-134.

Rendu, J., 1981, "An Introduction to Geostatistical Methods of Mineral Evaluation", South African Institute of Mining And Metallurgy, Monograph Series, Geostatistics 2, 84 p.

Sullivan, J., 1985, "Non-parametric Estimation of Spatial Distributions", Ph.D. Dissertation, Dept. of Applied Earth Scienes, Stanford University, 368 p.

Chapter 4

MINIMIZING ERRORS IN DRILL SAMPLING FOR GOLD

by

Frederick P. Schwarz
Senior Supervisory Geologist

Freeport-McMoRan Gold Company
Elko, Nevada

Abstract: Sample preparation and analysis can impose devastating errors on assay data. Minimizing these errors by testing and procedural redesign may be critical to assure representative assays and project success. Replicate samples, taken at any volume-reduction and/or analytical step in a sampling procedure, generate replicate assay values whose relative differences constitute a set of data. Variance of this data judges the addition to total sampling error imposed by the step generating the replicates. A variance more than $(11.5\%)^2$ is excessive for a single step. Sample volume increase and/or comminution at a step will reduce variance for that step. Rejects from well mineralized intercepts can provide bulk samples for testing.

EFFECT OF ERROR

Sampling procedures introduce errors into the reliability of an assay value at every step in sample preparation and analysis. These cumulative errors cannot be totally eliminated, but they can be controlled. Control of error is critical to a project since less error in data generates more reliability in a resource estimate based on that data.

Sample splitting steps generally impose the worst procedural errors. The most serious splitting errors are commonly generated at the drillsite. Analytical procedures can also impose significant errors. Errors not from splitting or analytical steps are usually trivial... unless processing is carelessly done.

A single sampling procedural step can generate so much error, especially in gold exploration, that a reported assay value does not reflect true assay value for the sample. Such error is excessive. Excessive error can unmeasurably modify a given sample so that, notwithstanding the effect on the sample of any subsequent subsampling, assaying, and/or extrapolation, the sample and its assay are ineradicably non-representative.

Non-representative data which results from excessive error is characterized by an assay value distribution dependent only on sample inhomogeneity. This distribution is in effect superimposed on the true distribution of assay values. Combined true and superimposed distributions can not be characterized reliably by classical statistics. In extreme cases, geostatistics may not help.

A common type of superimposed assay value distribution in gold deposits causes most assay values to report lower than true, and a very few to report much higher than true. Thus, deposit grade above cutoff tends to be underestimated.

Assays from a non-representative data set are individually meaningless. Since these assays can indicate fluctuations of grade which are not real, reasonably accurate zoning of the deposit by grade may be impossible. Thus, with excessive error, the true assay distribution and a reasonable characterization of the orebody might never be known.

MINIMIZING ERROR

For a single step of a sampling procedure, a practical limit for excessive error is considered to be when that step generates a data set of relative assay differences whose variance (the conventional statistic) exceeds $(11.5\%)^2$. With this limit in mind, the explorationist can test, judge, and minimize error.

A careful examination of sampling procedure is desirable early in the drilling stage of property evaluation. Such examination will reveal those procedural steps which may cause unacceptable error and, therefore, problems in assay reliability. These steps can be redesigned so that error is minimized. Minimizing error will improve accuracy of project evaluation work, including: geologic conclusions, grade estimates, economic studies, and, eventually, mine planning.

ESTIMATING ERROR USING VARIANCE OF REPLICATE ASSAYS

Assays of replicate samples isolated at any sample processing or analytical step can provide a data set comprised of relative differences between replicate sample assay values. Variance of this data set can be used to judge the magnitude of error which is generated by the step. Replicates from sample splitting of ore-grade mineralization usually are most useful for testing, since volume-reduction steps commonly generate highest variances.

The population of relative differences between replicate assays is a data set which shows a normal (Gaussian) distribution of values. A data population of actual assay values usually does not show a normal distribution. Thus, estimating error by use of the variance of a population of relative differences avoids many problems associated with the statistics of non-normal distributions.

The relative difference between assay values is shown by the relationship:

$$\% \text{ difference} = 100 \frac{(O-R)}{(O+R)} \qquad (1)$$

for original (O) and replicate (R) assay values. Thus, variance of a population of relative differences is expressed in units of percentage-squared (or $(\%)^2$).

TESTING AN ESTABLISHED PROCEDURE

Goals and General Approach

A drillhole sampling procedure is tested to discover if any processing step generates an excessive variance (more than $(11.5\%)^2$ for replicate assays), so that the step can be modified. Also, if variance of a step is substantially less than $(11.5\%)^2$, the procedure might be redesigned to reduce processing cost.

Once a sampling procedure has been optimized, reliability of data from later drilling is insured, and earlier samples from critical intervals, reprocessed and reassayed, become more useful. Thus, all the drillholes will provide reasonably representative assay data. The practical goal of testing an established procedure is to assure that reasonably representative assay data is generated routinely at reasonably minimized cost.

In the beginning stages of a drill evaluation program, samples for testing come from bulk reject drillhole cuttings. Assay values from these cuttings have confirmed interest in the project by returning "ore-grade" intercepts. Considering the value at risk in decisions which depend on reliable assay values, temporarily saving all bulk rejects from exploration drilling can be justified.

Example Procedure

For the first few holes in a drilling program, sampling procedure is typically based on convenient logistics, standard laboratory procedures, company tradition, etc.. After substantial and continuing interest is confirmed, initially expedient procedure should be validated.

Figure 1 shows an example procedure which begins with 56 kg of sample, the material from a typical 5 ft run of rotary reverse circulation drilling. The sample is split 50-50 on each pass through riffle splitters, and reject material is set aside. Retained material is comminuted at various stages. Finally, a pulp provides the analytical sample.

Figure 2 illustrates a test of the sampling procedure from Figure 1. Reject drill cuttings, saved from the original process, supply replicate samples at several splitting steps as follow:

[A] at 50 kg of cuttings;

[B] at 12.5 kg of cuttings;

[C] at 3120 g of -10 mesh material;

[D] at 780 g of -80 mesh material;

[E] at 32 g (or 1 assay ton (AT)) of -200 mesh pulp.

In practice, every sample splitting step of the established procedure could be similarly tested. At a minimum, several 50 kg samples should be tested.

Samples from each pair of procedurally equivalent "branches" in Figure 2 are treated independently. It is important to carefully document sample lineage, thus allowing comparisons of appropriate replicate assays. Assays resulting from testing the procedure are as follows:

a) The 50 kg of material is split, at [A], into a pair of replicate samples leading to assay #1 along one branch and assays #2, #3, #4, #5, and #6 along other, procedurally equivalent branches.

b) The 12.5 kg of material is split similarly, at [B], leading to assay #2 along one branch and assays #3, #4, #5, and #6 along others.

c) The 3.12 kg of material is split similarly, at [C], leading to assay #3 along one branch and assays #4, #5, and #6 along others.

d) The 780 g of material is split similarly, at [D], leading to assay #4 along one branch and assays #5 and #6 along others.

e) Analytical samples, run on -200 mesh pulp, are selected at [E], leading to assays #5 and #6.

Errors generated by the selected procedural steps of Figure 2 are judged by variances of populations of differences between replicate assay values. These variances are calculated, per formulation (1) discussed above, for pairs of values considered as follows:

a) Assay #1 versus assays #2, #3, #4, #5, and #6 produce variance [A]

b) Assay #2 versus assays #3, #4, #5, and #6 produce variance [B].

c) Assay #3 versus assays #4, #5, and #6 produce variance [C].

d) Assay #4 versus assays #5 and #6 produce variance [D].

e) Assay #5 versus #6 produces variance [E], the analytical error.

REDUCING THE VARIANCE

Reduction of excessive variance in part depends on available equipment and other logistical considerations, budgets, and project objectives. However, in order to avoid potentially devastating problems, a troublesome sample preparation step can be replaced with a step which reduces sample grain size, increases sample volume, and/or both. (Analytical variance might be reduced by using 2 or 3 AT; ie: "increasing volume".)

A minimum of several samples can be reprocessed, using the modified procedure. After accumulating paired replicate assay values from modified steps, the general approach as described above is repeated. Variance for differences between replicate assay values generated by volume-reduction steps should not exceed $(11.5\%)^2$. Establishing this fact will test if modifications have effectively optimized sampling procedures. The optimized sampling procedure should spread increments of error more-or-less evenly among the several steps of sample preparation and analysis.

After establishing an effective procedure, sample preparation and analysis should be routinely monitored for precision and accuracy during the drilling program. Excessive error from careless personnel, laboratory mistakes, changed ore characteristics, etc., otherwise may flaw assay data unexpectedly.

CONCLUSION

In preparation and analytical procedures for drillhole samples, ineradicable errors can destroy representativity of a sample. Such errors, which may fatally flaw a project, typically come from sample volume reduction and from assay procedures. Drillsite splitting steps commonly cause very substantial errors.

The key to optimizing sampling procedure and reasonably minimizing error resides with an explorationist who understands the sampling process. Comparison of procedurally identical replicate samples from bulk rejects of ore-grade intercepts provides a way to examine error.

Replicate samples, taken at a volume-reduction step in a drillhole sampling procedure, generate replicate assay values whose relative differences constitute a population of data. The variance of this population of differences can be used to judge error generated by the step which inaugurated the pair. A variance of more than $(11.5\%)^2$ for such data from any single step is considered excessive.

A procedural step can be redesigned if it generates non-representative assay data due to excessive variance. For a sample, excessive variance is minimized by increasing volume or by reducing particle size at a troublesome step. The modified procedure is tested for effectiveness and routinely monitored throughout the program to ensure continuing reliability of assay data.

An explorationist should avoid the unpleasant consequences of discovering, at the end of a drilling program, that assay data does not reasonably reflect samples taken. It may then be too late to solve the problem.

Acknowledgments: Thanks are due to Freeport McMoRan Gold Company for permission to publish this paper, and to Francis Pitard of Lakewood, Colorado, for teaching that a procedural step which increases variance of the assay value population by more than $(16\%)^2$ can lead to non-representative data.

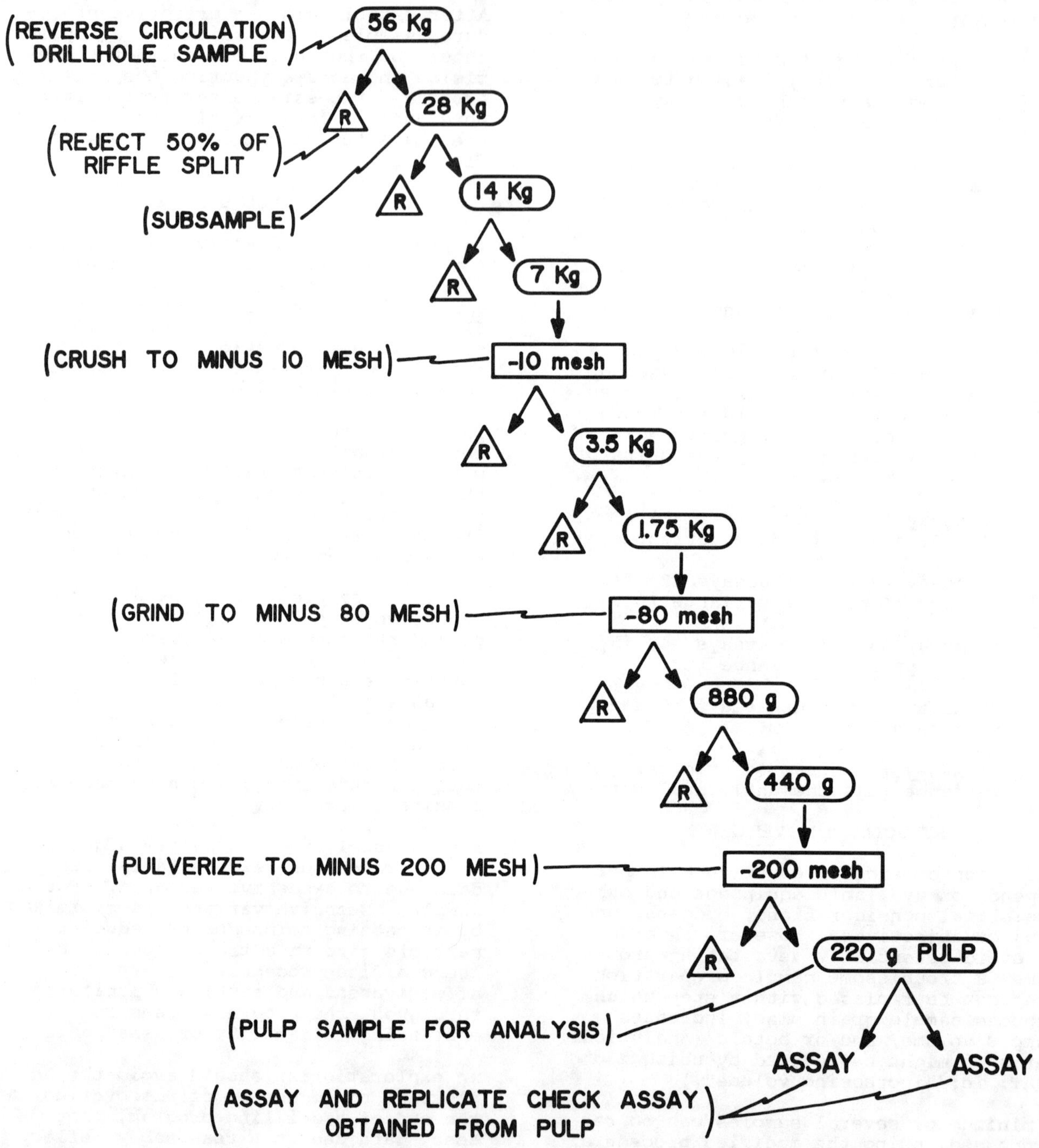

Figure 1. REVERSE CIRCULATION DRILLHOLE SAMPLING and ANALYTICAL PROCEDURE

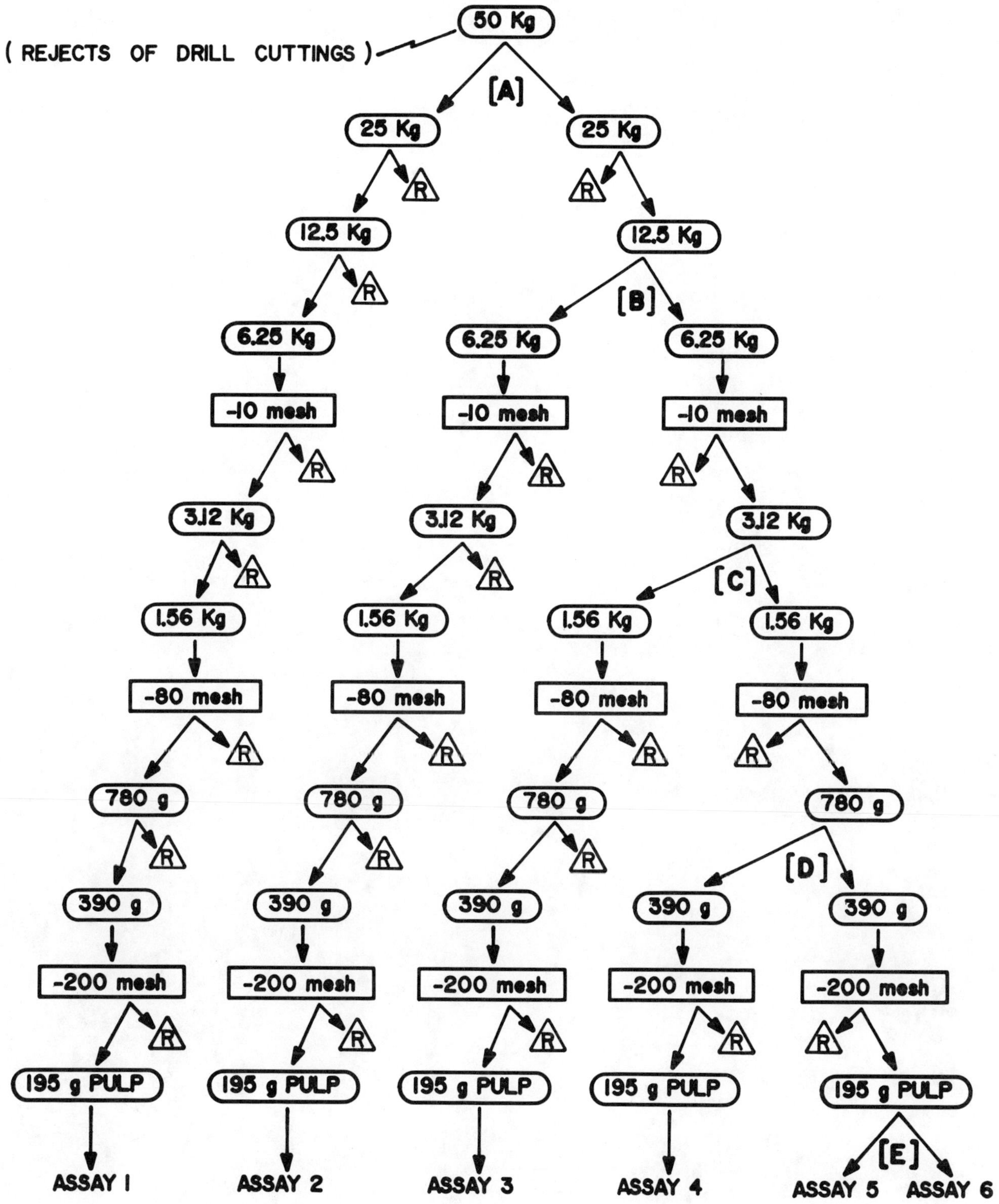

Figure 2. ASSAY STUDY OF REVERSE CIRCULATION DRILLHOLE SAMPLING and ANALYTICAL PROCEDURE

2

Geology/Exploration II Geophysical Exploration

N. Muncaster
Echo Bay Mines
Denver. Colorado

G.A. Travis
Ranger Exploration
Applecross, W. Aus., Australia

Chapter 5

HIGH RESOLUTION AIRBORNE GEOPHYSICS FOR EPITHERMAL GOLD EXPLORATION IN NORTH–EASTERN QUEENSLAND

S.S. Webster[1], R. Henley[2] and L. Harris[3]

ABSTRACT

High resolution airborne magnetic and radiometric data have been shown to optimize exploration for epithermal gold deposits in the vicinity of Charters Towers, northeastern Queensland.

The radiometric data allow extrapolation of poorly outcropping geology to provide a cost-effective mapping technique, even at exploration scales (eg. 1:25,000).

At a regional scale (eg. 1:100,000) the magnetic data reflect the regional tectonics and divide the area into domains for the application of different genetic models.

For example:

1. the Ravenswood Batholith is characterized by multiple intrusions dating from Ordovician to Permian age. The polarity of the magnetic anomalies implies an age for the intrusions and indicates a deposit style for exploration.
2. extensional and wrench tectonics along the northern margin of the Drummond Basin are interpreted from the magnetic data and indicate favourable blocks for detailed exploration.

Genetic exploration models may be tested in favourable environments by the recognition of geophysical signatures. These signatures reflect structural, lithological and alteration patterns arising from controls on ore deposits and can be applied at regional or detailed scales.

INTRODUCTION

Gold deposits in north Queensland have been classified by Morrison (1988), based on geological setting, mineral style and alteration. The diverse geophysical parameters of these deposits can be used to supplement this convenient classification, and provide regional signatures to assist exploration.

The cost benefit of airborne geophysical surveys is realised because data coverage of an area is complete and not hampered by lack of outcrop. Radiometric data give "analyses" of rock or detrital soils which can easily be transformed to a geological interpretation. Magnetic data reflect properties of outcropping or sub-surface rock, which may also be interpreted as a geological map, and may be enhanced by correlation with the radiometric data.

The integrated geological-geophysical interpretation derived from airborne geophysics provides a basis for targetting potential ore environments (e.g. calderas, structures, alteration zones, etc.). The aim of follow-up geology, geochemistry and surface geophysics is to localise mineral concentrations within these target areas.

REGIONAL CHARACTERISTICS

As in gold mineral districts elsewhere, the most common geological characteristic of the varied deposit types is that of deep-seated structures, which can provide zones of weakness for subsequent intrusion or channel-ways for hydrothermal, mineral bearing solutions. This factor dictates the scale of regional geophysical surveys used in exploration. The data set must cover a sufficiently large area to recognise disconnected segments of structures, crosscutting features and localised splays - all of which can control mineralisation. Structural interpretations may localise dilational positions within such deformation zones.

Examples of four mineralising environments described by Morrison (1988) are listed in Table 1 with their respective geological and geophysical signatures. This table also lists contour characteristics for the recognition of these geophysical signatures.

PLUTONIC

The rich plutonic gold vein deposits at Charters Towers are hosted by granodioritic rocks and are extensive, tabular, quartz reefs in fissures, or lenticular anastomosing quartz bodies in shear zones. Lode directions in the main mining area at Charters Towers generally strike E-W or N-S (Peters, 1987). with further ore shoots at the intersection with NW-SE and NE-SW structures.

The geophysical data (Fig. 1) available over the Charters Towers deposits include regional (BMR) and high resolution (Aerodata) detailed airborne magnetic and radiometric data (plus regional BMR gravity data at 11 km station spacing). The deposits at Charters Towers lie at the intersection of two major structural features :– the E-W trending Mosgardies Shear Zone and the NE trending Charters Towers Lineament, both of which are clearly expressed as linear zones in the regional geophysical data. WNW-ESE and NW-SE linears are also apparent in the Charters Towers area.

1. General Manager, Austirex International Ltd., 27 Merriwa St. Gordon, NSW 2072
2. Consulting Geologist, Epithermex Int. Pty. Ltd., 4 Juad Pl. Aranda, Canberra 2614
3. Lecturer in Structural Geology, Geology Dept., University of W.A. Nedlands 6009

TABLE 1

GEOPHYSICAL PATTERNS OF GOLD DEPOSITS IN NORTH QUEENSLAND

STYLE	TYPE EXAMPLE	SIGNATURE	AIRBORNE MAP RECOGNITION
PLUTONIC	CHARTERS TOWERS (Qld.)	– Deep seated structure – Broad pattern of intrusive	– Alignment of contours – Broad circular/elliptical contour pattern
PORPHYRY RELATED VEIN	RAVENSWOOD FAR FANNING (Qld.)	– Deep (and shallow) structure – Shallow intrusive pattern	– Alignment of contours – Offset of trends – Sharp circular/elliptical contour pattern
PORPHYRY RELATED CALDERA (DIATREME BRECCIA)	PEAK HILL (NSW) MT. LEYSHON (Qld.)	– Intersecting structures – Porphyry pattern . . . magnetic aureole . . . negative anomaly – Potassium enrichment	– Offset of trends – Circular/elliptical contours with low centre – Circular/elliptical contours with "strong" "negative" core – Radiometric aureole
EPITHERMAL VEIN AND DISSEMINATED	TEMORA (NSW) PAJINGO (Qld.)	– Magnetite destruction – Shallow structure/offsets – Potassium enrichment	– Broad "weak" magnetic low (circular pattern) – Sharp, elongated contours with offsets – Radiometric aureole

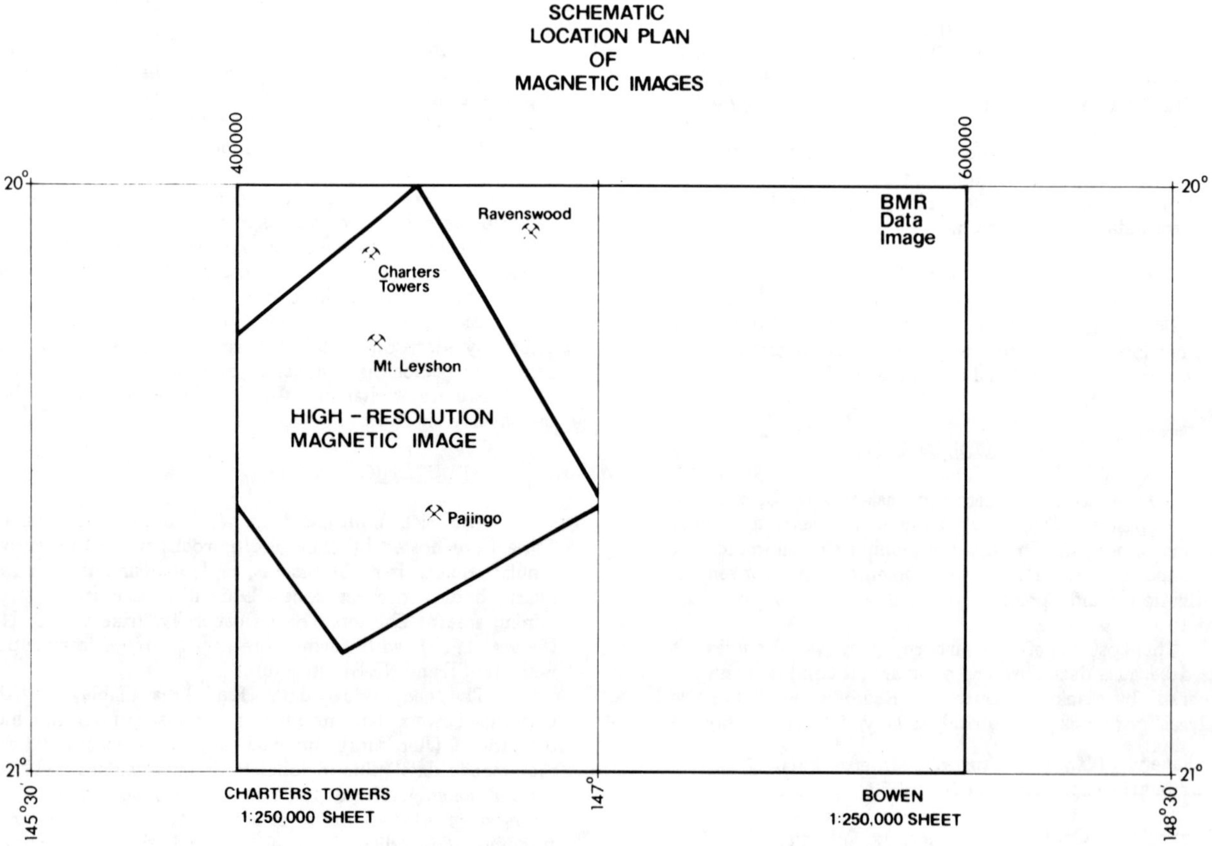

Figure 1: Location diagram showing coverage of multi-client data set within the Charters Towers 1:250,000 sheet.

VOLCANIC – SEDIMENT HOSTED

Porphyry related - vein

Two examples of porphyry related vein mineralisation are the fissure veins of the Ravenswood type and shear related veinlets at Far Fanning. The geophysical signature of the Ravenswood mineral district can be illustrated by the BMR data (Figs. 2 and 3) which show a circular, dipole anomaly some 10 km in diameter, located at the intersection of at least three major structural elements. The deposits cluster within the magnetic anomaly and are localised by structures. They occur along the NW-SE feature and, where this structure intersects the Mosgardies Shear Zone, along a divergent splay.

Porphyry related - diatreme

High level volcanic-hosted deposits, subdivided in Table 1 as porphyry related (vein, diatreme breccia) and epithermal types, are commonly associated with major regional structures (Henley, 1989). Linear features are clearly visible in the magnetic data as alignment of contours, anomaly offsets, disruption of trends etc.

Located along these linear trends are circular or elliptical magnetic patterns, the areal extent of which reflect the caldera size. Also evident are broad zones of increased radiometric response, particularly in the potassium channel, due to alteration associated with the mineralising solutions (Irvine, et al., 1989). For this deposit style the pattern is either

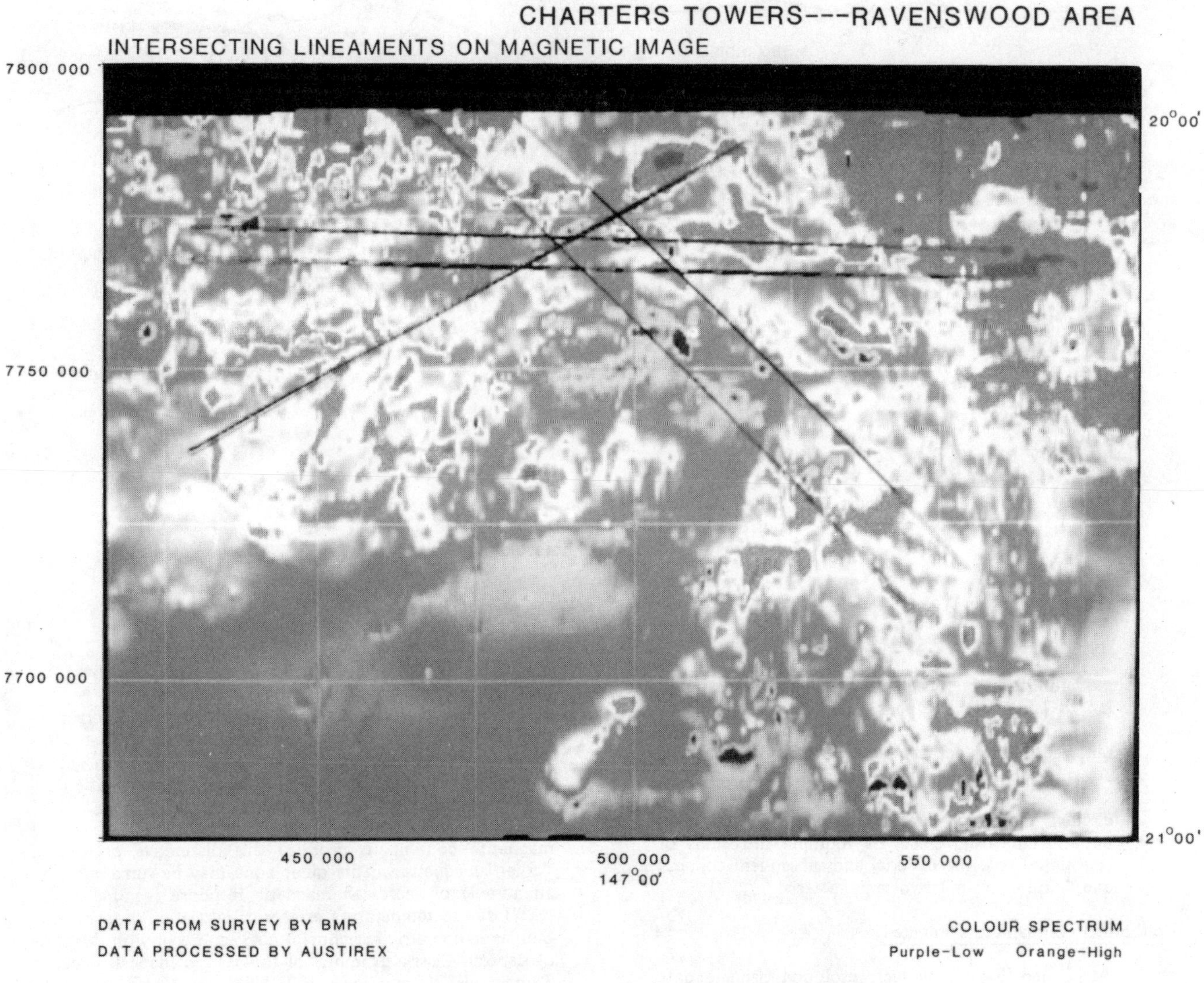

Figure 2: Enhanced image of BMR magnetic data for Charters Towers and Bowen 1:250,000 sheets showing some linears of importance for mineral occurrence.

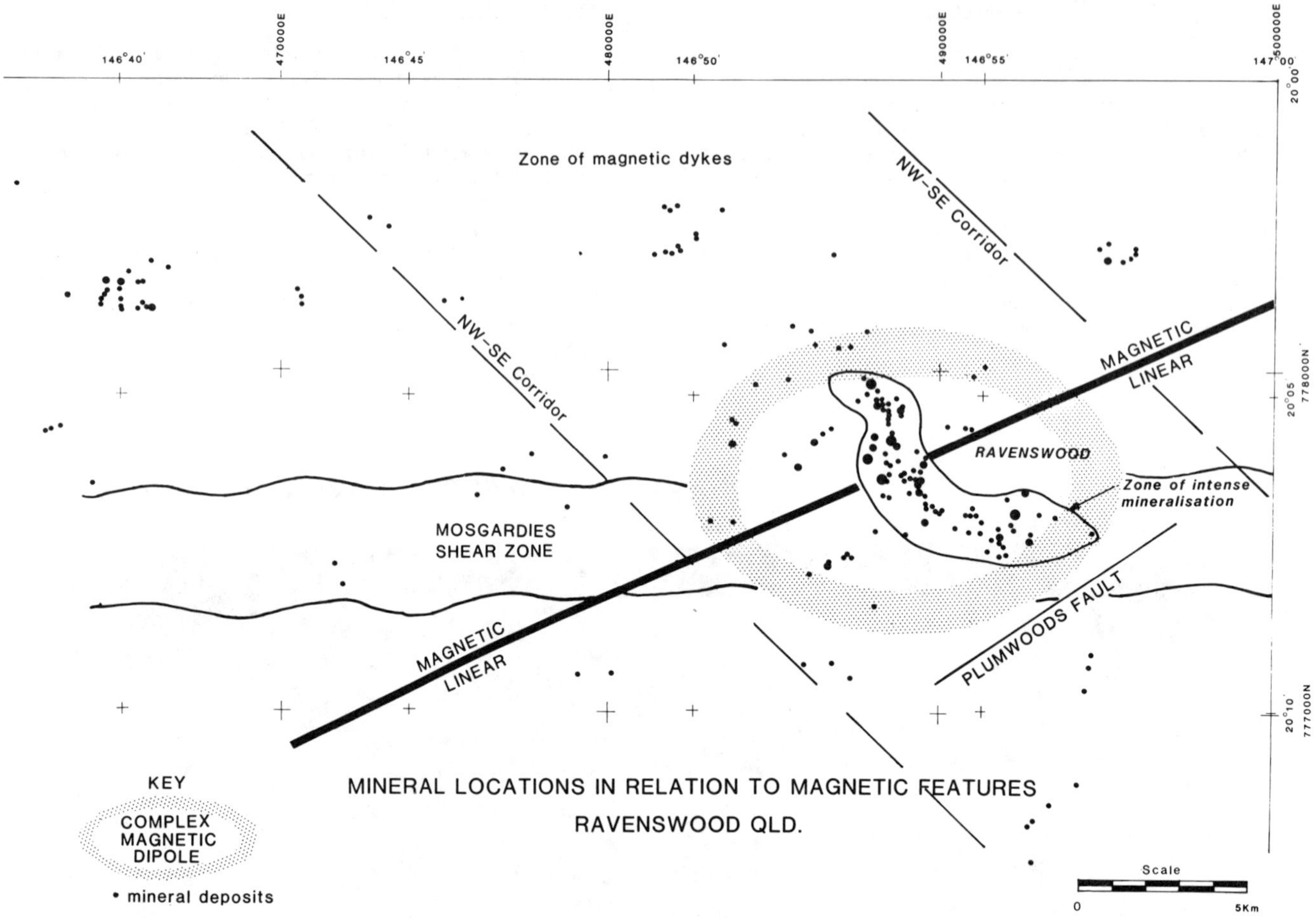

Figure 3: Schematic interpretation of Figure 2 showing position of mineral deposits within a magnetic feature and related to intersecting structures.

1. moderate to strong magnetic "negatives" corresponding to late magmatic phases, intruded at a time of magnetic field reversal. These intrusions indicate renewed igneous activity (eg. at Mt. Leyshon, Fig. 4) along the controlling fault or conduit.

2. circular magnetic aureoles of simple or complex pattern indicating single or multiple intrusions of (magnetic) volcanic material and subsequent collapse and/or fill by a similar volcanic material.

Epithermal - vein and disseminated

At Pajingo (Fig. 5) the high resolution magnetic data are essential to resolve the geological framework in an area of little outcrop. An E-W trending structure terminates a block of poorly exposed magnetic volcanics adjacent to a non-magnetic sedimentary sequence. A deep circular magnetic feature is observed within these sediments. Strong structural elements within the volcanics may be interpreted to indicate favourable dilation zones which coincide with the Scott Lode Vein and other mapped vein systems.

A geophysical signature, only recently observed in this north Queensland data set, are mineral environments where the geophysical expression is one of quiet magnetic response. Epithermal fluids (mixtures of deep magmatic fluid and convecting groundwaters with a reduced sulphur (H_2S) concentration (Henley, 1989)) deplete host rocks of magnetite content to produce this distinctive pattern of "quiet" magnetism. This quiet zone may be surrounded by an aureole of increased magnetic response (eg. Mt. Hope, NSW) due to temperature effects or relocation of magnetite. Similar radiometric enrichment also can occur with this style of deposit. Classic examples of this type of deposit occur at Temora, NSW, and Peak Hill, NSW. At Martha Hill near Waihi in the Coromandel area of New Zealand, the magnetic volcanic units exhibit highly disturbed magnetic anomalies. However, in the vicinity of mineral occurrences there are large areas of quiet magnetic response, which probably reflect magnetite depletion and constitute a distinctive geophysical signature.

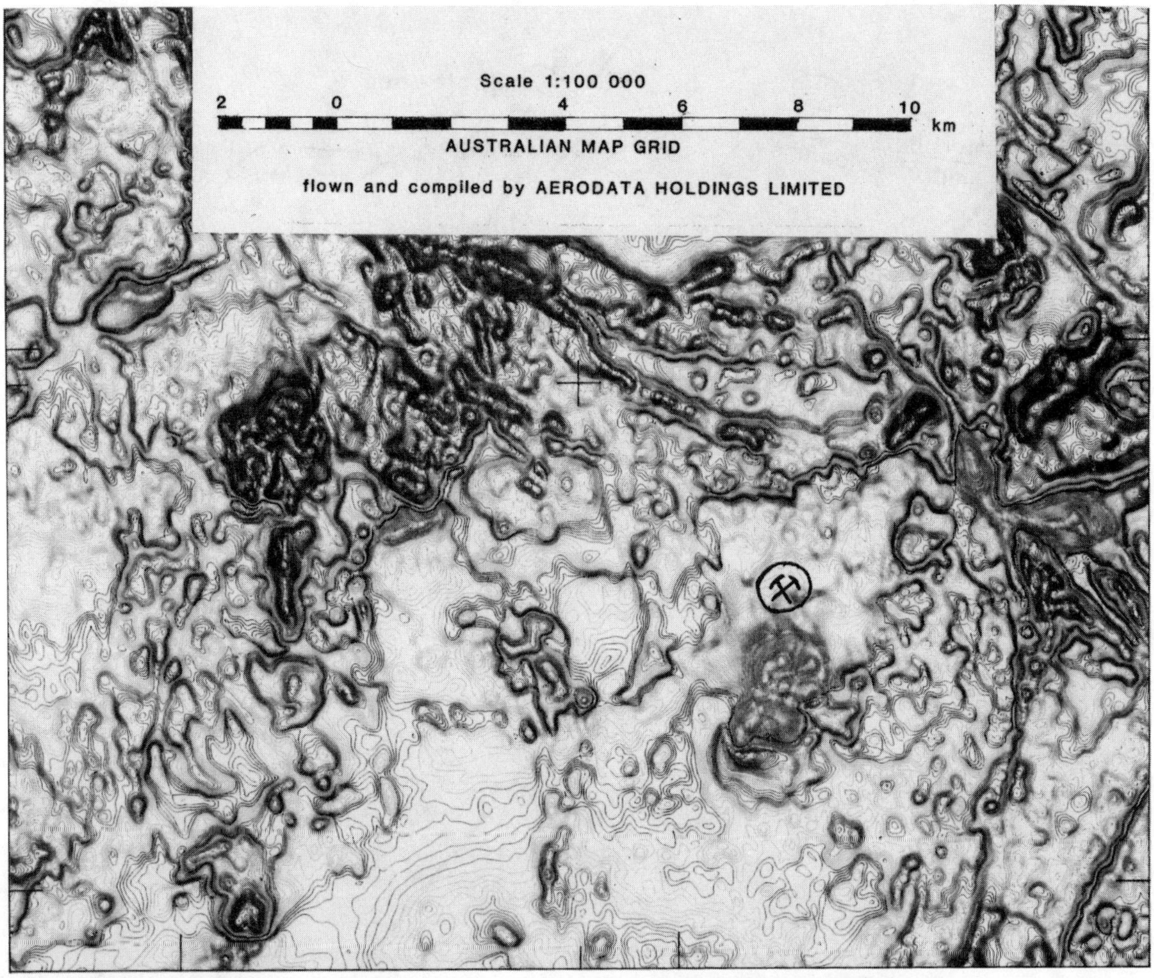

AEROMAGNETIC CONTOURS - LINE SPACING 200m

Figure 4: Multiclient data set in the vicinity of Mt. Leyshon showing a strong negative anomaly and a flanking magnetic high.

CONCLUSIONS

High resolution aeromagnetic and radiometric data have the potential for a major impact on exploration stategy. They provide cost effective data at regional scales which may otherwise be unavailable due to lack of earlier geological mapping or to poor exposure. Interpretation of signatures in such data provides definitive exploration target areas which may be followed up immediately by field geology, geochemistry and geophysical programmes, to locate buried gold deposits.

REFERENCES

Henley, R. W., 1989. Epithermal gold deposits in volcanic terranes,. in Foster R. P. (ed.), Metallogeny and Exploration of Gold, Blackie, (in press.).

Webster S.S. and Henley R.W. 1989
Application of high resolution airborne geophysics to epithermal gold exploration: in northeast-Queensland and Coromandle, New Zealand, ASEG Conference Melbourne, September, 1989.,

Irvine, R., Smith M.,
Geophysically significant characteristics of Epithermal environments, (in prep.).

Morrison, G., 1988. Gold 88 Abstracts, pp 91-101.

Peters, S. G., 1987. Geology and lode controls of the Charters Towers goldfield, northeast Queensland, EGRU Contribution 19, pp 117 (James Cook University of Townsville).

Porter, R., 1988. The Pajingo Gold Mine, Bicentennial Gold Excursion Guidebook, EGRU Contribution 29, pp 23-34.

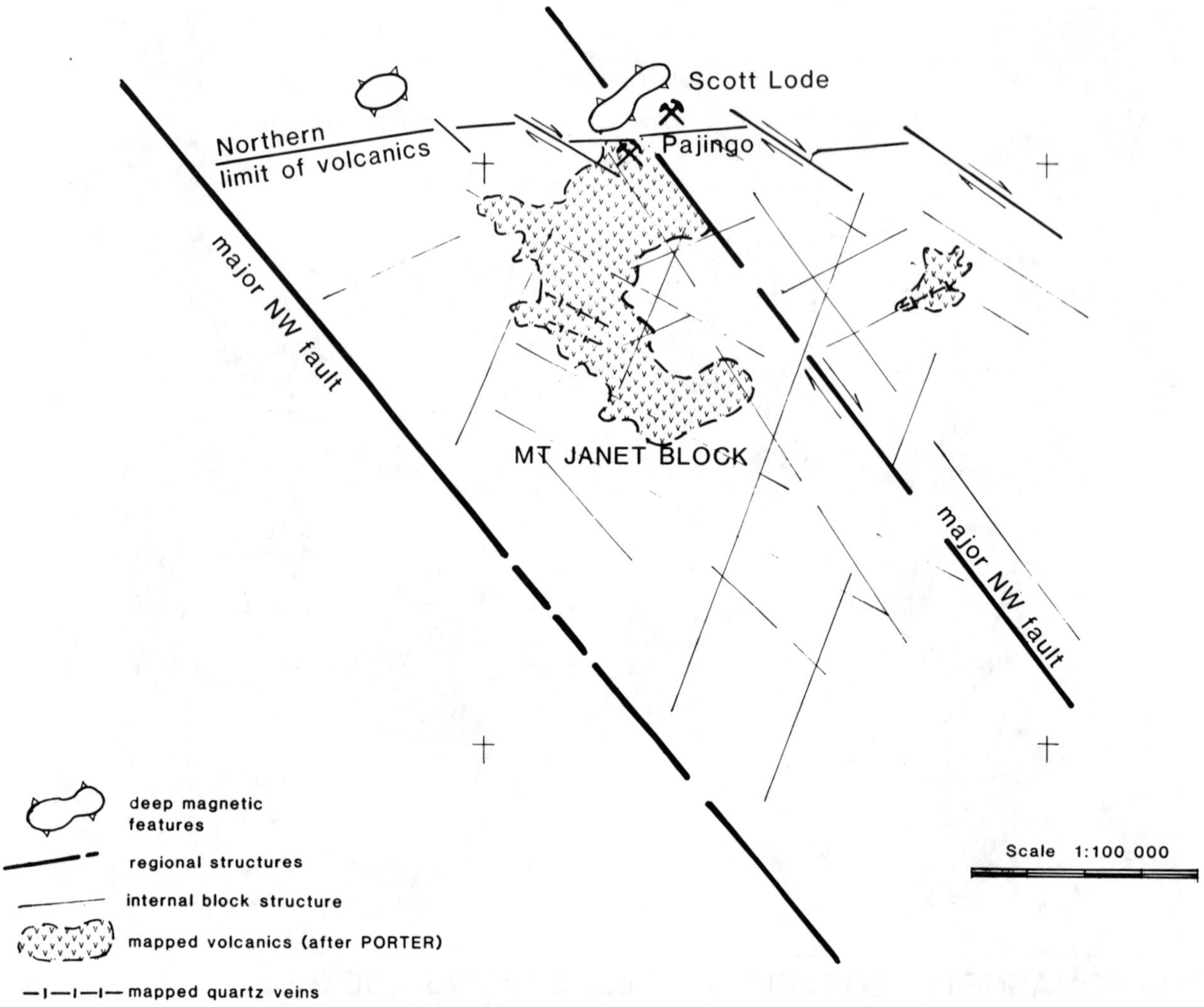

Figure 5: Structural elements in the vicinity of Pajingo.

Chapter 6

THE APPLICATION OF GEOPHYSICS TO INTEGRATED GOLD EXPLORATION IN CANADA

S.J. Bate [1], M. Boisvert [2] and M. Robb [3]

(1) MPH Consulting Limited, Toronto, Ontario, Canada
(2) Minnova Inc., Chapais, Quebec, Canada
(3) Giant Yellowknife Mines Limited, Yellowknife, NWT, Canada

ABSTRACT

The never-ending quest for gold-bearing ore deposits is a primary industry in Canada, employing people of widely differing expertise. The exploration models for gold deposits are generally geologically complex and the targets have relatively small cross-sectional areas. These problems are compounded in Canada by the minimal outcrop exposure in most mineral provinces and the high percentage of water-covered areas. Naturally the chances of successfully delineating an economic gold deposit are very much greater if all parties involved can act as a team of explorationists constantly conferring amongst themselves particularly as the applicable disciplines generally overlap and complement each other.

Three case histories (see Figure 1) are presented to illustrate this philosophy with emphasis placed on the role of the geophysicist in translating the exploration model as defined by the geologist into physically measurable parameters and identifying target areas with similar characteristics. The Conwest-Miminiska property in Ontario is a grassroots exploration play. Additional ore reserves needed to be delineated on the Salmita property in the Northwest Territories. At Laura Lake, near Chapais, Quebec geophysical methods were employed to trace thin high-grade mineral intersections under deep overburden and severe cultural interference.

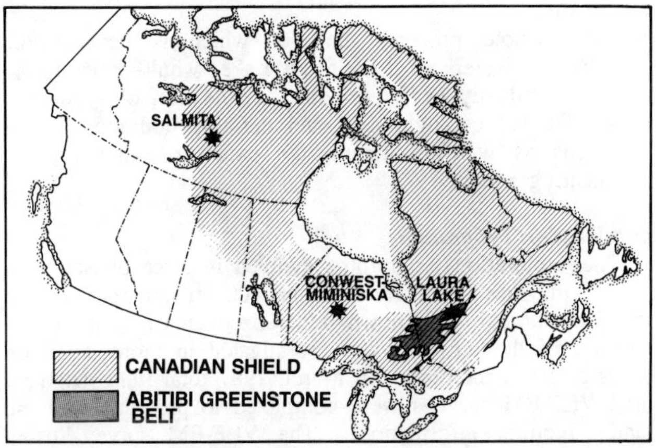

FIGURE 1. *Property locations in Canada*

INTRODUCTION

Gold, an elusive mineral much in demand, is the principal economic mineral in many deposits and also commonly occurs as by-product mineralization in base metal deposits. Discovering gold deposits could be described as looking for a needle in a haystack except that the search can be focussed constructively by the various geological models which have been developed and are evolving from studying known deposits. C.J. Hodgson (1987) summarizes the current philosophies:

> *In the past five years, there has been a major swing back to traditional views of gold deposits being epigenetic, structurally controlled and late in the geological development of greenstone belts ..... At present, the main battle is between those who profess a magmatic origin for the ore solution and those who profess a metamorphic origin.*

Opinions on the principal exploration criteria are therefore quite diverse. The prevailing economic, environmental and political climates will also have a bearing on when and where exploration will be undertaken. However, the general pattern of development from grassroots exploration to mineable deposit falls within, and rarely deviates from, a time-honoured pattern. A general area is first selected and often, though not necessarily, a multi-parameter airborne geophysical survey is completed. Priority areas are then established and subjected to a cyclical programme of geologic mapping and evaluation, geophysical surveying and, if warranted, diamond drilling. If the results dictate, underground development and, eventually, commercial mining operations commence.

At each stage the exploration risk is reduced by combining and evaluating the results from a variety of disciplines to produce the best fit of all the available parameters which are indicative of a particular ore deposit type or geology.

The scope of this paper is confined to the ground exploration phases where the results from several geophysical techniques will provide a broad band of parameters which can be screened with the available geologic information and multi-element geochemical results. The discussions in the three case histories presented are weighted in favour of the geophysical techniques employed and the relative success of the results. In each case the regional geology and exploration model proposed at that time are described and the reasons given for adopting particular geophysical techniques. The results of subsequent geological evaluation are then presented. Assay results are presented in grams per metric tonne and equivalent ounces per short ton unless otherwise stated.

The case histories are confined to hydrothermal gold mineralization within the Canadian Shield (Figure 1) but the ideas and exploration philosophies presented are pertinent to other deposit-types and areas around the world.

CONWEST-MIMINISKA PROPERTY

Regional Geology

The Conwest-Miminiska property, 115 km east-northeast of Pickle Lake in northwestern Ontario, lies within the Uchi Subprovince, an east-trending, predominantly metavolcanic-metasedimentary belt (Figure 2). With the exception of a few north-trending dikes of Middle Precambrian diabase, all the rocks in the area are Early Precambrian in age. Quaternary deposits of glacial till, sand and gravel now cover much of the bedrock.

Property Geology

A broad geologic understanding of the property was derived partly from regional government mapping and largely as the result of an exploration programme on the land portion of the claim block in 1944 and 1945 by Conwest Exploration Company Ltd.

A major unit of banded oxide facies iron formation trends eastwards across the north portion of the property. A number of subsidiary iron formation units, occurring to the south of the main unit, are actually composite clastic sedimentary/chemical sedimentary units consisting of chert-magnetite iron formation beds, generally less than 50 cm thick, interbedded with beds of fine wacke or mudstone in roughly equal proportions. The clastic interbeds often themselves contain up to 15% magnetite and may be highly magnetic. The iron formation units may be highly chloritic, a feature usually associated with later quartz or quartz-sulphide mineralization.

The balance of the property is underlain by a generally very well laminated epiclastic metasedimentary sequence comprising interbedded wacke, quartzitic wacke and mudstone. All units dip steeply. These rhythmically bedded sediments would appear to be part of a distal turbidite sequence.

Gold Mineralization

The gold values occur mainly in structurally-controlled, chloritic sulphide replacement zones within, or closely associated with, the subsidiary magnetite iron formation units. Pyrrhotite is the most common sulphide with lesser arsenopyrite and pyrite. Total sulphide content rarely exceeds 25 percent and more usually averages ten percent or less. Gold values appear to be directly related to the content of fine-grained disseminated pyrrhotite.

These disseminated to banded zones of mineralization are

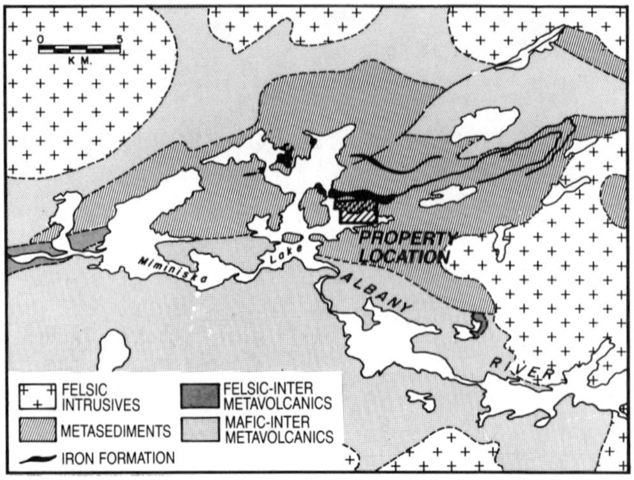

FIGURE 2. Regional geology of Miminiska Lake area

cut by later quartz-carbonate veins containing variable pyrite with lesser arsenopyrite and pyrrhotite. The quartz-sulphide veins usually do not carry appreciable gold values although visible gold has been reported.

Previous Work

Conwest drilled 28 holes totalling 8,630 feet which resulted in the definition of three gold zones in 1945. These values were considered of little interest at the time although drill intersections in portions of the zones, particularly the west end of the "A" zone - 5 g/t Au over 6 m (0.15 oz/ton over 20 ft), the central part of the "B" zone - 3.7 g/t (0.11 oz/ton) Au over 1.5 m and the east part of the "C" zone - up to 5.5 g/t Au over 3.1 m (0.16 oz/ton over 10.2 ft) are of economic interest (Figure 4).

Exploration Model and Objectives

A concerted evaluation of the entire property was not carried out until 1987. The geologic setting of the primary gold mineralization appears to be directly analagous to that at the Dona Lake deposit near Pickle Lake which was recently brought into production by Placer Dome. The basic criteria established were that:

(i) Precious metal sulphide replacement deposits associated with iron formation were the primary exploration model;
(ii) The whole property, half of which is beneath the shallow waters of Miminiska Lake, would have to be systematically explored with the intention of producing a detailed geologic map and defining additional target areas as well as tracing the extent of known gold-bearing zones.

Exploration Programme

The exploration programme evolved in three phases on a grid with north-oriented lines at 100 m intervals and a baseline through the known mineralized areas. The central portion of the property was investigated in more detail on lines at 50 m intervals. In March 1987 total field magnetic and VLF-EM surveys were completed to provide lithologic and structural information. The VLF-EM survey, using Cutler, Maine as the transmitting source, outlined numerous conductors of potential interest. These conductors were screened by horizontal loop electromagnetic (HLEM) surveying using a 100 m coil spacing and transmitting frequencies of 444, 1777 and 3555 Hz. Some detail work with a shorter and therefore higher resolving 50 m cable was completed to delineate multiple conductors.

Following this, mapping, trenching and soil and humus geochemical sampling were completed in September 1987 to provide ground truth to the geophysical interpretation, check previously reported assays and prioritize the geophysical anomalous responses, with a higher priority assigned to those with coincident geochemical signatures. Analyses for Cu, Zn, As and Au were completed using standard geochemical procedures. The geochemical surveys focussed attention directly to the area in Figure 4. However, while highly anomalous areas were delineated, geochemical signatures were not obtained at all sites where high gold values were noted from trenching and drilling (see Baseline Zone). No explanation for this is immediately apparent. In January, 1988 the priority targets were tested by diamond drilling.

Results

A strong regional gradient was removed from the magnetic dataset which is dominated by the major iron formation at the north edge of the survey area. The residual map produced (Figure 3) tends to enhance near surface, shorter wavelength features and enable slight magnetic lows indicative of magnetite depletion and sulphide replacement in the iron formation to be more readily interpreted.

The magnetic surveying and geological mapping defined a large anticline-syncline drag fold pair in the subsidiary iron formation units in the central portion of the property (Figures 3 and 4). It is uncertain whether or not the two

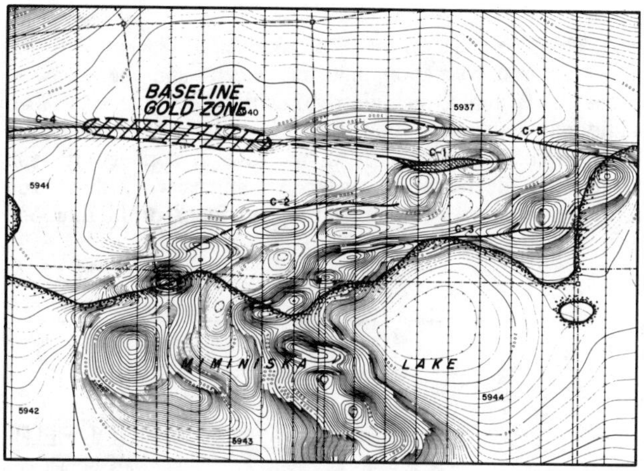

FIGURE 3. *Residual magnetic contours and EM conductor axes - Conwest-Miminiska*

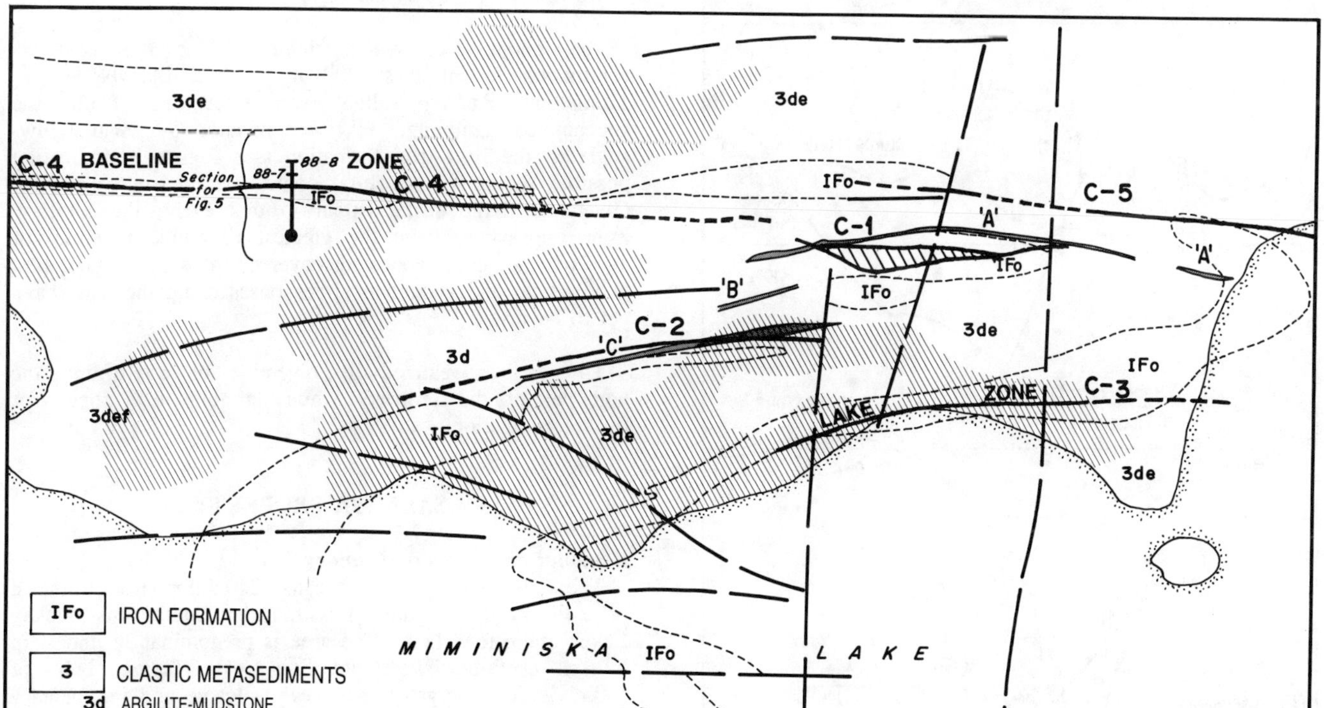

FIGURE 4. *Geology, geochemistry and geophysics compilation - Conwest-Miminiska*

iron formation units represent the dislocated and folded remnants of what was once a single unit or if two stratigraphic units are present although the latter would appear more likely at this point. Narrow shear zones can be discerned near the baseline and appear to be axially planar to the drag fold. These shear zones are typically rusty and chloritic and may be mineralized with quartz and sulphides. Similar features are known to be present elsewhere on the property based on geophysical data and previous drill results.

A close association of EM conductive zones with the magnetic iron formation units is evident and the Conwest zones A, B and C are clearly identified (Figure 4). Two additional conductors, both primarily on land, fit the Dona Lake exploration model of an increase in conductivity and a relative decrease in magnetic intensity within iron formation. They were the subsequent focus of 11 diamond drill holes (4,438 ft).

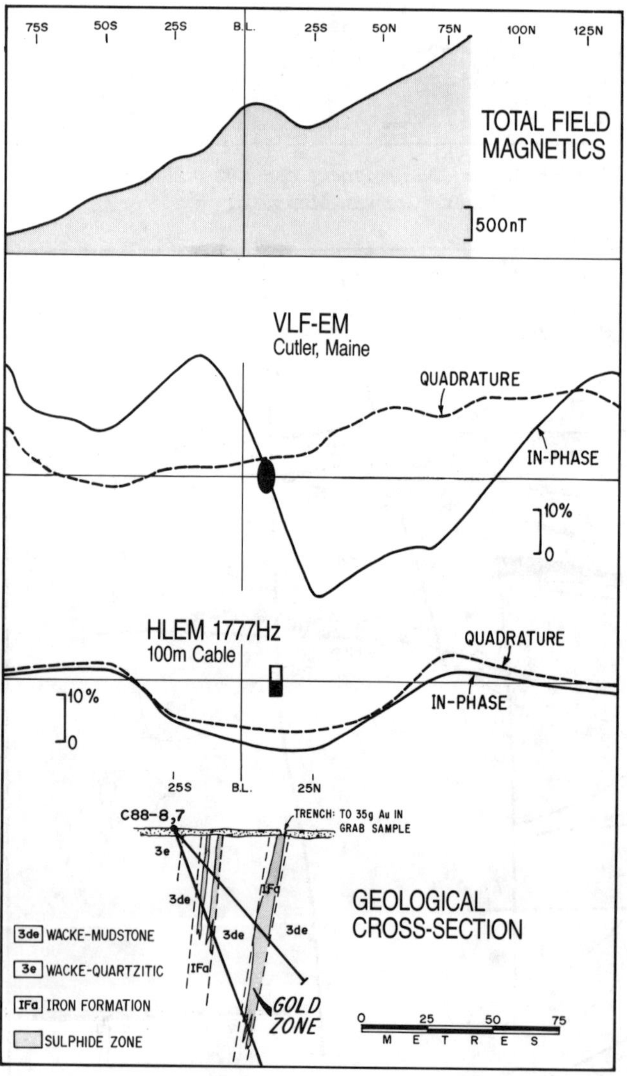

FIGURE 5. *Geophysics and geology section of Baseline Zone: line 16+00W - Conwest-Miminiska*

The ***Lake Zone***, close to the shore of Miminiska Lake, represents a narrow sulphidized shear zone of moderate to good conductivity with distinctly anomalous geochemical Au-As values.

The ***Baseline Zone*** is located immediately west of the known mineralized horizons. Both the VLF-EM and HLEM surveys delineated a strongly conductive bedrock feature coincident with a slight magnetic low possibly reflecting magnetite depletion within the iron formation. Diamond drilling of HLEM conductor C-4 outlined two mineralized iron formation units separated by 10-15 metres, with the bulk of the gold values in the more northerly unit (Figure 5). On line 16+00W the conductor is coincident with the northern iron formation (Figure 5) where the higher sulphide concentrations and better gold grades, up to 9.31 g/t (0.27 oz/ton) over 2.29 m in hole C88-7, are recorded. The southern iron formation did not display any distinct electromagnetic signature in all probability due to its narrow width and the low tenor of sulphides accompanying this largely cherty unit. A gold-bearing quartz vein is present between the two units in several of the holes.

It is of interest to note that on line 16+00W only background Au and As geochemical values were recorded. However, three grab samples of rusty sulphide-rich material, taken from a trench over the northern iron formation, assayed 0.23, 1.008 and 0.29 oz/ton Au (7.89, 34.56 and 9.94 g/t, respectively).

The geophysical results indicate that the zone is continuous for at least 200 m to line 14+00W where anomalous Au/As values were recorded from the geochemical sampling. HLEM conductor C-4 is at a low-angle to the host iron formation (see Figure 3) indicating possible secondary structural control of the mineralization. Given the large strike extent to this event, the Baseline Zone appears to have the greatest economic potential of any of the gold zones discovered to date. The best geophysical and geochemical responses along the zone have yet to be tested.

The project realized its objective of confirming the potential of the property while defining new zones for further evaluation.

SALMITA PROPERTY

Regional and Property Geology

The Salmita property is located 240 kilometres northeast of Yellowknife (Figure 1) within the Courageous-Mackay Lakes greenstone belt. The area is predominantly underlain by a north-northwest trending, east facing belt of Yellowknife Supergroup Archaean volcanic and sedimentary rocks (Figure 6). These units are invaded by younger granodiorite and diorite batholiths and stocks in the west and east along north-south linears. Sills of meta-gabbro and irregular masses of felsic porphyry intrude Yellowknife supergroup strata. Proterozoic diabase dikes cut all lithologies.

Mafic and intermediate volcanics occur in two cycles and form the bulk of the volcanic pile which ranges in thickness from 600 m to 5000 m in the north. Felsic volcanics are generally confined to two discrete cycles within the succession. The lower unit of cycle I is up to 60 m thick. The upper felsic unit of cycle II, where the majority of the gold zones occur, is more complete and extensive with a maximum thickness of 1800 m in the Matthews Lake area. This cycle has a lateral extent of some 25-30 km and may be a possible vent area. Sediments attain a thickness of at least 1500 m and generally conformably overlie the volcanics although locally may be interbedded with them. The volcanic and sedimentary beds have near vertical dips and form a series of steeply plunging isoclinal folds in a north-south direction.

North-south (strike) faults generally display left-lateral displacements ranging from one to 300 m in the horizontal plane with unknown vertical components. Two regional east/west linears are present while numerous minor cross-structures offset stratigraphy from one to several tens of metres.

Gold Mineralization

Most of the gold showings occur within 700 m of the steep easterly dipping volcanic/sedimentary contact zone, the mineralization being in three main settings associated with:

(i) ***siliceous vein like structures*** containing 3-5% sulphides and located within the uppermost rhyodacitic unit component of the upper felsic unit of cycle II. The Matthews (Tundra Mine) and B vein (Salmita Mine) structures were of this type with free milling gold occurring within highly siliceous zones averaging 1.0 to 1.69 m in width. Alteration within the host lithologies immediately surrounding the gold zone in the Salmita mine is intense and includes the development of silica, carbonate, muscovite/sericite and chlorite with minor graphite, epidote and tourmaline. Arsenopyrite is the major sulphide with traces of marcasite, chalcopyrite, pyrrhotite and galena.

(ii) ***finely disseminated sulphides within sericitic altered upper felsic tuffaceous units.*** The Noranda/Getty "Fat" deposit currently being placed into production is an example.

(iii) ***stringers and lenses of quartz with minor sulphide in narrow shear zones mainly within the overlying sediments***: for example, the Olsen and Horsehead showings.

Previous Production and Exploration

The Salmita Mine was brought into production by Giant Yellowknife Mines in August, 1983 based on ore reserves of 140,000 tonnes to a depth of 225 m grading 26.6 g/t, the highest reserve grade in the country at that time. Production continued to August 1987 totalling 215,295 tonnes at a millhead grade of 26.95 g/t Au, from which about 179,730 oz Au were recovered.

Prior to 1983 most airborne and ground geophysical surveys completed in the general area were focussed on

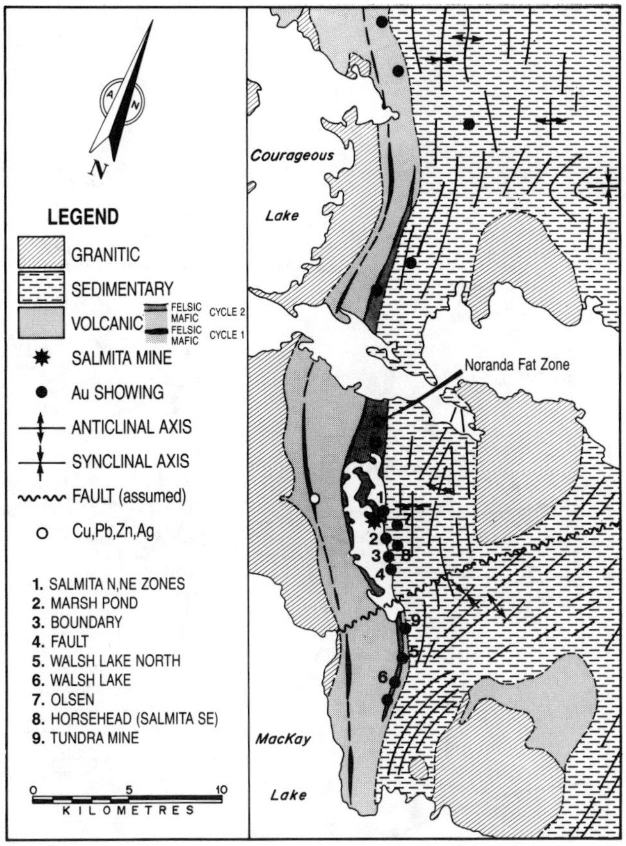

FIGURE 6: *Regional geology of Courageous Lake area*

potential base metal massive sulphide mineralization. VLF-EM and magnetics had successfully delineated two main conductive trends in the volcanic pile:

(i) Massive, stringer and disseminated sulphides within the lower felsic cycle to the west assayed erratic base metal values and appear to be devoid of gold mineralization.

(ii) Intermittent massive sulphide development along an essentially linear trend, possibly associated with sulphide facies "iron-formation" within the upper felsic tuffaceous units to the east, are barren of gold but spatially related to gold mineralization at Salmita.

Exploration Model and Objectives

The *objectives* of continued exploration were the delineation and diamond drill testing of potential gold mineralization targets and the production of a comprehensive geological-geophysical database from which a lithologic and structural overview of the property could be derived. The main obstacle to exploration on the Salmita property is permafrost which is prevalent in the northern mineral provinces of Canada and causes unique technical problems.

Disseminated sulphide and associated gold mineralization are known to be closely associated with north-south linear structures within a 1.5 km area about the

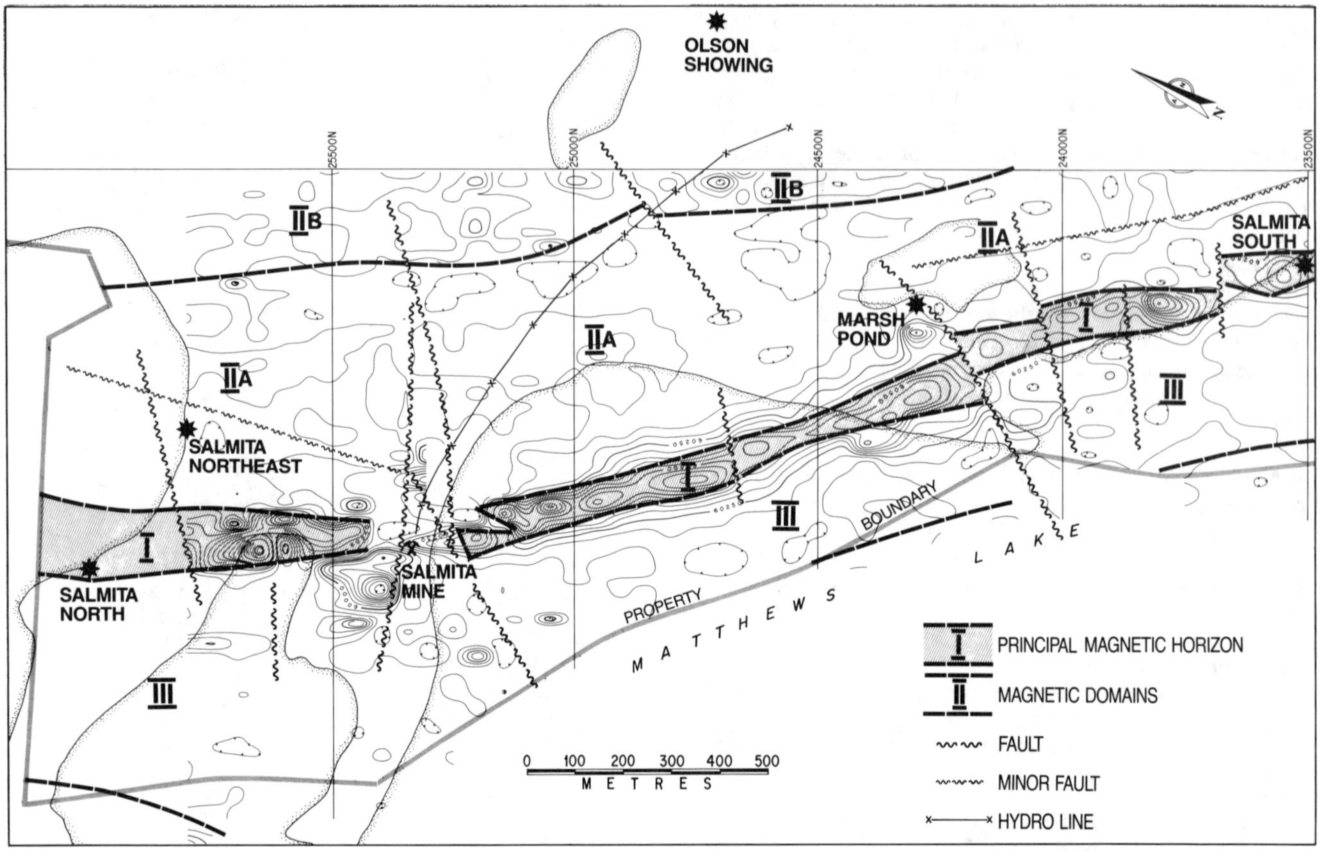

FIGURE 7. *Total field magnetic contours with interpretation and location of gold zones - Salmita*

volcanic/sedimentary contact. The *primary exploration model* is where these linears and the contact are cross-cut by essentially east-trending structures. There is a definite association between these areas of structural complexity and the presence of economic gold mineralization at the Salmita Mine (Ransom, 1983) where horizontal lateral displacement of the lithologies and north-south linear structures has occured. Sulphide-enrichment is common about the cross-cutting structures and, where arsenopyrite is the primary sulphide, similar scenarios for economic gold mineralization can be expected to occur.

Exploration Programme

Due to the permafrost in the ground and the relatively low percentage of outcrop on the property, geophysical techniques are the most cost-effective general methods available to efficiently delineate the volcanic-sedimentary contact, map polarizable sulphide-rich horizons and locate areas of cross-faulting conforming to the exploration model.

In consequence a grid was established centred over the volcanic/sedimentary contact with crosslines at 50 m intervals and the baseline parallel to the dominant stratigraphic trend. Total field magnetic readings were taken at 25 m intervals with a higher density of readings in areas of high magnetic gradient. A time domain IP/resistivity survey was completed using a dipole-dipole array with a dipole spacing of 50 m and dipole separations of 1 to 4. A Huntec 2.5 kw generator/transmitter system was employed with a 0.125 Hz transmitting cycle. The 60 Hz notch and low pass filters and delay and interval times of 100 msecs were used on the Huntec MK IV receiver.

Because of their narrow width the actual deposits are probably not directly detectable by geophysical techniques: an assumption which could not be confirmed as cultural development at the Salmita and Tundra minesites prevented any direct surveying. However, their alteration envelopes, which are more extensive and carry minor sulphides, should have detectable IP and resistivity signatures. The magnetic results, supplemented by the resistivity and IP data, were expected to map lithologies and alteration envelopes and provide information on crosscutting structural features.

Given the exploration model defined one might reasonably expect a combination of magnetic and resistivity signatures to be adequate for delineating targets. However, the permafrost conditions prevalent at this latitude and the semi-coincidence of the lake shore with the volcanic/sedimentary contact combined to present exceedingly difficult conditions for the resistivity survey. Over the land portions of the survey area, resistivities are generally quite high, in part reflecting the permafrost, with values at the smaller dipole separations often in excess of 30,000 ohm m (Figure 9). In contrast, over the lakes, where the overburden is generally unfrozen, resistivity values at similar dipole separations are considerably lower.

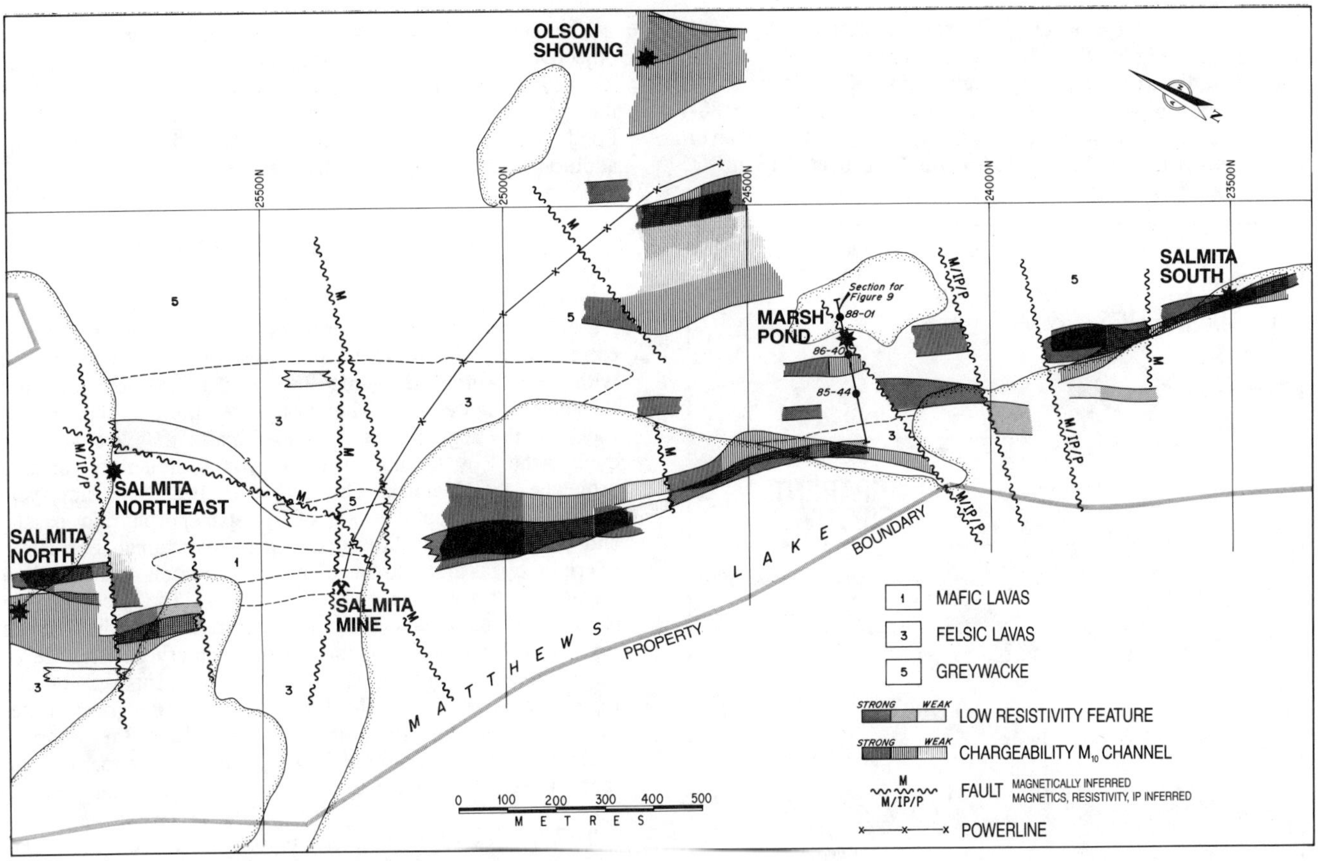

FIGURE 8. *Simplified geology, IP/resistivity interpretation and location of gold zones - Salmita*

These abrupt near surface lateral changes in conductivity give rise to anomalous resistivity responses themselves with the result that the apparent resistivity dataset cannot be interpreted with confidence in such situations.

The IP measurement, however, is normalized and therefore less sensitive to large lateral conductivity changes at the land/lake interface. The purpose of delineating polarizable horizons is thus threefold: supplementing the resistivity results in complex situations and supporting the resistivity interpretation elsewhere as well as indicating areas of greater sulphide content - possible sulphide enrichment zones.

Results

The magnetic response pattern (Figure 7) maps the general lithologies described above. The principal magnetic feature is a narrow linear zone which transects the property at, or close to, the shoreline of Matthews Lake. This zone marks the volcanic/sedimentary contact and is clearly disrupted by cross-faulting at a number of locations, in particular at the Salmita deposit. Strike-slip faults, known to exist with northerly orientations, cannot be confidently inferred.

The resistivity survey delineated a number of narrow conductive zones, generally conformable to the stratigraphy, many of which are disrupted by crosscutting faults. The majority of the moderate to strongly polarizable, narrow linear zones interpreted from the IP results correlate closely with these low resistivity features and the known gold zones (Figure 8). The volanic/sedimentary contact is clearly traced across the property as a zone of variable conductivity and a semi-coincident series of polarizable anomalies. Disrupting crosscutting faults are inferred from both datasets.

The permafrost, by blocking almost all paths of ionic conduction, frequently causes the bedrock conductive zones to have a geophysically inferred depth that is probably considerably greater than their actual geologic depth. The apparent polarizability of a given bedrock source is also affected. Where the host rock is highly resistive the permafrost may not allow a current to reach the target and, even if fairly conductive, it will therefore not have a detectable IP signature.

The results from all the datasets therefore have to be interpreted with reference to each other bearing in mind the possible physical constraints and distortions to individual datasets and in the context of the exploration models. First priority targets are where all three geophysical datasets correlate and indicate cross-cutting structures with lateral displacement of the volcanic/sedimentary contact.

A number of areas conformed to the exploration model, one of which is the Marsh Pond zone (Figure 8). In 1986, hole 86-40 intersected 24.68 g/t (0.72 oz/ton) Au over

1.13 m at the 150 m level by the felsic volcanic/sedimentary contact (Figure 9). However, subsequent holes 86-41 and 86-42 failed to trace any continuation of this encouraging mineralization a short distance away. In 1988, hole 88-01 was targeted to intersect the horizon at the 400 m level and returned only 8.40 g/t (0.25 oz/ton) Au over 0.18 m.

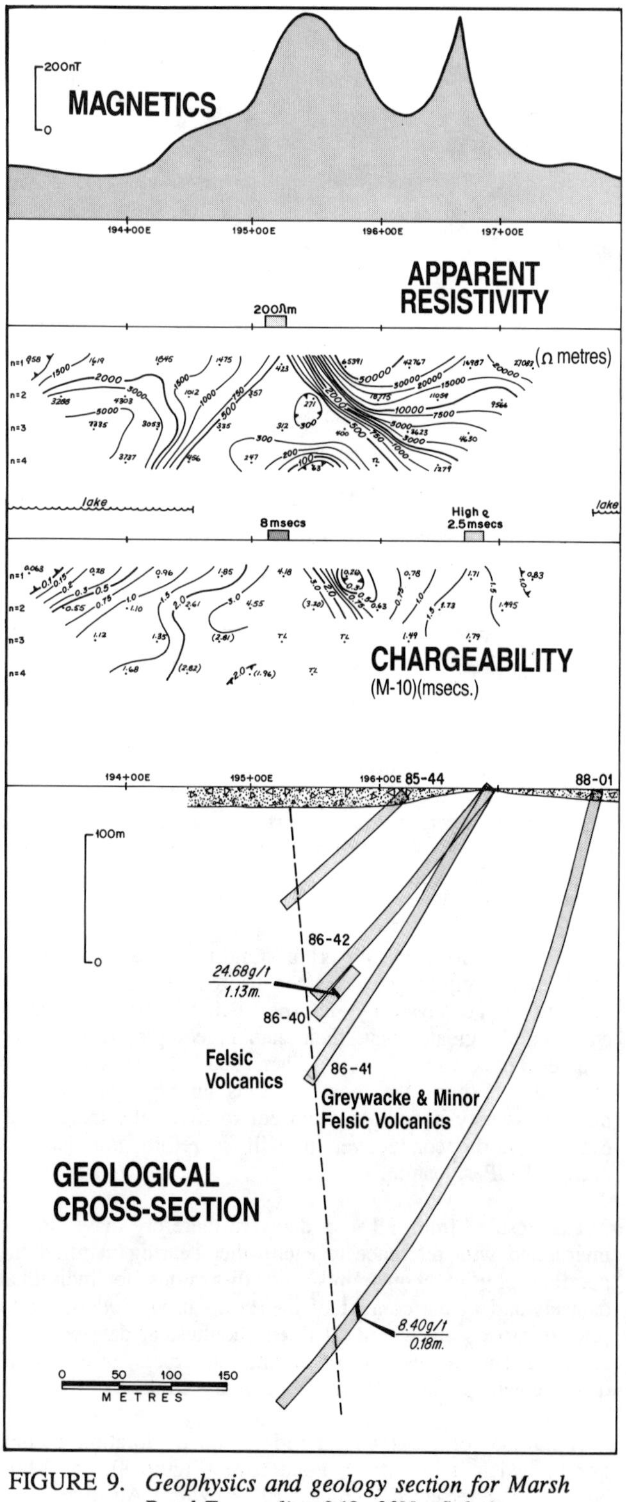

FIGURE 9. *Geophysics and geology section for Marsh Pond Zone: line 243+00N - Salmita*

The geophysical programme succeeded technically in its objectives of producing a lithologic and structural map and identifying Salmita-type exploration targets. Unfortunately, the grades, widths and continuity encountered at Marsh Pond and other zones were not sufficient to prove-up additional ore grade mineralization.

LAURA LAKE JOINT VENTURE

Regional Geology

The Laura Lake property is located at the eastern extremity of the Chibougamau-Matagami greenstone belt within the Abitibi Orogenic Belt of the Superior Structural Province (Figure 1). The early Precambrian supracrustal rocks of the region consist of two mafic to felsic volcanic cycles, the Roy Group, overlain by a sedimentary-volcanic sequence, the Opemiska Group (Figure 10). The first cycle of the Roy Group consists of feldspar porphyritic basalts and co-magmatic gabbro sills of the Obatogamau Formation overlain by a thin sequence of felsic pyroclastics and iron formation of the Waconichi Formation. The second cycle consists of mafic volcanics and gabbro sills of the Gilman Formation overlain by a volcano-sedimentary assemblage of the Blondeau Formation. The major intrusives in the region are Early to Middle Precambrian granodiorite-monzonitic or tonalitic-dioritic rocks. Late Precambrian, north to northeast-trending diabase dykes are numerous and range in width from 0.3 to 10 m.

Two phases of folding are reported: broad open folds with north-south trending axes, which were subsequently isoclinally refolded about east-trending axes (Allard, G. et al, 1984). Fault systems with east-west, northeast and northwest strikes have been identified in the Chapais/Chibougamau area. The oldest faults generally strike east-west to west-northwest and are characterized by steeply dipping, closely spaced zones of intense shearing, carbonatization and silicification. Sulphide mineralization associated with these faults is common.

The main regional fault system trends northeast and has three associated fault sets:

(i) The most prominent of these fault sets trends *45-60° east of north*. The regional Gwillim Lake Fault is the primary example, characterized by a vertically dipping shear zone which locally attains widths up to 100 m and transects both the supracrustal and intrusive granitic rocks. The lithologies in the Chapais region are displaced approximately 1 km in a left-lateral sense about the Gwillim Lake Fault.
(ii) The *20-35° east of north* set is generally local in extent and has horizontal displacements up to 100 m.
(iii) The *10-15° east of north* set mainly occur close to the Grenville Front and tend to be narrow with occasional slight shearing.

Property Geology

The portion of the property relevant to this case history is primarily underlain by east to east-northeast trending basalts believed to be of the Obatagamau or Gilman Formations and co-magmatic fine-grained gabbroic sills (Figure 11). The basalts are typified by the presence of

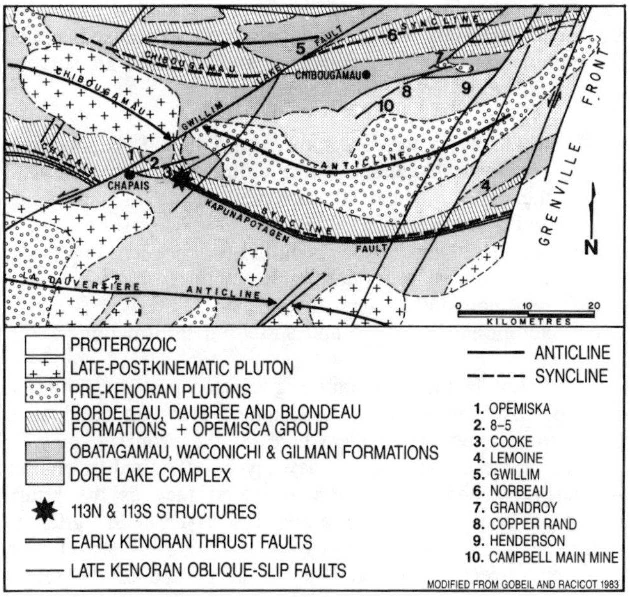

FIGURE 10: *Regional geology of the Chapais-Chibougamau area*

epidotized, albitized and locally silicified flow breccias intercalated within massive flows. The gabbros locally contain magnetite and are slightly coarser than the massive flows. Intermediate to felsic lavas are present on the north shore of Laura Lake. A major east trending leucogabbro dyke with an average thickness of about 800 ft intrudes the basalts and gabbros. A crude pyroxenite to dioritic differentiation is locally observed in the leucogabbro.

Two major east trending structures, the Gaudreault and Kapunapotagen Faults, are 1500 and 3500 ft south of the area, respectively (Figure 11). The Gaudreault Fault in particular indicates where the volcanic rocks of the Roy Group are in discordant contact with the mainly sedimentary rocks of the Opemiska Group.

Economic Gold Mineralization

A wide variety of ore deposit types and geological settings occur in the Chapais and Chibougamau base and precious metal mining camps and can be placed in three general categories:

(i) The majority of economic (Cu-Au) orebodies in both camps are vein-type sulphide deposits associated with differentiated basic intrusive complexes and closely related to 45-60° east of north structures: in Chapais the Gwillim Lake Fault and in Chibougamau the Dore Lake Fault. In Chapais the Cummings Lake Intrusive Complex, within the Blondeau Formation, hosts the Springer, Perry and Cooke deposits.

(ii) Volcanogenic massive sulphide deposits are rare in the region and those which were economic, the Lemoine and Coniagas Mines and the 8-5 zone in Chapais, have exhausted their reserves.

(iii) Two deposits are hosted by mafic volcanics: the Gwillim Mine (Au) in the Gilman Formation and the Chibex Mine (Au, Cu) in the Obatagamau Formation.

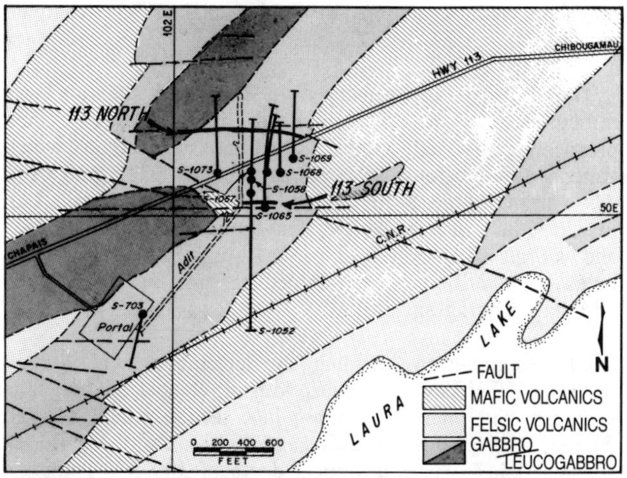

FIGURE 11. *Simplified geology of the 113 Structures and diamond drillhole and underground ramp locations - Laura Lake*

The Cooke deposit, the major gold producer in Chapais with more than 1.7 m tonnes at 5.1 g/t (0.15 oz/ton) Au (Gagnon, 1989), is hosted by epidiorite of the Bourbeau Sill within the Cummings Lake complex. Gold mineralization, accompanied by disseminated sulphides occurs in two parallel fracture zones oriented approximately N120E and dipping 65-85° northwest. The sulphide veins, which range in width from 6 ins to 10 ft, are hosted in a deformation zone up to 15 ft wide. The two fracture zones are chloritized and rich in carbonate with local quartz veining. Sulphide content locally varies from 1-75% with an average of 10%, the principal minerals being chalcopyrite, pyrrhotite, arsenopyrite and pyrite.

Exploration Model

Structural control of the gold mineralization is apparent in the deposits, which are spatially related to northeast faults. The mineralization is normally hosted in chloritic shear zones trending east or northwest. Sulphide mineralization may be narrow massive or semi-massive zones and/or as disseminated stringers. The gold potential of major shear zones such as the Kapunapotagen and Gaudreault Faults is not confirmed but considered possible (Gagnon, 1989). The *primary exploration model* is Cooke-style gold/copper mineralization.

Phase 1 Exploration

Late in 1984 the first systematic evaluation of the Laura Lake area was initiated. Much of the property is covered by 50 ft or so of overburden. Total field and vertical gradient magnetic and VLF-EM surveys were therefore completed at 300 ft line spacing and 50 ft station intervals on north-south lines. These results, together with the geology, prospecting and geochemical results, formed the basic database. A general lithologic and structural interpretation was completed and the VLF-EM conductors evaluated. Those VLF-EM conductors displaying coincident magnetic signatures are considered priority targets for Cooke-style mineralization. The vertical gradient magnetic data was extremely useful in refining the structural interpretation, often resolving two or more individual magnetic features from a single total field magnetic response.

The property was remapped in 1985 and attention was refocussed on the area outlined by Figure 11 when prospecting defined several structures in outcrop which contained anomalous gold values (180 to 9221 ppb Au). Also, nearby hole S-703, drilled by Opemiska Mines Ltd. in 1973, had intersected two one-foot sections grading 5.5 and 4.1 g/t (0.16 and 0.12 oz/ton) Au. Several follow-up holes were drilled in the immediate area in April, 1985 without encouraging results.

On recommencing drilling in September, hole S-1052 tested a VLF-EM conductor with a coincident magnetic signature and intersected a two-foot section which assayed 51.8 g/t (1.51 oz/ton) Au, 28.1 g/t (0.82 oz/ton) Ag, 0.27% Cu and 0.54% Zn at a vertical depth of 100 ft. The zone is now known as the 113 South Structure. A follow-up hole one hundred feet to the east failed to intersect this feature. Interpretation of the surface geologic and magnetic data and inspection of the drill core indicated that the structure may have been displaced by a left-handed fault. Hole S-1065 was drilled to test this hypothesis and a new vein, the 113 North Structure, was intersected 500 ft north of the 113 South Structure. The vein assayed 3 g/t (0.087 oz/ton) Au, 88.1 g/t (2.57 oz/ton) Ag and 6.81% Cu over 1 ft. Both structures are hosted by the gabbros with pyrite and pyrrhotite being the primary sulphide in the 113 North and 113 South structures, respectively. The 113 North Structure was not detected by the VLF-EM survey as interference from the power lines masks all electromagnetic responses up to several hundred feet away.

Phase 2 Exploration

Numerous occurrences of massive to semi-massive sulphide (py-po) veins and stringers with variable quantities of chalcopyrite and sphalerite were identified from a continuing drill programme in the general vicinity of the 113 structures. All anomalous to economic gold values in the area are found within these sulphide zones. The available surface geophysical data and drill core information suggested that a system of NNW striking faults may be present, a hypothesis not confirmed by the limited underground evaluation to date. Apparent horizontal displacements could be in the order of as little as twenty to thirty feet up to a few hundred feet.

In early 1986 additional ***magnetic and VLF-EM*** surveys were completed at 100 ft line separations in an attempt to refine the lithologic and structural interpretation of the immediate region and trace the extent and orientation of known sulphide zones. The results were of limited success due to the close proximity of Highway 113 and the railroad, both of which are highly anomalous conductive/magnetic features. They did, however, confirm the suspected presence of gabbros, in contact with basalts to the west, bounding the eastern extent of the 113 structures - an interpretation supported by subsequent geophysical and diamond drilling results.

The sulphides intersected in the 113 structures, while often massive, are of very narrow widths and suspect continuity and do not constitute a good electromagnetic target although inspection of the core revealed a degree of electrical continuity within and between mineralized zones. A combination of ***mise-a-la-masse and gradient IP*** surveys was therefore undertaken to determine:

(i) whether known sulphide zones could be mapped geophysically under severe cultural interference,
(ii) the trend, dip and lateral extent of the 113 structures and other zones,
(iii) whether there is any continuity between individual zones across cross-cutting structural features,
(iv) new potential targets, if any, and
(v) additional lithologic and structural information.

A 7.5 kw Huntec transmitter/generator system was used for the surveying as it was anticipated that thin conductive clay layers, encountered elsewhere on the property, may be present in the overburden. Clay layers effectively diminish the target response amplitude at the surface as the return signal is partly dissipated along the layer. In order to minimize the effect it is necessary to excite the target with a larger current source so that the amplitude of the return signal is correspondingly higher. A Huntec MK IV receiver, with the low pass and 60 Hz notch filters in and delay and interval times of 100 msecs, was used for both surveys.

Mise-a-la-masse pole-pole surveys were completed on the 113 structures to determine the lateral and updip extent of the sulphide mineralization. No confident estimates of downdip continuity can be made with this method. An infinite current electrode was established at 9+75N on line 138+00E and an infinite potential electrode at 102+00N on line 113+50E. These were used as the infinite reference points for all the mise-a-la-masse surveys.

The *113 North Structure* was investigated first. A current probe was placed at the deepest intersection of the structure in hole S-1065 and potential measurements were made in adjacent drill holes and on the surface. The surface results, presented as contoured measured potential amplitudes in Figure 12, indicate that the 113 North Structure, which trends N95°E, has a possible strike length of 450 ft. Powerline distortion of the response character is clearly evident but is at an acceptable level such that an interpretation can be presented. Indeed the mineralized horizon was interpreted to continue westwards across Highway 113 but the lower amplitudes in the order of 4 volts compared to over 6 volts suggested a degree of electrical discontinuity between the east portion and western extent of the 113 North Structure.

Hole S-1073 tested this hypothesis and intersected semi-massive sulphide mineralization which assayed 6.9 g/t (0.202 oz/ton) Au, 31.5 g/t (0.92 oz/ton) Ag, 0.81% Cu and 1.13% Zn over 5 ft. The current probe was subsequently placed in the sulphides in hole S-1073 and surface measurements collected (Figure 13). The results indicate that the 113 North Structure has a total strike extent of at least 950 ft with a possible NNW trending fault, with minimal lateral displacement, explaining the electrical discontinuity near 107+50E. Later underground exploration did not support this geophysical interpretation but did indicate a distinct lack and pinching out of the sulphide

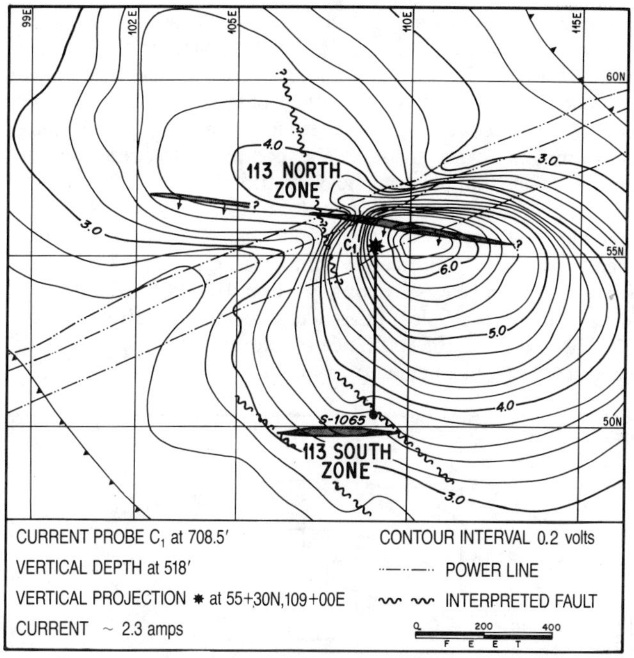

FIGURE 12. *Mise-a-la-Masse measured potential contours for current electrode in hole S-1065: 113 North Structure - Laura Lake*

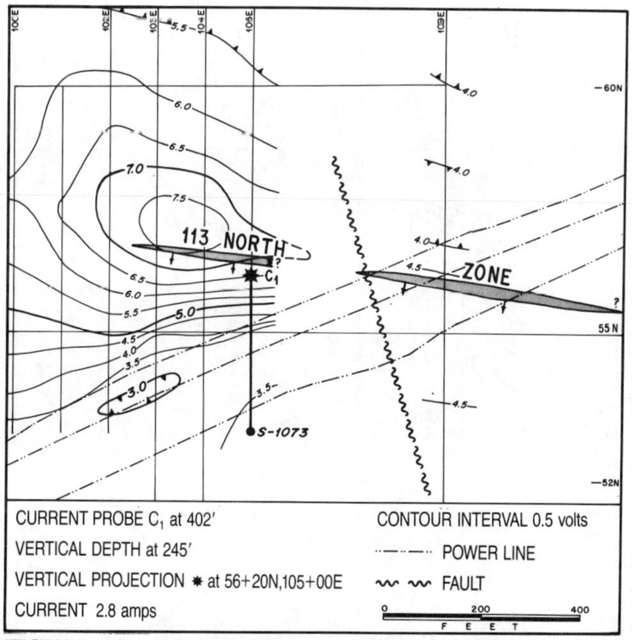

FIGURE 13. *Mise-a-la-Masse measured potential contours for current electrode in hole S-1073: 113 North Structure - Laura Lake*

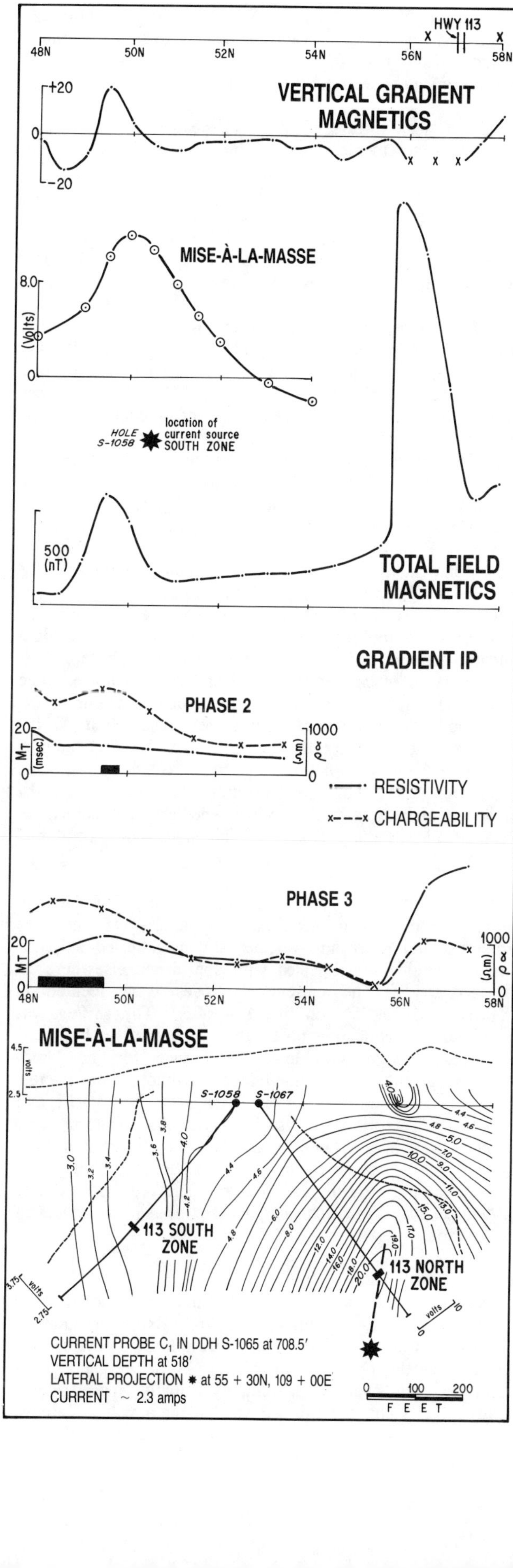

FIGURE 14. *Geophysical compilation of line 108+00E including Mise-a-la-Masse measured potential contours for hole-to-hole and hole-to-surface surveying: 113 Structures - Laura Lake*

mineralization. The 113 North Structure, although continuous, therefore appears to display distinct sulphide zonation such that the east and west portions of the structure are not electrically continuous.

Dips of near vertical to 80°S were estimated from hole-to-hole measurements (see Figure 14) and the lateral electrical continuity of much of the zone was also confirmed in the absence of cultural interference. The amplitudes recorded in hole S-1058 while the 113 North Structure was energized (Figure 14) also support the inference, drawn from the surface datasets, that the 113 North and South structures are not electrically connected.

Profile surveying on the surface over the 113 South Structure, while the current probe was in hole S-1058 (Figure 15), indicated that the zone is oriented almost due east and has a strike extent of about 250 ft. Dips are near vertical to steeply north. The coincident magnetic response (Figure 14) reflects the massive pyrrhotite content of the sulphide mineralogy. An attempt was made to preferentially enhance the anomalous response by calculating a pseudo-resistivity map (Figure 15). The processing involves removing the response from a point source in a homogeneous isotropic earth from the dataset and then calculating the ratio at each station of the amplitudes of a point source and the residual potential. An arbitrary half-space resistivity was used in the calculation as its absolute amplitude has no effect on the residual potential pattern. The routine must be used with care as it also serves to enhance topographic and cultural noise. The results, however, clearly identify the electrical extent of the 113 South Structure and suggest the presence of WNW structures. These were later confirmed as N130°E features by underground exploration. Definite knowledge on the actual controls on the sulphide mineralization is still the subject of debate as it is not clear whether the structures or changes in lithology constrain the potentially economic mineralization.

A ***test gradient IP*** survey was subsequently completed to augment the known geophysical signature of the 113 structures and determine whether similar untested sulphide horizons could be delineated with confidence elsewhere on the property. The current electrodes were located at 38+90N and 75+75N on line 114+00E. The survey was successful in delineating polarizable signatures for both 113 structures, though with lower confidence for the North structure, as the cultural interference is much more discrete than that encountered in the magnetic and VLF-EM datasets. The IP and resistivity response patterns also support the previous interpretation of an en echelon fault system with an axial trend of ENE underlying the immediate area. The results were therefore positive and influenced the approach adopted for the next phase of exploration.

Phase 3 Exploration

Evaluation of the area hosting the en echelon fault system continued in September, 1986. Extensive gradient IP surveys were completed and the results incorporated into the geological/geophysical database. As before the current electrode pairs were approximately equally spaced about Highway 113 and the railroad so that there was an almost continuous orientation of cultural effects with respect to electrode locations. As is often the case with gradient array surveys it was not possible to produce uniform contoured presentations of the datasets from multiple current electrode pairs, even though overlapping measurements were made. The complex structural regime ensures that the electric field distribution is different for any two current electrode pair locations. The IP/resistivity signature of a given anomalous bedrock horizon may therefore vary significantly for the different energizing fields. The repeat surveying of line 108+00E (Figure 14) when the current electrodes were located at 30+50N and 79+50N on line 105+00E neatly illustrates this problem of consistency from one setup to another.

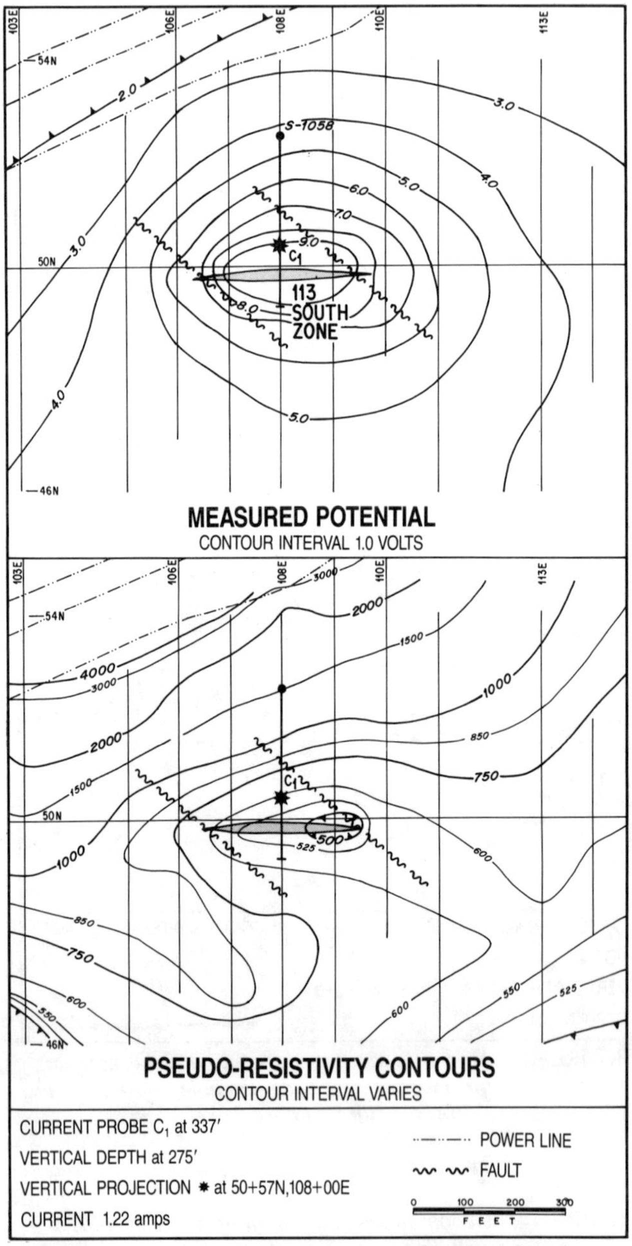

FIGURE 15. *Mise-a-la-Masse measured potential and pseudo-resistivity contours for current electrode in hole S-1058: 113 South Structure - Laura Lake*

Variations in weather conditions will also produce discrepancies from line to line. Heavy rain was encountered on 18 May, 1986 and, as is standard operating procedure, some of the readings were repeated on subsequent days until confidence in repeatability was achieved. The results on line 12000E (Figure 16) indicate that the sudden increase in surface water had the effect of decreasing the apparent resistivity amplitudes and increasing the total chargeability amplitudes. The general response pattern, however, remains consistent and the anomalous amplitudes with respect to background are relatively consistent. As the surface water drained on subsequent days, the total chargeability/apparent resistivity amplitudes decreased/increased, respectively. This illustration argues strongly in favour of a line-by-line analysis of gradient IP data and cautions that extreme care should be taken in the interpretation.

Diamond drilling of the priority targets was technically successful but failed to delineate potential economic mineralization on the scale of the 113 structures.

Phase 4 Exploration

Subsequent to further surface diamond drilling a decision was made to evaluate the ***113 North Structure*** underground for grade and tonnage at about the 300 ft level. Between February and June, 1988 a 1900 ft ramp was driven to the 113 structures which were explored via drifts (see Figure 11). The general east-west trend of both structures was quickly confirmed. The mineralized veins, generally subparallel to the shear zones and the deformation envelopes which enclose them, are often discontinuous and sometimes highly folded or boudinaged. There is clear evidence of internal deformation by shear movement in the sulphide textures. Chloritic alteration predominates.

The ***113 South Structure***, as known from the drilling results and confirmed by sampling, is uneconomic at the level where investigated. The structure was investigated, however, to provide valuable information on the continuity and mineralogical content of the sulphides and the structural relationships between the 113 structures. Mapping in the drift revealed that the massive sulphide veins or lenses are composed of 50-65% po, 10-30% py, minor qtz, cal, chl, and up to 5% sph. and 3% cpy. The deformation envelope dips 75°N-90° and structural interpretation suggests an oblique movement of the lithologies with a left-handed lateral component and a more important inverse vertical component: that is uplift of a north block over a south block.

This interpretation is supported by evidence from the ***113 North Structure*** (Figure 17) where vertical normal movement is evident from striae on the walls. The deformation envelope enclosing the mineralization is 8-9 ft wide and dips 70°S-90° - the shallower dip being towards the eastern extent of the structure. Modest right-handed lateral displacement is inferred. The most abundant sulphide is pyrite occurring either in disseminated stringers (1-20%) or as massive veins (50-90%). Other sulphides in lesser amounts are chalcopyrite which is more abundant in the west drifts and sphalerite in the east. Minor pyrrhotite and galena were mapped. Gold values are uniformly low in heavily disseminated (15-25%) or stringer sulphides. All

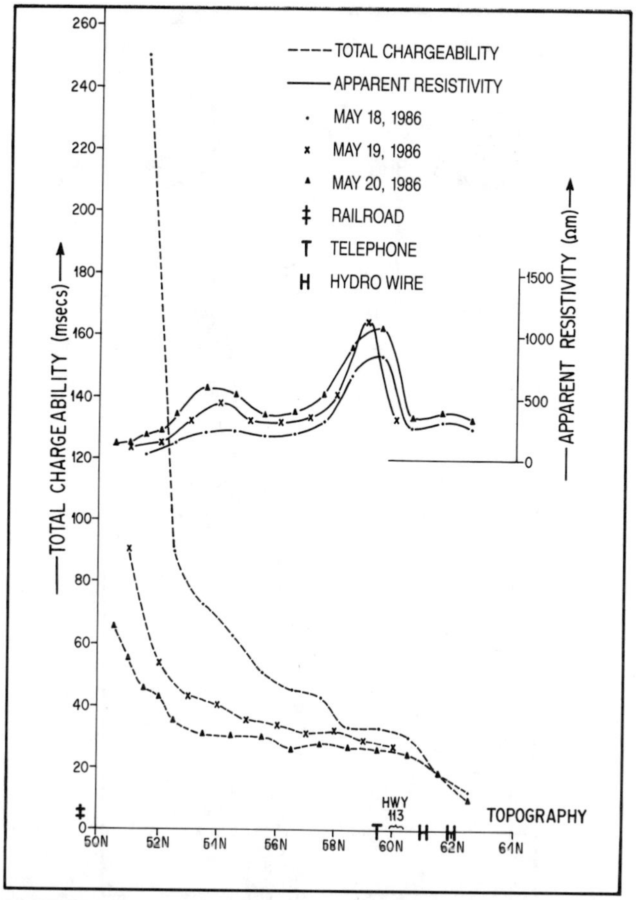

FIGURE 16. *Effect of surface conductivity on measured IP/resistivity amplitudes: line 120+00E - Laura Lake*

samples assaying 0.1 oz/t Au or more correspond to massive sulphide veins and sometimes to sulphide-rich quartz veins. Gold grades over 1.0 oz/t occur in massive sulphide lenses. Multiple sampling confirmed that a nugget effect is not present.

Grades in the *east drift* were inferior to and more erratic than those to the west: an 855 tonne bulk sample had a grade of 2.19 g/t (0.064 oz/ton) Au. In the *west drift* a 1017 tonne bulk sample from an area of 151.5 x 9 ft - the width of the deformation zone and therefore containing some uneconomic ore - had an average grade of 4.1 g/t (0.120 oz/ton) Au.

Studies by the Opemiska Engineering Division at Chapais concluded that the mineral grades in the west drift could support a mining operation: for example, in the west drift a 112 x 5.4 ft lens would have had a grade of 7.92 g/t (0.231 oz/ton) Au, 21.94 g/t (0.64 oz/ton) Ag and 0.30% Cu. Unfortunately, the study also concludes that the tonnage thus delineated, combined with the drilling results, does not justify additional underground development at this time.

Each phase of geophysical investigation successfully met the stated objectives at that time. The geological-geophysical database was enlarged and revised continually

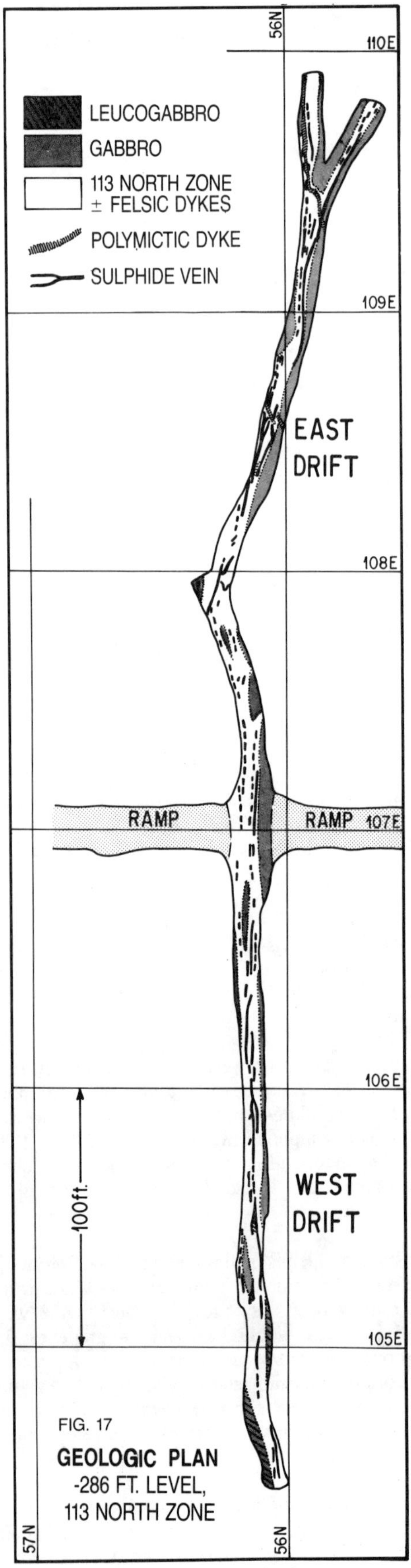

FIG. 17

GEOLOGIC PLAN
-286 FT. LEVEL,
113 NORTH ZONE

during each phase and continues to be extensively utilized in continuing surface exploration of the property.

CONCLUSIONS

The three case histories described in this paper clearly illustrate the efficiency of a concerted multi-disciplinary approach in translating an exploration model into physical reality. The process of outlining economic ore reserves involves a continuous cycle of evaluation, input of data and re-evaluation sometimes from a new perspective or even with a radically different exploration target in mind. A crucial function of each team member is the ability to communicate findings and ideas to each other. Geophysicists tend to address this problem by presenting data in compact forms. The best tool for this process is colour as the human eye can rapidly assimilate a great variety of information via this medium. For example, image processing of the geophysical data from all three case histories produced additional subtle information not immediately apparent from a black-and-white presentation. The fact that all the available data is referenced to the same grid system allows immediate accessibility of information and great flexibility in the correlation of results.

ACKNOWLEDGEMENTS

We would like to thank Conwest Exploration Company Limited, Giant Yellowknife Mines Limited and Minnova Inc. (formerly Corporation Falconbridge Copper) for releasing the data and for permission to publish this paper. W.E. Brereton, D. Jones and T. Kraft provided valuable discussions and comments.

SELECTED REFERENCES

Allard, G.O. and Gobeil, A., 1984, "General Geology of the Chibougamau Region" in: *Guha, J. and Chown, E.H. (eds.) Chibougamau - Stratigraphy and Mineralization.* CIM Special Volume 34.

Gagnon, J., 1989, "Travaux de Cartographie et d'Echantillonnage dans une Rampe d'Exploration, Projet Laura Lake J.V., PN 119, Canton Levy", Minnova Inc., Division Exploration, Chapais, Quebec.

Hodgson, C.J., 1989, "Paper 3, Uses (and Abuses) of Ore Deposit Models in Mineral Exploration"; in *Proceedings of Exploration "87: Third Decennial International Conference on Geophysical and Geochemical Exploration for Minerals and Groundwater*, edited by G.D. Garland, Ontario Geological Survey, Special Volume 3, pp. 31-45.

Ransom, A.H., 1983, "Paper 8, History and a Preliminary Update of the Geology of the Salmita Gold Mine, Courageous Lake, Northwest Territories", September, First District Five Meeting, CIM, Calgary, Alberta.

Vincent, R. and Drouin, M., 1986, "Laura Lake Surface Exploration Proposal", Corporation Falconbridge Copper, Exploration Division, Chapais, Quebec.

Chapter 7

HIGH RESOLUTION HELICOPTER EM-MAG SURVEY OF THE ORTIZ MOUNTAINS, SANTA FE COUNTY, NEW MEXICO

David A. Smith[1] and Craig J. Nelsen[2]

[1] Consulting Geophysicist, Salt Lake City, Utah
[2] LAC Minerals (USA), Inc., Reno, Nevada

ABSTRACT

In 1987, LAC Minerals (USA), Inc. conducted a high resolution helicopter electromagnetic (EM) and magnetic geophysical survey on the Ortiz Mine Grant, New Mexico. The primary purpose of this survey was to assist in geologic mapping of a major structural zone, the Tijeras-Canoncito Fault System (TCFS). This structure is a very important ore-controlling feature in the area. Geophysical signatures over known deposits was a secondary goal of the survey, but direct detection of mineralization was not anticipated. The TCFS was revealed to be a complex, branching and anastomosing system of faults which locally form the bounding structures of a graben. The geophysical survey guided subsequent geologic mapping by redirecting field investigations, and geophysically redefining some features which previously had been only partly defined.

INTRODUCTION

The Ortiz Mountains, located at latitude 35°20'N and longitude 106°10'W, occupy the center portion of the 100 square mile Ortiz Mine Grant, about 30 miles (50 km) northeast of Albuquerque, New Mexico (Figure 1). The Ortiz Mountains are formed by Oligocene stocks, laccoliths, sills and dikes which intrude Pennsylvanian through Tertiary sedimentary rocks (Maynard and others, 1989). The mountainous parts of the Ortiz Mine Grant are rugged and sharply dissected by erosion. They rise approximately 2000 ft (625 meters) above a gently sloping erosion surface which is mantled by a thin veneer (generally less than 50 ft (15 m)) of pediment gravel and sediments.

Gold mineralization in the Ortiz Mountains occurs in a number of deposit types including skarns, porphyries, veins, shears, breccia pipes and placers. Known recorded production since 1833 is approximately 350,000 ounces (11,000 kgs). The majority of this was from the mining of a breccia body at Gold Fields Mining Corporations' Cunningham Hill Mine from 1979 through 1987. Mineralization is generally controlled by a regional structure, the Ortiz Graben (Maynard and others, 1989), which is part of the approximately 60 mile long (93 km) Tijeras-Canoncito Fault System (TCFS). This fault appears to control intrusive activity in the region and has a well documented geologic history with numerous periods of recurrent movement (Kelley and Northrup, 1975; Lisenbee and others, 1979; Woodward, 1984). A detailed knowledge of geologic structure and intrusive body geometry was considered essential to a successful exploration program in the region.

LAC Minerals (USA) Inc. began an exploration program in late 1983 on two prospects within their holdings on the Ortiz Mine Grant. These geographically isolated programs consisted of geologic mapping, rock and soil geochemistry, ground magnetics, IP-resistivity and drilling. The widespread evidence of mineralization and extensive land holdings (20,600 acres or 8337 hectares) prompted LAC to undertake a more regionally extensive exploration program. In late 1986, the decision was made to conduct a complete and integrated exploration program on the property which included detailed geologic mapping, regional and detailed geochemical studies, and detailed geophysics. This program was intended to gather orientation signatures over known deposits and extrapolate into covered areas. Our understanding of deposit types known at that time, did not indicate that direct detection of mineralization by airborne geophysics would be possible. This paper will summarize the geophysical aspects of this program and describe how these data are used in the on-going exploration.

GEOPHYSICS APPROACH

LAC's earlier but restricted geophysical

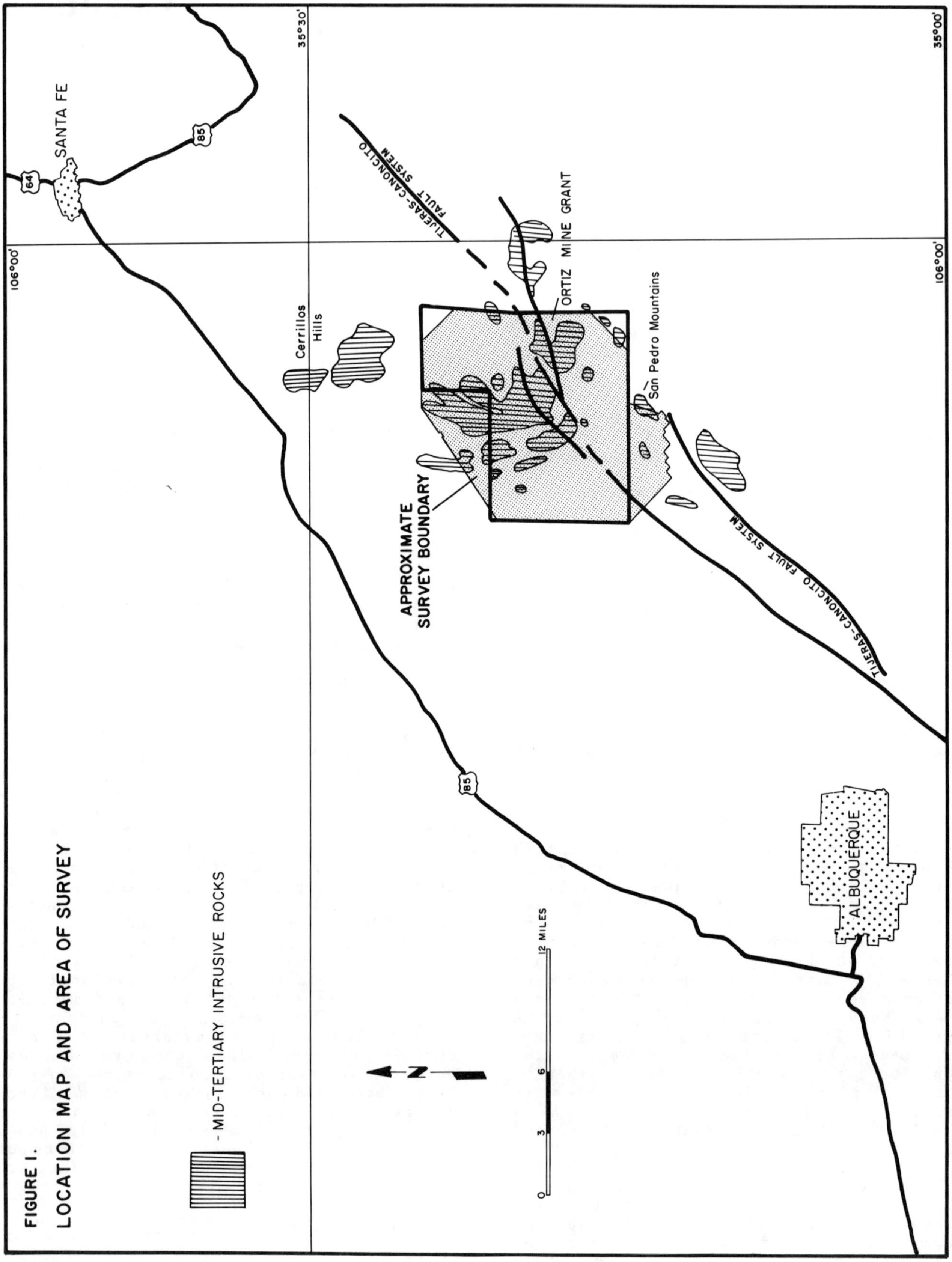

FIGURE I.
LOCATION MAP AND AREA OF SURVEY

studies and a previous property owner's extensive wide-spaced IP-resistivity and high altitude airborne magnetics comprised the existing data base. These data suggested that intrusive contacts and geologic structure could be mapped by a combination of magnetics and resistivity. The large area involved, the relative subtleties between intrusive bodies, and monotonous sediments indicated that the geophysics technique should be both close-spaced and detailed, yet extensive in coverage. Because of the extremely rugged terrain, a helicopter-borne geophysics system was selected. A multi-frequency electromagnetic (EM) system was chosen to provide resistivity information and to detect conductive structures such as faults and shears. The EM system was also equipped with a cesium vapor magnetometer of high sensitivity and sampling rate. At helicopter speeds, the results generally exceed the specifications of most ground magnetic surveys. VLF-EM was added for structural information in resistive areas. The specifications of the airborne system are presented in Table 1.

The survey was flown in two segments; one in April 1987 and a second extension in September 1987. Approximately 1,375 line miles (2,200 line kms) covering approximately 80 square miles (210 sq kms). The survey was flown in a northwest-southeast direction approximately perpendicular to the regional structural trend. The helicopter flew at a nominal height of 200 feet (60 m) with an EM sensor height of 100 feet (30 m), a magnetic sensor height of 150 feet (45 m), and a VLF sensor height of 170 feet (50 m). Line spacing was nominally 330 feet (100 m), controlled by a video recorder and electronic navigation. Altitude was measured by a radar altimiter. Electromagnetic, magnetic and altimeter data were recorded in analog and digital form. The total cost of the surveys was approximately $200,000.00 resulting in a unit cost of approximately $2,500.00 per square mile ($950 per sq km). Estimates of equivalent ground based surveys approached $50,000.00 per square mile ($20,000.00 per sq km).

RESULTS

The surveys were combined and presented at two scales; 1:24,000 and 1:12,000. The products received in the final report from the contractor were presented as computer generated color plots and black and white copies on topography. Figures 3 and 4 are reduced-definition, small-scale contour maps of total field magnetics and apparent resistivity (4600 Hz). Figure 2 shows the generalized geology of the area with some of the salient geologic features superimposed on Figures 3 and 4 for reference.

LAC geologists began using the survey results immediately on receipt of preliminary maps. Magnetics mapped intrusive rocks with the anticipated high amplitude response. In many cases the magnetics were used to infer geometry of bodies which were ambiguous from geologic mapping. For example, the intrusive at Candelario Mountain was first mapped as a floored body (eg. sill or laccolith). Detailed geologic mapping suggested that the intrusion appeared to have vertical contacts. The magnetic response indicated a steep sided body with some depth extent, suggesting a stock. The widespread evidence of mineralization around the periphery of the stock is now being examined in light of this new geometry. The magnetic response also revealed intrusions which had not been previously known and helped to map those contacts under covered areas.

Magnetics and resistivity together provide valuable assistance in mapping even in areas of sedimentary rocks; where similarities of lithologies make the recognition of structures difficult. Terminations or breaks of magnetic and resistivity features indicate faults which can be examined in the field for key outcrops. Mapping along such a break helped to accurately place the Buckeye Hill Fault, the south bounding structure of the Ortiz Graben. Geologic evidence suggested that a fault was required in this area, but poor exposure hindered mapping its trace. The boundary structures of the Ortiz Graben are significant ore-controlling structures (Maynard and others, 1989) making understanding of their position essential. The major Tijeras-Canoncito Fault Zone,locally called the Golden Fault, was poorly documented in the area. It was thought to be a discrete through-going feature, but geophysics revealed that it actually consisted of several components and extended eastward under cover as a multi-branching system of faults.

Within the central part of LAC's project area a great deal of geologic mapping, soil geochemistry and ground geophysics has been completed. The working scale of these data was 1:1,200 or 1:2,400, with compilation at 1:4,800. After working with the airborne geophysical data at smaller scales (1:12,000 and 1:24,000) it was decided to enlarge a portion of this smaller area of interest at the compilation scale (1:4,800). This proved to be a particularly useful scale for geophysical compilation and was comparable, if not superior to, ground surveys done in the same area. The detailed flight specification, high sampling density and precise location, makes this type of survey useful at both small and large scales.

AIDS TO INTERPRETATION

Maximum benefit of high resolution

TABLE 1. SPECIFICATIONS OF AIRBORNE SYSTEM (AERODAT LTD.,1987)

Aircraft

Aerospatiale 315B Lama helicopter.

Aircraft nominal height above terrain: 200 ft (61 m)

Geophysical Equipment

Electromagnetics

Aerodat 3-frequency system.
Two vertical coaxial coil pairs at 935 Hz and 4600 Hz.
One horizonal coplanar coil pair at 4175 Hz.
Transmitter-receiver separation: 23 ft (7 m).
Sensor container towed 100 feet (30 m) below helicopter.

VLF

Herz Totem 2A, receiving two transmitters and measuring the VLF total field and quadrature components. Towed 25 ft (8 m) below helicopter.

Magnetometer

Scintrex Model VIW-2321 H8 cesium, with 0.1 nanoTesla sensitivity at a 0.2 second sample rate; 50 ft (15 m) below helicopter.

Radar Altimeter

Hoffman HRA-100, measuring height of helicopter.

Flight Path Tracking

By color video camera.

Analog Recorder

RMS dot-matrix real time recording of 14 traces, plus manual and time fiducial markers.

Digital Recorder

DGR 33 recording on magnetic tape.

Recording Intervals

Equipment	Interval, Seconds
EM system	0.1
VLF	0.2
Magnetometer	0.2
Altimeter	0.2
Navigation	1.0

Magnetic Base Station

IFG-2, clock synchronized with airborne system.
Navigation equipment: Motorola Mini-Ranger

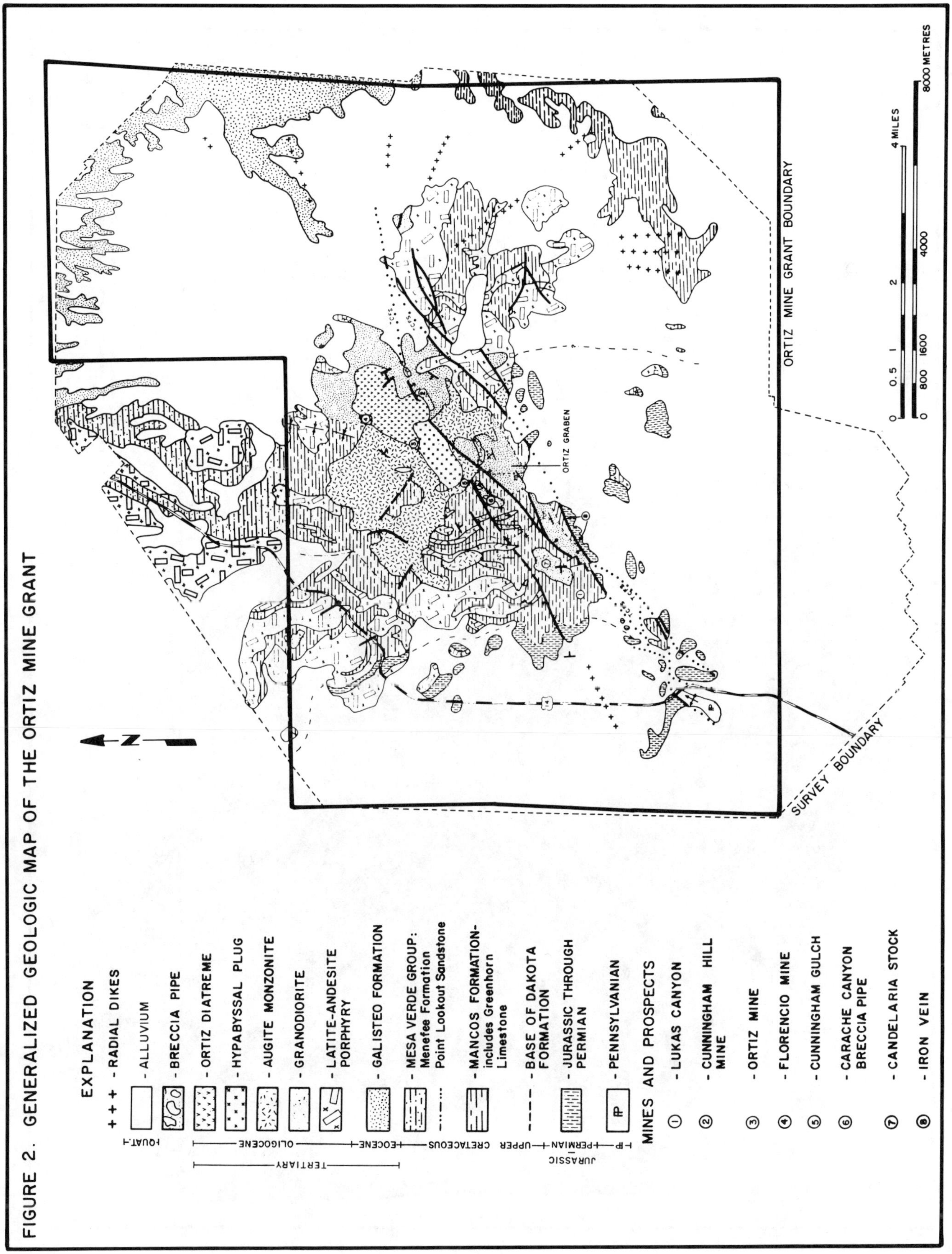

FIGURE 2. GENERALIZED GEOLOGIC MAP OF THE ORTIZ MINE GRANT

FIGURE 3.
TOTAL FIELD MAGNETIC CONTOUR MAP

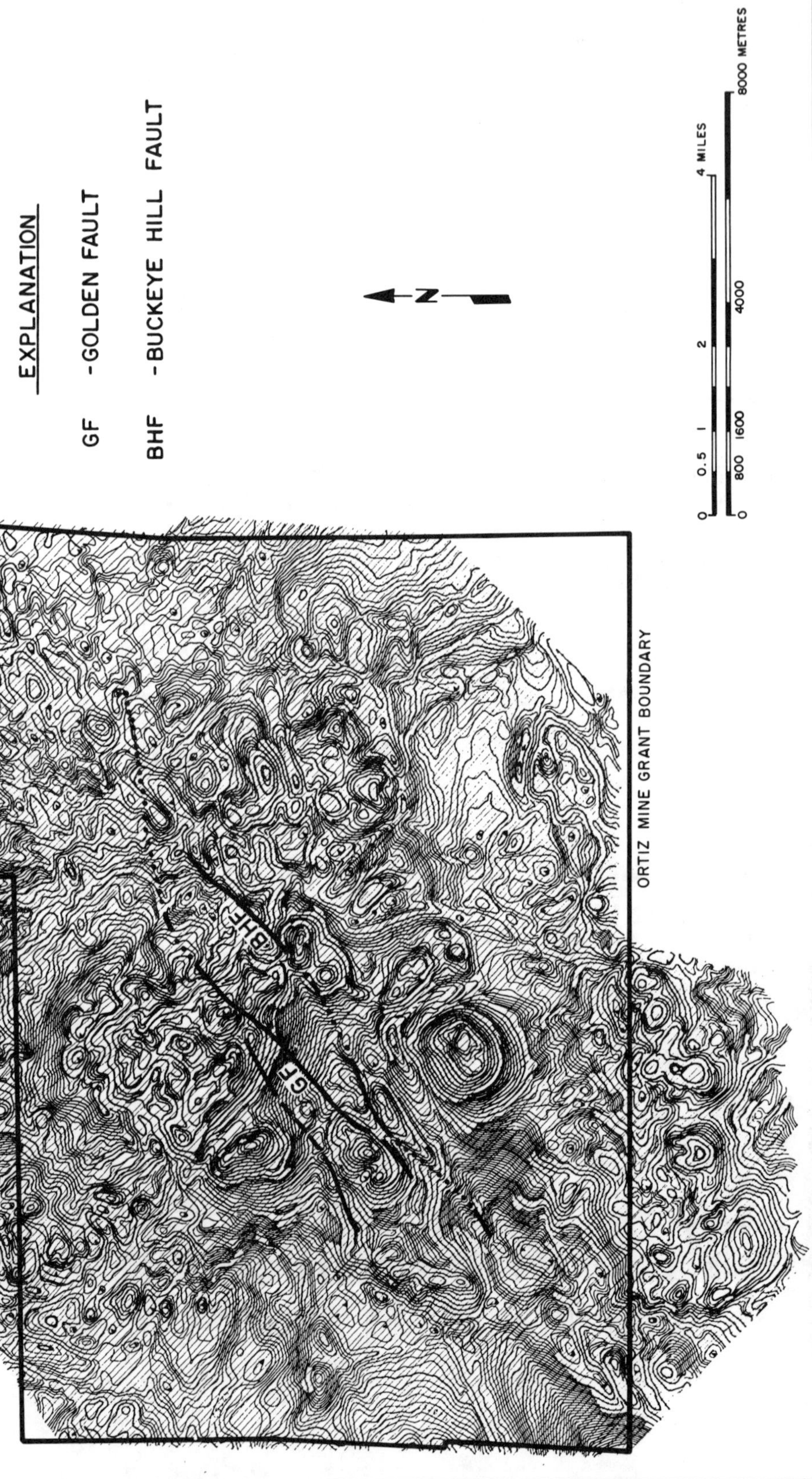

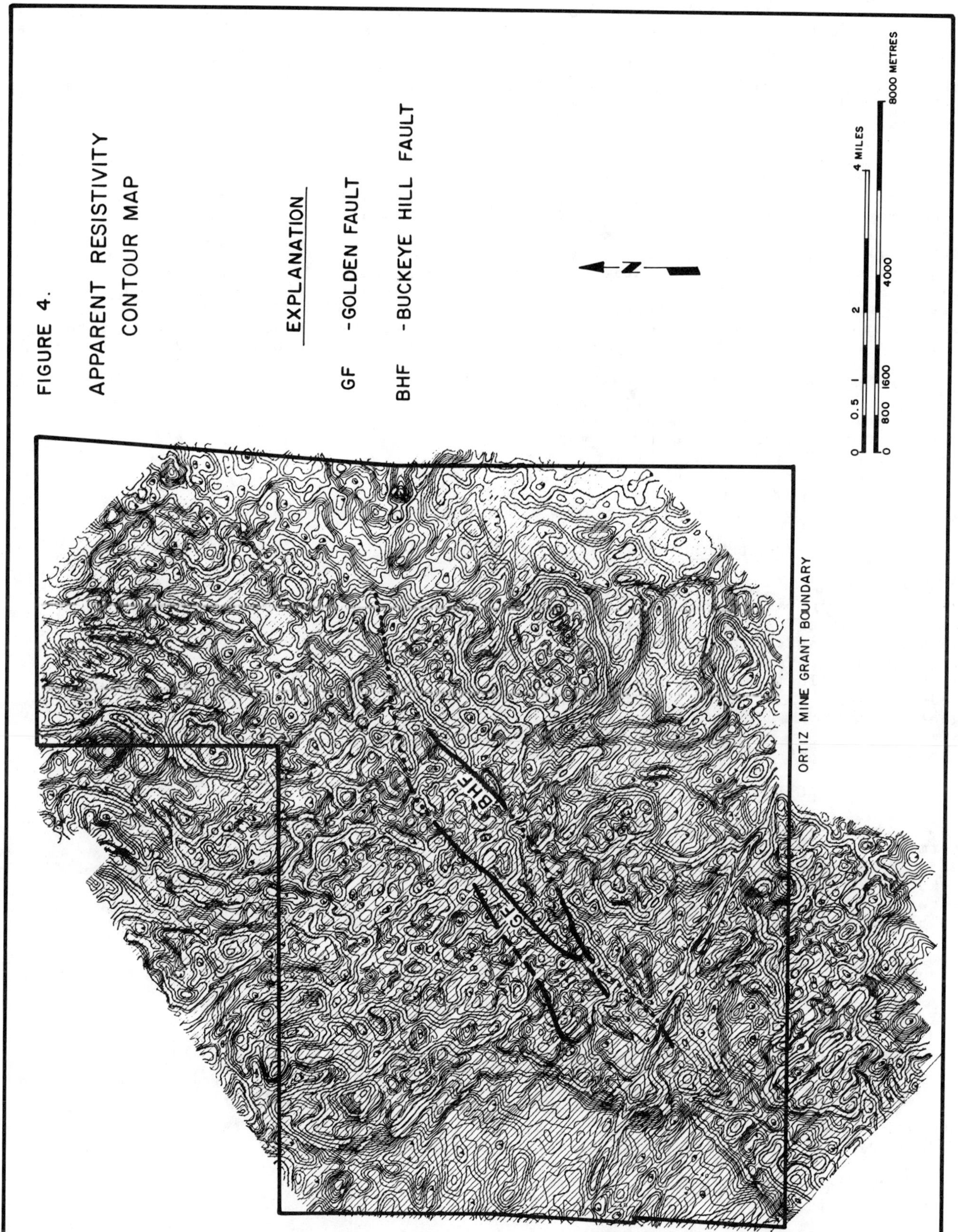

FIGURE 4.
APPARENT RESISTIVITY CONTOUR MAP

geophysical surveys is realized when the data are viewed utilizing some of the enhanced graphics presentation techniques currently available. Examples of these techniques include both the high resolution electrostatic color plots and real time imaging (RTI) PC-based computer graphics systems. It has long been recognized that color enhances ones ability to recognize very subtle features in data. Color computer plots are now able to combine various products onto one map, allowing interpretation of two products at once. This is even further dramatized with the use of RTI work stations which allow various combinations of shadowing and color imaging to be accomplished interactively with the field geologist or geophysicist. At this point, suspected features can actually be searched for in the data.

Other interpretive aids include products which are mathematical manipulations of the basic field data. These would include both first and second calculated vertical derivatives of total field magnetic data, calculated apparent resistivity data from multi-frequency EM data, and various filters or continuations of any of these data.

CONCLUSIONS

The high resolution helicopter electromagnetic and magnetic survey proved to be a cost-effective method to integrate detailed geophysical data within a regionally extensive framework. The survey helped in geologic mapping by focusing and reducing the time of exploration for geologic programs. These data can be used at a variety of scales and by individuals with a broad range of geophysical backgrounds.

ACKNOWLEDGEMENTS

We thank LAC Minerals (USA) Inc. for permission to use these data in this paper and the accompanying presentation, and also acknowledge with thanks the efforts of Aerodat, Ltd. in reprocessing the data and helping with the presentation.

REFERENCES

Aerodat Ltd., 1987, Report on the Combined Helicopter Magnetic, Electromagnetic and VLF Survey, Ortiz Mountain Range Area, for Long Lac Mineral Exploration (Texas) Inc.: author de Carle, R. J., 50 pp.

Kelley, V.C., and Northrup, S.A., 1975, Geology of the Sandia Mountains and Vicinity: New Mexico Bureau of Mines and Mineral Resources Memoir 29, 136 pp.

Lisenbee, A.L., Woodward, L.A. and Connolly, J.R., 1979, Tijeras-Canoncito Fault System - a Major Zone of Recurrent Movement in North-Central New Mexico: New Mexico Geological Society Guidebook 30th Field Conference: Santa Fe Country, pp. 89-99.

Maynard, S.R., Martin K.W., Nelsen, C.J.,and Schutz, J.L., 1989, Geology and Gold Mineralization of the Ortiz Mountains, Santa Fe County, New Mexico: Society of Mining Engineers Preprint No. 89-43, 10 pp.

Woodward, L.A., 1984, Basement control of Tertiary Intrusions and Associated Mineral Deposits along the Tijeras_Canoncito Fault System, New Mexico: Geology, v. 12, pp. 531-533.

Chapter 8

GEOPHYSICAL SURVEYS FOR PRECIOUS METAL EXPLORATION
IN THE BASIN AND RANGE, NEVADA

PIETER HOEKSTRA, JIM HILD AND MARK BLOHM

BLACKHAWK GEOSCIENCES, INC., GOLDEN, CO

Abstract. In the Basin and Range Province of Nevada, geophysics is used to assist exploration for bulk minable gold deposits, particularly over pediments. It is employed for mapping thickness of alluvium and Tertiary valley fill, for mapping covered structures, and for identifying alteration zones. In addition to potential field and seismic methods, three different electrical and electromagnetic methods are frequently used, resistivity-induced polarization, controlled source audiomagnetotellurics, and time domain EM. A review and comparison of the principles of these methods, illustrated by case histories, is the topic of this paper.

INTRODUCTION

The exploration for bulk minable gold deposits in Nevada is increasingly taking place over pediment surfaces. In these covered areas surface geophysics can be particularly effective in assisting exploration. With the schematic geologic sections of Figures 1a through 1c specific exploration objectives for geophysical surveys can be defined. They are

(1) Mapping thickness of alluvial and Tertiary valley fill. Developmental economics is influenced by the thickness of cover, and geophysics can map variation in overburden thickness over a claim block at modest cost. The required exploration depth in Nevada for this objective has varied from near surface to 3,000 ft.

(2) Structural mapping under cover. All three exploration models in Figure 1 are structurally controlled. Lineaments and faults are important for bounding tectonic provinces, and influence metallogeny by giving deep access to combinations of heat, metals, water and magma from various structural levels. Over pediment surfaces geophysics can be an important tool for mapping such structural features.

(3) Mapping of silicified zones and zones of argillic alteration. Ore is often associated with such zones.

Virtually all geophysical methods known have at one time or another been tested in Nevada for the above objectives. These methods include gravity, magnetics, seismic refraction and reflection, and measurements of electrical resistivity. This paper focuses on the uses electrical and electromagnetic geophysical methods.

The impetus for employing electrical and electromagnetic methods is that various units can often be delineated by their characteristic ranges of resistivity as illustrated in Figures 1a through 1c. It is also evident that any geologic interpretation based solely on a geoelectric section will have ambiguity, because (1) characteristic ranges of resistivity for various units may overlap, and (2) alteration may significantly change the resistivity of lithologic units. Such ambiguity may be partially removed by complimentary geophysical methods but can only be fully removed by drilling.

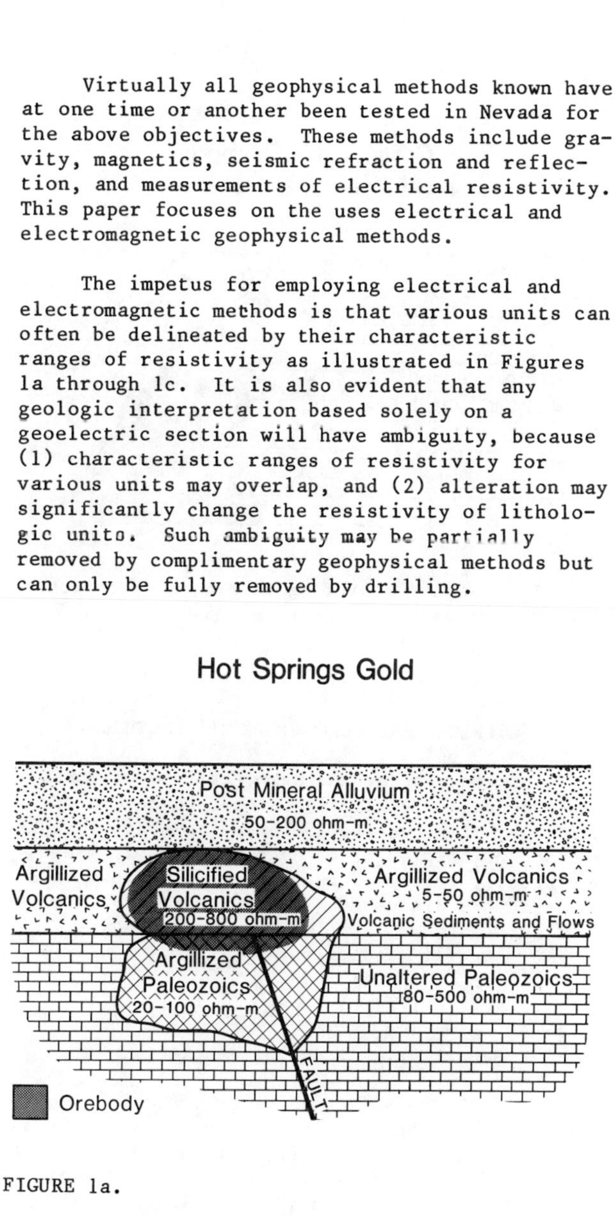

FIGURE 1a.

Fault Controlled Basin
Basin and Range

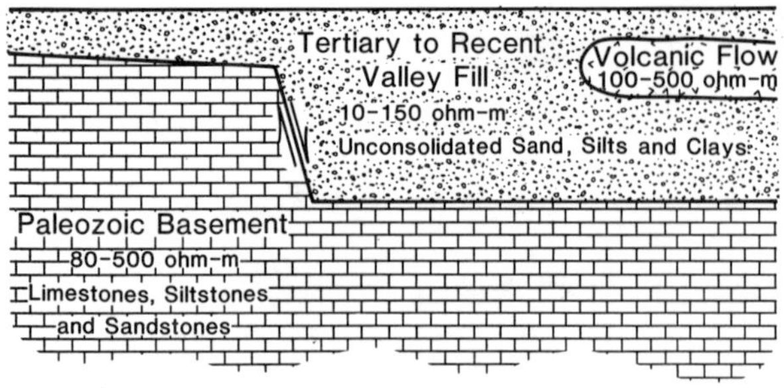

FIGURE 1b.

Carlin Type Orebody
in Paleozoic Rocks

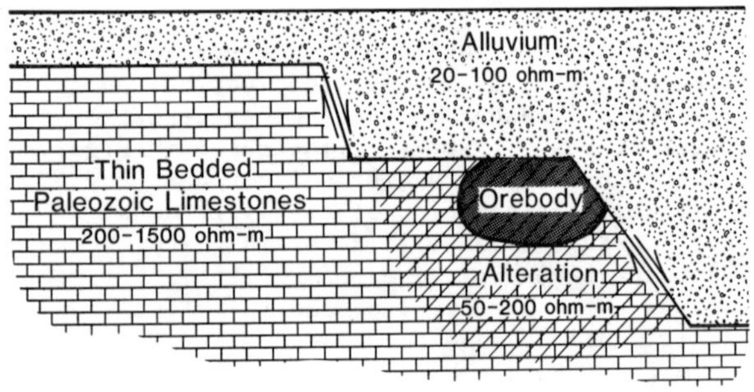

FIGURE 1c. Diagrammatic geologic and geoelectric sections of ore deposits in Nevada.

ELECTRICAL AND ELECTROMAGNETIC METHODS

Perhaps what makes the practice of electrical prospecting confusing is that there are a number of different technologies in use. In this section principles of methods now used in Nevada are briefly reviewed, and in comparing them three criteria are discussed

- lateral resolution
- vertical resolution
- sensitivity to geologic noise.

Geologic noise we define as variation in subsurface conditions not related to the exploration objective. For example, if the objective is to map thickness of Tertiary valley fill, geologic noise could be volcanic flows in the valley fill, changes in soil types, or changes in water salinity. All features commonly found over pediments.

The three electrical and electromagnetic methods now commonly used in Nevada are resistivity-induced polarization (Res-IP), controlled source audiomagnetotellurics (CSAMT), and time domain electromagnetics (TDEM). All three methods have in common that electrical resistivity is derived from measuring the resistance of the ground to flow of electrical current. They differ in the manner current flow is generated.

Res-IP

In Res-IP current is driven into the ground by galvanic contacts (steel probes). The resulting potential difference at the surface is measured by voltage electrodes. It is a characteristic of the method that effective exploration depth is a function of the separation between voltage and current electrodes. Typically, in a dipole-dipole array or a Schlumberger array, effective exploration depth is about 1/6 to 1/8 of the array length. To explore to a depth of 100 ft may require an array length up to 800 ft. Such large arrays make measurements sensitive to lateral variation in resistivity.

An important advantage of Res-IP is that in addition to resistivity, chargeability can also be derived. Chargeability is often diagnostic of disseminated sulfides.

CSAMT

In the two other methods, CSAMT and TDEM, current is made to flow in the ground by magnetic induction, i.e., induction caused by a time-varying magnetic field associated with a time-varying current forced through a transmitter. In CSAMT the current source is a grounded line, positioned a large distance from the survey area. The electromagnetic field vectors at the receiver station far away from the source are sketched in Figure 2. Over a near-horizontally stratified ground (pediment surface) they consist of a horizontal electric (Ex) and magnetic field (Hy), and a vertical electric field (Ez). Ex and Hy propagate near vertically into ground, attenuate with depth, and are reflected by boundaries of differing resistivity in the subsurface (Goldstein and Strangway, 1975). The intensity and phase of all three field vectors at the surface are influenced by the paths of wave propagation between source (grounded line) and receiver. The evidence is that all three field vectors are equally influenced by factors due to the path of propagation, but that only Ex is influenced by subsurface conditions below the receiver (Hoekstra, et al, 1975). Therefore, by measuring the ratio (Ex/Hy), called surface impedance, a quantity is derived independent of factors due to propagation path, but related to local resistivity stratification. The effective exploration depth is mainly a function of ground resistivity and frequency of operation. By stepping source and receiver synchronously through a range of frequencies the ground is sampled to a different effective depth, and a sounding can be obtained.

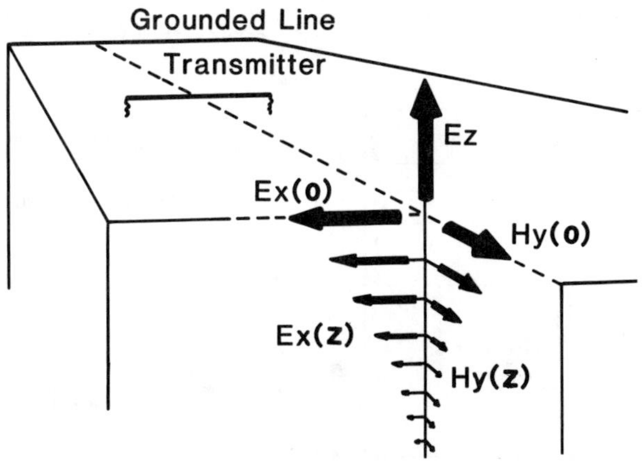

FIGURE 2. Electromagnetic field vectors at the surface in CSAMT.

Because the geoelectric section in CSAMT is derived from measurements of the electric field certain problems arise. There are two sources for electric fields, electrical charges and the time-varying secondary magnetic fields associated with current flow induced in the ground. Electrical charges physically form at boundaries of even minor discontinuity in electrical resistivity (Fig. 3). There are many causes of lateral and vertical boundaries of different resistivities where charges form (changes in soil types, boulders, depressions where water collects, arroyos, etc.).

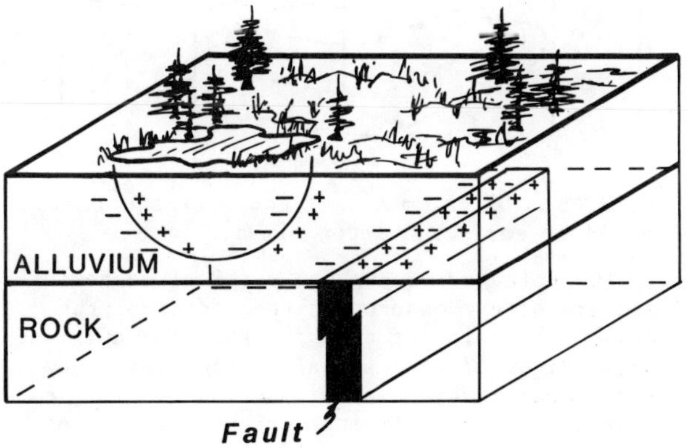

FIGURE 3. Formation of electrical charges at boundaries of discontinuity in electrical resistivity.

The effect of these charges manifest themselves in a major way in CSAMT data. The electric field due to these charges is a vector with a certain direction, and the receiver in CSAMT is an electric dipole (two probes in the ground), also oriented in a certain direction. Therefore, different readings can be expected with source and receiver dipoles oriented differently. That is indeed the case. Figure 4 shows CSAMT data measured at different orientation of source and receiver. Which is the correct curve? Which one is to be used in interpretation? Other independent electrical measurements are required to answer these questions. It is for these reasons that CSAMT is highly sensitive to geologic noise. At all lateral and vertical boundaries of changes in resistivity electrical charges form, influencing measurements at the receiver.

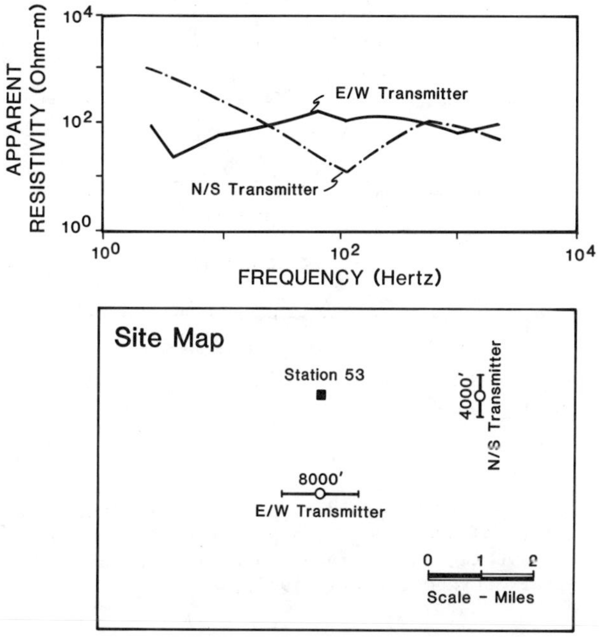

FIGURE 4. CSAMT apparent resistivity curves collected at WIPP at two different orientations of source and receiver.

CSAMT data are commonly displayed in pseudo-apparent resistivity-depth sections in which apparent resistivity is plotted versus frequency (Bartel and Jacobson, 1987). It is common in such sections to find near-vertical features cross-cutting such sections. It has been shown (Sternberg et al, 1988) that such events are often caused by near surface lateral changes in resistivity rather than by structures with depth extent. Figure 5 shows a pseudo-apparent resistivity depth section computed for a horizontally layered model with a thin near surface resistivity anomaly. The electrical charge formation on the boundaries of the near surface conductive anomaly is the cause of this apparent vertical structure.

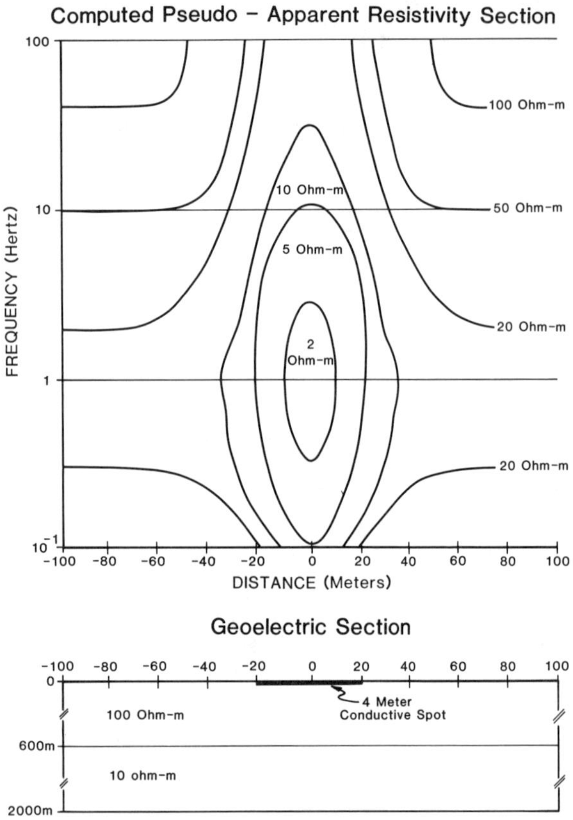

FIGURE 5. CSAMT pseudo-apparent resistivity section (top) computed for a geoelectric section with a thin near-surface conductive anomaly (after Sternberg et al, 1988).

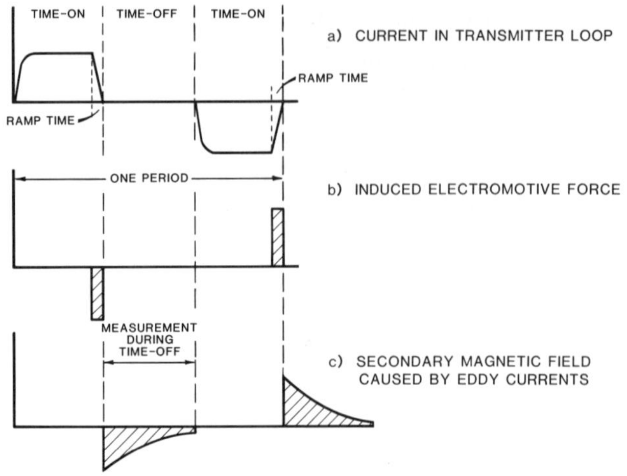

FIGURE 6a. Current waveform used in TDEM.

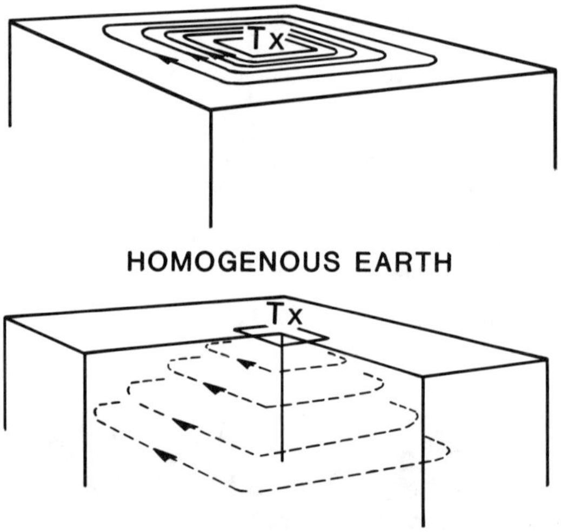

FIGURE 6b. Diagrammatic of eddy current flow induced at early and later times.

TDEM

In TDEM a source local to the receiver is used for the electromagnetic field (Mills et al, 1988). A common transmitter is a square non-grounded loop, and an air coil receiver can make measurements both inside and outside the transmitter loop. Through the transmitter loop an electrical current is driven with the waveform illustrated in Figure 6a. During the rapid current turn-off eddy current flow is induced in the subsurface. The intensity of these eddy currents is a function of the geoelectric section. Over a more or less horizontally stratified ground, the eddy currents are closed rings concentric about the transmitter. Immediately after turn-off these current rings flow near the surface directly under the transmitter, and with increasing time the currents are induced at greater depth (Fig. 6b).

The voltages due to the vertical magnetic field are often measured in the center of the loop as a function of time for sounding applications. Figure 7 schematically illustrates how voltage measurements, as a function of time, can be inverted in resistivity depth sections. At early time after turn-off eddy currents are mainly concentrated in near surface layers and a measurement of voltage at early time will mainly be sensitive to the electrical resistivity of near surface layers. With increasing time voltage measurements become progressively more sensitive to properties of deeper layer. Thus, in TDEM exploration depth is mainly a function of time of measurement after current turn-off. Apparent resistivities are plotted as a function of time and this data is inverted in true resistivities and depth. Because magnetic fields rather than electric fields are measured in TDEM, its measurements are less sensitive to near surface lateral changes in resistivity than CSAMT. Electrical charges effect magnetic fields in a secondary way (current channeling).

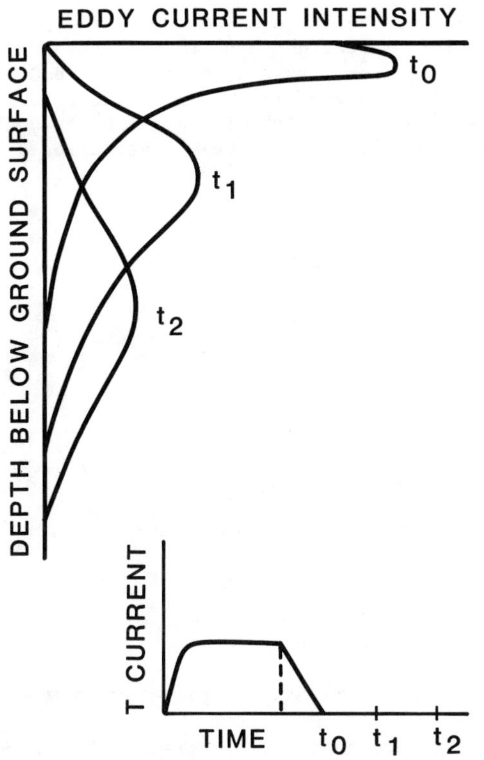

FIGURE 7. Schematic illustration of eddy current intensity versus depth at different times after current turn-off.

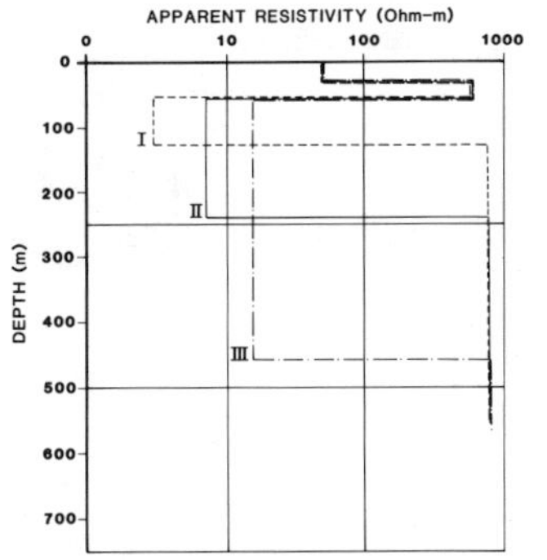

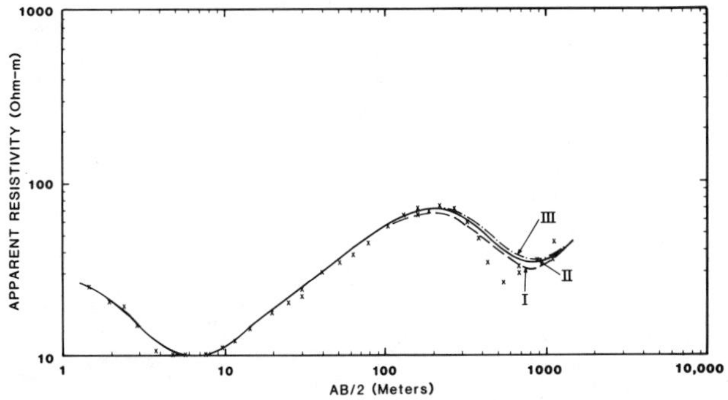

FIGURE 8. Results of Schlumberger soundings on coastal plains in Israel. Data are superimposed on three forward model curves. Depth to bedrock in drill hole was 168 m.

Thus, the two main differences between CSAMT and TDEM are (1) a transmitter local to the receiver is used in TDEM, and a transmitter distant to the receiver in CSAMT, and (2) subsurface resistivities are derived mainly from electric field measurements in CSAMT, and from magnetic field measurements in TDEM.

Next, the results obtained with the three methods are compared for a few case histories where measurements were made with more than one technique.

Direct Resistivity versus TDEM. On the coastal plain of Israel both Schlumberger (resistivity) and TDEM soundings were made near a drill hole. The geology consisted of quarternary coastal plain sediments (silts and clays) overlying a Mesozoic carbonate bedrock surface. Depth to bedrock in the drill hole was 168 m (550 ft). The resistivity soundings (Schlumberger array) are shown in Figure 8. Measurements were made to a AB/2 spacing of 1,500 m (an array length of 3,000 m). There is an increase in resistivity at the end of the spread likely indicating more resistive bedrock. To evaluate vertical resolution the data were superimposed on three models in which depth to bedrock was varied by several hundred meters. It must be concluded that vertical resolution is plus or minus several hundred meters.

Figure 9 shows the results of a TDEM sounding in the center of 200 m by 200 m transmitter loops. The TDEM data also are superimposed on three forward model curves, but the differences in depth to bedrock between the curves is now 20 m, and vertical resolution is evaluated to be ± 10 m. This reduction in equivalence and improvement in vertical resolution is an important advantage of TDEM.

The geoelectric section in the above example is similar to sections observed over pediments in Nevada, in that the Tertiary valley fill generally also has lower resistivities than the Paleozoic bedrock surface. Over pediments typically transmitter loops with dimensions about two-thirds of the required exploration depth are used. Thus, to explore to a depth of 250 m (750 ft), 500 ft by 500 ft transmitter loops are employed. Figure 10 is a typical section derived from center-loop TDEM soundings near Carlin, Nevada. In this section a drill hole encountered the welded volcanics and the depth to Paleozoics interpreted from TDEM was within ten percent of depth in a drill hole.

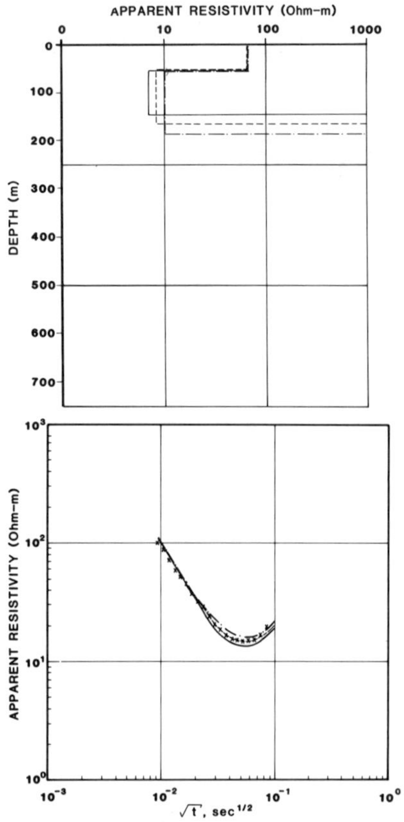

FIGURE 9. Results of TDEM soundings in 200 m by 200 m transmitter loop. Data are superimposed on three forward model curves.

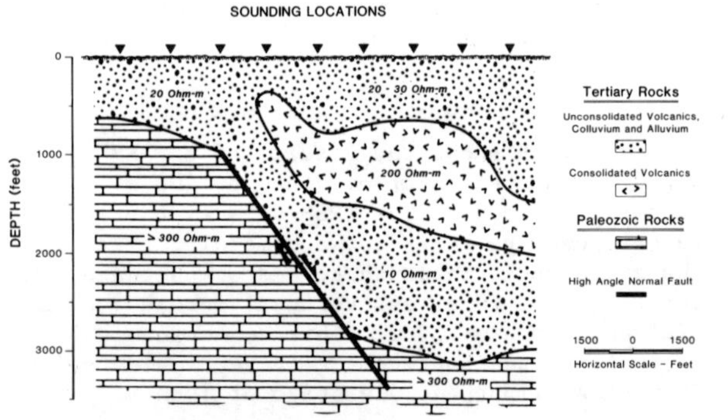

FIGURE 10. Section derived from center-loop TDEM soundings near Carlin, NV.

CSAMT versus TDEM. The most extensive comparison we have between CSAMT and TDEM is from the Waste Isolation Pilot Plant (WIPP) in Carlsbad, New Mexico. The CSAMT data set in Figure 5 was collected with transmitters in two different orientations so that the effect of electric dipole orientation could be studied. Without measuring with dipoles in two directions the true nature of the data may have never been known, and errors in inversion will result.

Figure 11 shows TDEM apparent resistivity curves and their inversions from WIPP. The sounding WIPP #12 is from about the same location as the CSAMT data in Figure 5. Also shown on Figure 11 is drill hole information. The main electrical marker horizons in this data is the brine encountered at a depth of about 800 m in WIPP #12 and deeper in DOE #1 (1,200 m). There is excellent agreement with drill holes. The prior extensive CSAMT survey never allowed a meaningful inversion of true resistivity and layer thicknesses.

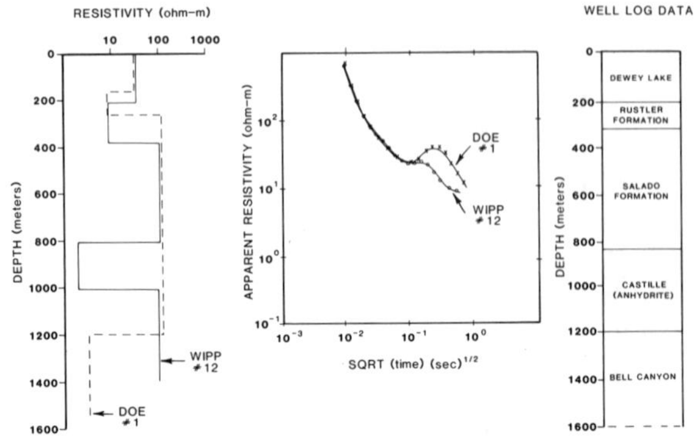

FIGURE 11. TDEM apparent resistivity curves and inversion from WIPP.

OTHER TDEM CASE HISTORIES IN NEVADA

Over pediments in the absence of alteration, the alluvium and Tertiary valley fill generally have ranges of resistivity substantially different from basement rock. Silicification and argillic alteration often cause major changes in the resistivity of these units. An example from north-central Nevada is given in Figure 12. Over the surface projection of the ore body the resistivity layering is reversed. Instead of the Tertiary unit having lower resistivity than Paleozoic rocks, silicification associated with mineralization in the Tertiary volcanic sediments, and argillization and fracturing in the Paleozoic rocks caused the Tertiary units to be more resistive than the Paleozoic units. Without drill hole information it would have been difficult to recognize the geologic significance of such a pattern. With drill hole information geophysics can (1) map the areal extent of the pattern, and (2) explore for similar patterns in the immediate vicinity.

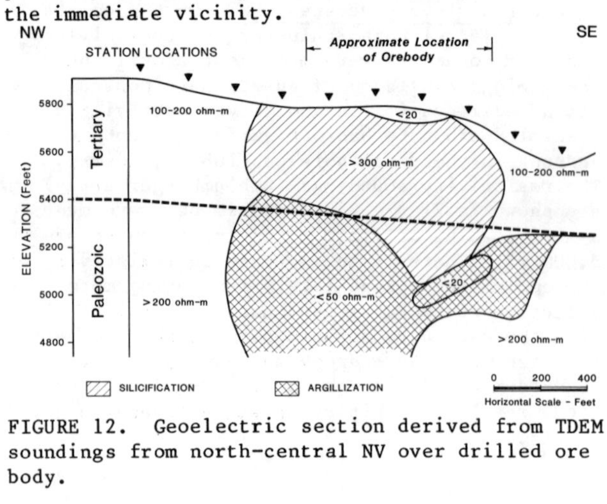

FIGURE 12. Geoelectric section derived from TDEM soundings from north-central NV over drilled ore body.

Another example of resistivity changes associated with mineralization is shown over the Betze deposit of American Barrick in Figure 13. An electrical marker horizon at a depth of about 500 ft is a carbonaceous unit. Over the surface projection of the Betze deposit, based on drill hole information provided, this carbonaceous unit is oxidized to greater depth. This is reflected in an increase in resistivity interpreted from TDEM. The correlation between oxidized and reduced carbonaceous units is further illustrated in a comparison of a drill hole log with a TDEM derived resistivity profile in Figure 14.

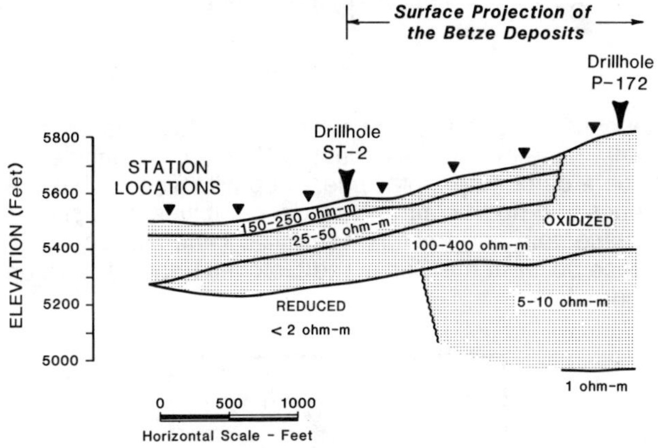

FIGURE 13. Geoelectric section derived from TDEM soundings over Betze deposit.

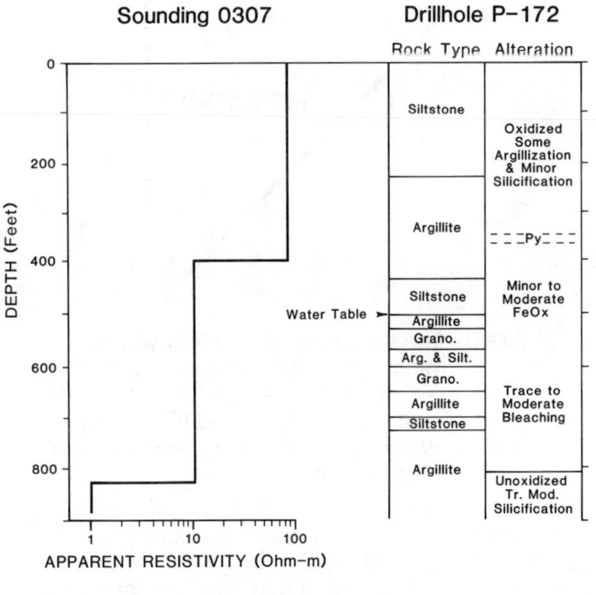

FIGURE 14. Comparison of drill hole log and TDEM derived resistivity-depth profile from near Betze deposit.

The advantages of TDEM in the case histories of Figures 12 and 13 are that the alteration zones of relative small dimensions (perhaps 500 ft maximum) can be mapped with good lateral resolution to a depth of about 500 ft. This would not have been possible with a Res-IP array; an array length in excess of 2,000 ft would have been necessary to map to a depth of 500 ft. However, a Res-IP survey may have indicated a "halo" of disseminated sulfides at shallow depth diagnostic of alteration.

CSAMT in principle has a good lateral resolution, but unless resistivities are measured with transmitting and receiving dipoles in two perpendicular directions (vector CSAMT), inversion of pseudo-apparent resistivity section in true resistivities and layer thicknesses is hazardous. When vector CSAMT data is collected, it often becomes apparent that an independent correction by another method may be necessary.

REFERENCES

Bartel, L. C., and R. D. Jacobson, 1987. "Results of a controlled-source and 10 frequency magnetotelluric survey at the Puhimau thermal area, Kilauea Volcano, Hawaii", Geophysics, V. 52, pp. 665.

Goldstein, M. A., and D. W. Strangway, 1975. "Audio frequency magnetotellurics with a grounded electric dipole source", Geophysics, V. 40, pp. 669.

Hoekstra, P., P. V. Sellmann, and A. Delaney, 1975. "Ground and Airborne Resistivity Surveys of Permafrost near Fairbanks, Alaska", Geophysics, V. 40, pp. 641.

Mills, T., P. Hoekstra, M. Blohm, and L. Evans, 1988. "Time domain electromagnetic soundings for mapping sea-water intrusion in Monterey County, California", Ground Water, V. 26, pp. 771.

Sternberg, B. K., J. C. Washburne, and L. Pellerin, 1988. "Correction for the static shift in magnetotellurics using transient electromagnetic soundings", Geophysics, V. 53, pp. 1459.

Chapter 9

GOLD EXPLORATION USING ILLITE POLYTYPES DEFINED BY X-RAY DIFFRACTION AND REFLECTANCE SPECTROSCOPY

P.L. HAUFF, F.A. KRUSE AND *R. J. MADRID

CENTER FOR THE STUDY OF EARTH FROM SPACE
Cooperative Institute for Research in Environmental Sciences (CIRES)
Campus Box 449, University of Colorado, Boulder, CO 80309
**CONQUISTADOR GOLD LTD.*
P.O. Box 1629, Palo Alto, CA 94301

ABSTRACT

Sub-microscopic gold grains in disseminated gold deposits in Nevada are often directly associated with the clay mineral illite. Delineation of gold ore-zones can be improved by accurate characterization of the associated illite varieties or polytypes. X-ray diffraction studies of clay minerals from several disseminated gold deposits demonstrate that illite polytypes reflect lateral and vertical zonation around the ore bodies. Polytypes act as geothermometers indicating temperatures of ore deposition and proximity to hypogene and supergene fluids.

Visible/near infrared spectral measurements using a field-portable reflectance spectrometer provide a reconnaissance technique for mapping the in-situ distribution of illites. Changes in the visible/near-infrared reflectance spectra are observed as illites progress from the lower temperature, less ordered 1M variety to the higher temperature 2M type. The combination of the reflectance spectroscopy with the X-ray diffraction data creates a powerful tool for detailed mineralogic mapping for gold exploration.

INTRODUCTION

Exploration for precious metals requires costly drilling and extensive assay determinations. These activities are usually guided by detailed geologic and alteration mapping and subsequent development of a model for the deposit. Alteration at the surface often acts as a guide to the ore, however, extensive laboratory and field work is required to produce an accurate map. This paper describes a technique under development that utilizes X-ray diffraction and reflectance spectroscopy for field identification of potential economic ore zones in precious metal deposits. The deposits used in this study are those termed "disseminated gold" as the gold found within them is distributed through a high volume of rock and is usually submicroscopic in grain size. A large number of these systems are found concentrated near or within the mountain ranges in the state of Nevada.

Numerous classifications have been used to describe these deposits. Commonly, they are divided into sediment hosted and volcanic hosted systems. This paper uses samples from two different sediment hosted deposits. The Carlin type, which takes its name from the Carlin Mine in northeast, central Nevada, is considered to be a carbonate replacement deposit. The Preble Deposit, in north central Nevada, combines both silica replacement and fracture controlled mineralization. These occur in the Carlin and Getchell trends shown on Figure 1.

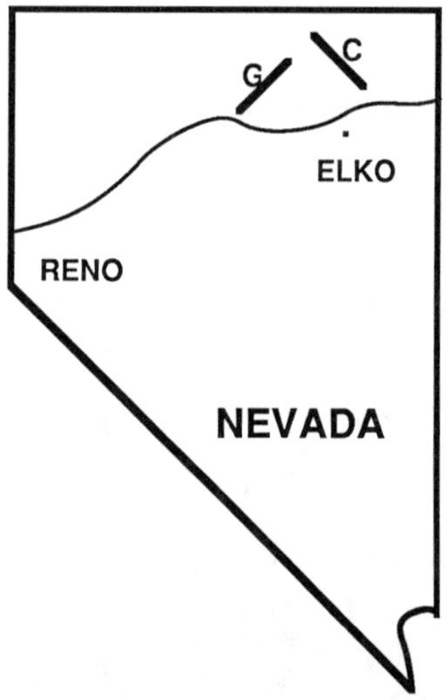

Figure 1. Location map of Nevada which diagrammatically shows two major structural trends, the Carlin trend (C) and the Getchell trend (G), along which many of these precious metal deposits occur.

The clay mineral illite is associated with the gold mineralization in many of these disseminated gold systems (Bakken et al, 1989). Illite is an important mineral species in disseminated gold deposits as it can reflect lateral and vertical zonation around gold ore bodies. Illites change crystal structure with temperature and may act as geothermometers indicating temperatures of ore deposition and proximity to hypogene and supergene fluids. The proposed method utilizes the structural ordering of the illite crystal lattice to predict certain zonal relationships within the ore body.

DISSEMINATED GOLD EPITHERMAL DEPOSITS

In the past five to ten years, the type of epithermal gold deposit termed "disseminated gold" has been discovered in large numbers in Nevada. The ability to bulk extract the gold using heap leaching technology has made exploitation of these deposits economically attractive. Extensive work by Bagby and Berger (1985), Bonham (1988), Buchannan (1981), Giles and Nelson (1983), Heald-Wetlaufer et al (1983), Percival et al, (1988), Radtke, (1985), and many others provide detailed descriptions of zonation and mineral distribution within the various volcanic and sediment hosted precious metal deposit types. There is no one model that appears to explain all deposits, however, many deposits share similar characteristics.

Nevada deposits appear to align along major structural trends within the state and are generally associated with igneous bodies and magnetic anomalies. Permeability often controls deposition rather than any paticular lithology. Carbonaceous material is present below the oxidized zones (and sometimes even within them) especially with shale and carbonate host rocks as at Carlin. Structural control is very important in many of the ore bodies with the tabular, stratabound and irregularly shaped ore localized within or near faults. Replacement and brecciation also occurs, again a function of the preparation and permeability of the host. Episodes of decalcification, silicification, argillization with later (?) calcite and quartz veining are indicative of the hydrothermal alteration stages. Jasperoids are common and important, but not always gold bearing. The contrast of the dark, carbonaceous, unoxidized ore against the reds, buffs and tans of the oxidized zones are a striking "hallmark" of these deposits. The gold is characteristically submicroscopic in size and associated with quartz and iron sulfides in the unoxidized zones and with iron oxides, sulfates and clays in the oxidized bodies. The trace elements As, Sb, Hg, Ba, Tl, +F, and +Ag are nearly always found in the alteration assemblages.

Figure 2 is a schematic cross section illustrating the two types of sediment hosted, epithermal, disseminated gold deposits discussed in this paper. The relative depths of these deposits are largely unknown, however, their approximate sizes and geometrics are depicted in Figure 2. The Carlin gold deposit is a replacement-controlled ore body developed in a sequence of silty

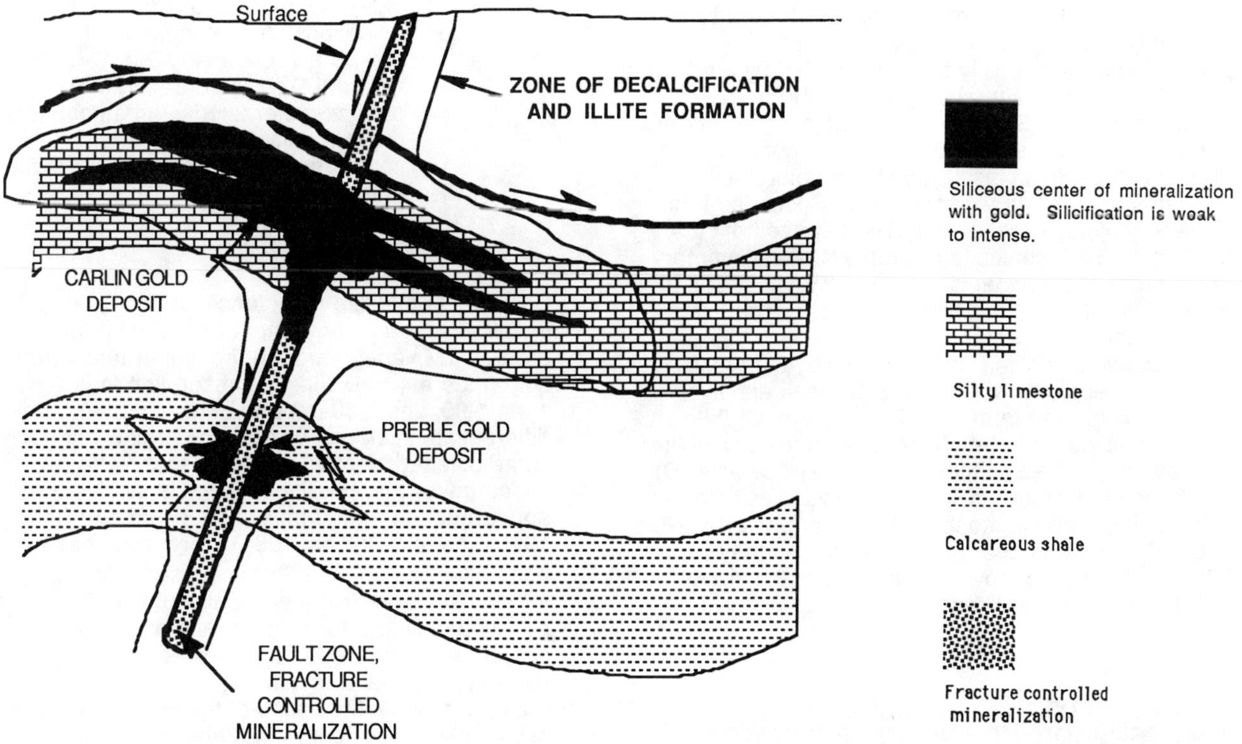

Figure 2. Schematic cross section of typical epithermal disseminated gold deposits developed in sedimentary rocks. Siliceous center of mineralization may be dominated by replacement mineralization as at the Carlin deposit or by fracture controlled mineralization as at the Preble deposit. In both cases a zone of decalcification accompanied by changes in and formation of different clay minerals, predominantly illite polytypes, is present in and around the areas affected by intense mineralization. Relative depths of deposits are unknown.

limestones. The Preble deposit is fracture controlled but also contains small replacement ore bodies. Both the Carlin and Preble deposits are accompanied by extensive zones of decalcification and illite occurrence in and around the most extensively mineralized portions of the ore bodies. Gold occurs predominantly in the centers of mineralization characterized by weak to intense silicification.

X-RAY DIFFRACTION DETERMINATION OF ILLITE POLYTYPES

Illites present within these deposits occur as at least two polytypes. Polytypes are varieties of the same mineral which differ in the stacking sequence of the compositional layers within the structure. The illite polytypes most frequently encountered are 1M, which is less ordered or symmetrical and 2M, which shows greater ordering of its structure. Polytypes can be used to learn about the environment of formation of the mineral.

The most common and effective analytical technique for the identification of illites has been x-ray diffraction. Figure 3 is a compilation of x-ray diffraction scans which shows the two most common illite polytypes. These range from the higher temperature, well ordered 2M to the poorly ordered, lower temperature 1M. Illite polytypes can now easily be differentiated by automated x-ray diffraction. Several diagnostic polytype peaks occur between 23-32 degrees two-theta. These peaks and their *hkl* values are marked in the figure. The $2M_2$ from Japan (A) is a classic high temperature hydrothermal illite provided by Dr. S. Shimoda of the University of Tsukuba, Japan. The 2M illite from the Marblehead, Wisconsin location (B) is of sedimentary origin and although identifiable as a 2M species, it is quite different from the Shimoda sample. There is considerably more water in its structure with interstratified smectite. This is shown by the broad, asymmetrical profile of its 10Å (8.9°) peak and the higher background from 2° - ~7°. Sample (C), Fithian illite, is considered to be 1M. It also contains smectite interlayering. The sedimentary Silver Hill sample (D) has been identified as a $1M_d$ which indicates disorder in its stacking. The diagnostic 1M, smaller peaks at ~25.5° and ~27° do not appear as well defined as they do in the Fithian. It becomes obvious that there is extensive variation among the illites.

LABORATORY REFLECTANCE SPECTROSCOPY OF ILLITE POLYTYPES

Infrared spectra were collected from the reference illites in the laboratory using a Beckman 5270 Spectrometer modified to measure at high resolution (<2nm). Measurements were taken from 0.4 to 2.5 µm. Spectra were measured using tungsten illumination and HALON as a reflectance standard. HALON is a highly reflective material with no absorption features in the 0.4 to 2.5 µm range.

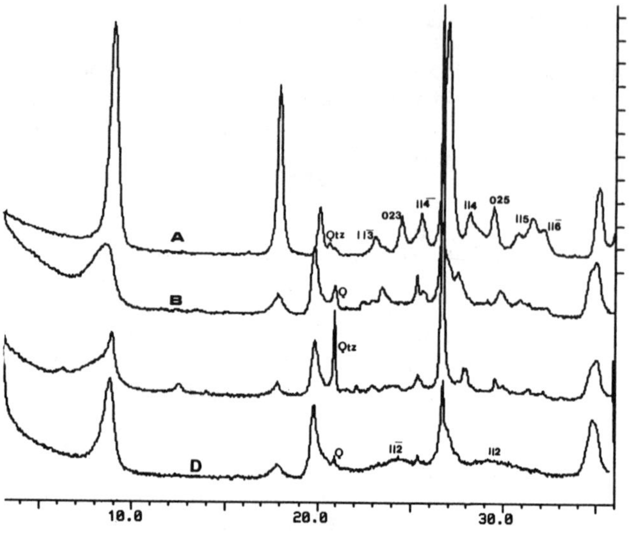

Figure 3. X-ray diffraction scans of reference illites illustrating polytype differences. Vertical scale shows intensities in counts per 2-theta step. Horizontal scale is in degrees two-theta. Peaks marked with *hkl* values indicate some diagnostic polytype reflections. **A** is a Japanese hydrothermal $2M_2$; **B** a sedimentary 2M; **C** a sedimentary 1M; and **D** a disordered 1M. The X-ray diffractometer used is an automated, low-angle modified Philips-Norelco with a copper target and the Nicolet Instruments automation package.

Changes in the illite reflectance spectra are observed as the illites progress from the lower temperature, less ordered 1M variety through the higher temperature 2M type (Figure 4). Distinctive absorption features near 1.40, 1.90, and 2.20 µm and between 2.3-2.5 µm differentiate interlayer water, structural water (OH-) and some octahedral layer characteristics. The reflectance spectrum for the 2M illite polytype is similar in appearance to that of muscovite (Hunt and Salisbury, 1970; Lee and Raines, 1984). It typically has a sharp absorption feature at 2.2 µm caused by OH^- in the mineral structure (Hunt and Salisbury, 1970). In the 1M polytype, this band is broader and asymmetrical towards longer wavelengths. The $1M_D$ spectrum from Silver Hill (C) suggests that a secondary absorption feature near 2.25 µm may be the cause of the absorption band asymmetry. Additional weak absorption bands at 2.35 and 2.45 µm in illites are better developed in the 2M polytypes. The presence of absorption bands at both 1.4 and 1.9 µm is indicative of molecular water as water of hydration or trapped in the mineral lattice (Hunt and Salisbury, 1970). If the 1.4 and 1.9 µm bands are sharp, as they are in the 2M polytype(A, B) this indicates that the water molecules are located in well defined, well ordered sites. When these bands are broad and asymmetrical, as they are in the 1M illite polytypes (C,

D), then they indicate that the minerals are relatively disordered (Hunt and Salisbury, 1970). The occurrence of the 1.4 μm band without the 1.9 μm absorption band indicates the presence of hydroxyl (OH^-) and that molecular water is not present.

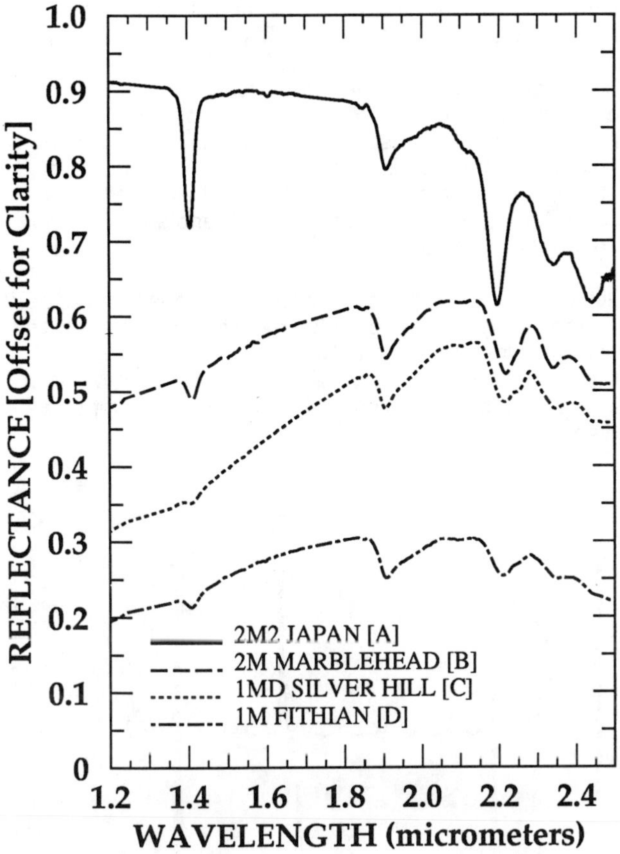

Figure 4. Reflectance spectra of illite reference standards showing polytype differences. Note how the absorption features at 1.4, 1.9, 2.2, and between 2.3 and 2.5 μm change with polytype. **A** is a higher temperature Japanese hydrothermal $2M_2$; **B** a sedimentary 2M; **C** a sedimentary 1M; and **D** a disordered 1M. Spectra were collected on a Beckman 5270 laboratory spectrometer with integrating sphere.

FIELD MEASUREMENTS

Field spectra were measured during 1988 and samples were collected from several localities within the East and Main Pits at the Carlin Mine, and from unaltered Roberts Mountain Formation outside the mine area. Reflectance spectra were measured using the Geophysical and Environmental Research (GER) Single beam Infrared Intelligent Spectrometer (SIRIS). The SIRIS is a grating spectrometer that measures reflectance from 0.4 to 2.5 μm at approximately 4 nm band centers. This instrument is portable and is controlled by a "laptop" computer. Spectra were measured using solar illumination and HALON as a reflectance standard.

THE CARLIN DEPOSIT

The Carlin-Trend disseminated gold deposits of northeast-central Nevada were chosen for this study of alteration assemblages because the geology is generally well known and gold assays already exist for many areas. The Carlin mine in Eureka County, Nevada, was selected for the initial study because it is recognized as the type locality for the original sediment hosted disseminated gold deposit. Acidic, gold-bearing hydrothermal solutions have extensively modified the host carbonates at Carlin. High grade ore is concentrated in the weakly to moderately calcareous, weakly to moderately silicified Roberts Mountains Formation rocks (Bakken and Enaudi, 1986).

Preliminary work at Carlin indicates that illites vary with proximity to ore bodies. Figure 5 contrasts the X-ray diffraction data for a 2M illite (upper scan) from a high gold bearing zone with a 1M illite (lower scan) from a barren, silica flooded alteration zone at the Carlin Mine. The contrast between the polytypes is striking and shows that with the automated diffraction equipment, they can be differentiated.

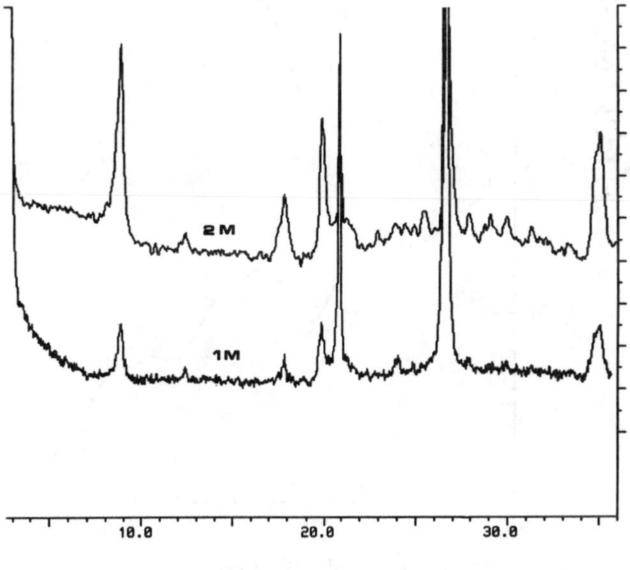

Figure 5. X-ray diffraction patterns for samples from the Main Carlin Pit. Vertical scale shows intensities in counts per 2-theta step with a maximum intensity of 5000 cps. Horizontal scale is in degrees two-theta. Sample C-24, a well developed 2M illite, is from an ore zone and shows high gold values (+10,000ppb). Sample C-47 is a 1M disordered illite from a barren, silica flooded zone.

The reflectance spectra for two illite polytypes taken with a portable reflectance spectrometer in the main Carlin pit are shown in Figure 6. Although the features are not as well defined as those observed on the laboratory spectra in Figure 4, the differences are apparent. Field spectrum C10885 is from alteration zone 3 (moderately silicified laminated beds) of Bakken and Einaudi (1986), a typically high gold zone (+10,000ppb). The well developed absorption bands at 2.2, 2.35, and 2.45 µm indicate that it is a 2M polytype. Field spectrum CAR88B is from alteration zone 5 (strongly silicified clastic beds) of Bakken and Einaudi (1986), a zone with silica flooding and jasperoids. There are "pods" of high gold, but low gold values (<1000 ppb) are common. These spectra substantiate that the polytypes can be identified and mapped in the field and that there is potential for this method as an exploration and mapping tool.

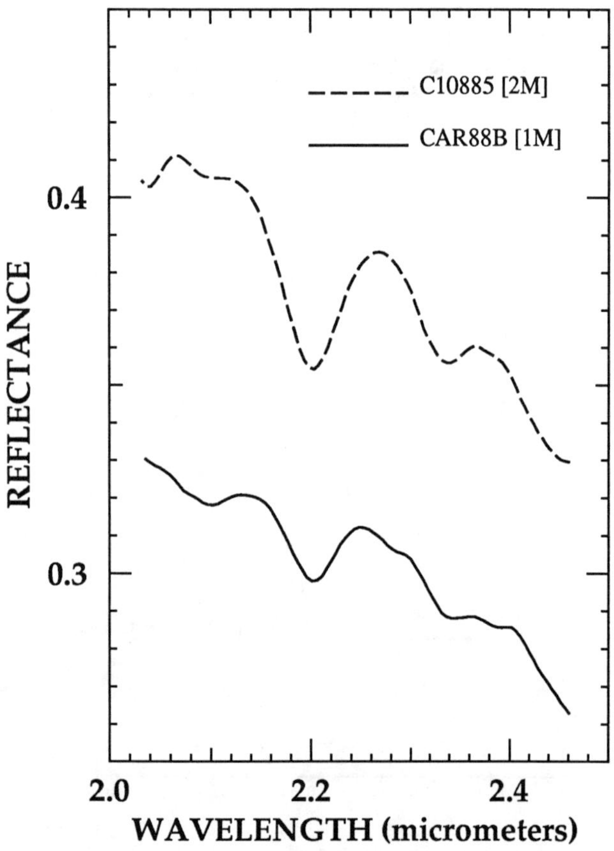

Figure 6. Illite reflectance spectra collected with a portable reflectance spectrometer in the Main Carlin Pit. The top spectrum is a 2M illite while the bottom is a 1M. Note the differences in the definition of the features at 2.2µm and from 2.3µm to 2.5µm.

THE PREBLE DEPOSIT

The Getchell trend is located to the West of the Carlin trend near the eastern margin of the Osgood Mountains (Figure 1). Gold deposits occur along a structural belt known as the Getchell Fault System. At the Preble deposit gold mineralization is hosted by the Middle Cambrian Preble formation which consists of regionally metamorphosed lower greenschist grade shales and massive to thinly bedded, silty, limestones. Silicification has occurred in the thin-bedded, carbonaceous, calcareous, silty, phyllitic shales that host the mineralization. (Kretschmer, 1988; Percival et al, 1988). Gold mineralization is post-metamorphic and occurs both in silica replacement bodies and along silica coated fractures within a major high angle normal fault zone (Madrid et al, 1988).

Illites are also found throughout this deposit and appear to vary in crystallinity or polytype with the alteration and hydrothermal temperatures. Outside of the ore zone they can exhibit poor crystallinity as illustrated by scan A in Figure 7 which shows an X-ray diffraction plot of a disordered illite.

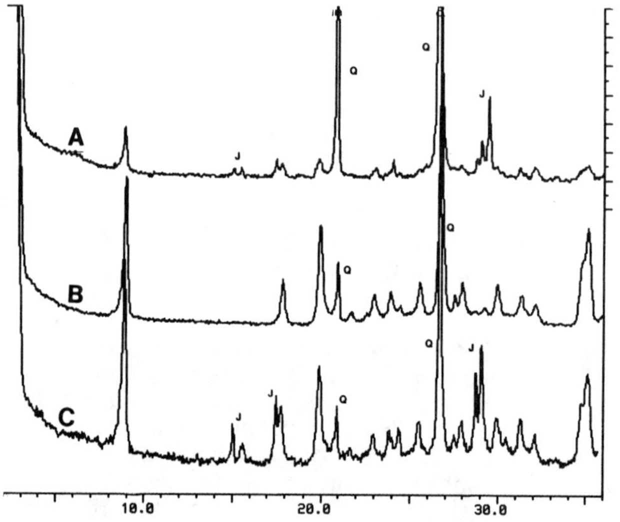

Figure 7. X-ray diffraction patterns of illites from the Preble Deposit. Scan "A" is a disordered illite from outside the ore zone. Scans "B" and "C" are from samples carrying greater than an ounce of gold per ton and show classic 2M ordering. They were obtained from core provided by Pinson Mining Company. The sample in scan "C" comes from a zone of intense fracture controlled mineralization that has been thoroughly oxidized and also contains jarosite(J). All samples contain silica as quartz (Q).

This lack of order is also shown in the reflectance spectrum. Curve A in Figure 8 has broad, poorly defined features. Part of the broadening of the 2.2 micrometer feature is due to the presence of jarosite which has an absorption feature at 2.26µm. Within the ore zone, however, ordering of the illite structure increases to a 2M polytype as shown by well defined peaks in the 23-32 degrees 2-theta range on the XRD pattern (scans B and C of Figure 7). The reflectance spectra (curves B and C of Figure 8) also change, showing definitive features in the 2.3-2.5 micrometer range which can be compared to A and B on Figure 4. The 2M illites found in the ore zone are quite distinctive with all eight XRD peaks (23-32° 2-theta) sharp and well defined. This lack of order is also shown in the reflectance spectrum (Curve A, Figure 8) which has broad, poorly defined features.

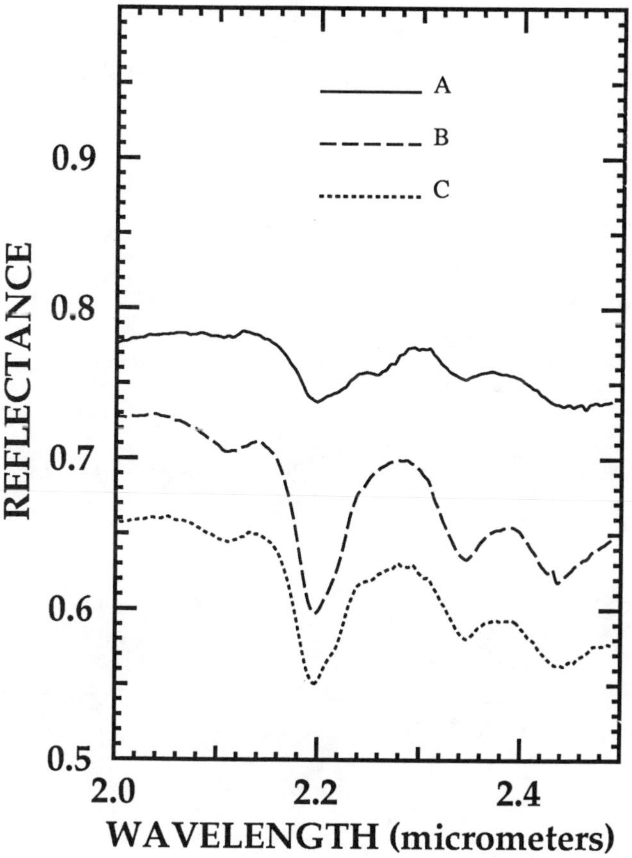

Figure 8. Laboratory illite reflectance spectra of Preble mine samples collected using the Beckman spectrometer. Spectrum A is a disordered illite which also contains jarosite. Curves B and C are well ordered 2M illites from the ore zone. 2M illites have sharp features at 2.2, 2.35, and 2.45 µm while 1M polytypes have broad features at those positions. Sample C contains jarosite.

The Preble ore zone illites appear to be slightly better ordered than Carlin ore zone illites indicating perhaps higher temperatures of formation. The economic values associated with these Preble illites are greater than an ounce of gold per ton.

DISCUSSION

The combined x-ray diffraction and reflectance spectroscopy study of illite polytypes has established that sufficient spectral differences exist to allow identification using near-infrared reflectance spectroscopy. Field and laboratory spectrometers have successfully been used to identify polytypes. Field reflectance spectroscopy can be used to assist exploration efforts by providing detailed mineralogical information real-time at the field location. The potential exists, given sufficient spectral and spatial resolution and instrument signal-to-noise to identify these polytypes from the air with an imaging spectrometer.

ACKNOWLEDGEMENTS

The authors would like to acknowledge the gracious cooperation and interest of the Newmont Gold Company in allowing us access to the Carlin property; Ed Kretschmer and the Pinson Mining Company for samples from the Preble Deposit; and Barbara Bakken of Stanford University for her knowledgeable assistance and loan of samples from Carlin.

REFERENCES

Bagby, W.C., and Berger, B.R., 1985, Geologic characteristics of sediment-hosted, disseminated precious metal deposits in the western United States, *in* Berger, B.R., and Bethke, P.M., editors, Geology and geochemistry of epithermal systems: Reviews in Economic Geology, vol. 2, Society of Economic Geologists, Chapter 8, p. 169-202.

Bakken, B. M., and Enaudi, M.T., 1986, Spatial and temporal relations between wall-rock alteration and gold mineralization, main pit, Carlin Gold Mine, Nevada: *in* Macdonald, A.J., ed., Gold '86 Proceedings, Willowdale, Ontario, Konsult Internat., Inc., p 388-403.

Bakken, B.M., Hochella, M.F., Jr., Marshall, A.F., and Turner, A.M., 1989, High-resolution microscopy of gold in unoxidized ore from the Carlin mine, Nevada: Economic Geology, vol. 84, No. 1, p. 171-179.

Bonham, Jr., H.F., 1988, Models for volcanic-hosted epithermal precious metal deposits: *in* Bulk Mineable Precious Metal Deposits of the Western United States, Schafer, R.W., Cooper, J.J. and Vikre, P.G., editors, Symposium Proceedings, The Geological Society of Nevada, A. Carlisle & Co.,Reno, NV, p.259.

Buchanan, L.J., 1981, Precious metal deposits associated with volcanic environments in the southwest: *in* Relations of tectonics to ore deposits in the southern Cordillera, Dickerson, W.R., and Payne, W.D., editors, Arizona Geological Society digest, vol. 14.

Giles, D.L., and Nelson, C.E., 1983, Principal features of epithermal lode gold deposits of the circum-Pacific rim: Transactions of the third Circum-Pacific energy and minerals resource conference, Circum-Pacific Series, AAPG.

Heald-Wetlaufer, P., Hayba, D.O., Foley, N.K., and Goss, J.A., 1983, Comparative anatomy of epithermal precious and base-metal districts hosted by volcanic rocks: U.S. Geological Survey Open File Report 83-710, p. 16.

Hunt, G. R., and Salisbury, J. W., 1970, Visible and near-infrared spectra of minerals and rocks: I: Silicate Minerals, *Mod. Geol.*, v. 2, p. 283-300.

Kretschmer, E.L., 1987, Geology of the Preble gold deposit, Humboldt county, Nevada, *in* Bulk Mineable Precious Metal Deposits of the Western United States, Guidebook for Fieldtrips, Johnson, J.L. and Abbott, E., editors, p.348, A. Carlisle & Co.,Reno, NV.

Lee, Keenan, and Raines, G. L., 1984, Reflectance spectra of some alteration minerals--a chart compiled from published data 0.4μm-2.5μm: *U.S. Geological Survey Open-File Report 84-96,* 6 p., 1 chart.

Madrid, R., Bagby, W.C., and Kretschmer, E., 1988, Gold occurrence and its relation to vein and mineral paragenesis in selected sedimentary hosted Carlin type deposits in Nevada: Geologial Society of Australia, Extended Absrtract series, P. 506.

Percival, T.J., Bagby, W.C., and Radtke, A.S., 1988, Physical and chemical features of precious metal deposits hosted by sedimentary rocks in the western United States: *in* Bulk Mineable Precious Metal Deposits of the Western United States, Schafer, R.W., Cooper, J.J. and Vikre, P.G., editors, Symposium Proceedings, The Geological Society of Nevada, A. Carlisle & Co.,Reno, NV, p. 11.

Radtke, A.S., 1985, Geology of the Carlin gold deposit, Nevada: U.S. Geological Survey Professional Paper 1267, 124p.

3

Geology/Exploration III International Activity

J.E. Frost
Frost International Minerals
Houston, Texas

Chapter 10

Mining Gold in Northeast China

I. S. Parrish

Manager ***DMBW, Inc.*** Golden, Colorado

ABSTRACT

Gold deposits in Hebei and Shanxi provinces in northeast China fall into three general classes:

- Placer deposits
- Mesothermal vein deposits
- Disseminated and oxidized volcanogenic deposits

Past production dating back thousands of years has all but exhausted the placers.

The mesothermal vein deposits, incorrectly considered in the past as "greenstone" type, are small to moderate in both size and grade. Production from these granulite and amphibolite hosted veins, at present greater than from the newly discovered disseminated deposits, is expected to wane.

Production from the disseminated deposits achieved only in the past few years is expected to equal or surpass that from the other classes of deposits. At least one "world-class" deposit is being developed.

INTRODUCTION

In 1986 and again in 1987, the author had the opportunity to review first hand gold mining exploration and mining in the Peoples Republic of China. The 1986 trip took him through central China, including the provinces of Sichuan and Anhui. The 1987 trip concentrated on Hebei province with an excursion to Shanxi province (Figure 1). As each visit was for a period of nearly two months, there was sufficient time to make more than a cursory examination and gather more than an initial impression.

The commission in 1986 was primarily to review exploration procedures related to disseminated "Nevada"-or "Carlin"-type deposits. In 1987 the concentration was on mining geology and in particular on deposits which many considered to be "greenstone"-type lodes.

This paper is derived principally from the 1987 trip, although some of the background and observations can be related back to the 1986 visit.

China as a Gold Producer

One might ask why mining people in North America should be overly curious about gold in China. The answer lies in the sheer volume of production. China does not provide official statistics and, due to the great number of unmonitored operations, accurate estimates are virtually impossible to come by. China is now considered as the fifth or sixth largest gold producing nation in the world. Present production, estimated at 60 tonnes in 1987, is expected to double in the next few years bringing China close to the U.S. and Canada in annual gold production.

The bulk of the gold comes from small mines. The average size is on the order of 500 tonnes per day, with the largest being on the order of 1,000 tonnes per day. Of the government owned mines most (80%) are underground operations. In addition to the government owned mines there are an unknown, but sizeable, number of private or "People's" mines. Most of these are placer operations, but there are small open pit and underground operations as well. For a description of the various levels of mining in China the reader is referred to the July, 1988 issue of Mining Engineering and an article entitled "Mining in China - A personal perspective" by I. S. Parrish and Barbara Parrish.

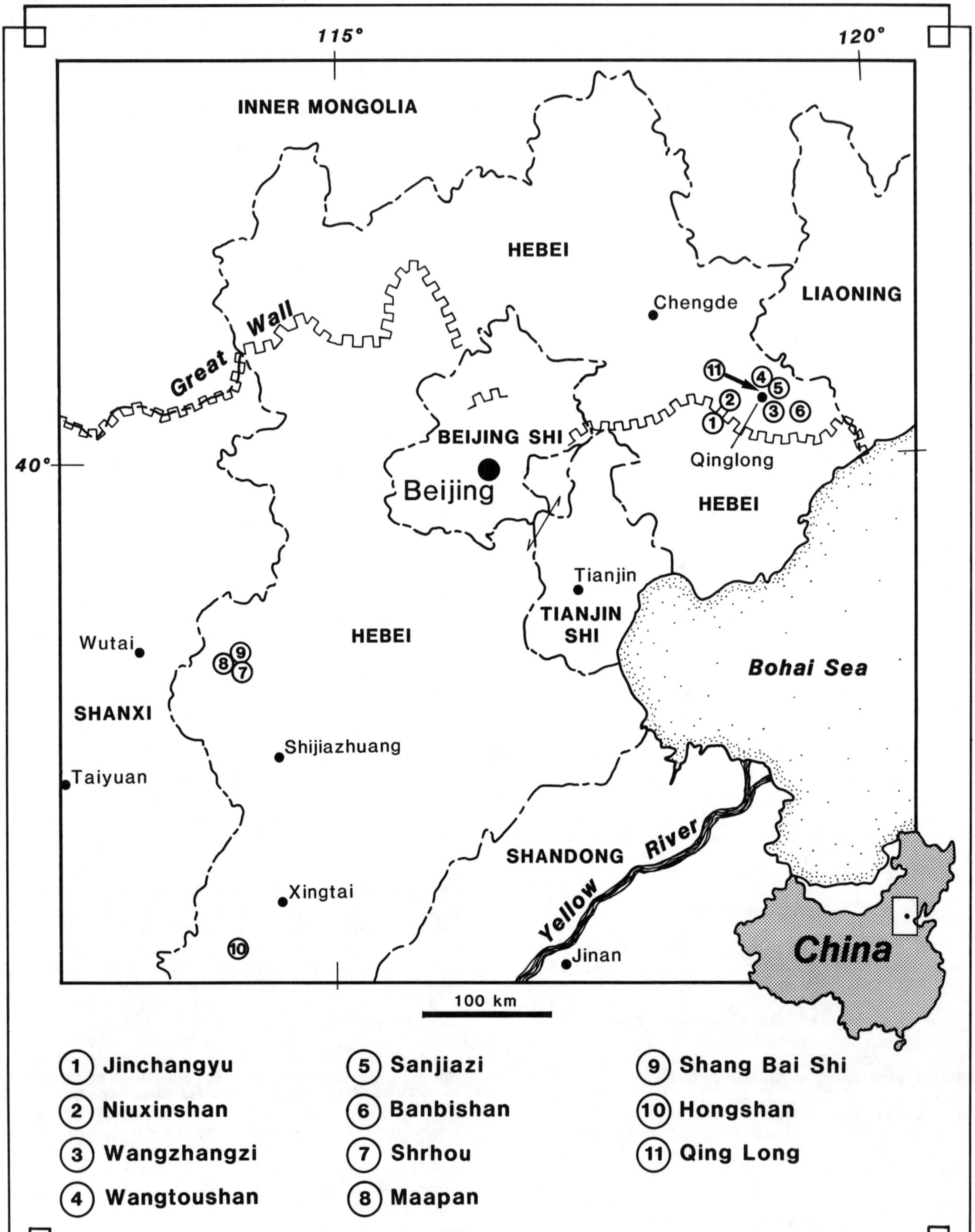
115°
120°
40°
INNER MONGOLIA
HEBEI
LIAONING
Chengde
Great Wall
BEIJING SHI
Beijing
Qinglong
HEBEI
Tianjin
TIANJIN SHI
Wutai
HEBEI
Bohai Sea
SHANXI
Shijiazhuang
Taiyuan
SHANDONG
Yellow River
Xingtai
Jinan
China
100 km
1 Jinchangyu
2 Niuxinshan
3 Wangzhangzi
4 Wangtoushan
5 Sanjiazi
6 Banbishan
7 Shrhou
8 Maapan
9 Shang Bai Shi
10 Hongshan
11 Qing Long

Figure 1

GOLD DEPOSITS OF NORTHEAST CHINA

The greatest number of gold deposits in northeast China fall into one of three classes. They are either:

- Placer deposits
- Mesothermal fracture controlled "vein" deposits
- Disseminated ("Nevada"-type) deposits

A very few skarn deposits are present. It is probable that other classes are represented as well, but the size and number of such "other" deposits is small to insignificant.

Placers

The placers are worked for the most part by individuals and communes (Figure 2). Methods are primitive, consisting of wooden rockers, riffles, and pans. In only one instance was there an indication that diesel-powered mechanical assistance (a nearby idle bull-dozer) might be used to augment the manual labor force.

Nowhere did I see the use of dredges, or suction pumps. Individual miners simply worked with inclined riffles and wooden pans. Commune efforts included hand sinking of pits 3-5 meters wide and up to 3 meters deep. There was no concerted effort to sink the pits to bedrock, possibly due to inability to handle water inflows. There was no attempt to sink to bedrock using timbered shafts, or to drive adits along the bedrock contact.

The placers have been worked for centuries. The gold now being recovered is universally as very fine grains, powder and flour. The streams are mature with broad flood plains, gravel bars, and a general absence of terraces. The streams are draining terranes of Precambrian rocks intruded by Mesozoic plutons. It is these upland areas that host the mesothermal vein deposits.

Production from the placers is difficult to quantify. There is no doubt that at some distant time in the past, placer production was paramount in the area. Today this is unlikely and it probably represents only a small portion compared to that of the vein deposits.

An unknown portion of this placer production is sold illegally through the "back door", finding its way to Hong Kong. China has a multi-tiered gold-pricing structure. Beijing sets a domestic gold price substantially below the London Market Price. The price paid to the producers varies, however, depending upon the class of producer, be it a Federal mine, Provincial, County, Commune, or People's (private) mine. The highest price is awarded the Federal mine descending to the lowest price being allotted the private miner. Hence, the illicit flow of gold out the "back door" to Hong Kong.

Mesothermal Veins

The most prolific gold mines of Hebei province are working vein deposits in Precambrian rocks. The mines range from "dog holes" developed by local peasants to complexes (Figure 3) such as the 550 m deep Jinchangyu mine developed through a series of portals and an internal shaft.

The Jinchangyu mine is located in Archean gneisses ranging in composition from magnetite-rich amphibolite to pyroxene granulite. Extensive migmatization imparts a striped or streaky appearance to the rocks. The deposit lies in the core of an anticlinorium in which compressive forces have caused numerous fracture zones. it is in these zones that the gold-bearing veins have formed. Nearby Mesozoic intrusives ranging from quartz diorite to granite in composition may have been the source of the Jinchangyu veining.

Six ore-bearing zones have been recognized. They are 500-6,000 m long, 10-130 m wide, and have been traced by drilling to a depth of 900 m. Alteration is principally by sericite, chlorite, pyrite, and carbonate. Sericite is the most important prospecting guide, although pyritization bears the most direct relationship to gold.

There are 16 ore bodies within the 6 zones. The average grade is 10 g Au/T. Gold occurs principally as native gold of about 850 fineness. Electrum is also present.

Producing since 1967, Jinchangyu is exceptional in its size and development. Most of the vein deposits are much smaller.

At Niuxinshan the veins reach a maximum of 2 m and average about 40 cm in width. Grade varies from 3-7 g Au/T to over 100 g Au/T. Quartz is the main vein material with accessory pyrite. Fluorite and kaolin are present. Sericitization is the principal alteration, but fluoritization is also important. Host rocks are upper Archean (2,700 m.y.) amphibolites which have been migmatized. Reserves are quoted in the Chinese fashion as 4 tonnes of gold.

Figure 2: Placer Mining Hebei Province, People's Republic of China

A. Placer miners reworking river gravel using picks, shovels, and wheelbarrow.

B. Miners inspecting gravels at base of 2 m deep pit. Tools of the trade include wooden scoop (center photo), riffle box and removable wooden riffles (foreground), and wooden handled pan (bottom left).

C. Panning of heavy minerals to concentrate gold. Local "pan" is of wood and has an open half funnel shape. Handles ease rocking, washing motion.

D. Washing gravel using wooden riffle box.

D

A

B

C

A

D

B

E

C

Figure 3: People's Mines, Hebei Province, People's Republic of China

A. The Great Wall of China from dump of a People's Mine near Niuxinshan.

B. Miners on break, Niuxinshan. Drift is 1.3 meters high. Working areas are not as commodious.

C. Shallow dipping massive sulfide vein is mined by hand using picks or by hand steel and explosives. Ore is mined separate from waste.

D. Near Wangtoushan ground support and electrical services are provided, but in a simple almost primitive manner.

E. Near Qing Long a new gold processing plant is under construction.

At Wangzhangzi Precambrian monzonitic granulite and sericite schist host three gold-bearing zones ranging up to 480 m long, but averaging only 1.5 m wide. Grades range up to 4.5 g Au/T over vein width. Metallic sulfides are rare.

A small granitic intrusive of Mesozoic age intrudes Archean hornblende plagiogneiss near Wangtoushan. Low grade (0.7-1.1 g Au/T) assays have been returned from narrow (0.1-0.6 m) sulfide veins filling fractures in the plagiogneiss. Silver to gold ratio is 1:4.2 One skarn deposit is present where the granite is in contact with a local dolomitic unit.

At Sanjiazi more than 20 gold-bearing veins with 6 tonnes of gold reserves have been discovered. Here too Mesozoic acidic intrusives have upwarped pre-existing Archean gneiss and migmatite. Sulfide and quartz sulfide veins have formed in fractures related to the Mesozoic intrusion. In places gold mineralization is found within and also along dikes radiating from the central intrusion.

The principal vein is more than 1,300 m long and 500 m deep, but averages only 0.6 m in width. Grade is reported as averaging 28 g Au/T. The main sulfide is pyrite with lesser chalcopyrite, argentite, and galena. Alteration is principally through sericitization, silicification, and pyritization. The Chinese reported that laboratory work indicated the veins formed at about 280-300°C.

At Banbishan Precambrian conglomerates (Figure 4) contain low grade gold and nearby lower Proterozoic mica granulites host veins grading over 8 g Au/T. The veins are being mined underground by open stopes. The ore is taken to surface where laborers reduce the broken muck to about 10 cm width cobbing out the more favorable pyrite-rich fragments and discarding the remainder. Pyrite and arsenopyrite are the principal sulfides. Silicification, potassium enrichment, and sericitization are the principal forms of alteration. Tourmaline has been reported from this locality.

Figure 4: Precambrian conglomerate near Banbishan, Hebei Province, P.R.C.

Near Shrhou in an area covering parts of western Hebei and eastern Shanxi province, eight separate ore bodies have been prospected. Local geologists estimate gold reserves of 7 tonnes in rocks grading 10 g Au/T, 16 g Ag/T, with significant amounts of lead and zinc. Ore has been found in a Mesozoic porphyritic granite at Maapan and in hornblende plagiogneiss at Shang Bai Shi and Shrhou. The gneiss-hosted quartz veins occur in a generally radial pattern to the south and east of the Maapan intrusive.

Reserves - The Chinese were reluctant to discuss reserves and those figures that were supplied must be considered with caution. Reserves are quoted as contained gold and not as so many tonnes at such and such a grade. Reserves are too often extended to depths well below the present mining capabilities. Also the reserves are based principally on surface sampling with only a very few widely scattered drill holes. For the most part, the Chinese reserves would be rated as no greater than "Possible" under North American standards. This is not a reflection on the potential of the areas, but only on the paucity of data.

Diamond drilling using standard core rigs (no wireline) is not uncommon, but the equipment is reminiscent of that used in Canada during the 60's. The Chinese geologists also seem to have a tendency towards drilling a few widely spaced, deep (750 m) holes rather than more numerous closely spaced shallow definition holes. This may stem from the desire to have "quantum" "reserve" increases with each hole. The preceding is speculation.

Classification - Not speculation, but a firm geological conclusion is that the gold deposits of northeastern China which have been considered as "greenstone" type are in reality mesothermal fissure veins. For a review of the model for greenstone deposits the reader is referred to R. G. Roberts,

"Archean Lode Gold Deposits", pages 1-19, "Ore Deposit Models", Reprint Series 3 of the Geol. Association of Canada.

The host rock for a "greenstone" type deposit should be a metamorphosed rock of the greenschist facies - a relatively low grade metamorphism. Host rocks for the majority of the Chinese deposits are gneisses, granulites, or amphibolites, often migmatized - a relatively higher grade of metamorphism.

The predominant alteration accompanying "greenstone" deposits, chloritization, is all but absent in northeast China where sericitization is the most prominent alteration.

An almost constant association in belts of "greenstone" deposits are large bodies of carbonatized rock. Such bodies are near absent in the Hebei and Shanxi gold districts.

Intrusive rocks of the great Canadian and Indian greenstone belts are of Precambrian age, those of northeastern China are of Mesozoic age.

Iron formations are another common feature in typical "greenstone" belts. These too are lacking, or uncommon, in Hebei and Shanxi.

The mineralogy of greenstone deposits is typified by noticeable free, visible gold. In China gold is present as very fine-grained, rarely visible, free gold. The gold:silver ratio of "greenstone" ore is usually high ranging from 2:1 to 12:1. In China the silver content is high resulting in an average 1:2 gold:silver ratio. Associated minerals such as tourmaline, ankerite, and mariposite which are all but ubiquitous in greenstone deposits are all but absent in the districts visited. Associated sulfides, pyrrhotite, chalcopyrite, and arsenopyrite found with most greenstone deposits are rare in northeastern China. Pyrite and quartz are the only minerals common to both the Chinese deposits and those of typical greenstone areas such as Kolar, India and Porcupine, Canada. Lastly, experimental work has shown that greenstone deposits are formed at temperatures in excess of 275° and often as high as 350-450° C. Work by Chinese geologists indicate temperature of formation for the deposits in northeast China to be in the range of 150-300° C.

Taken together, the mass of evidence indicates that the gold-bearing veins of northeastern China are typical of fracture filling mesothermal emplacements and are not of the "greenstone" variety. The portent here is that, whereas, greenstone deposits are noted for great depth potential, mesothermal veining has a more restricted vertical range. The Chinese deposits then must be considered as being restricted in their depth potential, bringing into question whether the present deep drilling program is proper and also whether the Chinese projection of reserves to depth is warranted or justified.

Disseminated Deposits

If the depth potential of the vein deposits is limited, then the significance of the recently discovered disseminated deposits increases substantially.

Typical of these deposits is one near Hongshan in the southern part of Hebei Province. In an area that strongly suggests a collapsed caldera, gold is found in stringers, as fracture coatings, and disseminated in oxidized pyroclastics and volcanic rocks (Figure 5). Individual ore lenses are up to 1,900 m long and 50-100 m wide. The host units are of Carboniferous (Pennsylvanian) age. These have been intruded during lower Cretaceous time by basic rocks ranging from dacite to gabbro. Venting occurred and tuffs and trachyte flows were extruded. In Upper Cretaceous time a caldera formed with at least 11 cycles of tuff and flows being recorded. A series of syenite porphyry

Figure 5: Strongly sheared, veined, and oxidized sedimentary units below a shallow angle thrust contain disseminated gold near Hongshan, Shanxi Province, P.R.C.

dikes was emplaced and then the caldera began to sink or collapse. Metals such as copper, molybdenum, uranium, thorium, and gold were introduced. The copper occurs central to the caldera, molybdenum rims the copper, and gold occurs outside of and overlapping the molybdenum. The deposits are low in silver.

Brecciation is widespread. Alteration includes limonitization, argillization, and silicification. There is a regional enrichment in potassium. Fluorite and barite are uncommon.

The gold is present mostly (88%) as free gold with an average fineness of 926. Grain size is usually 5-10 microns, but may be up to 100 microns.

At the time of my visit there were at least four separate deposits being worked, all by open pit (Figure 6). Gold was being recovered by heap leaching at each of the deposits. One such leach pad was privately owned and measured perhaps 15 x 10 m. The home-made leach tanks were two concrete vessels 3 x 1 m x 1 m deep. There were two larger pads being operated by the nearby town. Here peasants were "oxygenating" the pad by using pick and shovel and wheelbarrow to turn the pad over. The largest of the operations was provincially operated and was a modified vat leach.

The deposits had only been discovered in the mid-1980's and leaching begun in the past year. No estimate of recovery rate was given.

The deposit at Hongshan has not been drilled. The depth potential is unknown. The chances for additional discoveries both in the caldera and in the region are excellent (Figure 7). It is entirely conceivable that production from disseminated deposits such as at Hongshan will some day outstrip that from both the placers and vein deposits.

CONCLUSIONS

During an 8 week study tour of gold deposits in Hebei and Shanxi provinces of northeastern China, mining and exploration of placers, veins, and disseminated deposits were viewed. The placers, having been worked for centuries, are all but exhausted. Reworking of old gravels will continue to yield gold, but in ever decreasing amounts. The vein deposits, presently the chief source of gold, are not of the deep rooted greenstone class and as mining extends to depth the grade and quantity of gold will diminish. Disseminated deposits, whose presence has only recently been discovered, will become an increasingly important source of gold as exploration and heap leach technology become more widespread.

Figure 6: Overview of mining operations at Hongshan, Hebei Province, P.R.C. Government operated vat leach in foreground. A commune-operated heap leach at center of photo is marked by white plastic underliner. Ore is recovered from open-cut scars on the hillside.

Figure 7: A pilot heap leach near Qing Long, Hebei Province, P.R.C.

ACKNOWLEDGEMENTS

The author would like to acknowledge with appreciation the opportunity afforded to him by the Ministry of Metallurgical Industries of the Peoples Republic of China to visit the gold deposits of Hebei and Shanxi Provinces. Much of the data presented in this paper was gleaned from discussions and material provided by the geologists and engineers of the Ministry. These have been augmented by my own personal observations and conclusions. Any errors or misinterpretations stem from my own deficiencies and do not reflect the high professional calibre of my Chinese co-workers. My special appreciation must go to the First Exploration Co. of Yinjiao and my companions Mr. Li Guo Hua and Dr. Wang You Zhi.

Chapter 11

THE LAVERTON GOLD PROJECT - WESTERN AUSTRALIA

JOHN D GOW
Manager of Operations

IAN W WALKER
Manager of Exploration

ASHTON GOLD LIMITED

ABSTRACT

The paper describes the exploration and development activity associated with a series of open pit gold deposits in the Laverton region of Western Australia with an aggregate resource base of 4 million tonnes at a grade of about 3 grammes per tonne.

In September 1988, Hill Minerals N.L., now a wholly owned subsidiary of Ashton Mining Limited, completed the purchase of the Cork Tree Well treatment plant from Austwhim. This CIP plant had a rated capacity of 235,000 tonnes per annum, and is located within the Hill Minerals Laverton Project area. Ore is trucked by haulage contractors from each open pit mine to the central treatment plant at Cork Tree Well.

Studies were completed and a plant expansion undertaken lifting throughput to 500,000 tonnes per annum. A contract was let for this expansion in November 1988, and construction completed in April 1989.

Concurrent with the above project development work, heap leach metallurgical studies are being undertaken on low grade mineralisation associated with higher grade ore zones at several of the open pit mines, notably King of Creation and Craiggiemore.

It is anticipated that heap leaching will provide a useful adjunct to gold production from milling higher grade ores.

INTRODUCTION

From 1982-1984, Hill Minerals embarked on a tenement acquisition programme in the Laverton district of Western Australia (refer location map). During the next 4 years, an exploration programme indicated sufficient gold ore reserves in the district to justify a mining operation.

Current mining reserves are distributed between three deposits; King of Creation, Craiggiemore and Lancefield North, with individual tonnages in the 450,000 to 650,000 range and grades around 3 gAu/t. In addition, a number of resources have been outlined in the general district and these will be progressively brought into the mining category. An active exploration programme is being maintained in the Laverton area and new resources are continuing to add to the inventory.

In early 1988, a development team was assembled and the necessary detailed studies undertaken. A gold plant became available from another company, and was purchased in September 1988. This enabled gold production to commence in October, some 6 months before production would have been possible had a new plant been constructed.

The open pits and plant are typical of many in Western Australia. Mining and ore cartage is by contract and the plant is a carbon-in-pulp (C.I.P.) type. The workforce is flown to site from Perth for periods of 2 to 6 weeks.

Mining is currently from the King of Creation deposit, located 25 km north of the treatment plant and camp at Cork Tree Well and from Craiggiemore mine, 45 km to the south.

In December 1988, Hill Minerals N.L. and its operating arm Hillmin Gold Mines Pty. Ltd. became wholly owned subsidiaries of Ashton Mining Limited.

HISTORY OF THE LAVERTON DISTRICT

Laverton is situated 736 km northeast of Perth, the capital of Western Australia, and 250 km north-northeast from Kalgoorlie, the thriving major gold mining centre for the Eastern Goldfields (refer location map).

Whilst quite an active Aboriginal area, little white settler interest was evident until the Depression period of 1885-1905 brought in waves of gold prospectors.

In March 1896 Len Crawford and a party of three or four others, found gold in Crawford's Gully, just north of Laverton town. These prospectors had walked 430 km from Coolgardie, a then thriving centre. Others in their wake came from Coolgardie by horse and cart, pack-horse, horse-back and with wheelbarrows. Conditions were extremely harsh with summer temperatures above 40 deg. centigrade and annual rainfall less than 200 millimetres.

G. McOmish and T. Potts were the first to discover gold in the Laverton District at British Flag, some time in 1896. They took it to Dr. Laver, a medical practitioner in Coolgardie, who had "grub-staked" them. His interest aroused, he rode his bicycle along the prospector's road to the British Flag site where he camped and commenced prospecting. His subsequent success led to local fame and the naming of the town of Laverton.

Gold production for the region since that initial discovery, exceeds 40,000 kg or 1.3 million ounces. This historical production is dominated by Lancefield where 2,906,000t has been mined for 23,068 kg (741,603 ozs), Ida H where 292,000t were mined for 6,905 kg (222,026 ozs), plus smaller contributions from Burtville and Cox Find.

Since the early 1980's, significant new discoveries have included Western Mining Corporation's Lancefield Deeps (3.3 million tonnes @ 8.1 gAu/t), Placer Pacific/Delta Gold's Granny Smith (27.2 million tonnes @ 1.6 gAu/t), and Ashton Gold Limited's projects with combined reserves and resources in excess of 3 million tonnes at 2.9 gAu/t (300,000 ozs).

The more recent history, leading to Hill Minerals' involvement in the Laverton District, commenced when two prospecting syndicates (Burtville Prospecting Syndicate and R. L. Smith Prospecting Syndicate) pegged numerous tenements in the Laverton District in the late 1970's.

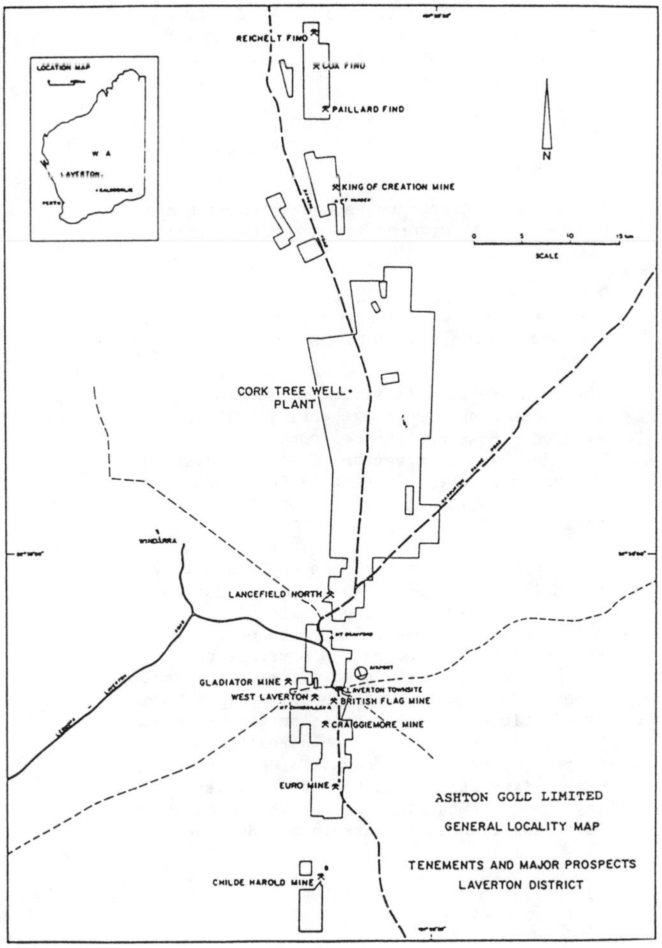

ASHTON GOLD LIMITED

GENERAL LOCALITY MAP

TENEMENTS AND MAJOR PROSPECTS
LAVERTON DISTRICT

Areas pegged included both the Craiggiemore and King of Creation open cuts presently operated by Ashton Gold Limited. As the tenement groups often formed interlocking land holdings, the two syndicates combined their tenements to make the package more attractive for potential joint venture partners.

Tenements were originally farmed out to Whim Creek Consolidated N.L. who carried out limited exploration at King of Creation and Lancefield North. Six vacuum holes were drilled at King of Creation. One hole intersected 22m at 5.70 gAu/t (KRC1 628m). At the Lancefield North area, vacuum drill holes were incorrectly logged which suggested the granite had intruded the northern extension of the Lancefield mineralized interflow sedimentary units. Whim Creek withdrew from the venture at this stage due to heavy exploration and monetary commitments in the Meekatharra area.

Morrison Petroleums Ltd (previously Consolidated Morrison Explorations Ltd) took over the expired joint venture under the same terms and conditions. They assigned 50% of the joint venture to Teck Explorations Ltd, who became the operator. This Joint Venture agreement was dated 26 June 1981. During 1983/84 Teck Exploration wound down their exploration programme in Australia and Hill Minerals was invited to farm in on the Laverton Joint Venture tenements and become operator.

More recently, additional tenements were pegged by Hill Minerals in their own right, and Joint Venture agreements signed with Peko Wallsend Operations Ltd (Lancefield Peko Tenements) and Nord Australex Nominees Pty. Limited (Laverton Downs North and South Tenements).

Tenement acquisitions and Joint Venture agreements are continuing to add to Ashton Golds' ground holdings in the Laverton District. By March 30, 1989, approximately 350 km^2 of tenements were under the Company's control, representing 100 km of strike, extending from 30 km south of Laverton to 70 km north.

The present phase of exploration commenced in December 1981 with RC drilling at King of Creation, and has continued until the present time. Major deposits located to date are King of Creation, Craiggiemore, Wedge (Lancefield North), West Laverton, Mary Mac and Bulldog.

REGIONAL GEOLOGICAL SETTING

Laverton is located in the northeast of the Archaean Yilgarn Block, and comprises part of the Eastern Goldfields Province. The base of the greenstone stratigraphy is best exposed in the core of the Margaret anticline where ultramafic rocks, including komatiite are overlain by quartz-magnetic iron formation. These are in turn overlain by thick tholeiitic and komatiitic basaltic piles that include narrow bands of interflow sediments and iron formation (including jasperlitic iron formation). The upper parts of the greenstone succession dominated by felsic to intermediate volcanics and coarse clastic sediments. A north northwest trending fault bounded graben filled with coarse clastic sediments appears to be the stratigraphically youngest unit.

Lower parts of the stratigraphy are intruded by mafic to ultramafic intrusives and porphyry complexes of felsic affinity. Sinuous strike conformable granodiorites and syenites intrude higher in the sequence. East-west and northwest trending mafic dykes of Proterozoic age intrude throughout the region.

At Laverton, and to the north, the greenstones have been extensively modified by north and north northwest trending regional structures including faults and ductile shear zones.

The most conspicuous are the north trending Crawford fault and the north northwest trending Admiral Hill - Ida H ductile shear zone. The trace of the Admiral Hill - Ida H shear is marked by the development of strong to intense shear fabrics in favourable lithologies and the presence of extensive linear quartz veins. This shear can be traced discontinuously over a strike length of at least 100 km, and probably continues northwards along the axis of the Duketon synclinorium. Apart from these structures that generally dip at moderate to steep angles, there have been delineated in the Laverton region a number of low to moderate dipping faults and shears that appear to be splays diverging from the regional structures.

Metamorphic grade in the greenstone succession ranges from greenschist to amphibolite facies.

EXPLORATION PRIOR TO THE DECISION TO MINE

From 1984 onwards, exploration in the Laverton district has involved a combination of geological and geophysical disciplines culminating in the successful delineation of mineable reserves at three localities; King of Creation, Craiggiemore and Lancefield North. Additional to these, at least seven other areas have been found to contain mineralized resources with good potential for generating additional mineable reserves.

Geological mapping on a regional and detailed scale, soil and rock chip sampling using detection limits as low as 1ppb, rotary air blast reconnaissance drilling, reverse circulation and diamond drilling for resource definition, ground and airborne magnetometry and various induced polarisation and electromagnetic techniques have all proved important in contributing to the discovery of mineralization in a range of diverse lithological and structural settings.

KING OF CREATION

In the period 1926 through to 1938, a total of 20,483 t of ore was mined for 140 kg (4,489 oz) of gold at a grade of 6.8 gAu/t. Underground mining was completed to a vertical depth of 44m and a limited open pit operation commenced.

The deposit is hosted within a north-trending transcurrent shear zone localised within a sequence of intensely deformed and altered quartz-poor sediments, banded cherts and intermediate to mafic volcanics. The shear zone, which in places exceeds 100m in width, dips 70° to 85° towards the west. Alteration assemblages within this zone comprise sericite, chlorite, pyrite and silica. The metamorphosed sediments and volcanics within the shear are flanked by felsic to intermediate intrusive porphyries, tholeiitic basalts, dolerites and gabbros that are relatively undeformed and weakly altered.

Mineralization occurs as a set of boudinaged quartz veins culminating in a breccia zone in the centre of the pit. Localization of the quartz veins appears to be in a mylonite zone developed along the contact zone (approx. 50m wide) of basaltic/doleritic rocks and metasediments consisting of shales banded iron units and cherts.' Some mineralization is also assocated with graphite schists, chloritic schists, banded cherts and metasediments adjacent to the quartz reefs. Within the ore zone the gold may be coarse grained and patchy in its distribution and is associated in many places with minor amounts of arsenopyrite and trace amounts of silver.

Dextral movements of the mylonite has caused the breccia pipe structure and boudinaged the mineralized quartz.

White (buck) quartz veins are later than the mineralized structures and are often mined as internal waste.

Surface weathering has had the effect of dispersing the mineralization laterally from the vein system, as goethite/haematite veining.

The main ore body is situated along a prominent ridge and is generally well exposed. However, further north it is obscured by a thin veneer of soil. Oxidation occurs to an average depth of 50 to 60m, deepening to 70m in places along the main shear zone.

The current reserve has been drilled out by a total of 203RC drill holes and 14 diamond drill holes over a strike length of 900m. Drilling was generally completed on 25m sections at 15m centres. However it did include infill sections to 12.5m in critical areas. Selected RC drill holes were twinned by additional RC drill holes or diamond drill holes. All holes were sampled using 4m composites and then assayed for gold by AAS with a fire assay finish or by fire assay. Any samples returning greater than 0.25 gAu/t were reassayed in 1m intervals. Selected intervals, both mineralized and unmineralized, were resplit in the field and submitted for check analyses.

CRAIGGIEMORE

In the period 1897 through to 1907, a total of 135,767 kg of ore was mined for 1,282 kg (41,214 oz) of gold at a grade of 9.44 gAu/t. Underground mining was completed to a vertical depth of 121m and a small open pit operation commenced. Mining between 1978 and 1984 produced a further 4,008t. of ore for 8.03 kg (258 oz) of gold.

The Craiggiemore gold deposit is associated with a banded iron formation in a thick Archaean mafic volcanic sequence. The iron formation can be traced northwards to Laverton, a distance of approximately 4 km. Other mines occurring along the same line are Mary Mac, Golden Pinnacles and British Flag. The main zone is approximately 20m thick and consists of finely laminated chert and magnetite ± carbonate ± sulphides. In the weathered zone to 80m vertical depth, hematite replaces magnetite and limonite replaces sulphides and carbonate.
In the area of the main Craiggiemore open pit the siliceous cherty bands have been desilicified and the iron formation occurs as fine bands of weakly cemented silica sand and limonite. To the west of the open pit, remnants of a surface silicified zone outcrop as hematite - jasperoid iron formation. A thinner (5m wide) iron formation occurs approximately 50m into the hanging wall, and a metasedimentary horizon comprised of chert and shale occurs 25m into the footwall.

The sequence has been extensively folded but maintains an overall dip of 70°to 80°to the east. On a broad scale the iron formation has been folded into a northerly plunging anticline and syncline that is exposed south of the open pit. On a detailed scale, the iron formation shows intense folding causing apparent thickening of the sedimentary unit. North of the Craiggiemore open pit a series of north-northwest trending thin shear zones cut the iron formation and appear to control the distribution of the desilicification in the iron formation. These north-northwest trending shears have upgraded the anomalous gold mineralization within the iron formation to form ore grade intersections.

Gold mineralization is generally confined to the main iron formation although some mineralization occurs in the weathered mafic volcanics (possibly associated with limonite filled fractures). Surface weathering has had the effect of lateral dispersion and depletion of gold values.

Regional shearing and mylonite development along the basalt-banded iron formations has formed the initial environment for the mineralization. Alteration and gold mineralization occurs within the mylonite zone and appears to favour the sheared BIF units as hosts (est. 60%). Subsequent block faulting to the N. E. (dipping at 60°) has offset the mineralization.
The interpretation at Craiggiemore is somewhat tentative as the majority of mill feed mined is mineralized backfill left by the miners of the 1900's.

Recent diamond drilling has located significant transitional and sulphide ore grade intercepts within the footwall tuffaceous sequence intercalcated with brecciated and boudinage cherts plus milky quartz in positions below the proposed open pit. The mineralized horizon has been detected between 75m - 125m vertical depth and appears traceable as two sub-parallel lodes, each of approximately 10m thickness pitching gently to the south and dipping between 70° and 80° to the east, over an interpreted strike length of approximately 200m.

Diamond drilling is continuing to define the extremities of this mineralization, currently open to both the north and south.

The mineralized zones exhibit intense pervasive sericite along with strong chlorite alteration. They appear carbonated with minor pseudomorph textures apparently after ankerite and sulphides. Foliation and bedding traces are obliterated by brittle fracture of the cherts and quartz occurring as intercalcations within the tuffs and minor graphitic interflow sediments.

This deformation appears to overprint a weak mylonite texture, and silicification of the sequence results in a microcrystalline fabric. The style of mineralization is principally veined and disseminated pyrite with associated specks of visible gold and minor arsenopyrite. Magnetite is confined to the thin banded chert horizons.

The current reserve has been drilled out with a total of 262 RC holes and 20 diamond drill holes over a strike length of 500m. Drilling and assay techniques followed the same procedures used at King of Creation.

LANCEFIELD NORTH

Lancefield North lies approximately 2km along strike from the largest historical gold producer in the Laverton district, the Lancefield Mine. However, no previous mining activity has been reported within the area to be mined at Lancefield North.

The deposit is hosted by an interflow sedimentary unit within a thick Archaean mafic volcanic pile. Other minor thin interflow sediments occur within the mafic volcanics. The principal interflow band or chert horizon is thought to correlate with the Main Lode at the Lancefield Mine. Topography is flat and the standing water table is at approximately 20m vertical depth. A laterite profile up to 20m thick has developed over the area, and weathering extends down to 40-50m in the volcanics and to 50-60m in the interflow sedimentary cherts. Below the zone of oxidation, the interflow sediments grade into black graphitic shales that contain pyrite and minor pyrrhotite.

The volcanic sequence has been folded and now dips at 40° to 45° south east. An increase in the amount of shearing near the interflow unit suggests regional stress has been concentrated along this unit, and has had a definite control on gold mineralization, especially near fold hinges or faults.

Mineralization occurs in the principal sediment with minor mineralization in the footwall mafic volcanics and some cross cutting quartz veins. A zone of supergene enrichment occurs between 15 and 60m, and above this, depletion of the gold has occurred. Below 60-70m, gold grades drop off in the sulphide zone, although one mineralized shoot has been traced down dip to a depth of at least 110m.

The current reserve has been drilled out with a total of 280 RC drill holes and 9 diamond drill holes over a strike length of 1,100m.

Drilling technique assay method and overall drill hole density was of a similar standard to both King of Creation and Craiggiemore.

ORE RESERVES

Mining ore reserves of the present and proposed operating pits are shown on Table 1. Also tabulated are the geological in-situ resources of seven other deposits in the Laverton district which are still in the process of evaluation and have not yet been classified as a mining reserve.

SUMMARY OF GOLD RESOURCES AND RESERVES

OF THE LAVERTON DISTRICT

as at December 1988

TABLE 1

Deposit	Ashton Interest %	IDENTIFIED MINERAL RESOURCES									ORE RESERVES				
			Measured & Indicated		Inferred Resources		Total Resources				Mineable, Proved & Probable				
		Cut off Grade (g/t)	Tonnes (000)	Grade (g/t)	Tonnes (000)	Grade (g/t)	Tonnes (000)	Grade (g/t)	Contained Gold (oz)	Equity Contained Gold (oz)	Cut off Grade (g/t Au)	Tonnes (000)	Grade (g/t)	Contained Gold (oz)	Equity Contained Gold (oz)
LAVERTON OPERATIONS															
King of Creation	100										1	658	2.7	57,627	57,627
Craiggiemore	100										1	438	3.2	44,517	44,517
Lancefield North	100										1	584	2.9	54,796	54,796
Sub Total															156,940
LAVERTON PROJECTS															
West Laverton	100	1	332	2.6	49	2.4	381	2.6	31,525	31,525					
Bulldog	100	1	54	4.3			54	4.3	7,465	7,465					
Mary Mac	100	1	237	2.8	61	2.5	298	2.7	26,250	26,250					
Childe Harold	100	1	77	3.1	26	2.9	103	3.1	10,108	10,108					
Paillard Find	100	1	57	3.9	13	4.3	70	4.0	8,937	8,937					
Lancefield North Workings	100	1	100	2.7	116	1.8	216	2.2	15,393	15,393					
Reichelt Find	17										1	312	4.0	40,094	6,816
Sub Total										99,678					6,816
TOTAL			857	2.9	265	2.3	1,122	2.8	99,678	99,678		1,992	3.1	197,034	163,756

THE DECISION TO MINE

Preliminary feasibility studies in late 1987 indicated the viability of a 500,000 tonnes per annum mining operation. The Board of Hill Minerals gave approval for a project team to be assembled to proceed with a definitive feasibility study. Detailed studies duly commenced in February 1988 (refer fig. I).

In late March, discussions commenced with Austwhim Resources for the purchase of their 235,000 tonnes per annum ore treatment plant, offices, camp and associated facilities, located at Cork Tree Well, approximately 40 km north of Laverton. Operations at Austwhim's two small pits, adjacent to the plant, were due to finish in August 1988. Agreement was reached on April 19, 1988, a deposit paid and takeover scheduled for September 1, 1988.

Studies indicated the plant would enable gold production to commence six months earlier than with the construction of a new plant. Cash flow benefits of an earlier startup and savings on construction and infrastructure costs justified the purchase, despite operating cost penalties in trucking ore into the plant from King of Creation (25km north) and the Lancefield District deposits (approximately 40 km south).

Feasibility studies confirmed that a plant expansion to 500,000 tonnes per annum provided significant economic benefits, and work commenced immediately on drawing up a Scope of Work and Invitation to Tender.
The final feasibility study was completed in July and a bank loan was secured during the next two months enabling full payment to Austwhim to take place in the first week of September, 1988.

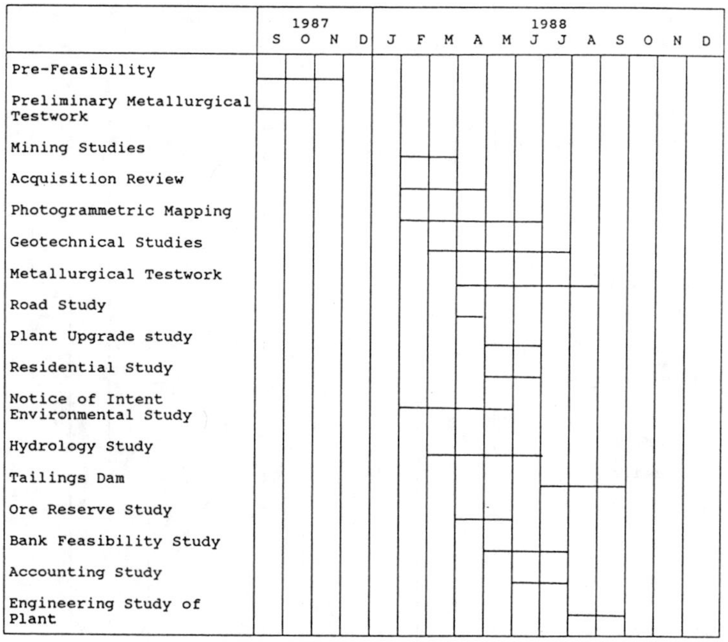

PRELIMINARY MINE STUDY SCHEDULE (Figure 1)

MINE DEVELOPMENT

Activities leading to the commencement of mining are shown in Figure 2. Development could be broken into four sections; recruitment of people, negotiating contracts, construction and production.

Recruitment was aggravated by the prevailing shortage of professional people in the mining industry. However, once the decision had been taken to operate a fly-in/fly-out roster from Perth, it became somewhat easier as the mine was then able to attract a mature, more experienced person.

Because of the then limited mine life of four years, and the requirement to commence immediate production, the decision was taken to use contractors for the mining, ore transport, catering and air transport. This involved several months of negotiation, contract preparation and adjudication by the project team.

During August, and prior to the mill handover in September, personnel moved to site and work commenced on clearing, stripping, roadworks and plant maintenance. Construction of roads, a tailings dam extension, and an airstrip continued after the pit pre-strips, as equipment (scrapers) became available.

Refurbishment of the Austwhim plant was also necessary subsequent to completing purchase arrangements on September 8. Ore milling duly commenced on September 23 and the first gold was poured on October 12, 1988.

Work commenced on expanding the treatment plant from its 235,000 tonnes per annum capacity to 500,000 tonnes per annum in late November 1988, with completion scheduled for the end of the first quarter 1989. The expanded facility was duly commisioned at the end of March and is currently operating above design levels.

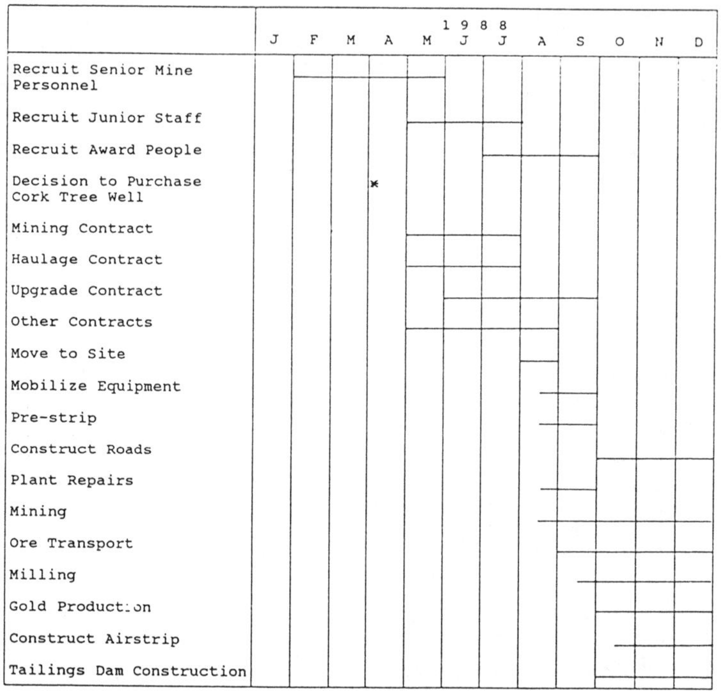

MINE DEVELOPMENT PROGRAMME (Figure 2)

MINING

PRODUCTION RATES

Consequent to the plant expansion, the Mill feed requirement has increased from 40 tph to 70 tph, and as a result, the two open cut pits currently being mined are each required to produce 25,000 tonnes of ore per month.

The King of Creation pit design is typical of many small open pit gold mines in Australia. Initially benches were designed at 20 metres in height, increasing in slope from 45° to 70° with depth. Maximum pit depth was 80 metres, berms at 5 metre width and the roadway was 1:10 and 15 metres wide. Falling gold prices prompted a review of the initial pit design to reduce mining costs. This led to a truncation at the north and south end of the pit of approximately 100 metres in length and the two lower benches had their slope increased to 70° with cable bolt support. Berms remained at 5 metres width but the roadway from 40 metres depth to the pit bottom was reduced to a single lane width of 8 metres. A passing bay was included in this design. Pit bottom remained at 80 metres. Pit dimensions are now 350 metres in length and 150 metres in width. The strip ratio has been reduced from 7:1 to 3.8:1.

Mining is by contract, initially using a Liebherr 984 excavator loading Cat 773 (50 tonne) trucks. With the ore zone running into stringers at the pit extremities, a two month trial using a Liebherr 942 excavator with a 1 metre wide bucket loading a single Komatsu 35 tonne truck was put in place. This left the 984 excavator to mine waste and low grade material at a rate of 4200 BCM/day while the 942 excavator mined ore, and a minimal amount of low grade material, at a rate of 700 BCM/day. Reduced waste dilution was the target with the smaller excavator exhibiting better control when mining the narrow stringers.

The revised pit design has necessitated the replacement of these two machines with a single Liebherr 974 excavator expected to mine 3000 BCM/day.

The Craiggiemore pit has undergone a similar redesign to King of Creation. Cable bolting has been used to increase bench slopes and reduce waste mined. The roadway below the 40 metre depth level has been reduced to a single lane with one passing bay. Due to the nature of foliation in the bedrock, bench slopes are unlikely to be increased beyond 60 on the west (foot) wall. Pit dimensions are 350 metres long by 170 metres wide by 80 metres deep. Stripping ratio is 6.9:1.

In all other areas, the Craiggiemore pit is similar to King of Creation, including the use of a Liebherr 974 excavator and Cat 773 (50 tonne) trucks and support equipment.

The ore is excavated in 2.5 metre flitches. A geologist remains with the excavator giving closer direction during ore excavation. Drilling and blasting is limited to the harder, tougher areas of the pit; commonly banded iron formations and chert zones. These areas are normally fired on a 3 m x 2 m pattern in the ore zone, and a 3 m x 3 m pattern in the waste zones. Holes are 5 m deep, 76 mm diameter, and are charged with ANFO and initiated with NONEL. As the pit deepens, the dolerite zones and some of the mafic volcanics require a "shake" in order to loosen up the zone to maintain budgeted digging rates and not disturb the ore. For this purpose a 4 m x 4 m pattern has been introduced and is working well.

Additional support equipment in the pit consists of a Cat D6 Dozer, Cat 14G Grader, and a 15 kL water truck. This equipment maintains the pit floor, all haul roads within and surrounding the pit and the stockpiles. They also assist in preparing the pit floor for grade control.
Four dewatering bores were installed prior to the pits reaching the water table to monitor water table level. If necessary, de-watering pumps will be installed in order to keep the pits in a relatively dry state and reduce sidewall pressure.

Four stockpiles are maintained - mill grade ore, (+1gAu/t), mineralized waste (0.5-1gAu/t), rehandle material and waste. The mill grade ore is separated into central zone ore (the main ore body) and stringer ore (offshoots north and south of the main ore body). Each flitch is also kept separate enabling mine grade and tonnes from each level to be reconciled with plant data.

Ore haulage from the mine sites to the process mill at Cork Tree Well is by contractors. Two Mack prime movers each pulling three 25 tonne trailers are used on a 24 hour per day, 6 days per week basis. From King of Creation the haulage distance is 23 km and the trucks are capable of hauling 2100 - 2400 tonnes per day. Craiggiemore to Cork Tree Well is 45 km and the subsequent daily haulage rate is 1500 tonnes per day. Ore from a single flitch is completely hauled into Cork Tree Well and batched through the mill to continue monitoring grade and tonnes. The haulage contractors are also responsible for maintenance of the public (gravel) roads from the mine sites to the mill. Road maintenance is carried out using a Cat 14G Grader and a 70 kL water tanker. Water is supplied from several dams along the two routes for the contractor's use.

In order to maintain mine production at 50,000 tonnes per month, it is envisaged that as production from the King of Creation pit declines, a third pit will need to come into production. Other pits are also being readied by the exploration team and the mining staff on site at Cork Tree Well. Outside consultants are regularly utilized for geotechnical, hydrogeological and pit optimization studies.

PIT GRADE CONTROL

Pit floor grade control is used to define ore bodies and control excavator movements. This is supplemented at times by blast hole sampling, face sampling and visual control by a geologist during excavation.

Benches are divided into 2.5 m high "flitches" which are sampled across the strike using either bulldozer rips or "Ditchwitch" earthsaw channels. Sample lines are 5-8 m apart and up to 0.5 m deep with samples of 2 - 3 kg taken at 1 m intervals. Sample lines are also geologically mapped.

Samples are processed by a laboratory at Cork Tree Well where up to 400 are treated daily. Turn around time for a sample is 3 to 4 days.

Data is plotted on 1:200 plans which are then interpreted using the assay and geology to a final ore pod design. Pods are marked out by the geologist who then supervises excavation. Veins as narrow as 1 metre are mined and internal waste greater than about 1.5 m is removed separately. Mining grade estimation is 13% below mill head grade for King of Creation and 2% above the mill head grade at Craiggiemore, when compared to the original feasibility study.

Use is made of a COMPAQ 386/20 computer with a HP Draftmaster I and GTCO 3648L Digi Pad running SURPAC and Cadsman software.

ORE TREATMENT

The treatment plant at Cork Tree Well had an installed capacity of 235,000 tonnes of ore per year as purchased. Following the successful commissioning of a second mill and additional leach tank in late March 1989, the plant is now treating in excess of 500,000 tpa. Total cost of the expanded facility is below A$10.5 million, making it one of the lowest capital cost plants in Australia.

CRUSHING

Run-of-mine ore is delivered to a 30 m^3 ore bin by front end loader (Caterpillar 966E) and feeds to the primary crusher by a vibrating grizzly feeder. The primary crusher is a Kurimoto 4032 NC Double Toggle Jaw Crusher. Product is conveyed to an Eljay 1.83 x 4.88 m triple deck screen with a closing screen size of 14 x 71 mm. Oversize ore is treated by an Eljay 54 inch rollercone crusher in closed circuit with the screen.

The crushed product is conveyed to a radial stacker conveyor which delivers it to two separate stockpiles each with a dedicated reclaim conveyor. The total capacity for crushed ore is some 14,000 tonnes.

MILLING

Crushed ore is reclaimed from each stockpile through three slot feeders. Each reclaim conveyor is dedicated to one mill and is fitted with a cross belt sampler and weightometer controlling a variable speed drive motor. Quicklime is added at this stage to maintain a pH of 9.8 - 10.0.

The No. 1 ball mill is a 3.20 m diameter x 3.96 m unit fitted with a 597 KW motor. With current ore hardness, a satisfactory grind is achieved at up to 40 dry t.p.h. feed.

The No. 2 ball mill commissioned in March 1989 is a second hand 3.8 m diameter x 6.0 m, fitted with a 1,300 kW motor. Mill capacity is approximately 75 tonnes per hour, although it is presently operating on a reduced charge at 45 tonnes per hour.
Both ball mills operate in closed circuit with nests of "Linatex" 375 mm cyclones, the cyclones being fed by one of two mill discharge pumps (Warman 6/4 DAH rubber lined casings).

Density control is achieved by controlling water addition rate, ratioed from the feed weightometer signal. A density of 45% solids is maintained.

It will be evident from the foregoing, that the grinding circuit capacity is considerably in excess of the nominated 500,000 t.p.a. It is anticipated that with continuing exploration success, further expansion of leaching capacity will be implemented to make full use of the grinding circuit.

LEACHING AND ADSORPTION

One 1,500 m^3 leach tank and four 653 m^3 leach tanks allow a residence time of twenty four hours at a feed rate of 75 tph, which is adequate for the current ores.

Cyanide in briquette or pellet form is premixed and added at a rate of 0.8 to 1.0 kg/tonne of ore to the first leach tank.

Leached slurry gravitates to the CIP adsorption circuit, consisting of six 58 m^3 tanks with countercurrent carbon movement carried out with air lifts. The carbon retention screens are cylindrical baskets of mixed wedge and woven wire (0.835 mm) and are mechanically wiped. Carbon concentration is maintained at 1.2 tonnes of carbon per stage. Typical loadings achieved are in the range of 3,000 - 4,000 gms/tonne.

ELUTION

An AARL elution system with a capacity of 1.3 tonnes is used to strip the loaded carbon. Eluate is stored in one of two 20 m^3 electrolyte storage tanks and circulated through one of two standard 9 cathode electrowinning cells.

Stripped carbon is regenerated through a first generation "Combustion Air WC 200" vertical kiln. Regeneration is carried out at ±750°C and the carbon is water quenched, screened and returned to the adsorption circuit.

SMELTING

The loaded cathodes are "direct smelted", i.e. the loaded steel wool is drained, dry fluxes added and smelted in the conventional crucible furnace. This procedure avoids acid handling and the inconvenience of residue filtration without detriment to dore fineness, which is typically +96%.

TAILINGS DISPOSAL

Tailings slurry discharges over a polyurethane carbon safety screen of 0.6 mm aperture, and pumped by Warman 4/3 AH centrifugal pumps to the tailings dam. Excess water is decanted and returned to the process.

SERVICES

Electrical power is supplied by Diesel generator sets consisting of two Caterpillar 3512 sets rated at 1 MW and three Caterpillar 3508 sets rated at 0.5 MW. The alternators produce power at 440 volts which is transformed to 3.3 KW for mill motors and transmission to the mine camp.

Process water is supplied by diesel pumpsets from a borefield 18 km away. Potable water and gold plant elution water is taken from bores adjacent to the mine site.

MANAGEMENT AND ADMINISTRATION

The mine operates autonomously except for the sale of gold and receipt of revenue. All accounting, payroll and supply functions are completed on site by an administration staff of six. Approximately 3,000 line items are held in stock with a value of about $800,000.

All employees and contractors reside at the Cork Tree Well camp, 40 km north of Laverton. There are currently eighteen staff who work a two week on, one week off roster and 36 award people who work six weeks on and one week off. In addition, 40 contractors reside in the camp. The camp has a capacity of 107, housed in portable duplexes and 6 bedroom houses.

Facilities at the camp include a wet and dry mess, gym and swimming pool. All accommodation units have TV with access to two channels and in-house videos.

Catering and cleaning services are provided under contract.

Figure 3 shows the labour structure of the mine.

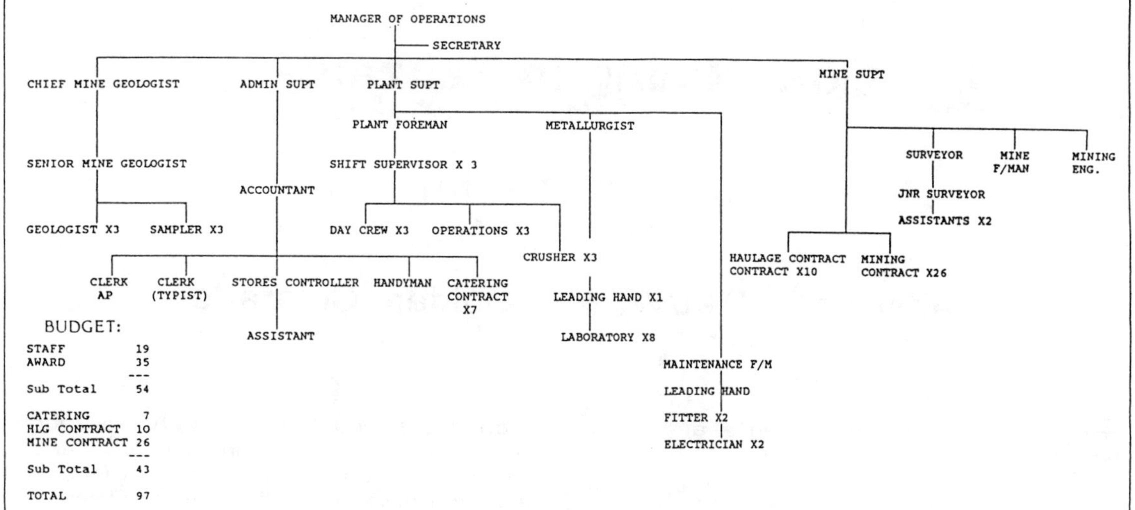

ASHTON GOLD LAVERTON OPERATION MANNING STRUCTURE (Figure 3)

FUTURE DEVELOPMENTS

Since commencement of operations, several developments have occurred which are likely to effect the future direction of the Laverton Operations. These include:

- Increase of ore reserves south of Laverton which will result in higher transport costs.
- Increasing number of small (± 100,000 tonne) low grade deposits scattered over a 100 km strike.
- Low gold price.
- The second-hand mill purchased for the upgrade is larger than originally specified.
- Greater quantities of mineralized (0.5 - 1 gAu/t) waste from the producing mines than anticipated.

The above factors have caused the mine to consider heap leaching for low grade ores. Extensive testwork is currently underway with varied recovery and agglomeration requirements.

Having a greater throughput capacity has resulted in production being pushed from the designed 500,000 tpa to 600,000 tpa. Consideration is being given to moving the smaller mill nearer to the southern deposits to save ore transport costs.

Exploration is being directed towards the ground in the vicinity of the plant. This is mainly grass roots, regional work at this stage. The smaller deposits will provide the mine with feed until a more substantial deposit closer to the plant can be developed. Because of the geology of the region, and the extensive land holding of the Company, expectations are high.

REFERENCES

Gee, R. D., Baxter, J. L., Wilde, S. A., and Williams, I. R. 1981, Crustal Development in the Archaean Yilgarn Block, Western Australia, Spec Publs Geol. Soc. Aust. 7

Hallbert, J. A. 1985, Geology and Mineral Deposits of the Leonora-Laverton Area, Northeastern Yilgarn Block, Western Australia, Hesperian Press

Chapter 12

Gold Mining in Tanzania

Brian W. Hester

Associate ***DMBW, Inc.*** Golden, Colorado

All gold occurrences in Tanzania are of Precambrian age and in geological settings similar to gold deposits in other parts the world. The three main areas of interest are perhaps the least prospected of areas of gold mineralization in Precambrian rocks in non-aligned, or market economy nations. Historic production exceeds 77,660 kg (2.5 million oz). Current results by small scale workers go unreported, but have been estimated to be as high as 11,000 kg (350,000 oz) annually.

Gold occurs in both lower and upper Precambrian rocks. In the older, gold is associated with greenstones and iron formations typical of many cratonic areas of Precambrian age. The Tanzanian Craton lies in the north west quarter of the country where it is flanked on both east and west sides by mobile belts involving upper Precambrian rocks of high metamorphic grade. Gold occurs in fractures in the western belt within two distinct areas.

INTRODUCTION

Gold was discovered at several localities (Figure 1) in Tanzania early in the present century. It was not until 1927 that consistent production began. Complete recording of gold sales began in 1935 soon after the widespread prospecting which started in 1932 when Great Britain abandoned the gold standard.

A peak level of production of gold of 4396 kg (141,000 oz) was reported for 1941. This level fell away and reached a maximum of only 3,110 kg (100,000 oz) in 1962 shortly before the major mines began closing. Underground mining of gold except for the unregulated activities of small miners virtually ceased in 1970 and has not yet been revived on its former scale. Total recorded production to 1970 was about 77,759 kg (2.5 million oz). Current annual production by small miners has been estimated in "Business Times" (1989) to be around 11,000 kg or 350,000 ounces.

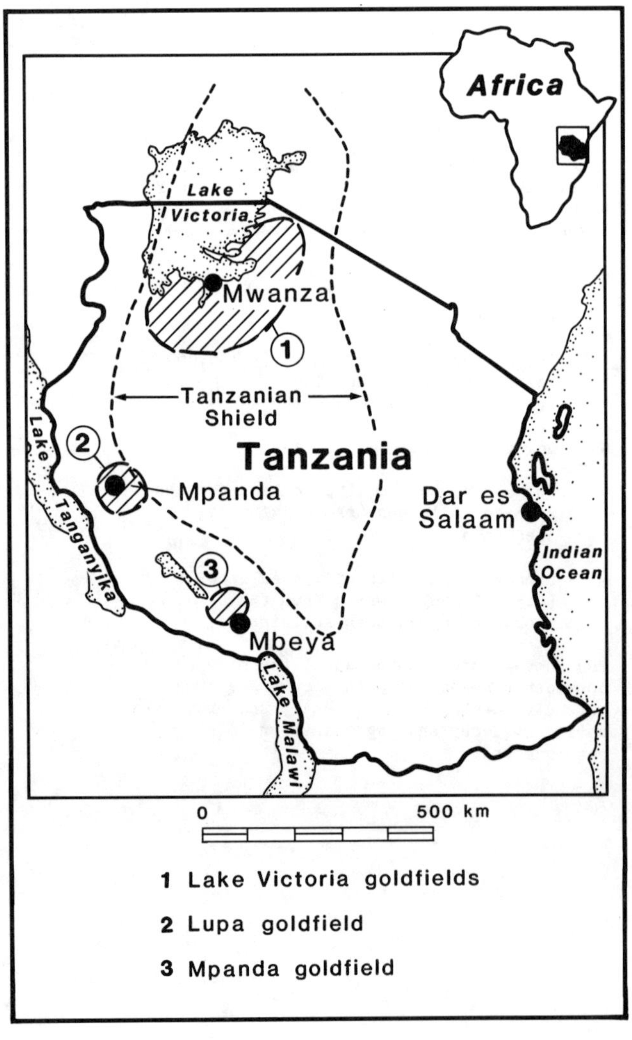

Fig. 1 Goldfields of Tanzania.

Of the three areas which have supported gold mining on a commercial scale, the one with lower Precambrian greenstone rocks of the Tanzanian Craton has contributed 75% of recorded production.

In recent years, many small scale miners have become active in areas of known gold occurrences. They employ primitive means to recover gold from outcrops of hard rock, placer, and most of all, eluvial material. Production from these small workers although not recorded is substantial, possibly three or four times the amount mined previously from underground mines.

The relation of the Tanzanian Craton to other gold bearing cratons of Archean age in the southern hemisphere is shown in Figure 2 with present coastlines in the positions occupied 200 million years ago before the break-up of the continent of Gondwanaland. Many of these cratons have been much more productive than the Tanzanian despite similarities of geology.

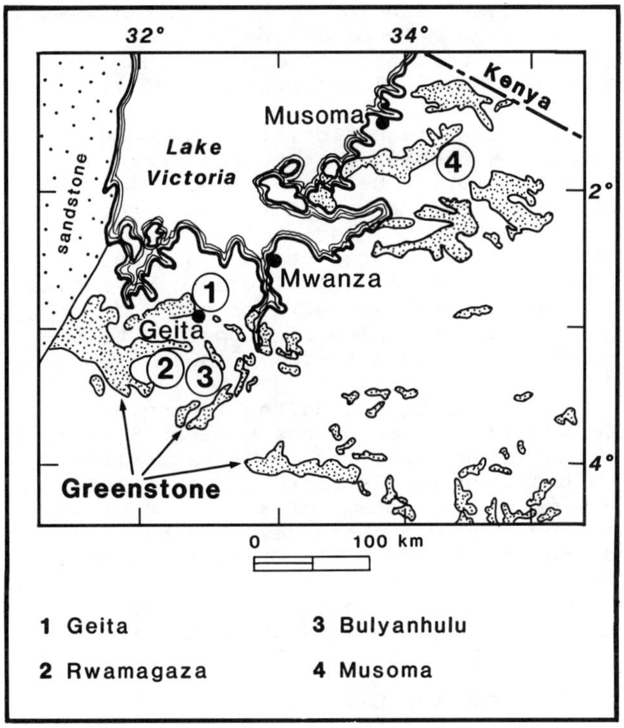

Fig.3 Goldfields and greenstones of Lake Victoria area. (modified after Van Straaten).

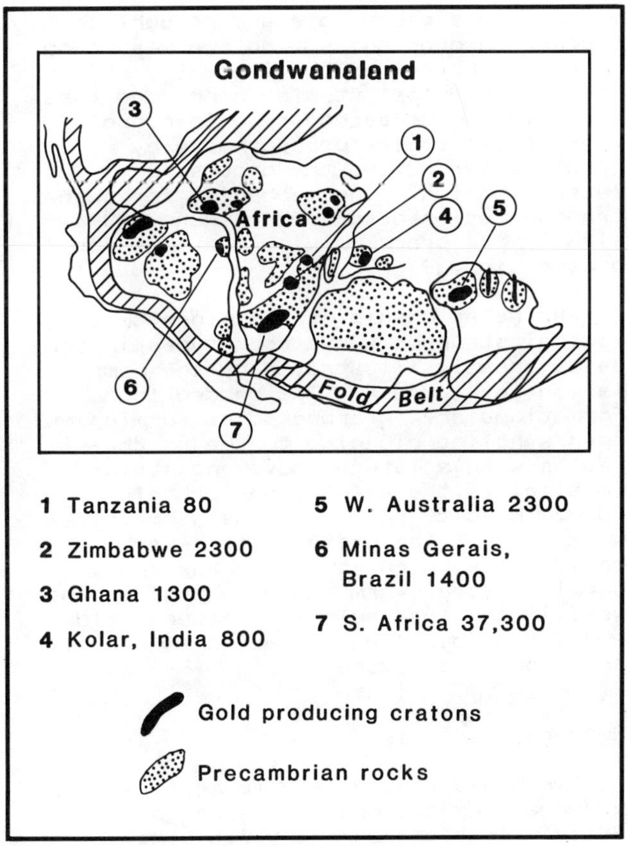

Fig.2 Gold producing cratons of Gondwanaland with estimates of gold production in tonnes. (Modified after Annhaeusser,1976)

GEOLOGICAL SETTING

Most of Tanzania's 937,062 sq km (362,800 sq miles) is underlain by rocks of Precambrian age. Rocks of later age are present as a cover of recent sediments close to the coast and as fresh water sediments of Permo-Jurassic age filling basins in the interior. These basins are often bounded by faults. Rift faulting in Quaternary time has disturbed wide areas in the western and central parts of Tanzania leaving down-faulted blocks now occupied by fresh water lakes of considerable size and depth. Hot spring and volcanic activity is associated with these faults. The Tanzanian Craton is exposed in the northwestern quarter of the country as the core of an area of Archean rocks. These are predominately granite with narrow remnants of sequences of volcanic and sedimentary rocks metamorphosed to greenschist grade. These sequences display characteristics typical of greenstone belts. These rocks are divided into a lower, predominately volcanic succession, and an upper, predominately sedimentary succession. The important gold deposits in the various greenstones and iron formations of the lower system are termed the 'Lake Victoria Goldfields' after the nearby lake. As in many greenstone terrains relationship with the surrounding rocks is not always clear. Age determinations on the granitic rocks date them as between 2500 and 2600 Ma (Bell and Dobson, 1981).

On its western side, the Tanzanian Craton is flanked by the upper Pre-cambrian mobile belt of the Ubendian System of sediments and volcanic rocks elevated to high grades of metamorphism (Figure 3) and dated at ca.1,800 Ma. (Van Straaten, 1982). These rocks strike generally north-south. The two goldfields of Mpanda and Lupa lie to the south within this system (Figure 1).

On its eastern side,the Tanzanian Craton is flanked by the Kibaran System, which is very similar in characteristics to the Ubendian in strike, rock types, and degree of metamorphism, but contains no known gold occurrences. It has been dated as 1200 Ma. (Van Straaten, 1982).

Both western and eastern mobile belts merge south of the craton into what is termed the 'Mozambique System'.

LAKE VICTORIA GOLDFIELDS

Known limits of several greenstone areas in the Lake Victoria goldfields have been extended by airborne geophysical surveys. Rock exposure in general is good in hilly areas, but in flat ground, bedrock is obscured by a thick cover of laterite. Because of this, the absence of detailed mapping, and the recent low level of interest in the area, the geology of these green-stone belts and the ore bodies they contain have not yet been interpreted in terms of modern volcanogenic concepts. Gold occurs throughout the many greenstone belts constituting the goldfields. Only some of the more productive sites are described in this paper.

Two sequences of volcanic rocks are present within some of the greenstone belts. Both are predominately mafic to ultramafic in composition and separated by a sequence of sulphide facies banded iron formation. Thickness of this unit has not been measured. The formation is best developed in the Geita gold field on the southern shore of Lake Victoria (Figure 3) and can be traced eastwards and southwards for 300 km (186 miles). There is no evidence of any extensive iron formation in the important greenstone belts lying to the east of Lake Victoria, although some do contain minor occurrences of this rock type.

Geita Goldfield

At Geita, the banded iron formation overlies a poorly exposed sequence of greenstones which contains minor deposits of gold. The Geita area proved to be the most productive with a total recorded production of 27,442 kgs (882,000 oz). Three mines operated in the Geita area. These were Ridge 8, Prospect 30, and Lone Cone-Geita-NW Extension. The last group consisted of three areas of ore within a strike length of four kilometers of iron formation and worked from a single adit system. At Ridge 8 mine, and at the Geita mines, ore occurred in fractures within the iron formation or at contacts between it and felsic, pyroclastic rocks. The iron formation is very resistant to weathering and forms ridges and plateaus.

Stope widths are reported to have varied from 1.5 meters (5 ft) to over ten meters (33 ft). Ground conditions in the stopes were excellent except at the Ridge 8 where fill was required.

Prospect 30 mine differs from the other deposits in several ways. Gold mineralization here is associated with banded iron formation, but not with any apparent structure. Widths were sufficiently great that the ore was mined from a glory hole and brought to surface through adits. No ore was found below the bottom of the present glory hole. Geological studies concluded that the gold was of secondary origin and transported to its present site by ground water during the process of lateritization. The mechanism for such transportation described by Mann (1984) clearly has practical implications for prospecting.

The relationship of the mines to geo-logical structures and rock types in the Geita area is illustrated by the compi-lation in Figure 4. The banded iron formations are interpreted as forming an open syncline plunging to the northwest. Faults with a lateral movement strike parallel to the axial plane of this fold. Results from airborne geophysical work suggest the presence of several basic dikes which strike northwards parallel to a swarm known further to the west. The close relation between gold and the banded iron formation is clearly shown on this figure.

Rwamagaza Goldfield

Two mines were worked formerly in the Rwamagaza greenstone belt south of Geita. That at Mawe Meru was simply a gold-quartz vein dipping steeply in basic volcanics close to a granite contact. Average grades of about 15 gms (0.5 oz.) per mt came from a single ore shoot of small dimensions. The second

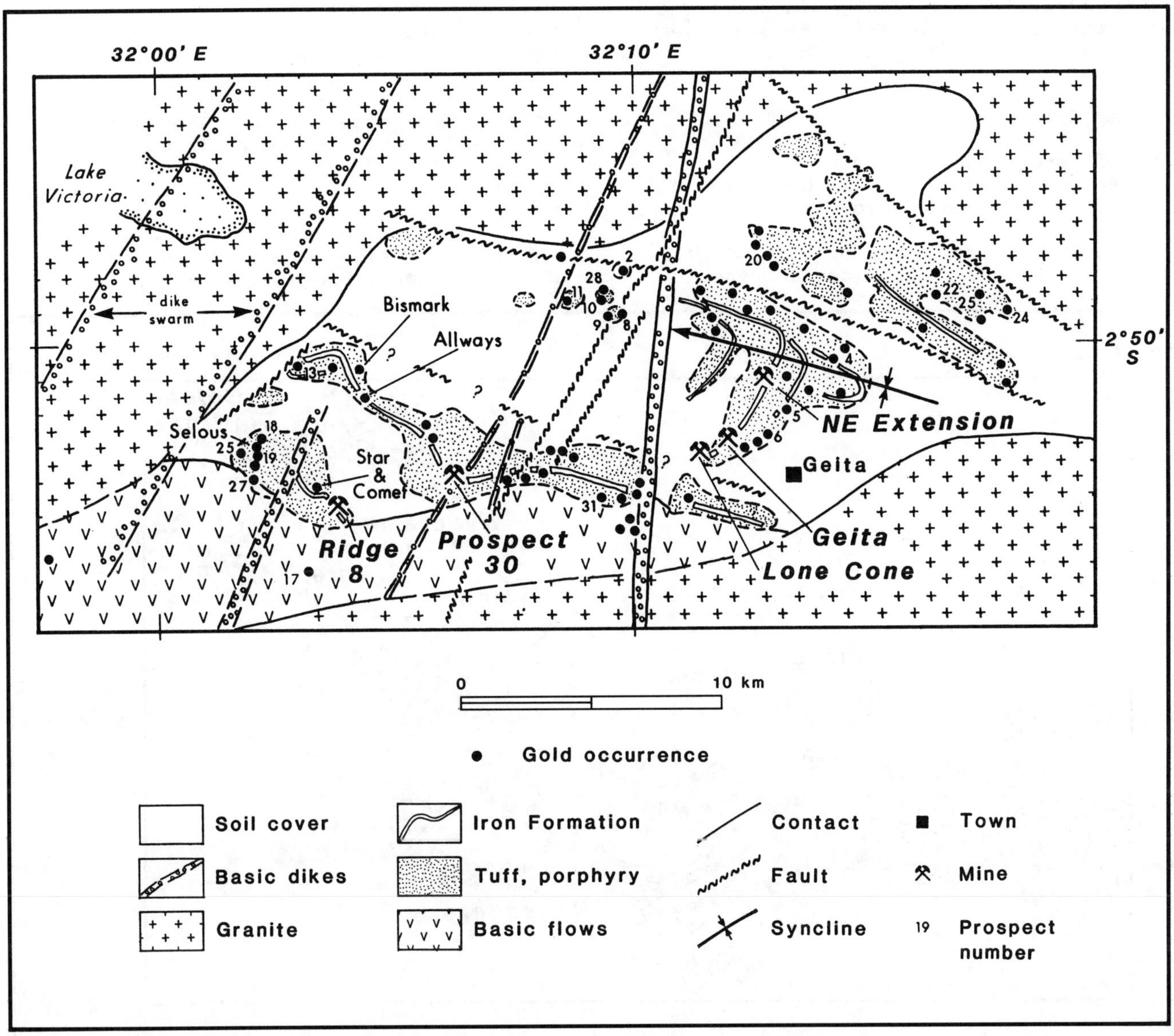

Fig.4 Geita goldfield. Compilation of geology and gold occurrence.

mine, the El Dorado is of more interest. Gold occurred in free form in a rubble zone about one meter (3 ft.) thick at the base of a laterite section itself up to 25 m (82 ft.) thick and formed over greenstone. This zone was mined by hand methods and the broken ore dropped through chutes to crosscuts driven in the underlying bedrock. The ore was hoisted to surface and washed to liberate the gold.

In 1984 prospectors found what developed into a major extension of rubble zone ore. This discovery attracted many independent miners who followed the zone for about 3 km (1.8 miles). The gold is free and in the form of nuggets which are rough in form with no evidence of being water worn. The gold appears to have been formed in place under the agency of laterite weathering as described by Mann (1984). At least one other deposit of this type has been discovered further west in this same greenstone belt. Specimens of bedrock from beneath these deposits include serpentinites with maraposite.

In a small area of greenstone east of Rwamagaza, small scale miners working at outcrop discovered important deposits at Bulyanhulu. Gold occurs here in at least three tabular bodies within felsic tuff. Fresh mineralization is associated with graphite. This occurrence has been incompletely tested with 22 diamond holes. These indicate the presence of about two million mt containing an average of 10 gm (0.3 oz.) gold per mt in one ore body. This is an exciting discovery and indicates what can be found here by even primitive prospecting techniques.

Musoma Goldfield

The largest greenstone belt in Tanzania and second only to Geita in quantity of gold produced, lies east of Lake Victoria close to the town of Musoma by which name the belt is known. It is 110 km (68 miles) long and from five to 30 km (3 to 19 miles) wide. Its more prominent geological features, presented here in Figure 5 have been compiled from published geological maps, geophysical work by various agencies, and personal observation. The eastern half of the belt is underlain by basic flows with some pyroclastics and felsic rocks.

In the western half of the belt, there is a well developed series of pyritic, cherty rocks possibly of

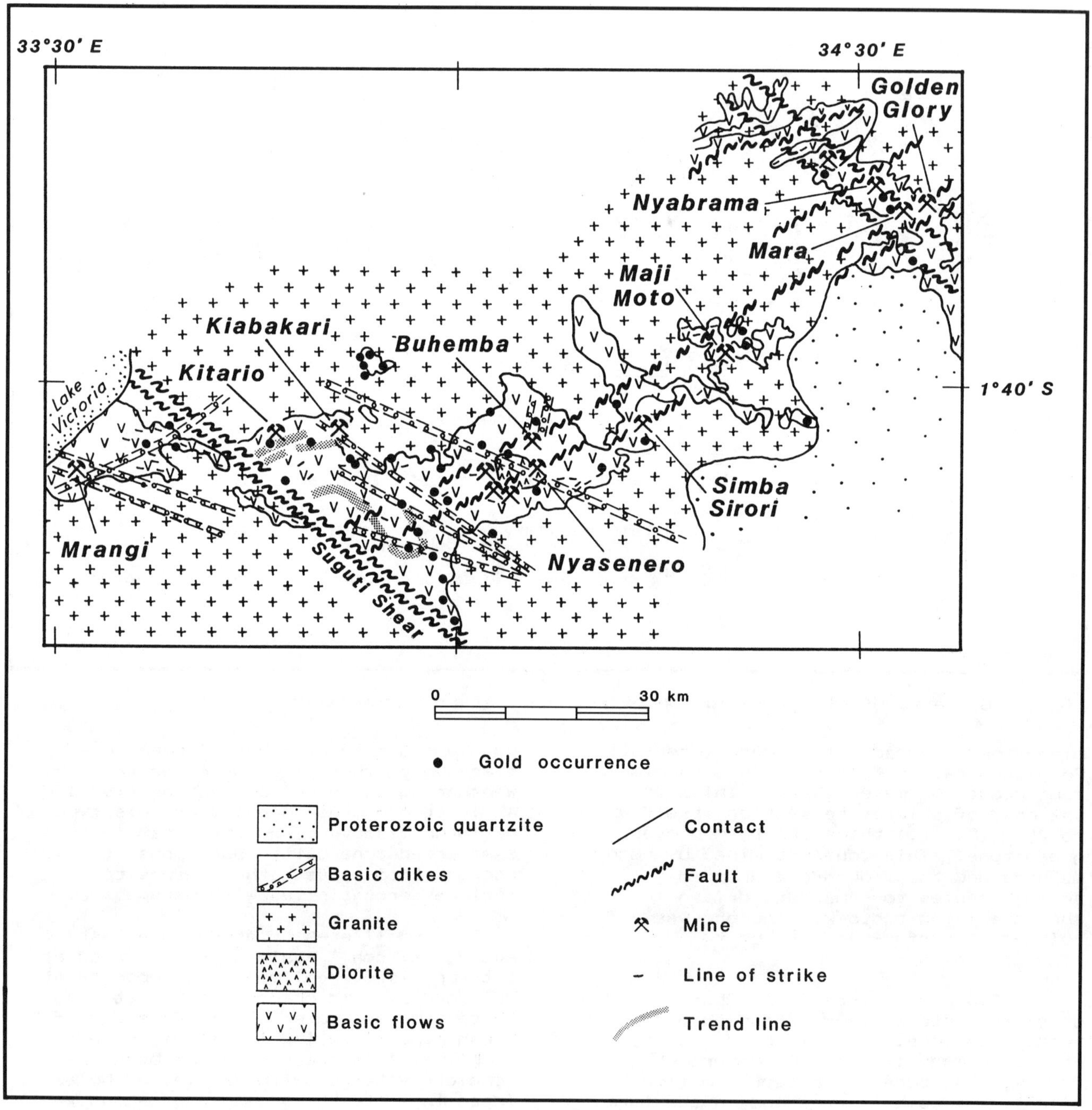

Fig.5 Musoma goldfield. Compilation of geology and gold occurrences.

exhalative origin. Results of airborne geophysics show certain horizons of these rocks to be either magnetic or conductive, and someplaces both. Directions of these features confirm strike directions recorded in the field. This series has been folded along a north-south axis. A number of gold prospects occur in these rocks, but production has been small except at Kiabakari mine. There is evidence of a positive relation between gold occurrences and stratigraphy.

The form of the various anomalies helps define the position and offset of a fault striking ENE along which a left-handed, strike slip displacement of about 8 km (5 miles) is interpreted. This fault is referred to here as the 'Buhemba Fault' (Figure 5). The strike extension of this fault was found by ground surveys by the United Nations in a cooperative effort with the Government of Tanzania.

Gold production from the Musoma greenstone belt is 21,836 kg (702,000 oz) of which nearly 97% came from two mines, Buhemba with 12,213 (393,000 oz) and Kiabakari with 8907 kg (286,000 oz). Production from Buhemba came from 1,067,000 mt of ore with an average of 11.45 gm (0.37 oz) per mt. Kiabakari produced 1,414,000 mt with an average grade of 6.3 gm (0.2 oz) gold per mt.

At the Buhemba mine, the original nature of the country rock is generally thought to be that of a pyroclastic of mafic composition, but original textures are obscured by intense carbonatization. Two ore bodies were mined. At the first, the original Buhemba mine, high grade ore was mined from several flat-dipping quartz veins by underground and open pit. At the second, or Nyasenero mine, 1.5 km (one mile) south of the Buhemba mine, ore was mined from a single vein for over a kilometer of strike length. The vein dips at about 45 degrees to the southeast and was up to two meters thick where stoped. Native gold occurs in the massive quartz with concentrations near the footwall. No ore was found below 200 m (650 ft) from surface. Near the NE end of this mine the strike of the vein swings gradually to north and then west before it terminates against a major fault. The form of the vein in the disturbed area is that of an anticline plunging east. Plans of the drifts show the vein to be crenulated in this section. The mine closed in 1970.

At the Kiabakari mine, gold was mined in a series of vertical open stopes in tabular bodies of a fine grained, siliceous rock, possibly of exhalative origin. The mine employed low cost, bulk mining methods. Grades were 6.3 gm (0.2 oz)per mt with continuity good enough to allow mining along 150 m (500 ft) of strike down to a depth over 450 m (1400ft). Ore continued downwards, but rising costs made the operation uneconomic. Altogether, 1.4 million mt were mined between 1959 and 1966. The ore bodies were found by diamond drilling beneath small scale workings at outcrop in the volcano-sedimentary rocks at the west end of the Musoma greenstone belt.

Only one of the many other gold prospects of Kiabakari-type has been diamond drilled. Opportunities to find more ore of this type appear encouraging.

MPANDA GOLDFIELD

The geology of Mpanda is poorly understood and little has been published. It is clearly of considerable complexity with several distinct series of rocks and both thrust and normal faulting. Placer gold was discovered here in 1936 and there was a short period of activity during which several gold prospects were explored underground. Base metal prospects were more attractive and an underground mine for lead and copper operated for about ten years in the forties and fifties. The ore occurred in a wide fracture system and was remarkable for its low zinc content. Production of gold totals 2164 kg (69,000oz) nearly 99% of which was as by-product from the base metal mining.

Gold mineralization occurs in several styles of fracture within an area measuring about 15 km (9 miles) by 15 km (9 miles). The southern limit of the field is not well defined and veins containing gold are known to continue for a considerable distance. Numerous veins are known and some drilled. Many of them are being mined at outcrop by small scale miners. The ore consists of free gold with base metal sulphides. Many of the veins are narrow and probably not of high enough grade to permit stoping underground. At a few prospects, mineralization occurs in stockworks of narrow veins. Some veins contain minor quantities of wolframite. Other veins are very persistent along strike and are up to 7 or 8m (22 to 26 ft) in width. The gold content of these is not known and none has been mined.

LUPA GOLDFIELD

The Lupa goldfield lies near the SW corner of Tanzania just north of the important town of Mbeya. Gold has been found over a triangular-shaped, fault-bounded area of 27,800 sq kms(10700 sq miles). The initial discovery in 1922 was of spectacular but small occurrences of placer. Gold was later discovered in bedrock and several underground mines established. The most important of these was the New Saza Mine which produced 7776 kg (250,000 oz) before closure in 1956. Total gold production from the district is 18,662 kg (600,000 oz).

The goldfield lies within the rocks of the Ubendian mobile belt. These consist of high grade metamorphic rocks of volcanic and sedimentary origin which have been intruded by several types of plutonic rocks. The structure is poorly understood. Ferruginous quartzites well exposed as ridges in parts of the district are thought to control the distribution of some of the gold mineralization and to be possible hosts for extensive low grade mineralization which has not been fully investigated.

Gold occurs in veins and fractures. At New Saza mine, near the western end of the district, stoping widths of 3 m (10 ft.) to 5 m (16 ft.) were attained. The ore occurs in distinct shoots which appear to be structurally controlled.

The distribution of the placer gold is influenced by the tilting and dislocations along several major faults of a rift system of Quaternary age. Elevations within the district range mostly between 2000 m (1.25 miles) and 2500 m (1.5 miles).

CONCLUSIONS.

1. Gold occurs in Tanzania in commercially interesting concentrations in three fields. All are of Precambrian age.

2. Over 75% of the gold mined so far has come from several greenstone belts in Lower Precambrian rocks near Lake Victoria.

3. International mining companies have shown little interest in recent years in the prospecting opportunities offered by these goldfields, yet by comparison with recent achievements in similar rocks elsewhere, these appear to be excellent.

REFERENCES

Anhaeusser, C.R., 1976,"The nature and distribution of the Archaean gold mineralization in Southern Africa," Minerals Sci.Eng.8.

Anon, 1989 Business Times ,April 28, 1989, Dar es Salaam.

Bell, K., and Dobson, M.H., 1981, "The geochronology of the Tanzanian Shield," Shield,". Geol.89.

Mann, A.W., 1984, " Mobility of gold and silver in lateritic weathering Proiles; some observations from Western Australia". Econ.Geol.79.

Van Straaten,H.P., 1982, "Gold Mineralization in Tanzania-A Review", Gold, '82 The Geology, Geochemistry, and Genesis of Gold Deposits. Proceedings of the symposium Gold '82 Univ. of Zimbabwee, May 1982. A.A.Balkema, Rotterdam.

Chapter 13

THE BANNOCKBURN GOLD PROSPECT - A NEW DISCOVERY IN EASTERN ONTARIO

Gordon D. House and J.B.P. Sawyer

Sawyer Consultants Inc.
Vancouver, British Columbia, Canada

ABSTRACT

The Bannockburn gold prospect is a new discovery in Ontario's oldest gold camp at Eldorado, eastern Ontario. Mono Gold Mines Inc. of Vancouver, British Columbia, acquired the property in 1980 and in 1984 discovered a previously unreported gold bearing quartz vein system to the northeast of the old Bannockburn Mine. Extensive diamond drilling carried out since 1985 defined a quartz vein system over 475 m (1,500 ft) of strike to a depth of over 105 m (350 ft). Limited underground exploration by decline to the 22.9 m (75 ft) level in 1988 confirmed the continuity and tenor of the quartz veins. Gold occurs as native metal with minor sulphides and bismuth tellurides in a quartz carbonate vein system within metasedimentary and metavolcanic rocks of the Grenville Province. Metallurgical test work on a surface bulk sample was favourable. Estimates of drill indicated reserves have been made as well as reserves based on the limited underground exploration.

INTRODUCTION

The Bannockburn prospect is located in southeastern Ontario, Canada, and lies 210 km (130 miles) west of Ottawa, approximately midway between Ottawa and Toronto. The property is centred on the village of Bannockburn in northern Hastings County, on Highway 62 approximately 16 km (10 miles) north of the junction with the Trans Canada Highway (Highway 7) at Madoc, a small market town lying 40 km (25 miles) north of the town of Belleville on Lake Ontario.

The Bannockburn property is owned by Mono Gold Mines Inc. of Vancouver, British Columbia, who acquired the property in 1980. Several adjoining claims were acquired in 1983 and 1985. The property consists of 9 full sized mineral claims, 2 partial mineral claims and the mineral rights to Lot 28, Concession V, Madoc Township, covering a total of about 700 acres.

PHYSIOGRAPHY

The property lies in the northern part of Hastings County underlain by Precambrian rocks which form part of the pre-Paleozoic peneplain sloping gently southwards from an average elevation of about 274 m (900 ft) above sea level north of Bannockburn to about 182 m (600 ft) above sea level near Madoc.

Figure 1. General Location Map.

Relief in the area is moderate being nowhere in excess of 61 m (200 ft). The area is drained by the Moira River which flows from north to south and southwest through the Bannockburn property. It empties through Moira Lake into Lake Ontario.

Forest cover includes a variety of hardwood as well as coniferous trees. The low lying, poorly drained areas consist of reedy willow swamps and beaver ponds between isolated rock ridges. A small part of the area is cleared farmland.

HISTORY

Madoc has been a mining centre for well over one hundred years. The discovery of gold in 1866 at the site that was to become the Richardson Mine near Eldorado, which lies less than 6.5 km (4 miles) south-southeast of Bannockburn on Highway #62, was the first such discovery in the province of Ontario and as such led to a "gold rush".

The Bannockburn Gold Mine appears to have been one of the earlier mines discovered in the period following the 1860s "gold rush". The exact date of the start of this operation is unknown but there is an account of it being well established and in operation, and subsequently being closed, prior to 1894. The property was reopened and operated again from 1894 to 1898.

Subsequently, the mine was farmed out to an American on a royalty basis, this arrangement enduring until about 1905. During this period a considerable amount of excavation work was done including sinking of a shaft to a depth of about 21.3 m (70 ft) and excavation of several other pits and trenches. During this period also a stamp mill was erected near the shaft to treat the ore from this property and other nearby properties.

Other gold prospects in the area include the Sophia or Diamond Mine which operated from 1896 to 1908 and again at a later date, the St. Joe Mine which was in operation prior to 1898, and the Craig Mine. Prior to the 1860s the early mining history of the area relates to the occurrences of iron ore including both hematite and magnetite on which a number of mines were operated until about 1910. Several other mines in the area were worked for base metals; these include the Eldorado Copper Mine which produced from 1903 to 1907, and a number of pyrite mines, including the Bannockburn Pyrite Mine which was finally closed in 1919.

PREVIOUS WORK

The first serious exploration effort on the Bannockburn property since 1905 appears to have been that carried out by local prospector Mr. Roland Belanger in 1965 under an agreement with the owner. All of Belanger's work was on Lot 28 and consisted of trenching, diamond drilling and sampling in the immediate area of the old workings. Belanger has reported assays of up to 103.0 g per tonne (3.0 oz per ton) gold but since all of the records from this work were destroyed in a house fire they cannot be verified. At least one major mining company examined the property during Belanger's work but declined to take an option on it.

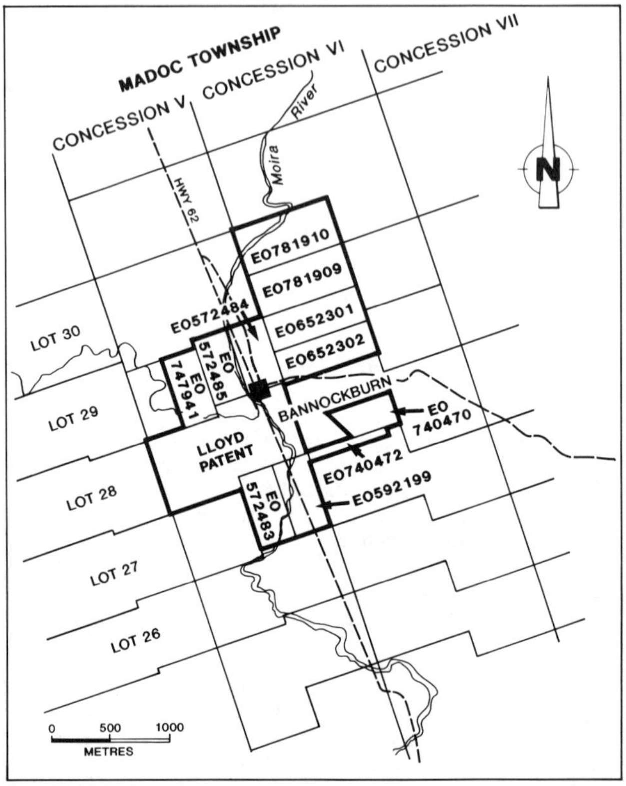

Figure 2. Claim Sketch of Property.

Previous geological work in Madoc Township and adjoining areas by government agencies includes early geological mapping at the end of the last century and several later maps and reports including those by Miller and Knight (1914); and Bain (1960). More recent work by D.F. Hewitt of the Ontario Department of Mines was published as ODM Geological Report 73 (1968). Several recent publications by the Ministry of Northern Development and Mines and the Ontario Geological Survey describe gold and base metal deposits in eastern Ontario, including Madoc Township.

Work by Mono Gold Mines Inc. since 1981

The first field work by Mono Gold Mines Inc. on the Bannockburn property was carried out in 1981 on the eastern part of the Lloyd Patent and adjacent claims to the north and south, centred essentially on the old Bannockburn Mine. Geological mapping revealed complex folding in metasedimentary and metavolcanic rocks adjacent to and immediately east of the Gawley Creek syenite intrusive but recognized the main target of economic interest in this district to be quartz vein structures carrying free gold, associated with shear zones along the intrusive-metasedimentary contact on Lot 28 on which the

Bannockburn Mine was based.

A small program of diamond drilling was carried out by Mono Gold Mines Inc. Several of these drill holes intersected the quartz vein structure confirming its existence. Geophysical work carried out as part of this initial program included magnetometer and VLF-EM surveys which in essence confirmed the geological picture. There was no field work on the claims from 1981 until 1984. Two claims were acquired on the northeast side of the property in 1983, and a further two claims adjoining on the north were acquired in 1985.

The cut line grid was extended to cover the whole of the property in 1984, and a geophysical survey consisting of ground magnetometer and VLF-EM was run over this expanded grid. Geological mapping and evaluation of geophysical anomalies was done in September 1984, and included the sampling of all quartz veins encountered during traverses over the cut-line grid. An assay of 33.1 g per tonne (0.966 oz per ton) gold was returned from a grab sample of a quartz vein in outcrop within an area of quartz veining and quartz float in the northeast part of the property.

A program of stripping and trenching was carried out in October-November 1984, with an assay of 175.7 g per tonne (5.124 oz per ton) gold returned from a larger grab sample of the same quartz vein outcrop, the Discovery vein. A diamond drill program was carried out in February 1985; visible gold was noted in quartz-carbonate veins in the core from seven of the eight holes drilled. Further diamond drilling was carried out in seven separate programs between 1985 and 1987 on the Northeast area.

The drill programs outlined a gold bearing quartz vein system over 457 m (1,500 ft) along strike with drill indicated reserves of 337,708 tonnes (372,154 tons) and a grade of 13.5 g per tonne (0.395 oz per ton) gold over 1.2 m (4 ft) width, to a depth of less than 107 m (350 ft). A total of over 12,500 m (41,000 ft) of diamond drilling was completed by March 1987.

A geochemical soil sampling survey was carried out over the whole property in late 1985, the samples were analyzed for bismuth as a tracer element for gold mineralization. Bismuth anomalies adjacent to the old Bannockburn Mine were subsequently drilled off, minor bismuth mineralization was encountered but no gold values were associated with it.

Two detailed geophysical surveys using EDA Omni IV instrumentation were carried out over the Northeast area of the property in 1986 and early 1987. There was no correlation between geophysical anomalies noted by the surveys and the mineralized quartz vein systems.

A program of surface stripping and ground sluicing was carried out during 1987 exposing about 183 m (600 feet) of the Discovery vein, which was geologically mapped and sampled.

At this point, Sawyer Consultants Inc., consultants to Mono Gold Mines Inc., recommended that the company should go underground on this prospect and outlined a first stage underground program estimated to cost approximately $1.8 million. To accomplish this work Mono Gold Mines Inc. entered into a Joint Venture Agreement in November, 1987, with Micham Exploration Inc. of Toronto, Ontario, for underground exploration of the gold bearing quartz vein systems. The underground program was carried out from January to April 1988 and consisted of 268.2 m (880 ft) of ramp to the 22.85 m (75 ft) level and cross-cutting to intercept three of the quartz veins with 97.5 m (320 ft) of drifting on the veins. An underground drilling program was carried out from three drill stations on the decline for a total of 1,737.3 m (5,700 ft) in 35 holes.

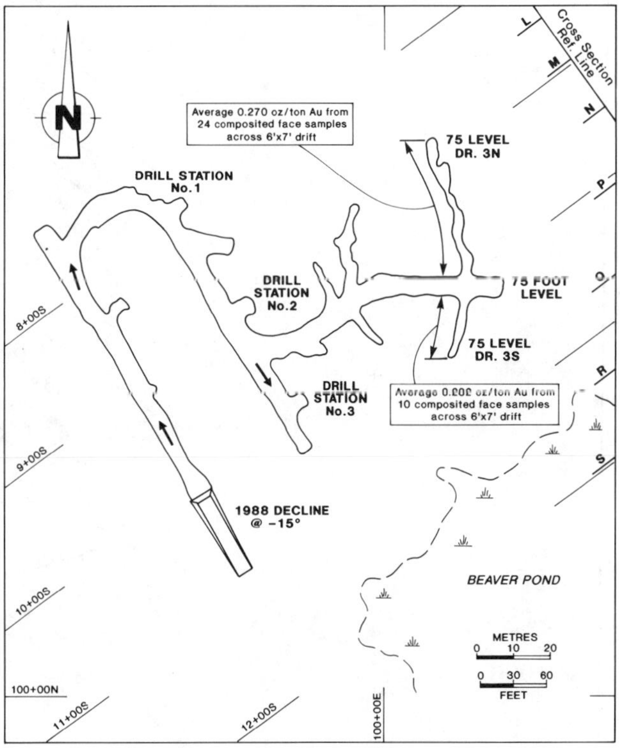

Figure 3. Plan of 1988 Exploration Decline.

A report on the underground exploration program for Micham Exploration Inc. outlined new reserve calculations utilizing the underground information. The reserves were classified as "Modified Geological Reserves" of 225,190 tonnes (248,160 tons) grading 9.15 g per tonne (0.267 oz per ton) gold (uncut) over a minimum 1.8 m (6 ft) width with 1.7 g per tonne (0.05 oz per ton) gold cut-off grade. These reserve calculations were restricted to specific areas of the quartz vein system in contrast to the earlier calculations of "drill indicated" reserves.

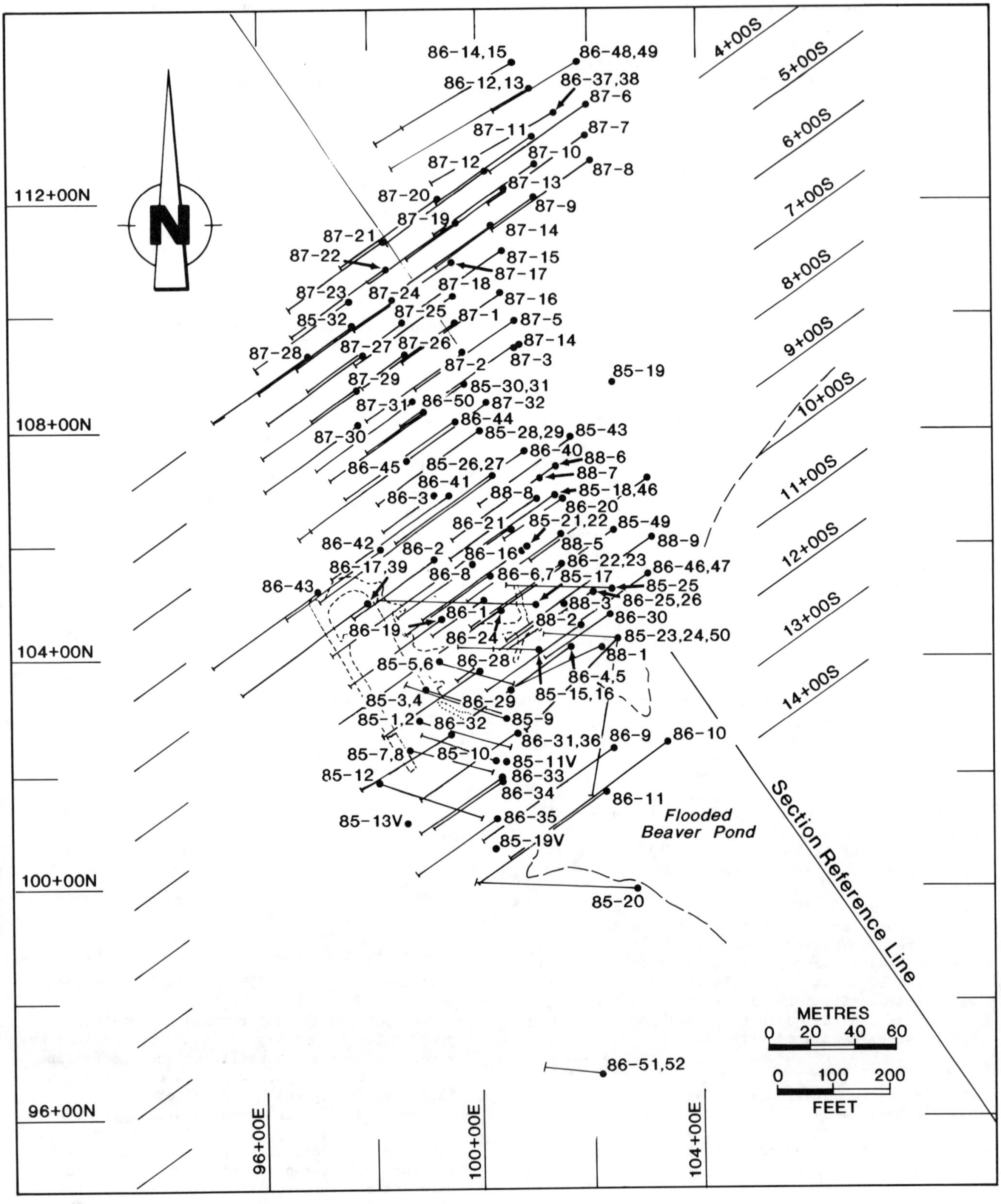

Figure 4. Drill Plan.

Recommendations for further underground exploration by drifting and raising with some additional drilling were not carried out due to financial constraints. Mono Gold Mines Inc. undertook a deep drilling program in October 1988 to explore the down dip extensions of the veins below the underground workings. The drill program was completed in December 1988 for a further 1,513.0 m (4,964 ft) of drilling in nine holes and has indicated a subvertical to vertical control on the gold mineralization within the quartz vein systems, related to a northeasterly trending mega-shear within the Grenville rocks of the district.

GEOLOGY

Regional Geology

The geology of Madoc Township and surrounding areas is characterized by good exposures of Precambrian greenstones, metasediments, intrusive rocks and relatively undisturbed, unconformable Phanerozoic sediments. Figure 5 is a regional geological sketch map illustrating the geological setting of the area.

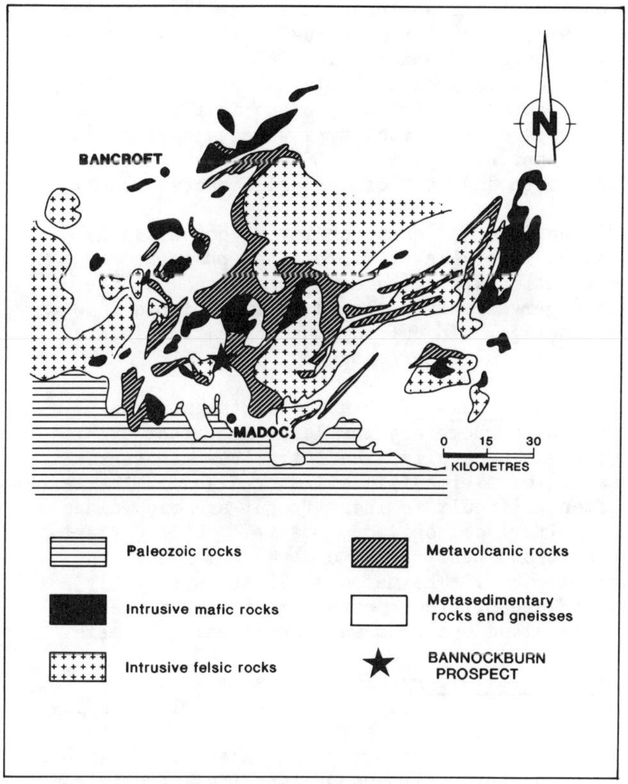

Figure 5. Geological Sketch Map of Part of Southeastern Ontario.

The oldest rocks in the region are mafic metavolcanics, predominantly of tholeiitic basalt type, of the Tudor Formation which occupy the base of the Hermon Group of supracrustal clastic to carbonate metasediments and greenstones. A second sequence of metavolcanic rocks, the "Madoc Volcanics" overlying the Hermon Group, are exposed in southern Madoc Township. These rocks range from andesite to rhyolite in composition and exhibit primary volcanic textures.

Overlying, and in some cases intercalated with, the Tudor Formation volcanics, are the Hermon Group metasediments which, in the Madoc region, are primarily impure marbles with some semi-pelitic and psammitic schists. Other sedimentary rocks include slates, and several bands of meta-conglomerate.

The major intrusive bodies in the area are the Deloro granite and the Gawley Creek syenite. The Deloro granite is a pink, medium crystalline granitic stock occupying several square miles of southwest Madoc Township and is associated with the Deloro gold occurrences in neighbouring Marmora Township. The Gawley Creek syenite body is located in the northwest quadrant of Madoc Township and extends into Marmora Township. This is a medium to coarsely crystalline biotite hornblende syenite which also includes a variety of granitic to dioritic differentiates. Figure 6 is a geological and structural interpretation of the Discovery area showing the main outcrop pattern.

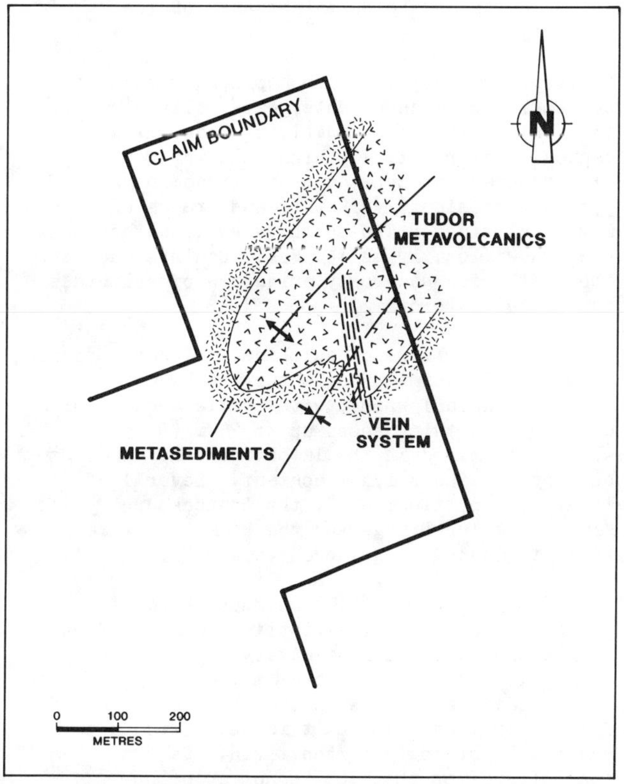

Figure 6. Generalized Geological and Structural Interpretation.

At least two major folds are present in the Madoc area, the Queensborough and Madoc synforms. The Queensborough structure has a northwesterly trend whereas the Madoc synform has a northeasterly trend similar to that of most Grenville structures.

Age determinations indicate these rocks were last deformed during the "Grenville Orogeny" which has produced highly deformed rocks of middle greenschist to lower amphibolite facies in the Madoc Township area.

Property Geology

The Northeast area containing the Discovery zone is underlain essentially by three principal rock types: mafic to intermediate metavolcanic rocks, semi-pelitic metasediments with sulphide-bearing semi-pelitic metasedimentary phases and minor intrusive rocks, all having altered equivalents. Paleozoic rocks have not been mapped on the property. Bedrock is for the most part well exposed with drift thicknesses of only several feet.

Metasedimentary Rocks

SEMI-PELITIC METASEDIMENTS: Much of the Northeast area is underlain by banded, fine to medium grained sericitic and siliceous semi-pelitic metasediments. The semi-pelitic metasediments are strongly deformed and isoclinally folded about the nose and axis of the southwesterly plunging antiformal structure which dominates the Northeast area of the property.

There are sulphide rich horizons within the semi-pelitic metasediments, especially close to the contact with the underlying Tudor metavolcanics. The relatively high sulphide content within these horizons consists predominantly of pyrite with minor pyrrhotite and traces of base metal sulphides. The sulphide enriched horizons are known locally as the Rusty Schists due to the weathering and rusty oxidation of sulphides on exposed outcrops.

Massive horizons of sulphide enrichment occur in semi-pelitic metasedimentary rocks to the south in Madoc Township with pyrite content of over 70% and thicknesses up to 28 m (90 ft) which were mined in the latter part of the last century for the sulphur content. Several of these old "pyrite mines", the Bannockburn Pyrite Mine at Keller Bridge and the Blakely Pyrite Mine at Queensborough, occur nearby.

GARNET SCHIST: A discontinuous band of garnetiferous, chlorite-biotite schist is found within the volcanic-sedimentary contact zone. This unit may range up to 21 m (70 ft) thick but is strongly deformed with the thickness probably fold controlled. The unit is sericitic in part and may be strongly carbonatized. Garnet content greater than 40% is not uncommon.

QUARTZ-SERICITE SCHIST: Quartz-Sericite Schists occur along the volcanic-sedimentary contact, and have been noted in the vicinity of the "Discovery Vein". These are apparently discontinuous lenses with fold controlled thicknesses, occurring within the semi-pelitic metasedimentary horizons closely overlying the contact with the volcanics.

Quartz-Sericite Schists are commonly associated with volcanic-sediment transitions within volcanogenic massive sulphide deposits where they are considered altered or metamorphosed exhalative cherts or rhyolites. Their presence along the main contact at the Northeast area may in fact be of similar origin - as cherty exhalatives.

Metavolcanic Rocks

MAFIC VOLCANICS: The most important rock types in the Northeast Area are the metavolcanics of the Tudor Formation. These rocks are generally mafic to intermediate, massive greenstones. They are moderately foliated, chloritic and exhibit several alteration types.

Distinguishing between individual flow or tuffaceous units within the metavolcanics is difficult, if not impossible. A number of features may be of use in separating similar sub-units, including the presence of possible amygdaloidal flows and sharp textural changes such as increased foliation or alteration.

FELSITE: Diamond drill holes in the northern portion of the Northeast area, drilled in 1985 to 1987, encountered intrusive related alteration zones, identified in 1985 as "felsite". Surface exposures of "felsite" were noted along drill access roads in 1986 and fairly extensive outcrops of this rock type were exposed by the bulldozer stripping in 1987 which extended to the north and west of the drill access roads.

The felsite zones consist of concordantly banded intervals of silicic and potassic alteration where the host metavolcanics have been pervasively replaced by a granitoid assemblage. In outcrop the zones form felsic "centres" over large areas.

Intrusive Rocks Within the Northeast area, intrusive rocks are the least important by volume. Generally, the intrusives are narrow, mafic to intermediate sills or dikes. These are often difficult to identify but are recognized by uniform composition, possible chill contacts, and strong penetrative cleavage or foliation development. The mafic sills are apparently very susceptible to carbonate alteration and often stand out from wall rocks on this basis.

Structural Geology The Northeast area is dominated by a series of northeasterly trending fold structures. The fold to the west of the Discovery trench outcrops as an antiform, with a minor synform composed of isoclinal parasitic folds on the eastern boundary. The antiform plunges to the south and southwest. Minor features are well expressed along the volcanic-sedimentary contact zone where ductility contrasts between the Tudor Formation and the Rusty Schists caused pronounced quasi-flexural folding.

The presence of these folds has resulted in the development of a strong, near vertical axial planar cleavage with associated axial plane shearing. Numerous parasitic folds have been noted, although much of the folding is complex and polyphase. This is especially noticeable in the Rusty Schists and semi-pelitic metasediments where refolded folds are obvious throughout the entire Bannockburn property. Within the metavolcanics folding is virtually invisible due to the lack of marker horizons but the units have undergone a more brittle deformation due to their low ductility.

Faulting in the Northeast area is dominated by the axial planar fabric of the antiform. Although a number of faults have been encountered in diamond drilling, these have not resulted in proven structural dislocation.

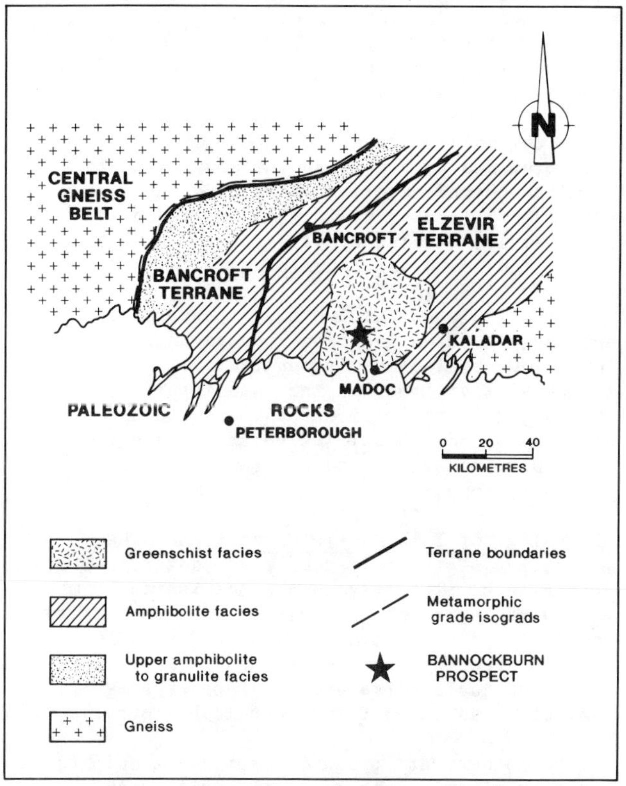

Figure 7. Grenville Province, Central Metasedimentary Belt Terranes, Metamorphic Isograds.

Metamorphism All of the rocks of the Northeast area have undergone Grenville aged deformation and metamorphism of lower to medium greenschist facies. The Madoc area appears as a window of greenschist facies metamorphism surrounded by amphibolite facies metamorphism.

The greenschist metamorphism produced chloritic greenstones in place of mafic volcanic rocks and siliceous, semi-pelitic schists in place of the original argillites, siltstones and tuffaceous sediments. The presence of abundant garnet in the more pelitic rocks implies the higher temperature ranges of low grade metamorphism, which is consistent with the occurrence of minor actinolite in greenschist and the development of biotite in both volcanics and sediments.

Alteration The Northeast area contains quartz vein systems both within and outside of altered host rocks. The alteration types include pervasive carbonatization, silicification and minor potassic alteration. In several cases the quartz veins are bordered by intense but localized carbonatization for several metres or feet, while in some drill holes the entire section exhibits varying degrees of carbonate enrichment.

Carbonatization is present throughout the entire Bannockburn property which makes it difficult to determine if regional processes or "ore" forming processes were responsible. Many major gold producing areas contain abundant carbonatized host rocks, considered by many workers as an effect of the metamorphic events which may have led to the deposition of gold.

Mineralization/Economic Geology

Gold occurs on the Bannockburn property as the native metal, in association with a minor sulphide assemblage, consisting of pyrite, minor marcasite, pyrrhotite, arsenopyrite, and other sulphides and bismuth tellurides, in quartz and quartz carbonate veins. The overall sulphide content in the vein structures is relatively minor and the tenor of the gold mineralization does not appear to be related to sulphide content. Gold may occur with or without carbonate in the veins, however it is sometimes concentrated near the vein margins where coarse calcite also occurs. Visible free gold has been frequently observed in drill core and, as a result of the stripping and sluicing program, in surface exposures of the vein systems.

The quartz vein systems in the Northeast area consist of a series of sheeted dilation veins which have a northerly to northwesterly strike and which dip to the east at between 40° to 55°. The quartz carbonate vein systems are comprised of at least four, and possibly ten, veins which crosscut folded Grenville supracrustal rocks. The veins are typically narrow, varying from 1.8 m (6 ft) to a few centimetres (inches) in true thickness in which "ore shoots" are developed. There does not appear to be a consistent relationship between vein width and grade of gold and frequently gold values even in narrow sections of the veins are high enough to support dilution to a 1.2 m (4 ft) minimum, or greater, mining width in continuous zones.

There are several types of quartz veins occurring within the property. All quartz veins intercepted during the drill program were sampled and assayed for gold, the results of these assays indicated that the gold mineralization was associated with a particular type of crosscutting quartz vein composed of coarse grained, glassy quartz with conchoidal fracture, and a bluish-grey colour, generally dipping at around 45° to 55° to the east. Carbonate in the form of calcite or rusty weathering ankerite is usually associated with the quartz veins and sulphides are generally present, although in minor amounts. The presence of chalcopyrite, galena, marcasite and arsenical pyrite - especially the botryoidal forms, is a good indication of the presence of visible gold or blue-grey tellurides.

Other types of non-auriferous quartz vein noted in drill core included milky white, cherty "bull quartz" veins, parallel to the foliation or axial plane cleavage and with the characteristics of metamorphic "sweats" or lensy veins. These veins are usually discontinuous, and where exposed on the stripped and cleaned surfaces can be seen as irregular, conformable layers of quartz within the axial plane cleavage and as folded pods of quartz within the tight isoclinal folds in the semi-pelitic metasediments.

Coarse visible gold associated with well developed tourmaline crystals has been noted in the sub-cropping quartz veins north of the Discovery trench area and was noted in quartz veins in several of the earlier 1985 drill holes drilled in this area. The occurrence of tourmaline in quartz veins is restricted to this area, and has not been reported from elsewhere on the property.

The gold bearing quartz veins have been traced in diamond drill holes and on surface exposure over a strike length of about 457 m (1,500 ft), and have been explored by drilling to a vertical depth of some 137 m (450 ft). Interpretation of drill core intercepts indicated there are from four to ten parallel veins within the quartz vein systems, however the surface stripping and sluicing programs exposed a sufficient strike length of the Discovery vein to show that the structural controls are more complex than had been anticipated.

Metallurgical Test

Samples of mineralized quartz vein material had been examined by petrographic microscopy during the earlier drill programs to determine the full suite of minerals occurring in the quartz veins, however it became evident that a proper metallurgical evaluation of the mineralized quartz vein material was necessary to ensure there would be no mineralogical surprises at a later date.

The Discovery vein was exposed over a strike length of 26 m (85 ft) by the stripping and sluicing program in 1987 and after diamond saw channel sampling at 1.5 m (5 ft) intervals the full 26 m (85 ft) was trenched by blasting. A total of 750 kg, or 1,650 lb, of broken quartz vein material was collected and sent to Witteck Development Inc. in Mississauga, Ontario, for metallurgical and mineralogical evaluation.

Metallurgical Evaluation of Bulk Sample The bulk sample was prepared for metallurgical testing by crushing to minus 3.2 cm (1 1/4 in) followed by coning and quartering. One quarter was split out and crushed to 100% passing 6 mesh, with representative samples of 500 gram and 200 gram weights riffled out for head analyses and subsequent test work.

Head Assays Gold assays were performed on the samples prepared for head analyses. The samples were reduced in stages through a 150 mesh screen until about 20 grams was left on the screen, both oversize and undersize were analyzed for gold. The average gold assay was 10.13 g per tonne (0.3257 oz per ton) gold, with a standard deviation of 2.73 g per tonne (0.088 oz per ton) gold. The large standard deviation can be attributed to "nugget effect", with about half of the gold reporting to the +150 mesh fraction, although this fraction comprised only 5% of the total sample weight.

An estimate of grade for the bulk sample had been prepared by weighting the assay results from the channel samples taken over the 26.0 m (85 ft) of Discovery vein which provided the bulk sample; a value of 14.4 g per tonne (0.463 oz per ton) gold was calculated.

A multi-element analysis was performed on a head sample which had been reduced in stages to 100% -150 mesh. A semi-quantitative scan indicated that iron was the main impurity present, with minor amounts of arsenic and antimony noted. Quantitative analyses showed that only iron and sulphur were present in high concentrations.

Mineralogy of Bulk Sample Concentrate The mineralogy of the bulk sample was determined by preparing samples of vanned concentrates from -48 mesh quartz vein material. The concentrate was prepared onto polished mounts for reflected light microscopy, with scanning electron microscopy and qualitative energy dispersive x-ray analysis used to confirm mineral chemistry.

The concentrate product comprised a sulphide dominant mineral assemblage of pyrite with lesser to trace quantities of pyrrhotite, arsenopyrite, marcasite, galena, sphalerite, chalcopyrite and covellite. Several bismuth minerals were also observed and included native bismuth, tellurobismutite, tetradymite, and bismuthinite. The gold mineralization consists of native gold with minor to trace quantitites of silver. The gold occurs as fine to coarse particles ranging in size from less than one micron to 100 x 130 microns with elongate grains up to 170 microns.

The gold may occur as fine particles associated with arsenopyrite or arsenical pyrite in which the gold follows intracrystalline boundaries and cleavages. Associations with bismuth minerals and commonly, with galena, are abundant. Gold occurs with hedleyite, and adjacent to tetradymite in association with galena, or partially intergrown with or as replacements of galena interstitial to tellurobismutite.

In summary, the gold mineralization occurs as fully liberated, fine to coarse grained particles of silver-poor gold and as numerous middling exposed particles commonly associated with arsenopyrite, chalcopyrite, several types of bismuth minerals, galena and gangue silicates.

Milling Test and Recoveries Representative samples from the bulk sample were submitted for metallurgical and milling tests, including grinding characteristics and a bond work index with mill power and size requirement determinations. Test work was conducted on representative samples to assess the potential for gravity concentration of gold values.

Cyanidation test work was conducted on a flotation concentrate sample to investigate the potential for direct gold recovery. The flotation concentrate was aerated for 24 hours, with lead nitrate added during aeration. Cyanidation was carried out in stages using increasingly harsher conditions, to determine minimum conditions for adequate gold extraction. The cumulative from staged gold extraction was only 82%, while gold extraction by intensive cyanidation alone was 56.6%. There were appreciable concentrations of tellurium, bismuth, arsenic and antimony remaining in the residue, with very little of these species present in the leach solutions. This indicates they were concentrated by flotation but were not readily leached by cyanide.

A series of bottle roll tests were conducted on representative samples to study the amenability of the material to direct cyanidation and to determine the optimum leach conditions. The parameters investigated included particle size, pre-aeration, lead nitrate addition and leach retention time. Gold extraction varied from 96.3% to as high as 99.7%, with cyanide consumption ranging from 0.15 kg per tonne (0.3 lbs per ton) to 0.101 kg per tonne (3.82 lbs per ton) depending on leach conditions. All tests were carried out at 50% solids, pH 11.0 and 1.0 gram per litre NaCN.

The high gold extraction achieved by direct cyanidation contrasts markedly with results from the cyanidation of flotation concentrates. There are references to the difficulty of leaching gold from high concentrations of sulphides and tellurides in the literature, and while both sulphur and tellurium are at low levels in the bulk sample, 0.3% and 21 ppm respectively, they are closely associated with the gold occurrences and are highly concentrated by flotation.

The effects of grind size and leach retention time were studied in different tests in order to optimize recovery conditions for the material. The grind size study indicated that a 77.5% minus 200 mesh grind provided recoveries of 91% extraction after 25 hours with 98.5% extraction after 96 hours retention. Cyanide consumption increased with decrease in particle size but the high cyanide consumption was of particular concern. A multi-element scan of leach solutions showed a high concentration of iron and sulphur; quantitative analyses revealed iron concentrations of 173 ppm Fe. It was calculated that the dissolved iron alone accounted for 65% of the measured cyanide consumption.

The effects of pre-aeration and lead nitrate addition on cyanide consumption was investigated in a series of tests which showed that pre-aeration did indeed reduce cyanide consumption from nearly 1.5 kg per tonne (3.0 lbs per ton) to approximately 0.25 kg per tonne (0.5 lbs per ton), with gold extraction increased slightly from 98.8% to 99.5%; a pre-aeration time of 8 hours was sufficient. Varying the pH of the solution or adding lead nitrate did not affect the metallurgical results.

The effect of leach retention time was studied with duplicate tests run for each of 24 hours, 48 hours, and 72 hours. Gold extractions were similar for tests with 48 hours or more retention time, resulting in about 99% extraction, with gold extractions for the 24 hour tests averaging 1.5% lower. The optimum leach time of 48 hours was in close agreement with the estimate of 50 hours leach time arrived at from mineralogical considerations. Cyanide consumption after 48 hours retention averaged about 0.25 kg per tonne (0.5 lbs per ton).

Chemical assays were carried out on the optimized leach solutions to allow for an environmental assessment of the effluent disposal, and suggestions were made for further chemical treatment methods for the destruction of cyanide and cyanide complexes prior to disposal to tailings.

Vacuum filtration tests were performed on a cyanidation leach slurry at 65% solids at various form times, in order to determine filterability. Design parameters were determined for drum filter size for specific rates of mill throughput, with the design subject to review after metallurgical process development.

Results of Metallurgical Test Work The metallurgical test work carried out on the quartz vein material from the Discovery vein shows that gold recoveries of over 98% can be achieved by direct cyanidation after 8 hours pre-aeration, with 48 hours retention time. Under these conditions cyanide consumption is about 0.25 kg per tonne (0.5 lb per ton).

The bond work index for the material was 15.62 kwh/t. below average for most gold mills, and grinding specifications of 80% minus 200 mesh appear optimum. No problems are anticipated in using the Merrill-Crowe process for gold recovery from pregnant cyanide solutions due to low base metal and tellurium content of the solutions. The filterability of the cyanide residue proved excellent.

Flotation concentration gave excellent results but cyanidation of the concentrates was unsatisfactory due to the presence of deleterious materials. Concentration by tabling, gravity concentration, was unsatisfactory even with cleaning, however consideration should be given to including a jig in the flow sheet, after the grinding circuit, to recover free gold.

CONCLUSIONS

The Bannockburn prospect is a new discovery in an old gold camp, brought to its present stage of development largely because of the management of Mono Gold Mines Inc. and their belief in the potential of the property. The gold bearing quartz veins were discovered by routine geological mapping and sampling following up on the geophysical survey, but without the continued efforts of management in raising funds to allow ongoing diamond drilling in 1985 and in 1986 there would have been little progress.

Diamond drill programs served to define the size of the deposit and indicated the high grade nature of the gold mineralization. Problems occurred in the assaying of samples of drill core containing coarse visible gold due partially to the "nugget effect," but subsampling problems as well as technical problems occurred at the commercial assay laboratories. Drilling served to define structure but it was necessary to go underground to define grades.

The underground exploration program was carried out towards the financial year end for the 1987 flow through funding, resulting in higher than expected costs. The exploration program could not be completed as recommended, but the continuity of the quartz veins was confirmed. Assays returned from the larger face samples and muck samples indicated that the assay results from the drill programs were conservative.

Metallurgical test work on a surface bulk sample was favourable and indicated the "ore" would be amenable to relatively low cost recovery methods. The mineralogy of the quartz veins has given an indication of both the age of mineralization and the origin of the gold. The presence of bismuth sulphides and tellurides are indicative of intrusive related sources for the mineralization.

The Bannockburn prospect is located near centres of population where there has been a mining tradition for over one hundred years, excellent infrastructure and electric power are readily available. The prospect has been opened up to the 22.85 m (75 ft) level and while there is still much to be done in the way of upgrading of reserves and mine engineering there is no doubt that this is a future producer.

There are tourist sign posts at Eldorado pointing out the location of Ontario's oldest gold mine, there could soon be sign posts at Bannockburn pointing to Ontario's newest gold mine.

REFERENCES

Bain, I., 1960, "Geology of the Grenville Belt through Actinolite, Ontario," unpublished Ph.D. Thesis, University of Toronto.

Carmichael, D.M., et al, 1978, "Isograds around the Hastings Metamorphic 'low'," Currie, A.L., and Mackasey, W.O., Ed., Toronto '78, Field Guidebook, Geological Association of Canada, pp. 325-346.

Cherry, M.E., et al, 1982, "Gold-selected References, Open File Report 5382, Ontario Geological Survey.

Gordon, J.B., et al, 1979, "Bannockburn Prospect," in Gold Deposits of Ontario, Part 2, Mineral Deposits Circular 18, Ontario Geological Survey, p. 34.

Hewitt, D.F., 1968, "Geology of Madoc Township and north part of Huntingdon Township," Geology Report 73, Ontario Department of Mines, 45p.

Holm, P.E., et al, 1986, "Geochemistry of Metavolcanic rocks and dykes from the central metasedimentary belt, Grenville Province, Southeastern Ontario, in The Grenville Province," Moore, J.M., Davidson, A.. and Baer, A.J., ed., Special Paper 31, 1986, Geological Association of Canada, pp. 255-269.

Hopkins, P.E., 1922, "Ontario gold deposits, their character, distribution and productiveness", Annual Report, Vol. 30, Pt. 2, Ontario Department of Mines.

Kindle, E.D., 1936, "Gold occurrences of Ontario, east of Lake Superior," Memoir 192, Geological Survey of Canada.

Kingston, P.W., and Papertzain, V.C., 1982, "Gold Development Potential in Eastern Ontario," Open File Report 5379, Ontario Geological Survey.

Meen, V.B., 1944, "Geology of the Grinsthorpe-Barrie area, Hastings, Lennox and Addington, and Frontenac Counties, Ontario," Annual Report, Vol. 51, Pt. 4, Ontario Department of Mines, pp. 1-50.

Miller, W.G., and Knight, C.W., 1914, "The Precambrian geology of southeastern Ontario," Annual Report, Vol. 22, Pt. 2, Ontario Department of Mines, pp. 1-21.

4

Geology/Exploration IV
New Discoveries and Developments

H. Hurst
FMC Gold Co.
Reno, Nevada

Chapter 14

TRANSPORT OF GOLD IN HIGH TEMPERATURE, LOW PRESSURE, GAS PHASES WITH Cl_2, CO, H_2S: A CLUE TO GOLD MOBILIZATION FROM THE LOWER CRUST

by

R. J. Puddephatt, S. Zhou and W. S. Fyfe

Departments of Chemistry and Geology
University of Western Ontario
London, Canada N6A 5B7

ABSTRACT

Gold is easily mobilized at temperatures within range 600-1100°C by gases such as Cl_2, CO, H_2S, in gas mixtures, at low pressures. Such gases are relevant to conditions of granulite facies metamorphism in environments of magma underplating. These observations suggest an efficient mechanism for moving gold to the upper crust during major thermal events. The efficiency of precipitation of gold at lower temperatures is related to available mineral surfaces and pyrite appears to be an excellent substrate for gold precipitation.

INTRODUCTION

It is now generally accepted that gold ore formation involves dissolution of gold from source rocks, where it is present in concentrations at the part per billion level, followed by transportation and redeposition of the gold. Enrichment by a factor of 10^4 is necessary and the deposit must contain 10^6-10^8 g of gold to be economically valuable. It follows that requirements for the successful formation of deposits must include: (1) An adequate supply of fluid containing reagents which can react with gold to give a soluble gold complex; (2) a sufficient quantity of source rocks containing gold, which must be accessible to the hydrothermal fluid; (3) an energy source to move the fluid, and (4) a mechanism for producing localized and selective precipitation of the gold from the fluid (Fyfe, 1984).

In the development of genetic models for hydrothermal metal deposits, it has been shown that large scale heat and fluid transport occur in geological situations such as cooling of igneous bodies (Lister, 1974; Norton and Knight, 1977), and the dehydration of rocks during metamorphism in subduction zones (Fyfe et al, 1978; Fyfe, 1986). For gold and platinum there is no doubt that transport can occur in an aqueous phase, but it is not known if the transport could occur in the gaseous phase.

Fyfe (1986) suggested that the fluid generated by the assimilation between crusted materials and underplate the magma bodies could be important in flushing gold from the lower crust. If the crust above the trapped underplate magma bodies contains dense materials, such materials will tend to sink and be partially assimilated. This might be the case for a typical deep greenstone belt with basalt or komatiite at amphibolite facies grade. The assimilation of such a body of rock, with water, carbonate and sulphide added at the oceanic floor cooling stage on the sea floor, will lead to massive gas production at temperatures exceeding 1000°C. In the final gas, species such as H_2-CO_2-CO-COS-H_2O-H_2S will be present, and the pressure of species such as CO could reach many bars. Under these conditions, it is possible that volatile metal carbonyl derivatives could be formed and that transport in the gas phase could occur. The possibility that CO and COS could act as ligands in hydrothermal transport of gold has been suggested previously (Fyfe, 1986; Fyfe and Kerrich, 1984) though several other ligands such as SH^-, S^{2-}, Cl^-, AsS^{2-} and NH_3 are also implicated in gold transport under the appropriate conditions (Seward, 1984; Fyfe and Kerrich, 1984).

For the deposition of gold from aqueous solution, several models have been proposed. These are based on changes of the solubility of gold due to changes in temperature or the concentration of ligands for gold (Fyfe and Henley, 1973) or on the adsorption and subsequent reduction of gold complexes on the surface of sulfide minerals (Jean and Bancroft, 1982, 1985; Bancroft and Hyland, 1988).

The aim of this work is to test experimentally the feasibility of Fyfe's hypothesis that gold could be transported in the gas phase by high temperature, high pressure fluids (Fyfe, 1986) and to determine if selective deposition of gold from the vapour phase onto sulfide minerals is likely to occur under geochemical conditions.

EXPERIMENTAL METHODS

Sources of Minerals

The minerals used in this study include pyrite (FeS_2), pyrrhotite (FeS), arsenopyrite (AsFeS), calcite and quartz. The minerals were cut into small cubes about 0.4 × 0.3 × 0.4 cm^3, and then polished. The polished samples were ultrasonically cleaned in an acetone bath and air-dried before experiments. Gases used in the experiments were CO, Cl_2, COS, H_2S, CO_2, and all were commercial products.

Gold Transport in the Presence of Gases

In this type of experiment, 0.02 g of gold powder was placed at one end of a pyrex or quartz tube (dimensions 20 × 1 cm), as shown in Figure 1. The tube was evacuated and then the desired gas or gas mixutre was added through a standard vacuum line. Then the tube was flame sealed and a temperature gradient was applied in a tube furnace. The end of the tube where the gold powder was situated was heated to 600°C-1100°C, while the other end of the tube was usually not heated. Gold was transported from the hot end of the tube and deposited at the cold end, usually as gold or a compound of gold.

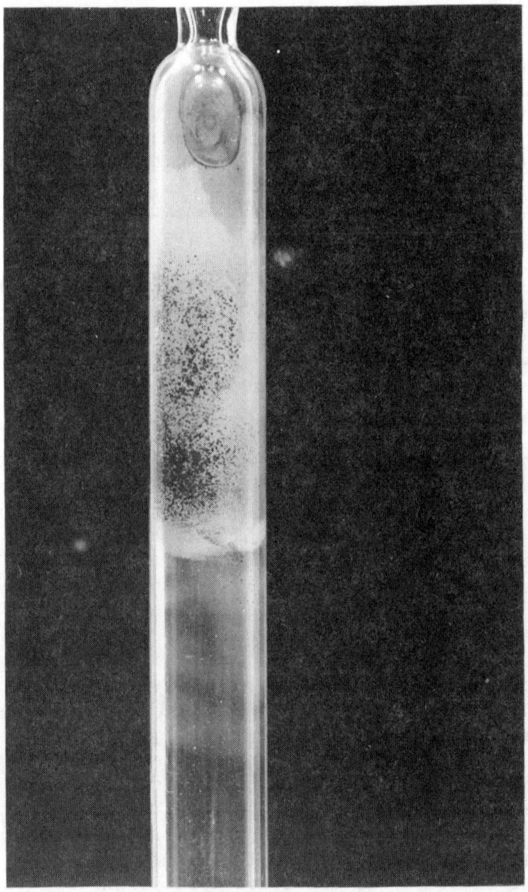

Figure 1. A photograph of a glass tube used for vapour phase transport, showing metallic gold deposits in the cold end of the tube, formed in "Au + $AuCl_3$ + CO" system.

Gold Transport and Deposition on a Mineral

In a second type of test, a mineral phase was present at the cold end of the tube, and all other conditions were the same as described above. Typically, gold was deposited in the metallic form on the surface of the mineral.

In addition to these two types of experiments, tests were conducted in a system with no gas present. Gold powder and mineral were placed at the hot and cold ends respectively of the tube. Tables 1-3 list all experiments which were carried out on gold transport.

Gold Transport Using Chlorine and Carbon Monoxide

Gold transport in the vapour phase using chlorine as carrier has been known for many years and the thermodynamics of the gold-chlorine system is also well-established (Biltz et al., 1928; Schafer, 1970, 1971, 1972, 1973; Puddephatt, 1978). Using chlorine alone at 650°C, gold could be transported and deposited either as AuCl (white) or $AuCl_3$ (yellow) in the cool zone of the tube (Table 1). Using a deficiency of chlorine (obtained by using a mixture of Au and $AuCl_3$ with no free chlorine) gold was transported and deposited in the metallic form as fine crystals.

Table 1

TRANSPORT OF GOLD BY CHLORINE AND CARBON MONOXIDE

Reagents	Temperature °C	Time h	Gold Transport
Au	650-1100	20	no
Au,Cl_2	650	6	AuCl or Au_2Cl_6[a]
Au, CO	650-1100	21	no
Au,Cl_2,CO	500	5	AuCl,Au_2Cl_6 or AuCl(CO)
Au,$AuCl_3$	650-1100	20	AuCl
Au,$AuCl_3$,CO	650-1100	20	AuCl or AuCl(CO)[a]

[a] The complex deposited depended on the pressure of Cl_2 or CO used. With a large excess of Cl_2, Au_2Cl_6 was the major product. Small, measured amounts of chlorine equivalent could be added by preforming $AuCl_3$ by reaction of Au with Cl_2 and adding a known amount of this solid.

Carbon monoxide alone did not give detectable transport of gold at temperatures from 650-1100°C. However, mixtures of Cl_2 and CO were more effective at gold transport than chlorine alone (Table 1). Using excess CO, the gold was deposited as either AuCl or [AuCl(CO)], and in all cases it is likely that transport occurs in the form of [AuCl(CO)]. The compound AuCl has a polymeric structure with bridging chlorides and $AuCl_3$ exists in both the solid or vapour as the dimer $[Au_2Cl_6]$, but [AuCl(CO)] forms a monomeric linear structure (Figure 2). It is therefore not surprising that [AuCl(CO)] is the most volatile and hence gives a better medium for gold transport. The various complexes can be interconverted according to the equations (1)-(3) (Puddephatt, 1978) and the complex which precipitates is then clearly dependent on the pressures of Cl_2 and CO.

$$AuCl_3 + CO \rightleftharpoons AuCl + COCl_2 \quad (1)$$
$$AuCl + CO \rightleftharpoons AuCl(CO) \quad (2)$$
$$AuCl \rightleftharpoons Au + 1/2Cl_2 \quad (3)$$

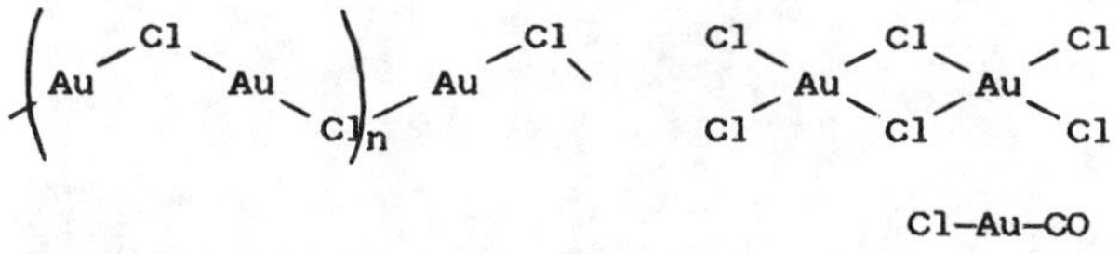

Figure 2. The structures of AuCl (polymeric), $AuCl_3$ (dimeric) and [AuCl(CO)] (monomeric).

The role of CO in this system is to convert polymeric, involatile AuCl to the more volatile, monomeric [AuCl(CO)]. It is easy to envisage how other gold salts could be mobilized towards vapour phase transport by similar reactions with CO.

Deposition of Gold on Sulfide Minerals

The results of the experiments in which gold was transported by Cl_2 and CO and then deposited on a sulfide mineral are given in Table 2. In general, experiments with excess chlorine present were complicated by reaction of chlorine with the sulfide mineral and so most experiments used a deficiency of chlorine or used preformed $AuCl_3$ as the chlorine source. Gold was selectively deposited on all of the sulfide minerals tested (pyrite, arsenopyrite, pyrrhotite) but was not deposited on quartz.

The case in which gold was transported and deposited from $AuCl_3$ onto pyrite was studied in most detail, but the other systems using chlorine as transport agent gave similar results. Islands were found on the pyrite surface which analyzed as elemental gold, elemental sulfur and iron(II) chloride respectively as determined by SEM analysis (Figures 3,4). Some sulfur was also present on the walls of the glass tube and this was shown by mass spectrometry to be S_8, the most stable form of elemental sulfur.

Table 2
GOLD TRANSPORT USING CHLORINE AND CARBON MONOXIDE WITH DEPOSITION ON MINERALS

Reagents	Mineral	Temperature °C	Time h	Gold Transport[a]
Au,Cl_2	pyrite	650	20	yes
	arsenopyrite	650	20	yes
	pyrrhotite	650	20	yes
$AuCl_3$	pyrite	500	3	yes
	arsenopyrite	650	16	yes
	pyrrhotite	650	23	yes
	quartz	650-1100	72	no
Au,$AuCl_3$	pyrite	650	7	yes
Au,$AuCl_3$,CO	pyrite	650	2	yes
$AuCl_3$,CO	pyrite	400	0.25	yes
Au,AgCl,CO	pyrite	1100	1	?[b]

[a] In all cases where gold was detected on the mineral surface, it was present as elemental gold.
[b] Only metallic silver was detected on the pyrite surface.

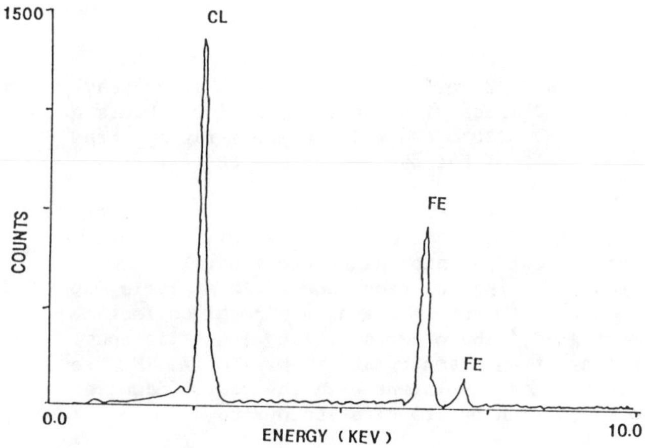

Figure 3. SEM x-ray spectrum of iron chloride on the mineral surface formed in the "Cl_2 + gold + pyrite" system.

As for the aqueous system (Jean and Bancroft, 1985), the sulfide mineral acts as a very efficient collector of gold and the gold is present on the mineral surface as small spheres and islands. A probable equation to describe the reduction of gold, which probably occurs after adsorption of the gold chloride at a sulfur donor site on the mineral surface, is shown in equation (4).

$$2AuCl_3 + 3FeS_2 \longrightarrow 2Au + 3FeCl_2 + 6S \quad (4)$$

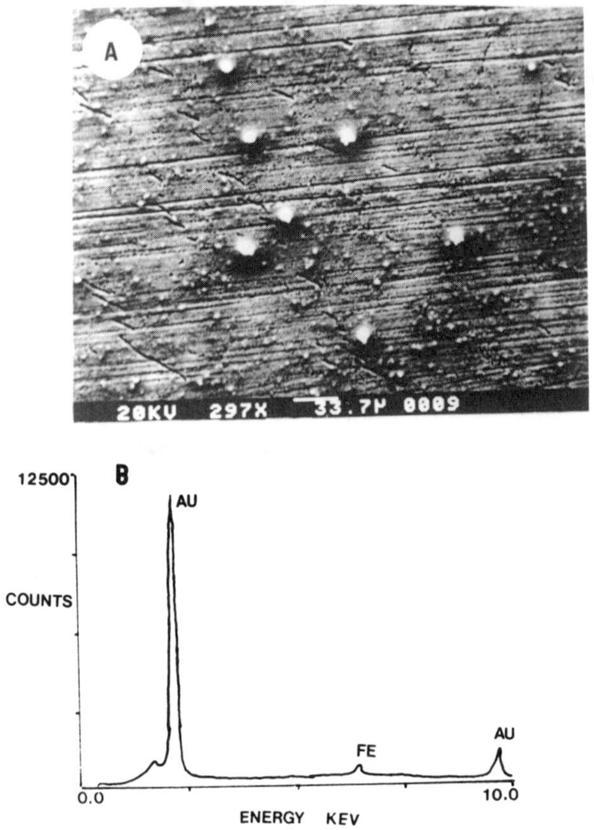

Figure 4. SEM micrograph of pyrite after reaction with $AuCl_3$ and Au for 7 hours at 650°C (A) and the SEM x-ray spectrum of the pyrite surface (B).

In order to test which mineral was the most effective collector of gold, experiments were carried out in which all three minerals were present in the reaction tube. SEM analysis was then used to determine which mineral collected most gold. The order of collecting efficiency was pyrite > arsenopyrite >> pyrrhotite, Figure 5, which is consistent with the common occurrence of gold with pyrite or arsenopyrite.

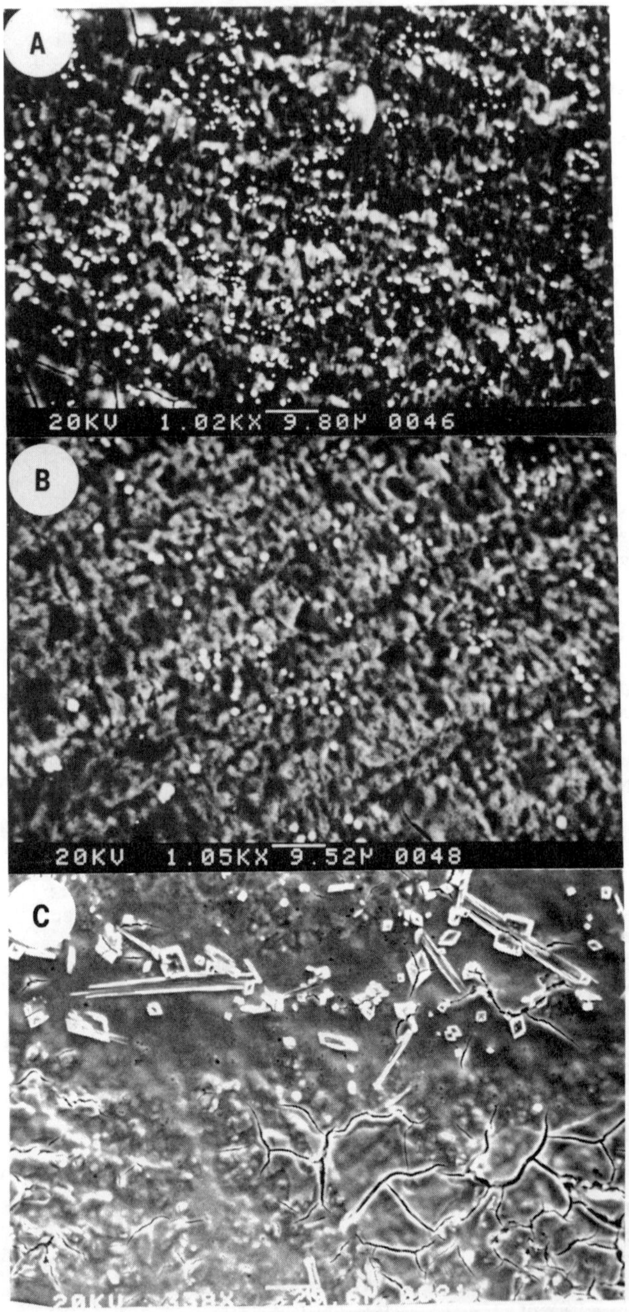

Figure 5. SEM micrographs showing regions of the concentration of gold found on the pyrite surface (A), arsenopyrite surface (B), and pyrrhotite surface (C) in "$AuCl_3$ + pyrite + arsenopyrite + pyrrhotite" system for 20 hours at 650°C. Pyrite provides best surface for the deposition of gold.

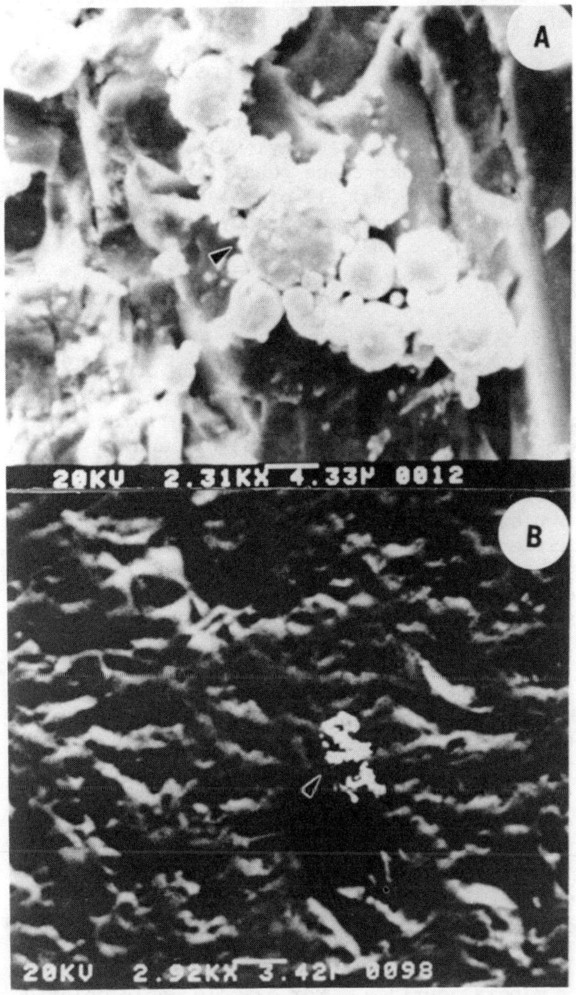

Figure 6. SEM micrographs of mineral after co-existing with CO, pyrite at 1100°C for 13 hours (A) and arsenopyrite at 1050°C for 51 hours (B).

Transport and Deposition of Gold Using Carbon Monoxide

The data for these experiments are given in Table 3. Gold was not transported by CO at 650°C but it was transported and redeposited on pyrite or arsenopyrite at 1100°C (Figure 6). Even at 1100°C the transport was slow and no gold was found on the mineral pyrrhotite under these conditions. However, the experiments were carried out with p(CO) ~ 0.2 bar whereas in geological systems it is possible that very high pressures of CO may be present and the transport using CO may then be much more efficient (Fyfe, 1986).

It is interesting to speculate on the nature of the volatile gold complex responsible for the gold transport, but it is not possible to determine this from the present data. Gold does not form a stable binary carbonyl complex but atomic gold reacts with CO in a low temperature matrix to give both Au(CO) and $Au(CO)_2$ (McIntosh

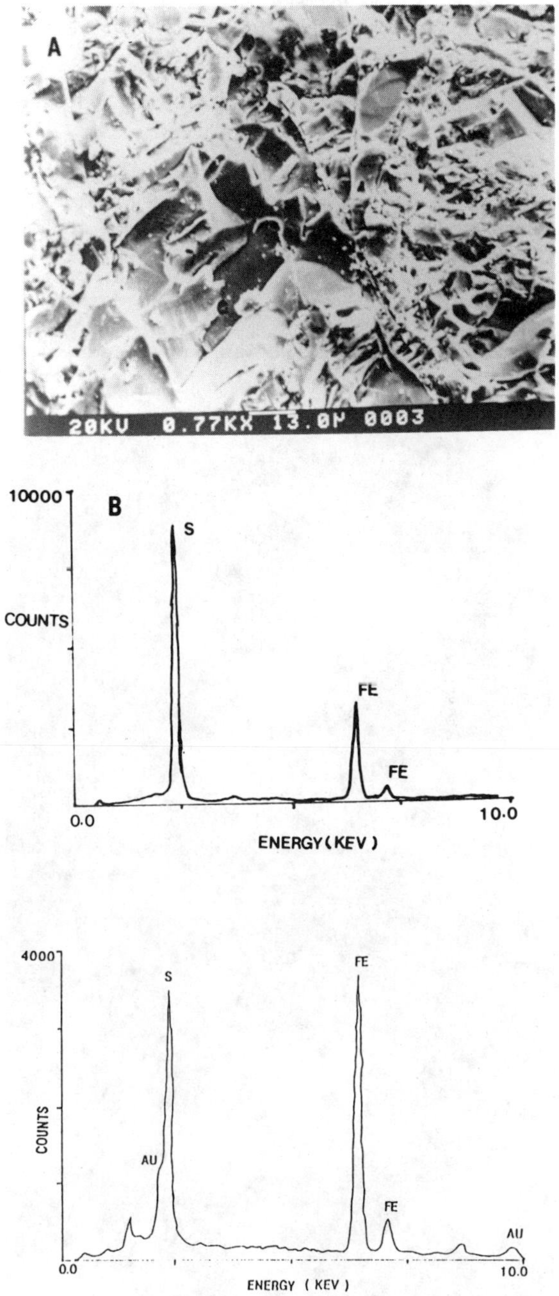

Figure 7. SEM micrograph of a blank pyrite (A) and its SEM x-ray spectrum (B), the ratio of sulfur and iron is approximately 2 to 1. This ratio decreased to 1 to 1 after gold transport by CO.

and Ozin, 1977) and it is possible that these complexes are responsible for the gold transport. The redeposition would then just involve adsorption on the sulfide mineral surface with loss of CO. Analysis by gas chromatography of the gases after transport at 1100°C showed only CO was present, thus ruling out the possibility that more reactive transporting gases might be formed by reaction of CO and pyrite or arsenopyrite. However SEM analysis of the pyrite surface did indicate some loss of sulfur (Figure 7).

Transport and Deposition of Gold Using Hydrogen Sulfide

H_2S at 1100°C gave transport of gold and deposition on pyrite or arsenopyrite (Table 3, Figure 8). Under these conditions, some decomposition occurred to give hydrogen and sulfur but H_2S is probably responsible for the transport of gold.

A mixture of CO and H_2S was more effective than either CO or H_2S alone for gold transport at

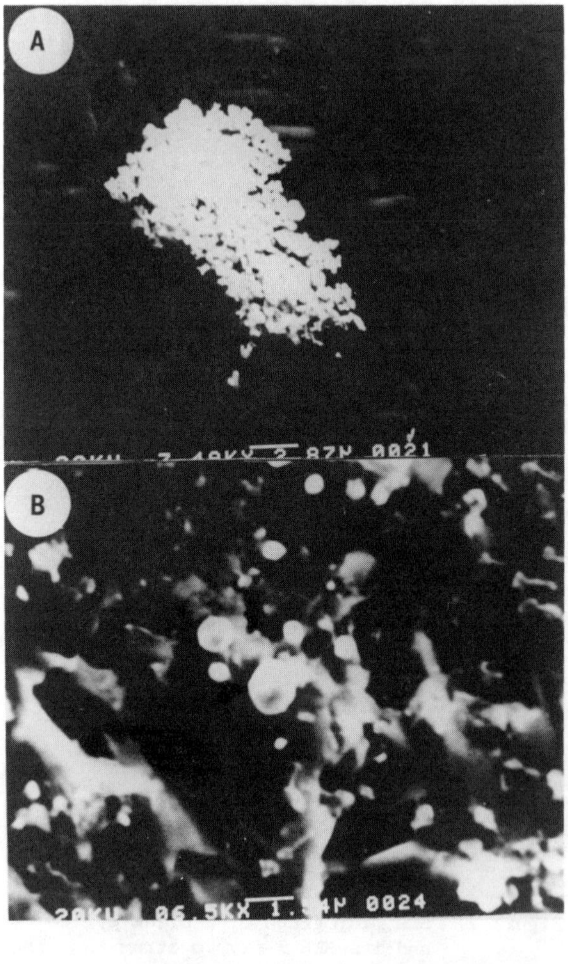

Figure 8. SEM micrographs of mineral reaction with H_2S and Au at 1050°C; pyrite, 6 hours (A), arsenopyrite, 18 hours (B).

1100°C. This is similar to the effect of CO on the transport of gold by chlorine which was discussed above, and it is likely that gold is transported as a carbonyl (sulfide) derivative such as $[Au(SH)(CO)]$ or $[Au_2(\mu\text{-}S)(CO)_2]$. Neither of these is a known complex and the nature of the volatile gold complex is obviously speculative.

Transport of Gold by CO_2 and COS

Neither CO_2 nor COS gave transport of gold at 1100°C (Table 3). Carbonyl sulfide decomposes to some extent under these conditions to give CO, S_8, CO_2 and CS_2 and all of these species could be detected by gas chromatography or mass spectrometry.

CONCLUSIONS

This research has addressed the question of whether gold could be transported in geological situations in the vapour phase by fluids which are rich in H_2, CO_2, CO, COS, H_2O and H_2S and which may be present at 1000°C and 100 bar. The experiments are oversimplified models for this complex system and, for safety reasons, have been carried out at much lower pressure (≤1 bar). Nevertheless they can give some indication of which gases may be involved in gold transport. The chief results are as follows:

1. Gold transport by chlorine is greatly enhanced by the presence of CO. Here carbon monoxide breaks down the structure of polymeric or dimeric gold chlorides to give the more volatile monomeric $[AuCl(CO)]$ and this leads to easier transport of gold. Although chlorine is not a significant constituent of geological fluids, this is important in showing how CO may react with other gold salts to give more volatile carbonyl derivatives.
2. The minerals pyrite and arsenopyrite act as very efficient collectors of gold from vapour phase transport. The gold compounds are adsorbed and reduced to metallic gold on the mineral surface. For example, gold chlorides react with pyrite to give gold, iron(II) chloride and sulfur. Pyrrhotite is a less efficient collector of gold and quartz and calcite are ineffective.
3. The gases CO and H_2S can give transport of gold and deposition on pyrite or arsenopyrite at 1050-1100°C and a mixture of these gases is more effective than either gas alone. The transport is slow under the conditions used (<1 bar) but, since geological pressures are likely to be much higher (up to 100 bar), it is clearly possible that vapour phase transport of gold by CO, H_2S could be geologically significant. We speculate that a carbonyl (sulfide) complex of gold is responsible

It is thus clear that under the conditions of granulite facies metamorphism (to 1000°C), and magma underplate conditions (to 1300°C), as gases in the system H_2-S-C-Cl_2-O_2 are common and normally CO_2-S rich, gold will be mobilized and available to be reprecipitated at lower temperatures. It is clear that the mineral phases in the precipitation zone can play a significant role in focussing precipitation.

Gold is a most reactive metal in many geological regimes. The new observations described here show the possible complexers at high temperatures. It is well known that under lower temperature water-rich conditions, that bisulphide is perhaps the most important ligand. Work on gold in near surface weathering conditions shows that the organics of living and degraded organisms have potential for gold transport (Fyfe, 1987). It is thus very clear that the search for gold deposits should not be restricted to any particular P-T regime of the crust. But perhaps the association of gold with greenschist facies rocks (300-400°C) is a result of the CO-H_2S gases generated at much higher temperatures.

Table 3
GOLD TRANSPORT USING CO, COS, H_2S, CO_2, CS_2 WITH DEPOSITION ON MINERALS

Reagents	Mineral	Temperature °C	Time h	Gold Transport[a]
Au,CO	pyrite	650	20	no
Au,CO	pyrite	1100	13	yes
Au,CO	arsenopyrite	1050	51	yes
Au,CO	pyrrhotite	1100	264	no
Au,COS	pyrite	1100	42	no
	arsenopyrite	1100	4.5	no
	pyrrhotite	1100	25	no
Au,H_2S	pyrite	1050	6	yes
	arsenopyrite	1050	18	yes
Au,O_2	pyrite	1100	22	no
Au,CS_2	pyrite	1100	23	no
Au,CO,H_2S	pyrite	1050	50	yes

[a] Gold was present in all cases in the metallic form.

REFERENCES

Bancroft, G.M. and Jean, G.E., 1982, Nature, Vol. 298, pp. 730.

Biltz, W., Fischer, W. and Juza, R., 1928, Z. Anorg. Allgem. Chem., Vol. 176, pp. 121.

Fyfe, W.S. and Henley, R.W., 1973, Miner. Sci. Eng., Vol. 5, pp. 295.

Fyfe, W.S., Price, N.J. and Thompson, A.B., 1978, Fluids in the Earth's Crust. Development in Geochemistry 1, Elsevier, Amsterdam, 383 p.

Fyfe, W.S. and Kerrich, R., 1983, "Gold'82: The Geology, Geochemistry and Genesis of Gold Deposits, Ed. Foster, R.P., pp. 99.

Fyfe, W.S., 1986, "Gold in the Western Shield", Ed. Laoyd, A.C., CIMM Special Vol. 38, pp. 156.

Fyfe, W.S., 1987. "Gold: One of the Most Mobile Elements", Gold Mining 87, Ed. Brawner, C.O., Post City Press, Baltimore, Maryland, pp. 3.

Huber, H., McIntosh, D. and Ozin, G.A., 1977, Inorg. Chem., Vol. 16, pp. 975.

Jean, G.E. and Bancroft, G.M., 1985, Geochim. Cosmochim. Acta, Vol. 49, pp. 979.

Lister, C.R.B., 1974, J. Roy. Abstr. Soc., Vol. 39, pp. 465.

McIntosh, D. and Ozin, G.A., 1977, Inorg. Chem., Vol. 16, pp. 51.

Norton, D., Knight, T., 1977, J. Geophys. Res., Vol. 81, pp. 3007.

Puddephatt, R.J., 1978, "The Chemistry of Gold", Elsevier, Amsterdam.

Schafer, H., 1970, Z. Anorg. Allg. Chem., Vol. 367, pp. 11.

Schafer, H., 1971, J. Cryst. Growth, Vol. 9, pp. 17.

Schafer, H., 1972, "Preparative Methods in Solid State Chemistry,", pp. 251.

Schafer, H., 1973, "Crystal Growth: An Introduction", North-Holland, Amsterdam, 143 p.

Seward, T.M., 1984, "Gold'82: The Geology, Geochemistry and Genesis of Gold Deposits", Ed. Foster, R.P., pp. 165.

Chapter 15

STRATIGRAPHIC CONTROLS ON GOLD MINERALIZATION CARLIN TREND, NEVADA

by
Odin D. Christensen
Newmont Gold Company
1700 Lincoln Street
Denver, CO 80203

Abstract Gold deposits of the Carlin Trend occur within sedimentary rocks deposited across the Paleozoic continental margin. Lithologically dissimilar units, juxtaposed along the Roberts Mountains Thrust Fault system, constitute three distinct stratigraphic packages: autochthonous miogeoclinal carbonates, allochthonous eugeoclinal siliciclastic rocks, and an overlying coarse clastic overlap assemblage. Gold deposits occur within all three stratigraphic packages, with economic concentrations localized at structural and stratigraphic settings similar to those which host petroleum traps. Along structural breaks, gold is selectively concentrated within favorable stratigraphic units. Host stratigraphy controls the character of the ore, including average grade, grade variability, physical properties and metallurgical response.

INTRODUCTION

The Carlin Trend is a 60 km long northwest alignment of epithermal, sediment-hosted, disseminated gold deposits. Since 1965, more than 250 tonnes (8 million troy ounces) of gold have been produced. During 1988, more than 32 tonnes (1,050,000 oz) were produced from six active mines; 1989 production is forecast to exceed 50 tonnes (1,630,000 oz). Currently identified geologic resources in 20 deposits exceed 1,600 tonnes (52 million oz). The Carlin Trend is the premier gold district in the United States and is currently the center of a major gold rush.

The Carlin deposit, located near the center of the Trend (Figure 1), has become the type deposit for the class of sediment-hosted, disseminated gold deposits. Discovered in 1960 by Newmont Mining Company geologists, it was the first significant gold deposit in the world to be exploited by bulk mining methods. Since that time, more than 30 separate centers of gold mineralization have been identified along the Carlin Trend. While all of these deposits possess similarities, there are significant differences between the individual deposits. Six subdistricts are distinguished, based upon geographic, structural, lithologic, alteration, geochemical, and mineralization characteristics (Christensen et al., 1987).

REGIONAL GEOLOGIC SETTING

The Carlin Trend is located centrally within the Great Basin geologic province of the western United States. Host rocks for the deposits include sedimentary units ranging in age from Ordovician through Mississippian, as well as a Jurassic-Cretaceous granodiorite stock.

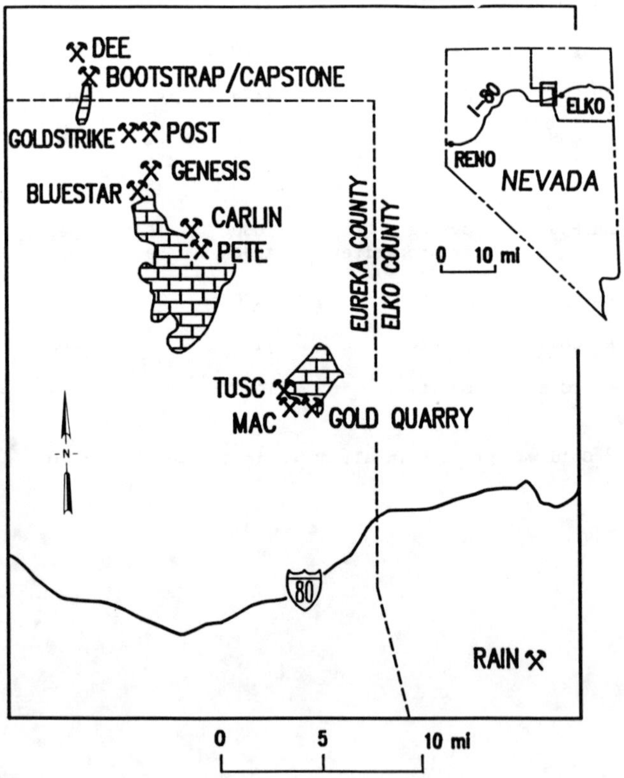

Figure 1. Major gold deposits of the Carlin Trend. Patterned areas are windows exposing Paleozoic carbonate rocks through overlying siliceous rocks (after Rota and Hausen, 1989).

During the Early Paleozoic, the area that is now eastern Nevada lay along the passive western margin of the North American continent. A westward thickening and deepening wedge of sedimentary units was deposited across the shelf and slope. Rock units deposited vary from eastern miogeoclinal continental shelf carbonates to western fine-grained siliceous clastic and cherty units (Stewart, 1980).

Tectonic activity associated with the Late Devonian to Early Mississippian Antler Orogeny resulted in the eugeoclinal siliceous rock assemblage being thrust eastward over time-equivalent transitional and miogeoclinal carbonate units along the Roberts Mountains thrust fault system. The leading edge of the allochthonous Roberts Mountains thrust plate formed the emergent Antler highland, from which siliceous clastic sediment, eroded from the overriding plate, was shed eastward into the adjacent foreland basin. Local terminology refers to the three major units of the geologic architecture as the eastern assemblage, the western assemblage, and overlap assemblage for the autochthonous shelf carbonates, the allochthonous siliceous lithologies, and the coarser clastic flysch deposits, respectively. In late Mesozoic time, plutonic activity was accompanied by doming and folding of the sedimentary units along the north-northwest trending axis of the Tuscarora anticline. Several windows, eroded through the overthrust western assemblage units along the broken trend of the anticline, expose eastern assemblage rocks of the lower plate. A Jurassic-Cretaceous granodiorite stock is exposed in the vicinity of the Post Mine.

Tertiary stocks and dikes of intermediate composition are common, but generally poorly exposed throughout the district. Cenozoic extensional faulting began about 17 Ma, creating the present Basin and Range topography characteristic of the area.

The age of gold mineralization is poorly constrained. A maximum age of approximately 145 Ma is indicated by the mineralized granodiorite of the Goldstrike intrusive. A minimum age of approximately 5 Ma is indicated by the presence of gold mineralized clasts in Pliocene Carlin Formation alluvial conglomerates overlying the Gold Quarry deposit. A middle to late Tertiary age of gold mineralization is most likely.

CARLIN-TYPE GOLD DEPOSITS

Common features of sediment-hosted gold deposits have been summarized by Romberger (1986). Sediment-hosted gold deposits occur throughout the western United States but are most highly concentrated in the Basin and Range Province, many occurring along regional trends characterized by coincident alignments of igneous rocks and geophysical anomalies (Percival and others, 1988).

High- and low-angle faults are important features in all deposits. Fracturing and brecciation associated with structural intersections facilitated optimum hydrothermal fluid introduction into, and interaction with, receptive host rocks.

Hydrothermal alteration varies from negligible to intense between deposits. Typically this includes decarbonatization, silicification, argillization, and baritization. Gold mineralized and altered rocks are characteristically geochemically enriched with silver, arsenic, antimony, mercury, and thallium.

Sedimentary host rocks can be of any lithology if the effective permeability was favorable for the introduction of hydrothermal fluids. Gold mineralization along the Carlin Trend is hosted within silty dolomite of the Roberts Mountains Formation (Carlin), massive fossiliferous limestone (Post and Bootstrap), interbedded chert and shale (Gold Quarry), and coarser clastic lithologies of the Mississippian Webb Formation (Rain).

Extensive exploration and mining experience has demonstrated that host stratigraphy, however variable, plays a critical role in gold ore localization and in the physical and chemical character of the ore. Three examples from the Carlin Trend provide useful examples.

Carlin Deposit

The Carlin deposit, centrally located along the Carlin Trend, lies near the northeastern margin of the Lynn Window, in which Silurian and Devonian carbonates of the lower plate are exposed through overlying upper plate Ordovician-age cherts and siliceous shales (Figure 2). Mining began at Carlin in 1965 and continued through 1987. Gold ore at Carlin occurred in three principal zones aligned some 2,100 m (7,000 ft) in an east-west direction, roughly parallel both with bedding and with the trend of the Roberts Mountains thrust. Although gold deposition was controlled by high-angle faults on a district scale, it was largely confined to replacement of favored permeable facies, resulting in conformable, tabular orebodies (Figure 3). During mining, the structural control on gold mineralization was evident from thickened ore zones in the vicinity of north-trending cross structures.

Most of the gold ore occurs within the upper 250 m of the Middle Silurian to Late Devonian Roberts Mountains Formation, with only minor occurrences within structurally or stratigraphically adjacent units (Radtke, 1985). The Roberts Mountains Formation, originally a platy, silty dolomitic limestone, has been strongly altered in mineralized areas. Carbonate removal, along with local bleaching and iron staining, left a porous low-density rock that ranges in composition from an argillaceous to dolomitic siltstone.

The upper part of the ore zone has been oxidized; conformity of the bottom of the oxide zone with the present topographic surface suggests

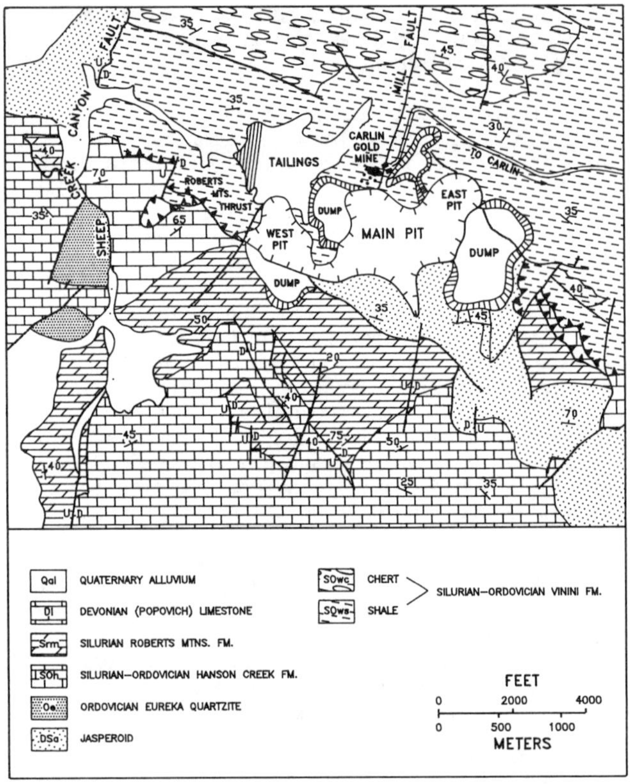

Figure 2. Geologic map of the Carlin gold deposit (after Radtke, 1985, and Rota and Adkins, 1984).

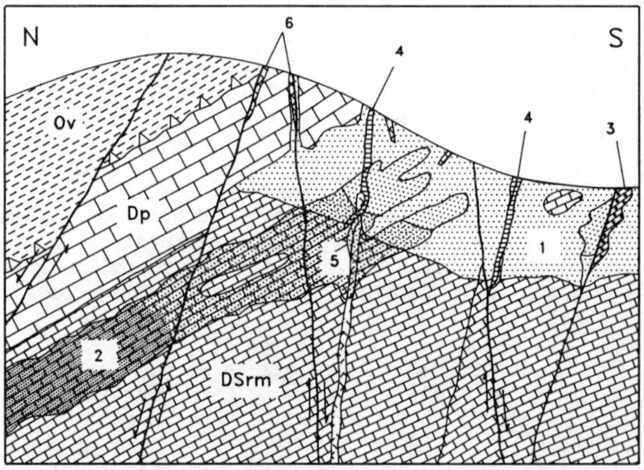

Figure 3. Generalized cross section of the Carlin gold deposit showing important geologic features: Ordovician Vinini Formation (Ov), Devonian Popovich Limestone (Dp), Silurian-Devonian Roberts Mountains Formation (DSrm), zone of leaching and alteration (1), main ore zone including lower unoxidized ore and upper oxidized ore (2), jasperoid bodies (3), barite veins (4), quartz veins (5), and calcite veins (6) (after Radtke, 1985, and Rye, 1985)

that the oxidation was at least in part a supergene feature. Oxide ores consist of fine-grained mixtures of quartz, illite, kaolinite and iron oxides.

Unoxidized deeper ores are characterized by abundant sulfide minerals, much carbonaceous material, and minor barite with residual calcite, dolomite, and illite. The carbonaceous material at Carlin has been characterized as pyrobitumen, the thermally altered residue of petroleum. Over 90% of the gold within the unoxidized ores is submicroscopic (less than 0.2 microns in diameter) (Rota and Adkins, 1984).

Orebodies in the Carlin deposit were formed as multiple pulses of hydrothermal fluids were focused upward along high-angle fractures, penetrated the structurally broken dolomitic beds of the Roberts Mountains Formation, altered the permeable lithology, and left behind a residue of gold. The ultimate geometry of the ore zones was significantly controlled by the stratigraphic distribution of permeable receptive lithologies within the Roberts Mountains Formation.

Gold Quarry

Gold Quarry is the largest sediment-hosted gold deposit discovered to date on the Carlin Trend. Proven reserves are estimated to be 223 million tonnes (246 million st) grading 1.5 g/t (0.044 oz/ton) and containing 290 tonnes (9.3 million troy ounces) of gold (Newmont Gold Company, 1989). Mining and milling at Gold Quarry began in 1985; dump leaching began in 1986. The Gold Quarry ore body is to be developed by an open pit that will eventually be 1,830 m (6,000 ft) long, 1,220 m (4,000 ft) wide and 300 m (1,000 ft) deep (Rota and Hausen, 1989)

Gold Quarry is located along the southwestern margin of the Carlin Window, a carbonate fenster roughly circular in outline and about 3 km (2 miles) in diameter (Figure 4). Steeply dipping normal faults and fractures associated with the window margins are the primary control on the distribution of mineralization. The two principal structures are the northeast-trending Gold Quarry fault and the northwest-trending Good Hope fault.

Extensive exploration and ore-definition drilling has demonstrated that the Gold Quarry deposit consists of two distinct ore zones, termed the Main and Deep West zones. The Main zone, which is dominantly structurally-controlled, is located above the dominantly stratigraphically-controlled Deep West zone with a gold-poor zone between (Figure 5) (Rota and Hausen, 1989)

Main ore zone gold is hosted by a 450 m (1,500 ft) thick sequence of siltstone, shale, sandstone, silty limestone, and chert which is interpreted to be part of a transitional sequence of strata, deposited on the Paleozoic slope between the eastern and western assemblage rocks. The siliciclastic sediments are typically thin-bedded and noncompetent.

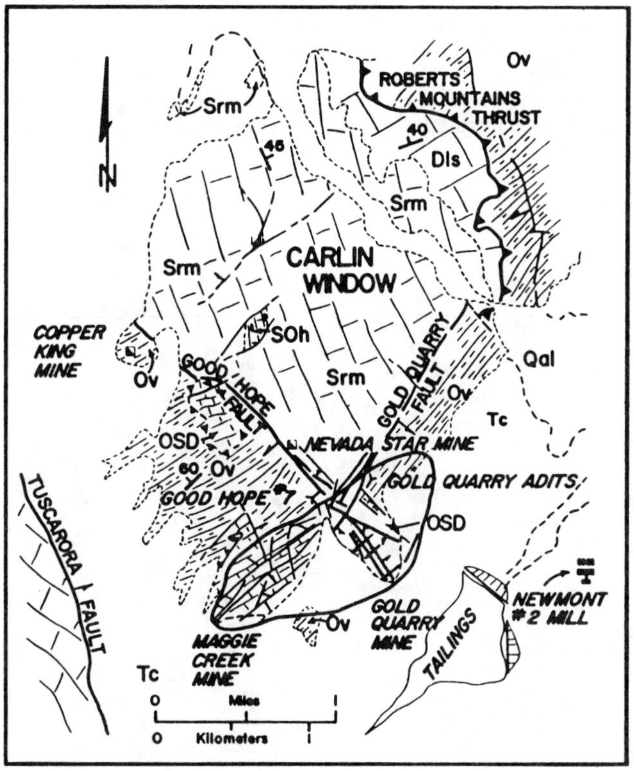

Figure 4. Geologic map of the Gold Quarry mine area: Quaternary alluvium (Qal), Tertiary Carlin Formation (Tc), Devonian limestone (Dls), Silurian Roberts Mountains Formation (Srm), Silurian/ Ordovician Hanson Creek Formation (SOh), Transitional sediments (OSD), Ordovician Vinini Formation (Ov) (after Rota, 1987)

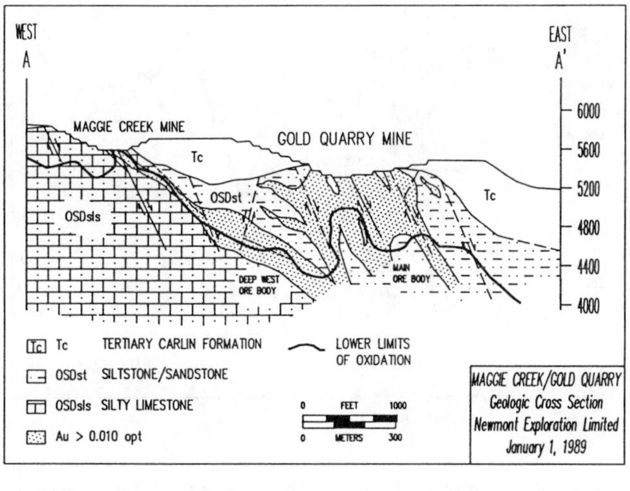

Figure 5. Generalized geologic cross section through the Maggie Creek and Gold Quarry orebodies (after Rota and Hausen, 1989)

Gold mineralization in the Main ore zone is principally structurally controlled. Tabular to irregularly shaped high-grade ore shoots are directly related to northeast and northwest trending faults and fractures. Intersections of structures often host significant, though discontinuous, ore pods. The intervening rock, generally less fractured and altered, constitutes the majority of the low-grade ore in the Gold Quarry deposit.

The deposit can be best characterized as consisting of pods of high grade gold ore distributed irregularly throughout a much larger body of lower-grade material. The Main ore zone strikes generally north-south and dips 45-50 degrees to the east and is continuous over a strike length of 600 m (2,000 ft).

Alteration is so extreme in Gold Quarry that ore types are distinguished more by intensity of alteration than by primary lithology, which is often not recognizable. Distribution of alteration, like that of gold, is highly variable and complex, being related to high-angle faults and fractures. Silicification is the most pervasive style of alteration, with several episodes of silicification and brecciation evident. Silica flooding has produced rocks which range from relatively unaltered siltstone to a cherty, flint-like, rock that is more than 97% silica.

Argillization and alunitization followed the main silicification event; baritization is locally common in the Main zone, especially in the vicinity of larger faults. Supergene alteration followed the formation of the Gold Quarry orebody. Primary, unoxidized, carbonaceous ore is present at depths greater than 120 to 200 m (400-650 ft) beneath the original ground surface.

The Deep West zone is a highly silicified, stratabound, replacement ore zone that averages about 75 m (250 ft) in thickness. It strikes approximately N20E, dips 30-35 degrees to the southeast, and is recognized over a strike length of 770 m (2,500 ft). The average gold grade of this zone is approximately twice that of the Main zone. Deep ores are carbonaceous and pyritic.

Deep West gold mineralization lies immediately above the lithologic contact between a lower limestone unit and the overlying transitional siliciclastic rocks. Both the upper and lower limits of the zone are marked by abrupt grade boundaries. Gold-barren, thin-bedded grey limestone lying beneath the Deep West has been interpreted to be Roberts Mountains Formation, although the intensity of alteration obscures the contact as sampled in drill holes.

The style of mineralization within the Deep West can be characterized as more passive, and more continuous, than the Main ore zone. In style, it is very similar to the Carlin deposit.

Mineralization at Gold Quarry is inferred to be of Tertiary age. The hydrothermal system responsible for gold mineralization was probably characterized by deeply convecting meteoric water, driven by an underlying heat source. The system appears to have been aerially extensive and long-lived. Primary down-dip structural conduits were inferred high-angle structures parallel to the Gold Quarry Fault. Fluid flow from these feeder structures was focused through the permeable lithologies at the carbonate-clastic contact, altering and mineralizing the host to form the Deep West zone. Rising further along structures in the overlying siliceous rocks, the fluids reached a level where boiling resulted in the alteration, brecciation, and gold mineralization of the Main zone.

Differing geologic characteristics of the two zones result in markedly different physical and chemical properties, which significantly affect the exploitation of the deposit. Geologic information is utilized daily to segregate different metallurgical ore types, to distinguish hard and soft milling ores for blending, and for grade control.

Rain Deposit

The Rain deposit is the southernmost of the producing deposits along the Carlin Trend, located 14 km (8 miles) south of Carlin, Nevada, near the crest of the Carlin-Pinon Range (Figure 1). Discovered in 1979, gold production began in 1988. (Thoreson, 1987; Knutsen and West, 1984)

The Rain orebody is located near the base of the Mississippian Webb Formation where it unconformably overlies the Devonian Devils Gate Limestone. The Webb Formation is part of the overlap assemblage, deposited within the Mississippian foreland basin near the then-emergent Antler highland. At Rain, the Webb Formation consists of a sequence of siliceous mudstones, claystones, siltstones, and interbedded fine to medium-grained sandstones.

An apparent high-angle reverse fault, trending west-northwest and dipping southwestward, is the dominant ore-controlling structure (Figure 6). In general, the main orebody is an elongate manto within the coarser basal Webb Formation in the hanging wall of the high-angle fault (Figure 7).

Hydrothermal alteration processes included oxidation, argillization, silicification, and baritization. While much of the ore-bearing host rock is argillized and oxidized, silicification is restricted to the main hydrothermal conduit structure and to the ore zone. Hydrothermal barite occurs within the jasperoidal fault breccia as fragments, as replacements of other rock fragments, and as massive white interstitial barite.

Other smaller gold deposits in the vicinity of the Rain deposit are similarly and exclusively hosted at this Devonian-Mississippian unconformity, a surface of substantial solution porosity, and within permeable coarser basal lithologies. Again, effective permeability played an important role in focusing the ore-forming hydrothermal fluids, and the combination of stratigraphy and structure controlled the location and geometry of the resulting deposit.

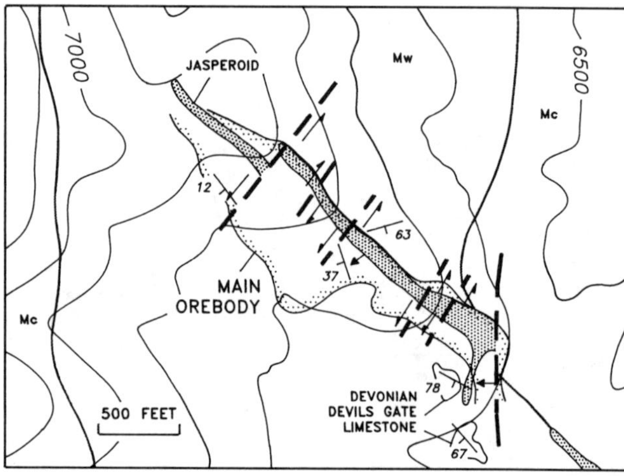

Figure 6 Geologic map of the Rain gold deposit: Mississippian Chainman Shale (Mc), Webb Formation (Mw) (after Knutsen and West, 1984).

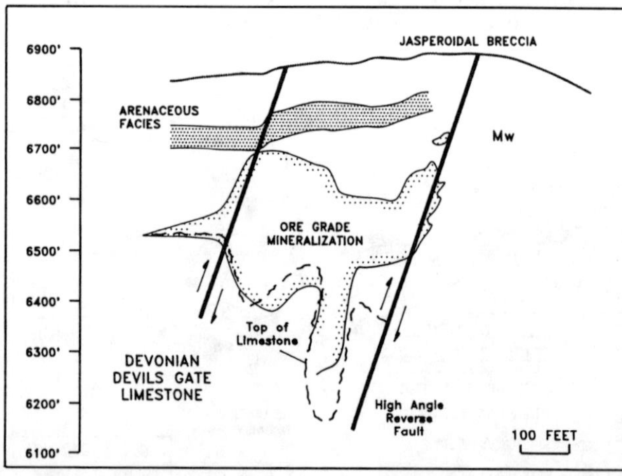

Figure 7. Generalized cross section through the Rain deposit (after Knutsen and West, 1984).

SUMMARY

Disseminated sediment-hosted gold deposits of the Carlin Trend represent the greatest currently identified gold resource in the United States. Twenty explored deposits contain a gold resource exceeding 1,600 tonnes (52 million troy ounces).

While each deposit is unique in its particular combination of lithologies, structural elements, alteration, mineralization style, and geometry, all possess common elements. Principal among these are the strong control exhibited by both structure and host-rock stratigraphy upon the flow course and chemical interaction of hydrothermal fluids during deposit formation.

Mineralization styles vary from the passive stratabound replacement within relatively permeable and porous silty limestone at the Carlin deposit, to the highly variable, structurally-controlled, Main ore zone at Gold Quarry.

Where permeability of the host rocks was relatively uniform, the resulting gold deposits are of relatively uniform grade and physical character; where the host rock was marked by physical inhomogeneity, as in the case of fracture permeability in siliceous rocks, the resultant ores are similarly highly variable. Recognition of the importance of stratigraphic and structural control upon the distribution of ore and upon the physical, chemical and metallurgical properties of the ores is critical to the efficient exploration for and exploitation of these deposits.

REFERENCES

Anon., 1989, "Goldstrike plans 28 t/a (900,000 oz per year) production," Mining Engineering, Vol. 41, p.78.

Christensen, O.D., Knutsen, G.C., and Ekburg, C.E., 1987, "Disseminated gold deposits of the Carlin Trend, Nevada," AIME Preprint 87-84.

Cunningham, C.G., 1988, "The relationship between some disseminated gold deposits, the western edge of the Precambrian craton, and paleothermal anomalies in Nevada," Bulk Mineable Precious Metal Deposits of the Western United States, Schafer, R.W., Cooper, J.J., and Vikre, P.G. (eds), Geological Society Nevada, Reno, pp. 35-48.

Knutsen, G.C., and West, P.W., 1984, "Geology of the Rain disseminated gold deposit, Elko County, Nevada," Arizona Geological Society Digest, Vol. 15, pp. 73-76.

Newmont Gold Company, 1989, Annual Report, 1988, 28 pp.

Percival, T.J., Bagby, W.C., and Radtke, A.S., 1988, "Physical and chemical features of precious metal deposits hosted by sedimentary rocks in the western United States," Bulk Mineable Precious Metal Deposits of the Western United States, Schafer, R.W., Cooper, J.J., and Vikre, P.G. (eds), Geological Society Nevada, Reno, pp. 11-34.

Radtke, A.S., 1985, "Geology of the Carlin gold deposit, Nevada," Professional Paper 1267, U.S. Geological Survey, 124 pp.

Romberger, S.B., 1986, "Disseminated gold deposits," Geoscience Canada, Vol. 13, pp. 21-31.

Rota, J.C., 1987. "Geology of Newmont Gold Company's Gold Quarry deposit, Eureka County, Nevada," GEOEXPO '86, Exploration in the North American Cordillera Elliot, I., and Smee, B. (eds), Association of Exploration Geochemists, pp. 42-50.

Rota, J.C., and Adkins, A.R, 1984, "General geology of the Carlin gold mine," Reno '84 Field Trip Guide, Association of Exploration Geochemists, pp. 17-22.

Rota, J.C. and Ekburg, C.E., 1988. "History and geology outlined for Newmont's Gold Quarry deposit, Eureka County, Nevada," Mining Engineering, Vol. 40, pp. 239-245.

Rota, J.C. and D.F. Hausen, 1989, "Geology of the Gold Quarry Mine -1988," Paper presented at AIME Annual Meeting, February 1988. Proceedings in preparation.

Rye, R.O., 1985, "A model for the formation of carbonate-hosted disseminated gold deposits based on geologic, fluid-inclusion, geochemical, and stable-isotope studies of the Carlin and Cortez deposits, Nevada," Geologic Characteristics of Sediment- and Volcanic-Hosted Disseminated Gold Deposits - Search for and Occurrence Model, Tooker, E.W. ed., Bulletin 1646, U.S. Geological Survey, pp.35-42.

Sporlede, L.E., 1989, "Newmont - new projects, new methods, new mines," Engineering and Mining Journal, Vol.41, p.N10-N21.

Stewart, J., 1980, Geology of Nevada, Special Publication 4, Nevada Bureau of Mines and Geology, 136 pp.

Thoreson, R.F., 1987, "The Rain gold deposit" Bulk Mineable Precious Metal Deposits, Guidebook For Field Trips, Johnson, J.L., and Abbott, E. (eds), pp. 269-270.

5

Mining I
Open Pit

D.L. Simpson
Echo Bay Mines, Ltd.
Reno, Nevada

R.J. Robinson
Ashton Mining Ltd.
Melbourne, Vic., Australia

Chapter 16

OPEN PIT PLANNING AND SELECTIVE MINING PRACTICES AT THE NEW CELEBRATION GOLD MINE

Barrie W. Sullivan

Resident Manager
New Celebration Gold Mine

Abstract. The New Celebration Gold Mine near Kalgoorlie in Western Australia was commissioned in late 1986 and the CIL treatment plant currently processes 1.7 million tonnes of gold ore annually.

The 1988 geological resource was 13.2 million tonnes grading 2.37 g/t Au in several open pittable deposits with the bulk of gold mineralisation occurring as a stockwork in Archaean quartz feldspar porphyry intrusives along a major regional shear zone.

Mine planning, scheduling and operating techniques have focused on minimising open pit dilution whilst maintaining production requirements.

This paper describes the successful achievement of this aim in extremely variable ground conditions and orebody widths.

OVERVIEW

The New Celebration Gold Mine is located 32 km south of the regional mining centre of Kalgoorlie in the eastern goldfields area of Western Australia (see Figure 1). New Celebration draws its name from the Celebration Mine which along with many other small underground and surface mines on the freehold block known as East Location 50 was worked on an intermittent basis from the early 1900's to the 1950's.

The current New Celebration Joint Venture, formed in 1981 between Newmont Australia Limited (60% and manager), Hampton Australia Limited (25%) and Mt Martin Gold Mines NL (15%) holds the East Location 50 block and other adjacent tenements which cover an area of approximately 5,000 hectares.

After sufficient exploration and resource definition drilling had been completed up to 1984, project feasibility studies commenced with final approval for construction of a 500,000 tonnes per annum carbon in leach (CIL) treatment plant announced by the Joint Venture partners in January 1986. Project construction commenced in May 1986 with commissioning and design production tonnage achieved in December 1986.

In December 1987 the Joint Venture partners announced final approval of an expansion of the CIL treatment facilities to 1.6 million tonnes per annum.

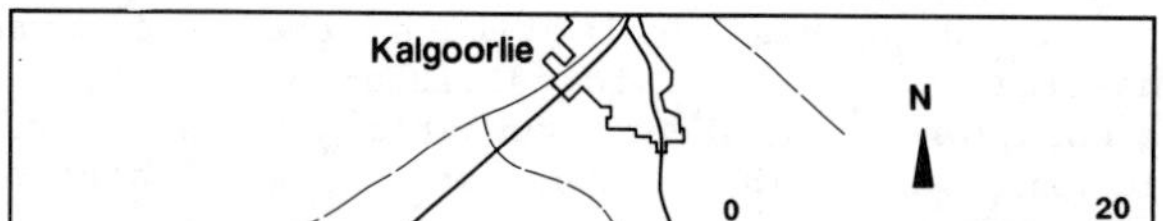

FIGURE 1. LOCATION PLAN

Commissioning and design tonnage of the plant expansion were achieved in August 1988 and the current mill throughput is 1.7 million tonnes per annum with an annual gold production of 120,000 ounces.

Open pit mining operations commenced in July 1986 to initially feed the 500,000 tonnes per annum mill using a mining contractor, and subsequently operations were increased in July 1988 to cater for the mill expansion. With a projected resource base and possible mine life in excess of ten years, a decision was made to go to owner mining in October 1988, to lower overall operating costs and tighten mining control.

The mining and treatment operations at New Celebration currently directly employs 200 personnel, 30 of whom are contractors. All employees live in either of the two nearby towns of Kalgoorlie-Boulder and Kambalda, and many of the mine's support and service facilities are drawn from these communities.

THE RESOURCE

The New Celebration Gold Mine tenements cover approximately 5,000 hectares of Archaean greenstone stratigraphy straddling a major regional geologic structure, the Boulder Lefroy Shear Zone (see Figure 2). This shear hosts three of the main New Celebration ore deposits and on a regional basis links many of the major gold deposits in the eastern goldfields region, including the famous Kalgoorlie Golden mile. Ten km of lateral movement on this shear has resulted in a zone of intense shearing up to 100 m wide striking 340°. At New Celebration this shear truncates a steep north plunging anticline known as the Celebration Anticline.

The bulk of the 1988 mineral resource of 13.2 million tonnes at 2.37 g/t Au is contained in several open pittable deposits associated with the Boulder Lefroy Shear Zone.

Mine Geology

Gold mineralisation is predominantly associated with porphyry intrusive rocks along the main Boulder Lefroy Shear Zone, and occurs as a stockwork mainly in the porphyry units. Four main orebodies make up the bulk of the mineral resource and these are known as Hampton Boulder, Mutooroo, Celebration and Pernatty (see Figure 2).

Typical of the shear related porphyry mineralisation is the Hampton Boulder porphyry body which is continuous over 600 m strike along the shear zone and can be up to 80 m thick. This body dips steeply to the west and divides a dominantly mafic hangingwall sequence from ultramafic rocks with numerous thinner felsic to intermediate porphyry bodies in the footwall (see Figure 3). The hangingwall sequence is dominated by chlorite schists derived from shear deformation of gabbroic rocks. Foliated talc carbonate ultramafic rocks form the footwall sequence. Mineralisation of economic significance is developed in several positions within the mine sequence, forming a stockwork of micro-veinlets particularly in the hangingwall and footwall endocontact positions of the Hampton Boulder porphyry. As a consequence, the orebody is one of a series of west dipping ore zones, varying in width from 2 to 30 m, in which steeply plunging higher grade ore shoots are developed. Pyrite is the dominant sulphide material occurring in fine veinlets and disseminations. The gold is spatially associated with it in clusters of fine particles less than 50 micron in size, hence is rarely seen in hand specimen.

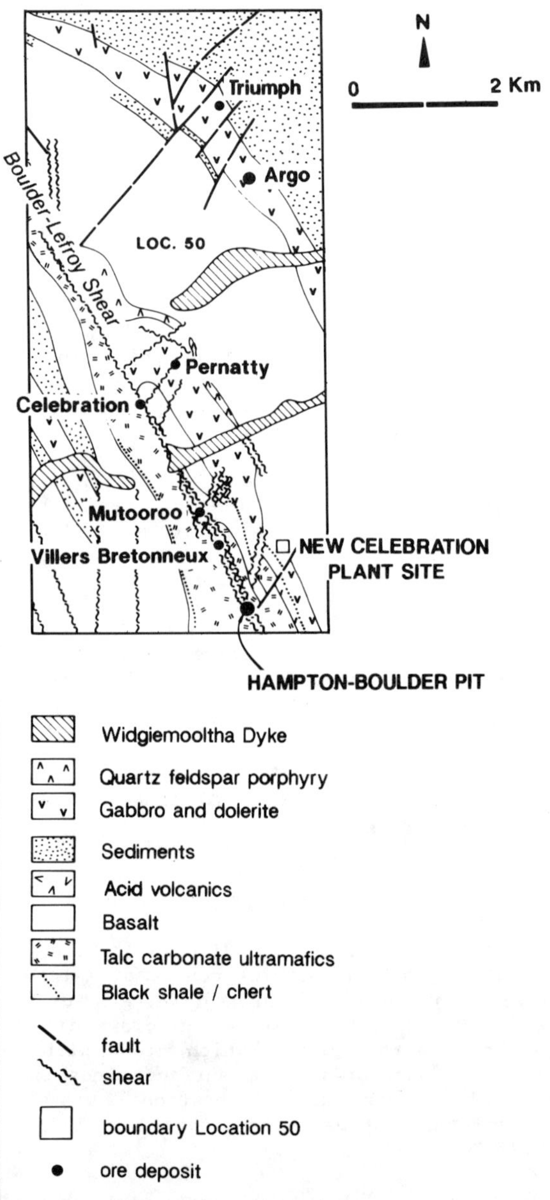

FIGURE 2. REGIONAL GEOLOGY

Depth of weathering varies considerably along strike of the main shear zone and can range from almost 0 m with the fresh porphyry outcrop at Hampton Boulder to some 40 or 50 m in the Mutooroo deposit. It is this depth of weathering combined with the horizontal width of mineralisation which determines the grade control and selective mining methods to be used. An example of the depth of weathering and variability of ore zone widths is the Mutooroo deposit where up to 32 ore zones of widths between 1 and 30 m can occur within a single pit outline (see Figure 4).

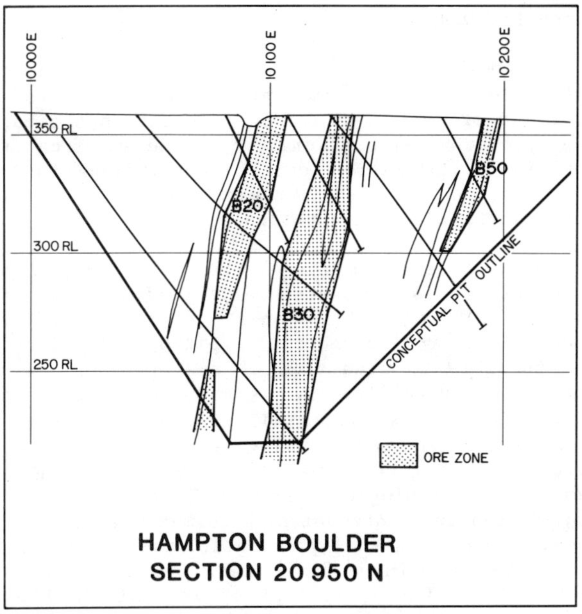

FIGURE 3. HAMPTON BOULDER DEPOSIT
TYPICAL VERTICAL CROSS SECTION

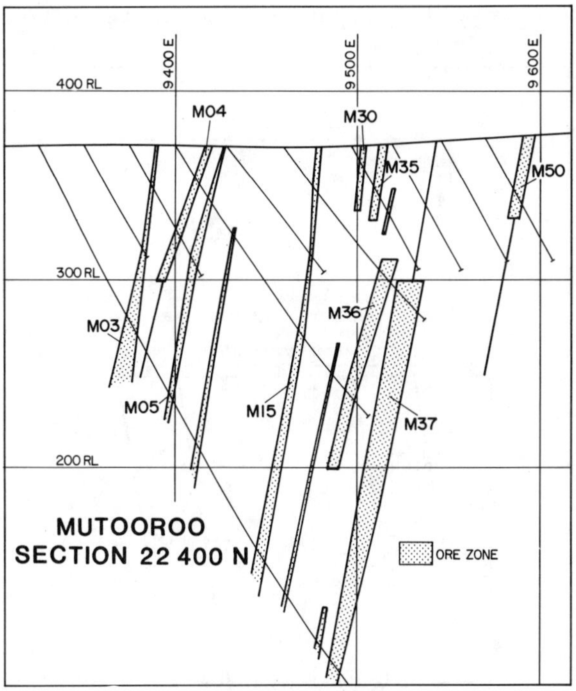

FIGURE 4. MUTOOROO DEPOSIT
TYPICAL VERTICAL CROSS SECTION

Orebody Definition

Exploration drilling to bring orebody definition confidence into the measured and indicated category is dependent on continuity but is usually done on 50 m or 25 m grids along strike with 25 m spaced 60° inclined reverse circulation (RC) and diamond drillholes across strike. Normal practice is to use RC drilling to water table depth and then continue to end of hole with diamond drilling. Traditionally drilling grids are placed on north-south AMG orientations except in one instance on the Pernatty deposit where a splay shear is oblique to the main shear zone.

Orebody Modelling

A geological resource is calculated from the drilling information after hole logging, diamond saw splitting and fire assay using gravimetric weighing.

All logging and assay information is compiled and stored using DATAMINE software on DEC computer hardware. Several custom and tailored DATAMINE packages allow data entry, compilation, storage and then sectional plotting of all drillhole information onto worksheets for geological interpretation and modelling.

Geological modelling is done manually and usually on section. After the ore zones have been interpreted geologically for structural and lithological continuity, ore outlines are drawn. Grade intersections have a universal cut of 30 g/t Au applied.

Based on the nature, width and weathering profile of the ore zones, the pure geological model in cross section now has mining dilution applied. For ore zones in the softer free digging material, a minimum mining width of 3 m of ore or waste is assumed with 1 m of downhole dilution on each of the footwall and hangingwall contacts (see Fig 5). For the ore zones in drill and blast material (generally below depth of weathering) a minimum mining width in ore or waste of 5 m is assumed with 1 m of downhole dilution on each of the hangingwall and footwall contacts (see Fig. 5). The dilution applied is the downhole assay in the area of dilution.

With all sectional information complete and continuity between sections established, orebody modelling and resource calculation using DATAMINE software is undertaken. In the case of New Celebration the DATAMINE model is built up on 5 m by 5 m by ore width blocks, using inverse distance squared for area of grade influence.

From this model a diluted mineral resource is produced which is either measured, indicated or inferred depending on drilling density and model confidence. This resource is then available for further evaluation to take it possibly into the mineable reserve category.

Ongoing Drilling

As deposits are mined, drilling for resource definition continues. This is usually done below or outside the boundaries of the existing pit designs where confidence in the ore zone needs to be increased. Resource definition drilling outside the existing pit boundaries is normally on a 25 m by 25 m spacing 60° incline drill pattern. Infill pit drilling for mine planning purposes may close up to 12.5 m square grid pattern.

This information is used to either upgrade

resource confidence or to increase the existing resource base.

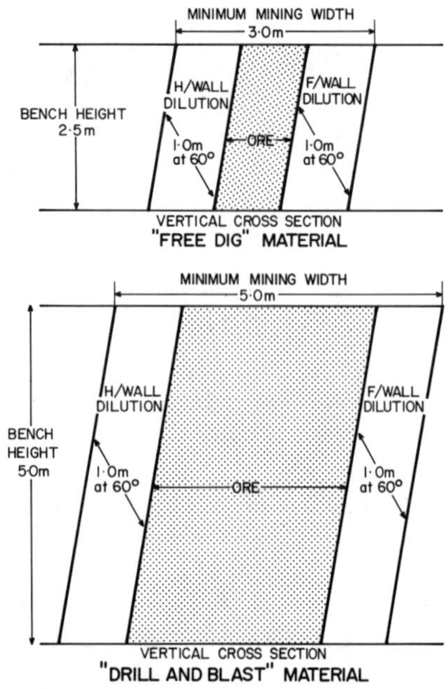

FIGURE 5. TYPICAL MINING WIDTHS FOR OREBODY MODELLING

MINEABLE RESERVES AND PLANNING

With the completion of a diluted mineral resource, the next step is to apply mining, metallurgical and economic parameters to the resource for determination of a mineable reserve. This mineable reserve forms the basis of all scheduling and production studies leading to operating and production budgets and cost information.

Reserve Calculation

The diluted mineral resource is presented in DATAMINE model form. To this model are added the following parameters:

- Ore and waste mining costs
- Milling costs
- Metallurgical recovery
- Gold price
- Pit wall slopes

As part of the DATAMINE modelling, a LERCHS-GROSSMAN (LG) pit optimisation software package is used to generate the mineable reserve and economic pit outline using a cone of influence in 3 dimensions. Ramp access and berms are added as fine tuning of the final pit outline and resultant mineable reserve with waste strip ratio. The mineable reserve is available in varying grade increments which allows flexibility for modelling at different cut off grades according to particular economic circumstances.

Open Pit Design

An important part of the LG optimisation is the parameters used for open pit design. Pit designs for the ore deposits at New Celebration are based on geotechnical information gathered from the original exploration drilling and then further information gathered from either ongoing drilling or practical mining experience.

The ability to stage all of the open pit operations is an extremely useful tool for designing final pit slopes. As part of the scheduling process and the LG pit optimisations it is possible to plan each of the open pits to be mined in several stages, encompassing a series of pushbacks to the final wall design. This staged approach to pit design allows some experimentation with wall slopes, haul roads, berm spacing and smooth wall mining techniques. There is no better way of testing geotechnical information, theory and design than by actually doing it, without the risk of a failure in a final wall which could have severe long term practical or economic problems. Staging of pit designs allows this to happen before commitment to a final design.

Mine Scheduling

Sequencing of ore deposit mining is greatly aided with the DATAMINE software and LG optimisations.

Mine scheduling is largely dependent on the type and grade of material available from each of the deposits. Of the 1988 mineable reserve of 8.5 million tonnes grading 2.38 g/t Au, over 60% of the ore is fresh porphyry rock from the Hampton Boulder deposit. The remaining 40% of material is essentially highly oxidised, high clay content material with varying milling characteristics and metallurgical recovery. From a process control aspect it is desirable to blend the fresh porphyry and high clay content material as much as possible. Grade balance is also an important objective and this requires careful scheduling from each deposit.

Apart from the staged pit approach to design and scheduling, 3D Linear Optimisation scheduling software has been used to optimise mining areas and cash flow.

Mine planning and scheduling is updated on a long term basis annually, when update of the geological resource and other critical parameters such as gold price, operating costs and hence cut off and breakeven grades are reviewed. This annual mine plan takes in the life of the mineable reserves, as updated and at a minimum is done for 5 years of the mine life.

The first year of the 5 year plan is scheduled on a calendar month basis and the second year quarterly, with annual scheduling of the remaining 3 years.

Short term planning looks at basically the next year of mining and breaks this into twelve calendar months. Each month, a monthly mining plan is produced, in detail for the first month and projecting into the second month. Re-forecasting of the annual budget plan is normally done at least once, but this is largely dictated by the degree of change of the fundamental parameters used in the annual budget plan.

GRADE CONTROL

Following on from short term scheduling is production scheduling which involves weekly and daily review of the mine operations, particularly in relation to grade control and selective mining of the orebodies. With a mineable reserve grade of 2.38 g/t Au and an overall strip ratio of 6 tonnes of waste to 1 tonne of ore, there is very little room for error in attempting to achieve the budget mill feed grade. Economics tells us that it costs the same to process a tonne of waste as it does a tonne of ore, so we are constantly aware of the need to minimise dilution through selective mining practices and tight grade control.

Sampling

As discussed previously the ore zones throughout the New Celebration deposits present themselves in extremely variable width and ground conditions. It is possible to generally identify two basic methods of grade control that have been adopted for New Celebration in the softer, "free dig" material (highly weathered rock) or the harder "drill and blast" material (fresh rock below weathering horizon).

"Free Dig" Material This material is typical in the Mutooroo deposit where sub vertical narrow highly weathered mineralised porphyry zones can be freely dug by hydraulic excavator to depths of 50 m below surface. Because of the narrowness and relative discontinuity along strike and down dip of these ore zones, grade control is carried out to allow for 2.5 m high mining benches. Ore zone positions are approximately determined from the initial resource drilling and the bench above. This information forms the basis of the grade control sampling grid laid out along the mining bench. The method of sampling is to use a mobile rubber tyred trenching machine, known as a Ditchwitch which cuts 200 mm wide slots up to 1 m deep into the ore zone, cutting across strike. A typical slot spacing would be 6.25 m along strike (see Figure 6). The material brought to surface by the Ditchwitch and deposited as a neat windrow beside the slot is then sampled in 1 m measured intervals across strike allowing for the cutting displacement of the material. Mapping and logging of this slot is also possible to ascertain structure and lithology.

"Drill and Blast" Material The subvertical fresh porphyry ore zones in the Hampton Boulder deposit typically represent this material. The geometry of the ore zones and the need to drill and blast the fresh rock established the ore mining bench height at 5 m. Blasthole drill patterns are laid out on a 2 m by 2 m offset pattern using outlines for the ore zone positions as indicated by initial resource drilling and the bench above. The blasthole pattern across strike is placed to maximise information at the estimated ore footwall and hangingwall contacts (see Figure 7). Nominal 75 mm diameter vertical percussion blastholes are drilled on this pattern to a depth of 5.5 m. Sampling of the blasthole cuttings is done only on the first 5 m as a composite. In dry ground sample collection is by way of dust collector cyclone fixed directly onto the drill rig which dumps the 5 m composite sample directly into a specifically designed 4 stage splitter mounted beneath the cyclone. This sample is nominally 4.5 kg in weight and is immediately bagged and labelled by sampling personnel stationed at the rigs. In wet ground where dry cyclone dust collection is not possible, a "pie plate" sampling device is used which basically collects the damp chips as they are deposited around the collar of the blasthole during drilling. This "pie plate" takes a representative 4.5 kg wedge of cuttings from the first 5 m composite of the blasthole, and is bagged and labelled by sampling personnel.

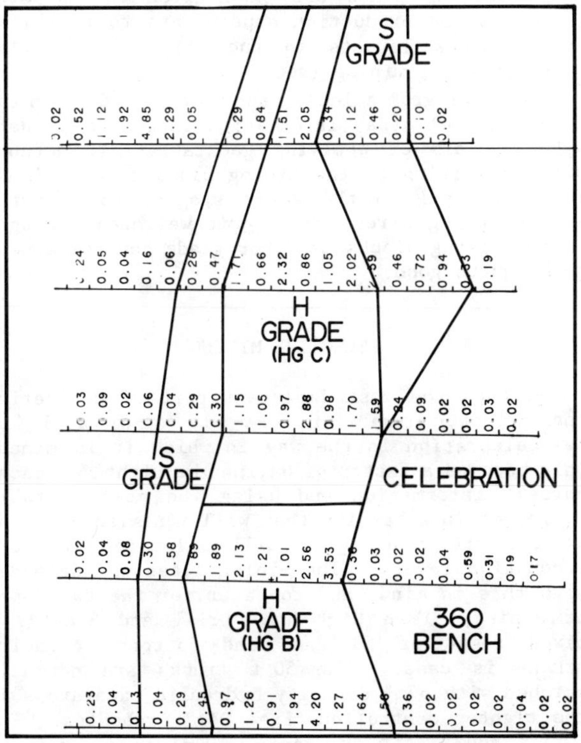

FIGURE 6. TYPICAL TRENCH SAMPLING PLAN

Various other methods of sample collection including a "rock saw" trenching machine and close spaced shallow RC drilling have been trialled, however the three basic methods outlined above have proved to be the most cost efficient and reliable. Total grade control costs for 1988 were $A0.82 per ore tonne mined or $A12.30 per ounce mined.

Assaying

All grade control samples are collected and sent by road daily to a contract assay laboratory in Kalgoorlie. Between 250 to 500 samples are sent daily for preparation and fire assay with gravimetric weighing. Average turnaround time is 18 to 24 hours. Check assaying is an ongoing process both at the contract laboratory and also with check samples going to other laboratories.

Data Collection and Compilation

Results from the grade control assays are sent from the contract laboratory via data link directly to the on site computer system at New Celebration. With the aid of some customised DATAMINE software, this assay information is then compiled and matched to survey information and a hardcopy plot produced for either the Ditchwitch slots or the blastholes (see Figures 6 and 7).

The grade control assay plans are then contoured by assay grade manually and this information then forms the basis for the production of daily cutting plans for mining control. The grade control cutting plan is designed to use the assay information, contours and geology to delineate practical mining blocks of ore and waste material. These cutting plans are then given to the mine surveyors and production supervisors to establish mining block layouts in the pit and effective guidelines for mining (see Figures 6 and 7).

Semivariogram calculations have been done on the grade control assay information generated to date with the aim of applying geostatistical methods such as Kriging to the mining block calculations. This is still in the early stages, however the results using direct arithmetic weighted averages on the mining blocks used for grade control appear to be quite good.

SELECTIVE MINING

Perhaps the most important part of delivering the right grade of ore to the mill at New Celebration is the way in which it is mined. This becomes a factor of having the right approach, correct information and using the most suitable equipment in a fashion that will minimise dilution of the ore zones yet still maintain practical production levels. The mining fleet was selected with this in mind, and to cater for the two types of mining, "Free Dig" and "Drill and Blast", a mixed fleet of 50 tonne and 85 tonne capacity trucks is used. The 50 t trucks are normally matched with 6 m³ capacity hydraulic excavators in the tighter controlled "Free Dig" orebodies. For "Drill and Blast" ore zones, 10 m³ capacity front end loaders can also be used matched with the 85 t or 50 t capacity haultrucks. Supervision plays a key role in all aspects of this process, and part of the justification to switch from contract to owner mining was to improve supervision and control.

"Free Dig" Material

Although the majority of the ore zones in this material are sub vertical in dip, this dip consistently tends to the west, hence dilution can be minimised by mining with the dip, ie working from west hangingwall to east footwall, across strike wherever possible. From the Ditchwitch slot information and the subsequent grade control cutting plan, the mine surveyor physically lays out the ore zone outlines on the floor of the mine bench using lime or flagging tape and pegs as markers.

Under geological supervision a 2.5 m high bench is mined with hydraulic excavator, normally working from hangingwall to footwall. Extreme care is taken when clearing to hangingwall or footwall contacts. Due to the practical consideration of bucket size and production requirements, the minimum width of ore or intervening waste that is flagged for mining is 3 m. Where structure and exposure allows, narrower zones have been mined. Bench level control is critical during mining and at the completion of a bench, the floor is carefully surveyed and smoothed with a road grader in preparation for the next lift.

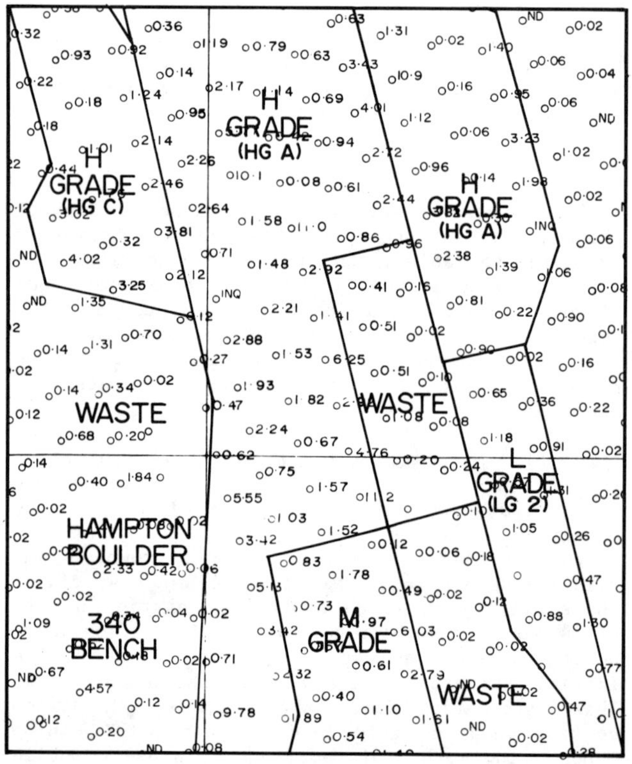

FIGURE 7. TYPICAL BLASTHOLE PLAN

"Drill and Blast" Material

Over 60% of the total material mined at New Celebration falls in this category, mainly from the Hampton Boulder pit. With the westerly dipping sub vertical ore zones, minimisation of dilution over 5 m high benches working from the west hangingwall to the east footwall is essential. The hardness of the fresh rock, particularly the porphyry with a compressive strength of up to 200 MPa means that blasting must be able to move and break the rock to ensure adequate fragmentation and diggability. This indicates that blasting must be done selectively without mixing ore and waste, particularly where powder factors of around 0.7 kg/m³ are used.

Once a blast has been designed using the grade control cutting plan, mining blocks are then charged and fired according to the mining direction and block outline. Wherever possible shots are fired to a free face to assist fragmentation and minimise backbreak. Open holes adjacent to blasting blocks are protected from collapse by the insertion of a 1 m long polyethelene pipe into the hole collar. Particularly on ore/waste contacts, each shot is cleaned out completely to a free face before further firing. In the case of the schist/porphyry contacts, the bench face is actually scraped down with loader or excavator to minimise dilution.

Because of the blasthole pattern size of 2 m offset, a minimum mining width of 5 m ore or waste is used. The practical requirements of production rates are also considered when designing blasts, hence the number of holes fired must provide sufficient broken material to match the scale of operations.

Cleanup and floor level control are considered essential and monitored continuously by the supervisors with the aid of laser levels. Normal tolerance is ±0.2 m.

Where there is a transition zone between free dig and drill and blast material, "freeze" or "paddock" blasting has been used with good results. This method involves firing a large area of ore and waste using a very light powder factor (0.3 kg/m³) to shake up the material with very little displacement. Ore zones can then be flagged on the blasted ground using survey co-ordinates, and then mined selectively.

Supervision and Stockpiling

For the grade control system to work effectively, good supervision and stockpile operation control is essential. All ore digging locations are constantly supervised by the mine geologist assigned to each particular pit, who directly liaises with the mine production supervisor and equipment operators. Two way UHF radios are fitted to all heavy and light vehicles in the mine, and are used extensively for communication.

Ore is selected and stockpiled according to grade, material type and metallurgical characteristics. Mill feed ore, typically >1.0 g/t Au is despatched to the run of mine (ROM) stockpile area, where ore is stockpiled in a series of parallel fingers (see Figure 8). These fingers are 6 m high and 25 m wide and are formed by truck tipping out from a preformed ramp. Each finger will contain some 30,000 tonnes of ore when full. These finger stockpiles allow similar ore types to be stockpiled together and precise control of the dumping and subsequent reclaim of ore from many different areas. A 10 m³ capacity front end loader continuously reclaims from these finger stockpiles according to a recipe which will supply a balanced material blend and correct feed grade to the mill. The composition of each finger stockpile is carefully recorded and monitored as it is dumped and then reclaimed. This then forms the essential information required for reconciliation of grade and tonnes back from mill gold recovered through to the stockpiles and initial reserves.

Mineralised material below 1 g/t Au is stockpiled separately away from the ROM stockpile area in grade ranges of 0.5 to 1.0 g/t Au, 0.2 to 0.5 g/t Au and waste less than 0.2 g/t Au.

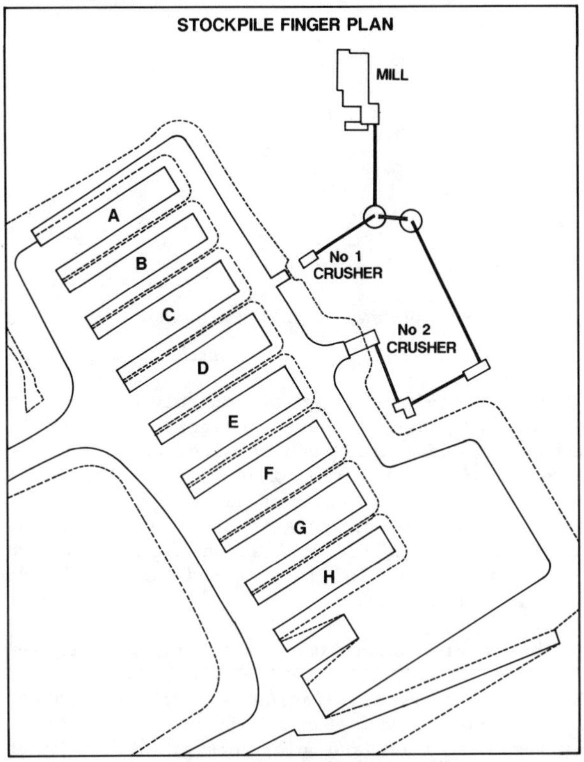

FIGURE 8. STOCKPILE FINGER PLAN

RECONCILIATION

As with any mining grade control system, at the end of the day the "proof is always in the pudding" and this directly relates to the amount of production achieved through the mill versus that which was calculated to be fed. Any system requires continual monitoring and checking, and the method of reconciliation used at New Celebration can be summarised as:

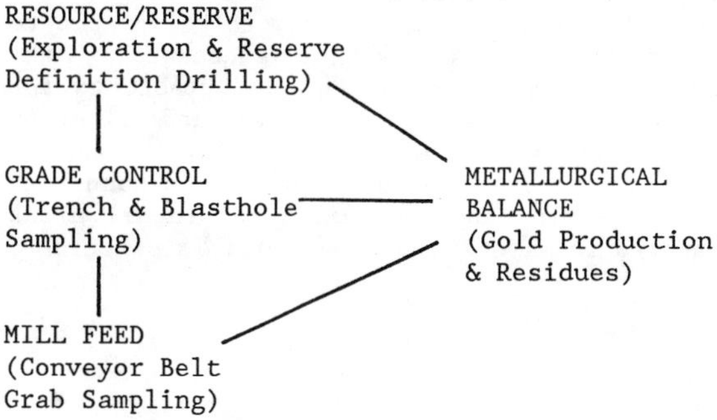

The relationship between each of the above areas in an ideal world would be simple, as each should give the same answer. However this is unlikely to be the case, and in fact the relationships can be extremely complex and varied. The answer to a

Table No 1

RECONCILIATION RESULTS

Period	Diluted Geological Reserve			Grade Control			Metallurgical Balance		
	Tonnes	g/t	Au (ozs)	Tonnes	g/t	Au (ozs)	Tonnes	g/t	Au(ozs)
1/8/86 to 31/12/87	681,767	2.68	58,744	650,253	2.76	57,700	650,253	2.61	54,565
1/1/88 to 30/6/88	358,761	2.68	30,912	353,903	2.77	31,518	353,903	2.72	30,949
1/7/88 to 31/12/88	690,064	2.51	55,598	683,027	2.46	54,021	683,027	2.41	52,753
TOTAL	1,730,592	2.61	145,254	1,687,183	2.64	143,239	1,687,183	2.55	138,267

successful mining and grade control system is to constantly update, measure and monitor the results of each of these areas and to alter or factor them as required. The basis of all reconciliation at New Celebration is the metallurgical balance to which all the other relationships are measured. Table No 1 shows reconciliation figures for the period since ore mining commenced in August 1986 through to December 31, 1988.

This table summarises the mining for all four pits and in detailed analysis each deposit has shown variable reconciliation performance according to numerous circumstances including depth of surface depletion and old underground workings. The overall result shows a slight overestimation in grade and tonnes for the initial reserve calculation and an approximate 3% overestimation in grade for grade control methods. This is a pleasing result and supports the reserve estimation, grade control and selective mining methods used at New Celebration.

CONCLUSION

The New Celebration Gold Mine has a gold resource base which could be considered medium to lower grade, in Australian terms, for a mainly hard rock open pit gold mine. The key to being able to mine and treat this resource successfully and economically has largely been attributable to the careful planning, tight grade control and selective mining practices coupled with the initiatives of economy of scale through process expansion. Constant checking and reconciliation in all areas is required to maintain and monitor this position.

ACKNOWLEDGEMENT

The author would like to acknowledge the Management of Newmont Australia Limited and the staff of the New Celebration Gold Mine for valuable assistance and support with preparation of this paper.

Chapter 17

LARGE SCALE OPEN PIT GOLD MINING AT GOLD QUARRY

Scott Santti
Ali Soltani

Newmont Gold Company

ABSTRACT

The Gold Quarry Mine is a large scale open pit mine located 11.3 km (7 miles) north of Carlin, Nevada. The operation consists of an open pit mine, two milling facilities, a leach crushing plant, and three dump leach pads. Gold Quarry is operated by Newmont Gold Company, which is 90.1% owned by Newmont Mining Corporation, a Denver based company. The property is the largest of the three gold producing properties operated by Newmont Gold Company on the famous Carlin Trend.

This paper will give a general overview of the Gold Quarry mining facilities and detail mine operating practices. Specifics on cost saving ideas being implemented and those under investigation will be discussed.

INTRODUCTION

Newmont Gold Company operates three operations management areas as determined by geographic location (Figure 1). All of them are located in northeastern Nevada along what is known geologically as the Carlin Trend. The North Area consists of three active mines, two milling facilities, and a leach crushing plant and pad. The Rain Area includes a mine, mill and leach pad. The Gold Quarry Mine is part of the largest management area, the South Area. In addition to the mine, two milling facilities, a leach crushing plant and three leach pads are operated here. The Gold Quarry pit currently produces 60% of the total material mined and 62% of the gold recovered by Newmont Gold Company.

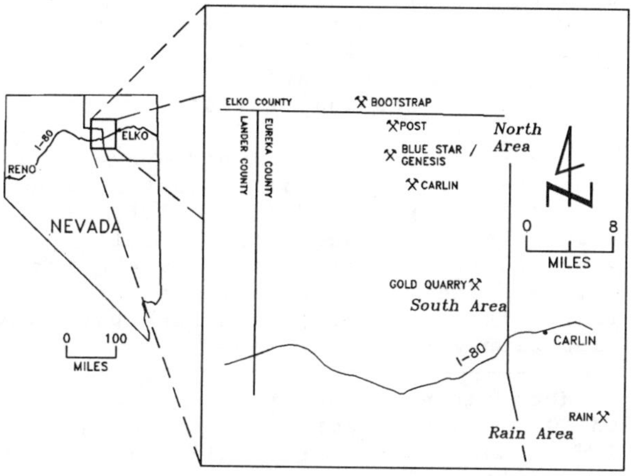

Figure 1. Newmont Gold Company

HISTORY

The Gold Quarry mine encompasses a sizeable portion of what is known as the Maggie Creek Mining District. Newmont Mining Corporation began mapping and sampling in the vicinity of what is now the present northern limit of Gold Quarry around 1960. At this time drilling was centered around and toward the southwest of a jasperoid outcrop where previous mining had taken place. The Maggie Creek leases were acquired in 1977 and an expanded exploration drilling program was started in early 1978. Exploration drilling continued into 1979 with good results. Evaluation of this drilling during 1979 resulted in the Maggie Creek deposit with a measured and indicated mineable reserve of 4.35 million tonne (4.8 million st) at 3.19 g/tonne (0.093 oz/st). Pre-stripping began in late 1979 from the Maggie Creek pit. A 2,270 tonne/day (2500 STPD) leach crushing plant and a 68.6 m x 253 m

(225 ft. x 830 ft.) asphalt heap leach pad were constructed beginning in July of 1980 and the first gold was poured in early 1981.

After 1979, exploration continued to the east of the Maggie Creek pit toward the present location of the Gold Quarry Deposit. When significant gold values were encountered in this area, the drilling program was accelerated to define the extent of mineralization. Between 1979 and 1983, 550 conventional circulation holes were drilled to an average depth of 183 m (600 ft.) on a 30.5 m (100 ft.) square grid. Bulk metallurgical test samples of mill grade and leach grade material were mined in 1982-83. As a result of the exploration drilling and metallurgical test work, a mineable reserve of 130.6 million tonnes (144.0 million st) at 1.68 g/tonne (0.049 oz/st) was estimated at the end of 1983.

Pre-stripping of the Gold Quarry Mine began in late 1983. Construction of a 6,350 tonne/day (7,000 STPD) mill, a dump-leach pad and support facilities were started in February of 1984. The mill was dedicated in September of 1985. The total investment for this expansion was $130 million.

Much expansion has taken place at the Gold Quarry facilities since the dedication of the first mill (Mill No. 2) in 1985. A chronology of the highlights is given below, most notable of which are the huge production increases in 1987-1988.

1986	The No. 2 Mill was converted from autogenous to semi-autogenous grinding to increase its capacity to 8,160 tonnes/day (9,000 STPD).
1987	The Maggie Creek leach facility was decommissioned.
1987	A gyratory primary crusher was commissioned to process the leach material with a rated capacity of 10.9 million tonne/year (12.0 million stpy).
1988	A secondary crushing facility for leach grade material was constructed and placed into service.
1987-88	The mining rate at Gold Quarry was increased from 72,560 tonnes/day (80,000 STPD) to 163,260 tonnes/day (180,000 STPD).
1988	A second leach pad was constructed to replace the original leach pad and a third was built to process leach material with a different royalty rate.
1988	19,614 Kg (630,000 oz.) of gold was produced at Gold Quarry operations.
Early 1989	The second mill facility (No. 5 Mill) was completed with a design capacity of 14,510 tonnes (16,000 st) per day. This brought the total nominal milling capacity to 23,580 tonnes (26,000 st) per day.
1989	Annual production from Gold Quarry is forecasted to be 26,905 Kg (865,000 oz.) of gold.

GEOLOGY

The Gold Quarry Deposit can be described as a sub-microscopic, sediment-hosted, epithermal, disseminated gold deposit. Paleozoic siltstone, shale, sandstone, silty limestone and chert host the gold. These sediments have been highly silicified and argillized, especially along the numerous high angle fault structures located in the deposit.

The ore is concentrated in two distinct zones - the Main zone and the Deep West zone. The Main zone overlies the Deep West and is the zone currently being mined. It strikes in a northeasterly direction and dips toward the east at 45° to 50°. Structurally controlled mineralization is continuous over 1,829 m (6,000 ft.) of strike. The majority of the Main zone is oxidized with only 10% of the mill grade material being carbonaceous-sulfidic refractory.

The Deep West strikes approximately at N20°E and dips 30° to 35° to the southeast. Its average thickness is 76.2 m (250 feet). Mineralization is known to extend for a distance of 762 m (2,500 ft.) along strike, but is open to the south, east, and at depth. This zone is a silicified replacement ore body with 30% to 50% of the material being of a carbonaceous-sulfidic nature. The grade of the Deep West is twice that of the Main ore zone.

The majority of the bedrock at Gold Quarry is covered by Tertiary material known as the Carlin Formation. The Carlin Formation (generally called alluvium) consists of tan-white to pink lacustrine claystone and reworked tuff, bentonitic clay, and calcareous silt. It ranges from 0 to 91.4 m (300 ft.) in thickness in the pit walls. The abundant montmorillonite and bentonite clays in the Carlin Formation promote unstable pit walls and difficult mining, especially during inclement weather.

MINEABLE RESERVES

The geologic resource at the Gold Quarry mine is estimated based on exploration drill hole information. Both a fire and a cyanide amenable assay determination was performed on a majority of the 1.5 m (5 ft.) samples collected. The assays were then composited in 6 m (20 ft.) intervals to reflect production bench heights. The gold in inventory in the deposit was calculated using ordinary kriging with a block recovery function to reflect the effects of mine selectivity. The Deep West and Main ore zones required different kriging parameters to account for the different geologically controlled mineralization. The Main ore zone model was compared to eleven benches of historical blast hole assays to verify its accuracy. No production data is available for the Deep West to verify the model for this zone. The model for the entire deposit includes an oxide-carbonaceous boundary. This outline is based on examination of exploration cuttings, core, and AA to fire assay ratios by the geologists. This surface is used to estimate the amount of refractory material. A bedrock-alluvium boundary surface was also used to prevent the extrapolation of gold values into the overburden. Future geologic models will attempt to predict sulfide refractory ore in addition to the carbonaceous refractory ore.

The mineable reserve was estimated utilizing a computerized floating cone algorithm to produce economic pit limits. Input parameters included: gold price, mining costs, leaching costs, oxide and refractory ore milling costs, milling and leach pad recoveries, royalty payments, refining costs and capital replacement costs. Estimates of overall slope angles which include allowance for haul roads and material density were also inputs to the computer program. The final engineered pit design, using the floating cone output as a guide, incorporates 45° inter-ramp slopes in bedrock, 35° slopes in the alluvium, and 30.5 m (100 ft.) wide haul roads with grades not greater than 8%. The measured and indicated mineable reserves using this technique for Gold Quarry at the end of 1988 were 223.2 million tonnes (246.1 million st) at 1.44 g/tonne (0.042 oz/st) or 320,365 Kg (10.3 million oz.) gold. The total material in the ultimate pit was estimated at 522.7 million tonnes (576.3 million st) which reflects a 1.3 to 1 waste to ore ratio. The final pit measures approximately 1.6 Km (1 mile) x 1.2 Km (3/4 mile) in size and 366 m (1,200 ft.) in depth.

MINE PLANNING

The primary tools used by the mine engineer for pit scheduling are powerful 3-D color graphics workstations. The workstation environment allows the engineer to view and design their mine plans interactively on a high resolution monitor. All the mine planning software for the workstation was developed in house by Newmont personnel. Improvements to the workstations and the software are continually being added to speed up the planning process.

Life of mine annual pit plans and monthly plans for the first year are produced annually for Gold Quarry. These plans are continuously being refined to produce the most economic schedule. Some of the parameters used to produce schedules for the pit include: (1) the layback width required - ranging from 36.5 m (120 ft.) to 91.4 m (300 ft.) in width; (2) inter-ramp slope angles required as determined by the geotechnical conditions; (3) a continuous supply of mill and leach ore; (4) royalty requirements; (5) bench access and road design for short and long term; (6) maintaining sufficient working areas for the seven loading units required to meet production; and (7) the blending of durable and soft ore to the mills. Once the plans are completed, they are reviewed with mine operations personnel for approval. If no problems exist, the plan is implemented.

In addition to pit scheduling, the Mine Planning group assists operations by designing and sequencing waste dumps and leach pads, producing stockpile plans, determining future dewatering requirements, calculating cutoff grades, analyzing equipment performance, reconciliations of predicted tons and ounces with actual, and performing ore control activities.

MINE OPERATIONS

The Gold Quarry mine is budgeted to produce 186,000 tonnes/day (205,000 STPD) of total material for an annual total of 67.8 million tonnes (74.8 million st) in 1989. This compares to 51.6 million tonnes (56.9 million st) produced in 1988. A complete comparison is given below:

GOLD QUARRY MINE PRODUCTION
(In Millions)

	1988 Actual Tonnes (st)	1989 Budget Tonnes (st)
Mill Ore	3.6 (4.0)	9.1 (10.0)
Leach Ore	16.3 (18.0)	19.6 (21.6)
Waste	31.7 (34.9)	39.1 (43.2)
Total Material	51.6 (56.9)	67.8 (74.8)
Waste to Ore Ratio	1.6:1	1.4:1

To achieve this production, the operation is scheduled three shifts per day, seven days per week. A total of 300 hourly and salaried people are employed in the Mine Operations group.

A list of the mining equipment at Gold Quarry is given in Table I.

TABLE I
MINE EQUIPMENT LIST
GOLD QUARRY MINE (1989)

2	DeMag H-285 Hydraulic Shovels
1	DeMag H-185 Hydraulic Shovel
3	LeTourneau L-1100 Front End Loaders
1	LeTourneau L-1000 Front End Loader
5	Caterpillar 992-C Front End Loaders
3	Driltech 60K Rotary Drills Track Mounted
4	Ingersoll-Rand DM-M Rotary Drills Track Mounted
1	Reedrill SK60 Rotary Drill Track Mounted
18	Dresser Wabco 510E Haulpak Trucks Electric, 120 ST

TABLE I (CONT'D)
MINE EQUIPMENT LIST
GOLD QUARRY MINE (1989)

5	Dresser Wabco 510E Haulpak Trucks Electric, 140 ST
15	Caterpillar 785 Trucks Mechanical
2	Caterpillar D10N Dozer
2	Caterpillar D9L Dozers
1	Caterpillar D8H Dozer
1	Komatsu D375A-1 Dozer
2	Caterpillar 834C Rubber Tire Dozers
2	Caterpillar 16G Motor Graders
1	Caterpillar 14G Motor Grader
5	Water Trucks: 3 - 16,000 gal. 1 - 11,000 gal. 1 - 4,000 gal.
1	Caterpillar 623B Scraper
1	Ingersoll-Rand SP150 Compactor

DRILLING

Bench heights at Gold Quarry are currently being increased to 7.6 m (25 ft.) from 6.1 m (20 ft.). The reasons for this are outlined later in this paper. Regardless of bench height, 1.5 m (5 ft.) of subdrilling is drilled. Drill hole patterns are most commonly designed on a 5.5 m (18 ft.) by 5.5 m (18 ft.) square grid, but are often reduced to as low as 4.3 m (14 ft.) by 4.3 m (14 ft.) centers in highly silicified areas. Because of the expansion in drill hole spacing, drilling the holes in the correct location is extremely important for ore control. Therefore, each drill hole is marked with paint or a stake in the ground by a survey crew. The driller can then position the drill exactly where the hole is required to be placed. The blast holes are drilled with a 20 cm (7-7/8 in.) diameter tri-cone rotary carbide button bit.

There are eight rotary drills available for blast hole drilling. Four are Ingersoll-Rand DM-M drills, three are Driltech D60KII, and the last is a Reedrill SK60II. All of the drills are track mounted, equipped with dust collectors, have heated water injection systems for positive dust control, and are equipped with pneumatic backflushing. The Driltech and Ingersoll-Rand drills have a maximum pull down of 27,200 Kg (60,000 lb.) while the one Reedrill has a 31,752 Kg (70,000 lb.) pull down capacity. All the drills are equipped with 39.9 m^3/min (1,200 CFM) capacity air compressors with the exception of one Ingersoll-Rand drill which has a 46.6 m^3/min (1,400 CFM) capacity. The drills utilize 10.7 m (35 ft.) long integral steel 15.24 cm (6 in.) in diameter with either an API or Beco threaded box on each end.

Efficiencies for the first half of 1989 averaged 27.5 m/hr (90 ft/hr) for the drill fleet. Mechanical availabilities for the same period was 76%. Five to six drills are operated each shift to meet the production goals.

The driller is responsible for the collection of an ore sample from each hole. Samples are collected for the first 6.1 m (20 ft.) or 7.6 m (25 ft.) for each hole in bedrock depending on bench height. The subdrilling is not sampled. Collecting of the sample is accomplished by lowering a plastic cylinder containing an open sample bag down a hole through the drill deck before drilling begins. The sample cylinder is 45.7 cm (18 in.) from the drill steel and the outside diameter of the cylinder is 10.2 cm (4 in.). With the cylinder near the ground, a representative sample is collected. A sampler collects each pre-labeled, bar coded sample bag, ensuring it has been correctly taken, and delivers it to the assay lab for analysis. Before the drill hole is blasted, its location is surveyed in by the survey crew.

BLASTING

Currently, 450 holes must be blasted each day to keep abreast with production. This number will reduce as the percentage of benches at 7.6 m (25 ft.) rather than 6.1 m (20 ft.) increases. To accomplish this requirement, Gold Quarry is equipped with two Amerind-McKissic 6.4 tonne (7.0 st) pneumatic prill trucks. A third 4.5 tonne (5.0 st) prill truck is available as back-up. A specially built high explosives truck is used to handle delays and primers. A total of ten people are employed on the blasting crew.

The blasting grade ammonium nitrate prill, used in the majority of blasting, is stored in two 54.4 tonne (60 st) bins. After the trucks load from these bins, the prill is pneumatically loaded into each blast hole at a rate of 204 Kg/min (550 lb/min). Diesel fuel is sprayed into the prill at a prescribed rate as it is being loaded. The amount of ANFO added to each blast hole varies with the hardness of the rock and the burden, generally ranging around 0.85 to 1.09 Kg/tonne (0.35 to 0.45 lb/st) powder factor. The alluvium overburden requires a much lower powder factor. Wet holes are loaded with ANFO bagged in 12.7 cm (5 in.) diameter waterproof bags or bagged slurry. When mining reaches the water table in the very near future, bulk loading of slurry will be required.

All the holes are primed with 0.566 Kg (20 oz.) directional cast primers. Detonation of the primer is accomplished with Ensign-Bickford 18-grain Detacord as downlines. Delaying of each hole in the blast pattern is done utilizing Ensign-Bickford Noiseless Trunkline Detonator surface delays. 17 millisecond and 25 millisecond delays are used between rows of holes. Blast patterns are usually shot in a Chevron (wedge) method. Two No. 6 caps and fuse, attached just prior to blasting, are used to initiate the blast. Blasting is done every day at 12:00 noon.

Crushed rock, usually mill reject material, is used for drill hole stemming whenever possible. The stemming confines the energy in the drill hole when blasting, thus maximizing rock fragmentation. When crushed rock is not available, drill hole cuttings are used. Investigations into an alternative economical source of crushed rock stemming is underway. The selection of a stemming truck to transport and mechanically stem each blast hole to eliminate labor intensive hand stemming is also in progress.

LOADING AND HAULING

The loading equipment available for production at Gold Quarry include: two DeMag H-285 and one H-185 diesel-hydraulic shovels, three L-1100 and one L-1000 LeTourneau diesel-electric loaders, and two 992-C Caterpillar diesel-hydraulic loaders. Three additional 992-C Caterpillar loaders are used at crushers to smooth out the feed between trucks and to blend ore from stockpiles. The mechanical availability for all twelve of these units for the first six months of 1989 was 80%.

The two H-285 DeMag shovels were purchased primarily for alluvium overburden removal. They have been used successfully in muddy conditions created by inclement weather. The buckets on these units have a 17.6 cum (23 cu. yd.) capacity. Productivity for the first half of 1989 for the shovels has been 1,415 tonnes (1,560 st) per operating hour. The third DeMag shovel, the H-185, has been utilized in both bedrock and alluvium overburden effectively. It has a 12 cum (16 cu. yd.) bucket and has been producing 1,125 tonnes (1,240 st) per operating hour. Although the bucket on the H-185 is 70% of the capacity of the H-285, the H-185 has faster cycle times which improves the productivity to 80% of that of a H-285.

The LeTourneau and Caterpillar front end loaders are used primarily in bedrock. Their ability to move quickly between mill and leach ore zones is essential for blending and maintaining a constant flow of material to the two mills and the leach primary crusher. Bucket capacities for the front end loaders and their productivities are presented as follows:

	Bucket Size	Per Operating Hour
L-1100 LeTourneau	16.8 m^3 (22 cu.yd.)	1020 tonnes (1120 st)
L-1000 LeTourneau	13.0 m^3 (17 cu.yd.)	940 tonnes (1040 st)
992-C Caterpillar	9.2 m^3 (12 cu.yd.)	680 tonnes (750 st)

Mine production trucks at Gold Quarry total 38. The fleet consists of eighteen 108.9 tonne (120 st) and five 127 tonne (140 st) Dresser-Wabco electric drive 510E Haulpaks. Fifteen 127 tonne (140 st) Caterpillar 785 mechanical drive trucks make up the remainder of the haul trucks. Each of the trucks, as well as each loading unit, dozer, grader and water truck are equipped with two-way radios to facilitate communication with the pit supervisor and dispatcher. The haul truck mechanical availability for the first six months of 1989 was 84%.

To optimize pit production, Modular Mining Systems, Inc. truck "DISPATCH" system was installed during late April of this year. An 8% to 10% efficiency improvement for the haulage fleet is expected with the fully automatic computerized dispatch system. This conservative estimate is based on productivity improvements which actually occurred at other mines where the system has been installed. It accomplishes this by reducing non-productive truck operating time caused by the following:

1. Queuing of trucks at loading units, dump points and service facilities.
2. Trucks waiting for loading units because of moves, minor repairs, and servicing.
3. Trucks proceeding to incorrect destinations.
4. Driver caused delays.

In addition to lowering the haulage cost per ton, other benefits which arise from the dispatch system are:

1. Loading unit productivity increases.
2. Ore control is improved by ensuring the different types of ore are dumped in the correct locations.
3. A more consistent flow of material and ore blending can be maintained to the crushing facilities.
4. Real time accounting for each shift is obtained.
5. Fast and accurate load counts are produced.

6. Eliminates the driver monotony of trucks locked to one loading unit.
7. Provides for operator involvement.
8. The system has simulation capabilities for determining short and long term equipment needs.

HAUL ROADS AND DUMPS

Approximately 27.4 Km (17 miles) of haul roads must be maintained at Gold Quarry. A minimum width of 30.5 m (100 ft.) and a maximum grade of 8% is desired whenever possible. These design parameters are used to maintain adequate clearances, prevent overstressing the trucks while on grade, reduce spillage and, therefore, minimize premature tire failures. To maintain these roads, two Caterpillar 16G and one 14G motor graders are operated. An Ingersoll-Rand SP150 compactor has also proven useful for haul road maintenance. The mechanical availability of the motor grader fleet has been 82% for the first six months of 1989. Consideration of the purchase of a fourth motor grader is under consideration.

Waste material is hauled to the waste dump at a rate of 118,000 tonnes (130,000 STPD). The dumps are designed at 21.3 m (70 ft.) lifts due to the unstable nature of the alluvium. In addition to the waste, 45,350 tonnes (50,000 STPD) is hauled to the leach pads. The leach dumps are designed in 9.1 m (30 ft.) lifts to maximize recoveries. To maintain all dumping areas and perform the additional pit work required, six track and two rubber-tired dozers are utilized. The fleet consists of two D10N, two D9L and one D8H Caterpillar track dozers. A Komatsu D375A-1 finalizes the track dozer list. The rubber-tired dozers are Caterpillar 834B's and are used primarily for loading unit pit floor clean-up and road maintenance.

Topsoil stripping ahead of advancing dumps and the reclamation of abandoned areas is additional work performed by the auxiliary equipment. The topsoil is stockpiled utilizing a Caterpillar 623B self-loading scraper. Track dozers and graders re-contour and scarify areas to be seeded as quickly as possible after being abandoned.

The control of fugitive dust is taken extremely seriously at Gold Quarry. A total of five water trucks are in use. Three of the trucks have 60,560L (16,000 gal.) capacity, one has a 41,635L (11,000 gal.) capacity, and the last has a 15,140L (4,000 gal.) capacity. Environmentally safe chemicals to assist in controlling dust and to improve the road base are also in use. Soil Stabilization's Earth Materials Catalyst 2 is the primary chemical being used to improve the road base. Magnesium Chloride is being used for dust control. Surfactants, asphalt emulsions, and others are being tested for their effectiveness in dust control.

OPERATION COST SAVINGS

The mine operators at Gold Quarry are constantly investigating ways to increase efficiencies and lower unit mining costs. Some of the methods implemented or under investigation which have been discussed include computerized truck dispatching, mechanization of blast hole stemming, and chemical treatment of haul roads for dust control and to improve road bases. Other projects in progress include: higher bench heights, wider drill hole spacing, reduction of material carried back in haul trucks due to the high clay content of the alluvium, and the use of crushed rock rather than run of mine material for haul road maintenance.

A study was completed during spring 1989 to analyze the economics if the bench heights and drill hole spacings were increased. Assuming the utilization of blast hole drills on site (no capital expenditure) and current blasting methods, an increase in bench height from 6.1 m (20 ft.) to 7.6 (25 ft.) and drill hole spacing from 4.3 m (14 ft.) to 5.5 m (18 ft.) was proposed by operations. A partial list of the advantages to this were:

1. A maximum 51% reduction in the drill holes required for the same tonnage in bedrock. This reduces blasting and assaying costs.
2. A maximum 42% reduction in the drill footage required in bedrock. This reduces drilling costs.
3. The large shovels and front end loaders can increase bucket fill factors in the higher bench height and reduce moves, thus increasing efficiencies.
4. Fewer benches, therefore fewer roads, will need to be maintained for the same production rate.
5. The work required for ore control and blast patterns by the survey crews will be reduced.

Disadvantages to increasing bench heights and drill hole spacing include:

1. Ore grade control loss - mill ore going to leach, leach ore to waste, and vice versa.
2. Misallocation of oxide and refractory ores.
3. Increased dilution.

Costs and cost savings were estimated for the above advantages and disadvantages. It was concluded that the mine operating cost savings exceeded the ore control loss cost by a significant margin. The blast hole spacing has been increased and the gradual change to the higher bench height is in progress.

Due to the "sticky" nature of alluvium, especially in wet conditions, carry back of material in haul truck beds is a major problem. Not only is the capacity of the haul truck reduced - cycle times are increased, delays due to having the bed cleaned out are increased, production time for a front end loader to clean beds is lost, and fuel efficiencies are decreased. The net result is a higher cost per ton. Various ways to reduce carry back are being investigated and include: periodic chemical treatment of the beds to prevent adhesion of the material to the bed, special plastic polymer liners, and special rubber liners on cables that eject the material on the bottom of the bed each time they dump. The evaluation is still in progress.

A good haulage road surface is essential to maximize safety, truck speeds, tire life, and production time, especially in slippery winter conditions. Having sized, crushed rock available to maintain good surfaces on roads is essential. A study into screening and washing leached spoils is in progress. Other sources of rock that would require crushing and screening will also be investigated.

CONCLUSION

Mine operations at Gold Quarry are committed to goals of efficiency and low cost while maintaining safety, quality, and the morale of its quality work force. This goal, shared by Newmont Gold Company as a whole, will ensure gold production for years to come.

REFERENCES

Rota, J.C., and Hausen, D.M., 1988, "Geology of the Gold Quarry Mine - 1988", As Presented at 1989 AIME Annual Meeting, Las Vegas, Nevada.

Smith, G.C., "Western U.S. Disseminated Gold, Typical Example - Gold Quarry, Nevada", 1988, to be included in revised "Mining Engineer Handbook - 2nd Edition".

Chapter 18

ALLUVIAL GOLD BACKHOE PLANTS IN NEW ZEALAND

John C Braithwaite

Consultant
Nelson New Zealand

ABSTRACT

Gold production commenced in New Zealand during the gold rush of 1861, and recovery to date is over 900 tonnes, almost 60% of which was alluvial. New Zealand pioneered development of bucket line dredges during the 1880's. The advent of the modern hydraulic excavator and increased gold price led to a resurgence in alluvial production with about one hundred plants in operation in 1988.

The processing plants are mainly floating, with throughputs of 50-100m^3/hr fed by backhoes of 20-40 tonne operating weight. Some larger plants with more sophisticated recovery systems have commenced over the last few years, with the largest using a 180 tonne backhoe.

INTRODUCTION

The river systems of Otago were the sites for the early discoveries of gold in New Zealand during the early 1860's. Gabriel's Gully was the nucleus of the first rush in 1861, and then the Americans Hartley and Reilly deposited over 30 kg (1000 ounces) of gold with the Dunedin gold receiver in 1862. They had obtained this from a beach in the Clutha river below Cromwell, and triggered a rush which revealed the full extent of the rich alluvials in the Clutha, Kawarau, Shotover and Arrow rivers. Hartley and Reilly recovered their gold in a few weeks using a cradle, while the river was particularly low during the winter. When the river was running at normal levels the rich gravel could only be reached by "blind stabbing" with a long handled shovel. This was a great incentive to invent mechanical contrivances to raise the gravel so that its gold could be extracted, and thus began the development of the gold dredge.

From the early spoon dredge through current wheelers, the design evolved to the steam powered "Dunedin" dredge in 1881. This "New Zealand Type" bucket line dredge was exported initially to California. Here American engineers improved many features and the modern close coupled bucket line dredge evolved from this beginning.

Following the Otago discoveries, the rush spread to the West Coast, where the alluvial gold had been brought down from the mountains by Pleistocene glaciation. Successive advances deposited gold in the moraines, which were reworked in the outwash gravels and recent riverbeds. Continuing tectonic uplift had preserved many deposits as raised terraces and these were favoured sites for sluicing operations in the early days. As dredging developed, most of the riverbed and terrace areas were worked.

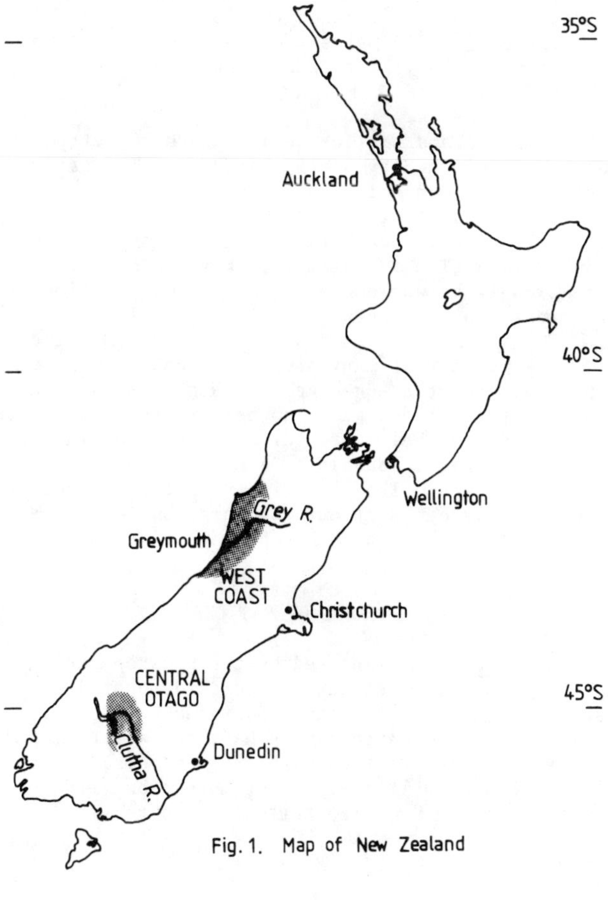

Fig. 1. Map of New Zealand

The author's interest in backhoe operations was aroused by a very experienced American alluvial mining engineer, W H Breeding, who visited New Zealand in 1976 to look at some of the early dredging sites. He suggested that many of the smaller areas of recent streambed and terrace could be mined using a backhoe to feed a floating plant as was being done in South America. About the same time, an engineer from the Kanieri dredge, J Acker, commenced operations with a small excavator and light floating plant.

When the gold price peaked in 1980, this was the impetus for a great increase in the number of plants operating on the West Coast. At one stage this was apparently as high as 130, but has now settled back to under 100. In Otago there are 20-30 operating at present. During 1989 the numbers were dropping for several reasons. Firstly the gold price had fallen to about US $380 per ounce, and the NZ dollar remains comparatively strong at US 62c. On top of this, the recent formation of the Department of Conservation has resulted in prospecting and mining licences taking up to three years each to come through, often with very onerous conditions which make working difficult.

EARLY DREDGING

During the dredging peaks around 1900 and the 1930s, the bulk of the larger areas were worked. Very large dredges such as the Rimu, Grey River, Ngahere, Arahura, Kanieri, Lowburn and Alexandra, digging up to 30 metres depth with bucket capacity of 0.34-0.51 cubic metres, worked the large river flats and suitable terraces. Annual throughput was about 3 million m^3 per dredge.

Shallower and smaller areas up to 10 metres in depth were worked by smaller dredges. The early units had wooden hulls and with the small buckets and light equipment could not handle ground containing large boulders. As even a small dredge took a long time to assemble, only reasonable sized areas could be worked to allow for amortisation of the installation costs.

This meant that many areas remained such as the upstream end of dredged ground when it became too shallow for the dredge. Another situation was when the outwash gravels were close to the moraine and not well sorted, containing a high proportion of large boulders. The versatility of the backhoe is such that it can sort out the large boulders and either stack them aside or work around them.

BACKHOE OPERATIONS

The screening plant fed by the backhoe can either float on pontoons, or be skid mounted and shifted around by the excavator or a bulldozer. Skid or wheel mounted plants have the water pumped in by pipeline and thus must stop and disconnect the line when shifting. A lot more rehandling of both fines and oversize tails is also required.

Floating Plants

These are found in all sizes from very small up to almost 400 tonnes weight. Floating plants are the most economical to operate as the pond in which the plant operates is the water supply, the settling pond, and the transport system for moving the plant around. The plant is moved frequently and over small distances, and always remains level. This reduces the amount of tailings levelling which has to be done, and means that feeding is continuous.

The plant basically consists of a feed hopper, trommel, and conveyor belt to stack oversize tailings. These units are all mounted on a pontoon structure which also holds the gold saving equipment, and diesel motor powering the pump and hydraulic systems. The feed hopper has a curved bottom corresponding to the shape of the trommel. This accepts the gravel from the excavator bucket and feeds it to the trommel, with water jets to assist sticky material. The deposits worked generally have low clay content, and scrubbing sections are not normally required in the trommel.

A few punched plate screens have been tried in trommels, but their greater weight and length has mitigated against their use. Wire woven screens are bolted into the trommel frame and usually have wear bars fitted to improve screen life, and this generally runs at 2000-4000 hours. Screening apertures depend on the size of gold, but vary from about 14 mm down to 6 mm. Vibrating screens have been used on several plants, and when fitted with rubber or polyurethane decks, have given extremely long life.

In the simplest plants the trommel undersize is led over riffled tables or sluices and discharged aft of the plant. On the larger plants it is generally pumped to jigs.

Fig. 2. Small Floating Plant

A small floating plant is shown in fig 2. This has a 1.2 m diameter trommel with a throughput of 40-50 m^3/hr, and is being fed by a 25 tonne backhoe. This unit weighs only about 10 tonnes, and can be dismantled and reassembled on another site in 2 days. It is useful for working small higher grade areas, or processing bulk samples of several thousand cubic metres to confirm grades and gravel type before committing to a larger plant.

The unit is powered by a diesel motor which direct drives the water pump, and also powers a hydraulic system which rotates the trommel and drives the stacker conveyor. Gold recovery is by riffled sluices. Capital cost of the plant would be about NZ$50,000, and total capital for an operation like this around NZ$350,000.

Fig. 3. Medium Scale Plant

A medium plant with a throughput of 60-80 m^3/hr is shown in fig 3. This has a 1.8 m diameter trommel and is fed by a 40 tonne excavator. It has two flow lines of primary jigs followed by a secondary jig and Knudsen bowl collection. The weight of this plant is about 60 tonnes. Capital cost for a plant of this type is approximately NZ$350,000. With a 40t excavator at NZ$350,000 the total capital involved would be about NZ$1 million.

Fig. 4. Larger Floating Plant

Fig 4 shows a larger plant on the Kawarau River in Central Otago. This handles 100-120 m^3/hr, and has a 2 m diameter trommel 6 m long. It has four flowlines of primary jigs with one secondary, followed by Knudsen bowls. Total weight with 160 kw generator is about 90 tonnes.

Because of the depth of ground in this river, a very large excavator was required to achieve the large digging depth needed to reach bottom. The unit in the photo is a Hitachi UH801 with a very long boom and dipper arm to dig to about 13m. The bucket is correspondingly small at 1.8 m^3.

Capital cost for this type of processing plant would be around NZ$500,000. It would normally be fed by a 60t excavator costing about NZ$600,000 giving a total project cost of about NZ$1.5-2 million.

Fig. 5. Very Large Plant

The very large unit shown in fig 5 was commissioned in 1988 at Rimu, south of Hokitika. The plant is working poorly sorted outwash gravels in the Rimu channel, which had been left by the famous Rimu Dredge which worked in the area from 1922-1952 and recovered almost 10 tonnes (320,000 ounces) of gold. The ground is deep varying from 12-23 m but can be worked due to the low water table.

The plant weighs about 380 tonnes and has a capacity of 250-300m^3/hr with a radial primary jig followed by secondary and tertiary jigs and Knudsen bowl clean-up. Twin vibrating feeder grizzlies scalp off the large material, and vibrating screens produce the minus 10 mm jig feed. The excavator is a 180 tonne Hitachi EX1800 with 5.3 m^3 bucket. Total project cost was NZ$6 million, with a processing plant cost of about NZ$2 million.

Land Based Plants

Skid plants have higher operating costs than floating operations, as a second machine (either bulldozer, wheel loader or excavator) needs to be on hand almost full time to keep tailings clear. Production time is lost each time the plant is being shifted, and water lines have to be relocated regularly, along with settling ponds and water pumps. Their use is common where the topography is broken, or old underground workings may make it difficult to retain water. They also have the advantage that 'bottom' can be seen and cleaned up thoroughly.

The plant shown on fig 6 has a 1.5 m trommel and processes about 60 m^3/hr. To keep the plant compact so that it is easy to shift is paramount, but this restricts the width of riffled tables which can be fitted and fine gold recovery suffers accordingly. Capital cost for the plant would be around NZ$ 100,000, depending on whether it had a swivelling stacker. Total capital cost would depend on what additional equipment was required for the particular location.

Fig. 6. Land Based Plant

One very large plant on the West Coast operated for about two years around 1984. This was the Goliath operation which aimed at 300 m^3/hr from a 120 tonne Demag excavator feeding a 3 m diameter trommel 13 m long, mounted on crawler tracks. Recovery was by riffled tables, and the company suspended operations when costs exceeded recoveries.

EVALUATION OF POTENTIAL AREAS

Drilling technology for placer deposits has advanced very little over the years. Hardrock drilling has progressed to sophisticated wireline core drilling, and a host of new reverse circulation methods are coming into use. The Bekker rig has proved effective in fine-grained alluvials, but for coarse bouldery gravels the old cable tool rig appears to give the only reliable results. With the comparatively slow drilling rate of 5-10 m per shift, the cost is high at about NZ$200 per metre, but there appears to be little alternative.

Some advances have been the use of 300 mm diameter drive shoes compared with the traditional 190 mm. This gives a larger sample, and often faster drilling as larger cobbles can fit inside the casing shoe. Several rigs are now driven by hydraulic motors, rather than the large open gears and clutches. Calweld holes up to 1 m diameter have been successful in parts of Central Otago with comparatively fine gravel and low water table. Up to 20 m per shift have been drilled, and a very large sample is obtained.

Most operators assess their potential ground with test pits dug by a backhoe. A bucket measured volume of 10-50 m^3 is fed through a small trommel plant as in fig 7. The 0.9 m wide riffled sluice under the trommel catches the concentrate which is then cleaned up and the gold weighed to give agrade figure. This recovery equipment is very basic, but gives a conservative value. A few more sophisticated sample plants have been used, but are not as rugged for positioning in difficult areas and cannot process a large sample as quickly. Where a lot of fine gold is present, tests on the fine tails show the extent of plant losses.

Fig. 7. Sampling Plant

The system is cost effective as it is quick, and a 6 m pit would cost around NZ$600, or half the cost of drilling. Difficulties are sometimes experienced below the water table with slumping of upper material and diluting of grades. Large samples are processed, and if the grades are reasonable the gold recovered makes a significant contribution towards the costs. As with drill holes, the pits are put down in lines across the direction of flow, with a spacing close enough to detect any channels which may contain high values.

A more detailed assessment to determine mining plans is often made with a skid mounted plant by digging 2-6 m wide channels right across the valley or area to be tested. The plant is washed up every 50 m or so along the length of the channel and the results recorded. This method gives an excellent assessment of an area, in that practical experience has been gained in handling the gravels for final plant design. It also has the great advantage in economic grades of producing sufficient gold to pay for the test programme.

RECOVERY METHODS

The early plants had very simple recovery systems with riffled sluices to retain the gold, as were used 100 years ago. Various types of riffles have been tried with vertical sloping and angle iron bars being tested. Overall most operators use vertical riffles about 30 mm high spaced 40-50 mm apart. If the water is dirty, the riffles are generally raked every hour to counteract packing of the bed. The upper riffles are washed up each shift and the concentrate taken away for clean up. Lower sections are washed up every few days when the operator judges it necessary.

It is acknowledged that there are losses from these simple riffle plants, the extent depending on the gold particle sizing in the deposit, and care taken by the operator. A series of systematic tests was done on several plants by Muir and Fricker (1986). This showed losses to be about 20%. The smaller operators generally prefer to keep their capital costs low with the plant simple to operate, and suffer this sort of loss.

Fig. 8. Large Sluice Plant

In general only the larger company operated plants have gone to heavier and more complex recovery systems. Fig 8 shows a typical larger riffle plant in the Shotover river, Otago. It has a 2 m diameter trommel, handles about 100 m^3/hr, and 6 m width of sluices. This plant was later fitted with jigs and spirals, and showed 10-14% increase in recovery, mainly in the minus 0.6 mm fraction (Braithwaite 1989).

Most of the larger plants are now fitted with jigs as these have shown to be the most reliable and effective system. The types in operation include Yuba, Pan American, and an ADL radial jig, as well as local variations. The secondary jig concentrate is generally handled by Knudsen bowls as these have proved to be the simplest and most effective unit for this job. The Knudsen bowl is built in Reno, and has been a great boon to the New Zealand alluvial gold industry.

The Knelson concentrator from Canada with its high speed and injected back pressure water has also been used. They did not catch on for some years due to their high price and requirement for very clean water, but are now starting to be used in beach-sand gold operations on the West Coast. This gold is very fine at about 60 microns in size, and recovery is apparently quite acceptable.

Hydraulic riffles were developed in Otago some years ago and are now quite widely used. They are deeper than normal riffles, and the lower section of each riffle is made of 25 mm hollow section steel. Low pressure water from holes in thishollow section keeps the bed of material between the riffles partly in suspension so that gold can settle in more easily.

OPERATING COSTS

Costs vary greatly from one operation to another depending on the amount of stripping and restoration to be done, size and abrasiveness of the gravels, whether a good water supply is available and many other factors. In general floating plants operate at a lower cut off grade than land based plants, and larger plants can operate at a lower cost per cubic metre than smaller units.

Little information has been published on costs, but a typical medium scale plant processing 80m^3/hr would have a bare operating cost of about 120mg/m^3. If deep and extensive stripping of swampy overburden is required, or restoration is particularly difficult this will be increased. On the West Coast a good profit can usually be made at 170mg/m^3 (NZ$3.30/$m^3$ and grades above this have shown excellent returns.

Many operators with small floating plants as in Fig 2 have made good profits on 170mg/m^3 with a throughput of only 50m^3/hr. This is because of lower capital costs, and care with the equipment reducing repairs and maintenance costs. No operating cost figures have been published for the very large plant at Rimu near Hokitika, but it should be lower than the figures given above.

In Otago rivers such as the Shotover, operating costs would be higher because of the more expensive equipment required, and significant downtime when the river level is high.

Land based plants would generally require values of 170-250mg/m^3 to be profitable. At Cromwell in Central Otago an operation based on terrace gravels making extensive use of conveyors and selective mining, has successfully worked grades of about 150mg/m^3.

CONCLUSIONS

The hydraulic excavator and the new breed of floating and landbased processing plants opened up a whole new phase of alluvial gold development in New Zealand during the 1980's. Production now appears to be declining with the falling price and continuing delays in issuing of mining licences. The size of excavator used has increased from the initial 20t units up to the 180t giant at Rimu, feeding 300m^3/hr. This has made possible the working of large volumes of alluvial ground at a comparatively modest capital cost.

A new scale could be about to start with the commissioning now underway of the R A Hanson Co - Giant Resources new Grey River dredge. This is a combined bucketline-suction cutter dredge with a total predicted annual output of 12 million m^3. If this is successful in processing this huge output, there could be many marginal areas which will become economic in the future.

REFERENCES

Braithwaite, J. C., 1989, "Gold Mining in the Shotover and Kawarau Rivers", Mineral Deposits of N.Z., AusIMM Monograph 13.

Fricker, A. G., and Minehan, P. J., 1986, "The Morphology and Recovery of Placer Gold", N.Z. DSIR Report No IPD/TSO/2024.

Jury, A. P., and Hancock, P. M., 1985, "Alluvial Gold Deposits and Mining Opportunities on the West Coast, South Island, N. Z", Proc. AusIMM N. Z. Branch, Annual Conference, 1985.

Muir, D. G., and Fricker, A.G., 1986, "A Semi-Quantitative Survey of the Metallurgical Efficiency of Alluvial Gold Operations in New Zealand", N.Z. DSIR Report No IPD/TSO/2025.

Salmon, J. H. M., 1963, "A History of Gold Mining in New Zealand", Government Printer, Wellington, N.Z.

Utting, B. S., 1988, "Gold Dredging in New Zealand", Proc., AusIMM Second Mineral Heritage Seminar.

Williams, G.J., 1974, "Economic Geology of New Zealand", Monograph Series, No. 4, AusIMM, Melbourne.

6

Mining II
Underground Mining

R. Armstrong
Echo Bay Mines, Ltd.
Edmonton, Alba., Canada

R.J. Harden
Fluor Daniel Australia Ltd.
Melbourne, Vic., Australia

Chapter 19

INVESTIGATION AND IMPLEMENTATION OF MINE DEWATERING SYSTEMS

by

Lee C. Atkinson
Principal
Hydrologic Consultants, Inc.
Lakewood, Colorado

Thomas M. Hanna
Project Hydrogeologist
Hydrologic Consultants, Inc.
Lakewood, Colorado

John W. Anthony
Project Hydrogeologist
Hydrologic Consultants, Inc.
Lakewood, Colorado

Abstract. Ground water can pose major engineering and environmental problems to the new generation of large, deep gold mines currently being developed throughout the world. It is important that the potential magnitude of these problems be assessed early in the planning of such projects and that any necessary dewatering and environmental remediation programs be designed at least preliminarily in the feasibility stage. The assessment should begin with acquisition of hydrogeologic data during exploration drilling and be followed by carefully designed and analyzed aquifer tests. Once a sufficiently detailed conceptual hydrogeologic model has been defined, state-of-the-art numerical models should be used to estimate ground-water inflows to proposed pits and underground openings, to design and evaluate the potential effects of various dewatering schemes, to define specific environmental impacts, and to predict pore pressures in highwalls relative to slope stability. The modeling results will indicate which method(s) of dewatering will be most effective both technically and economically. If wells are used, it is imperative that they be as hydraulically efficient as possible. There are new methods of drilling and completing dewatering wells that can help assure maximum efficiency.

INTRODUCTION

As mines become larger and go deeper to develop gold deposits, the chance of encountering significant quantities of ground water increases. Whereas many past mining operations simply stopped either at the water table or when pumping costs reached a certain limit, the goal of many modern mines is to excavate through hundreds of meters of saturated rock.

Mining "in the wet" can significantly increase the cost of mining. Direct increases in cost can include the need to use waterproof explosives; hauling and processing of wet ore; pumps, pipelines, and electrical power to run the pumps; the possible need to store and/or treat discharge; generally higher maintenance of almost all mining equipment; and periodic slowdowns or complete shutdowns when pumps fail or winter freeze-ups occur (Brawner, 1979). Other indirect costs include delays associated with obtaining permits to extract, store, or discharge water and the possible need to reduce slope angles because of high ground-water pore pressures.

To eliminate, reduce, or at least be prepared for some of these problems, it is important that the role of ground water in a proposed or expanding mining operation be defined as quantitatively as possible well in advance of the problems occurring. A gold deposit is, by definition, a geologic "anomaly". Therefore, if it is located below the water table, it is also usually part of a complex hydrogeologic system which requires a relatively comprehensive program of investigation. Fortunately, much important information can be obtained during the exploration drilling program at very little incremental cost. Together with more quantitative data from carefully designed pumping tests, these findings are used to define a conceptual hydrogeologic model of the mine area.

In a somewhat parallel development to that of current ore reserve and mine design models, relatively sophisticated numerical ground-water flow models (Durbin and O'Brien, 1987) have been developed which can incorporate even the most complex hydrogeologic setting. This type of model can simulate the results of various mining and dewatering scenarios and should be a part of the overall mine plan for any large-scale mining operation which will encounter significant ground water.

Recently developed drilling and well completion technology also needs to be incorporated into any mine dewatering system utilizing wells. The importance of hydraulically efficient wells can be realized in the millions of dollars that can be saved on a large scale dewatering project.

ACQUIRING HYDROGEOLOGIC DATA DURING EXPLORATION DRILLING

Semi-quantitative hydrogeologic data can be obtained during exploration and development drilling if air-rotary or air-hammer methods are used. The two main sources of data include ground-water production (and temperature in some hydrogeologic settings) with depth during drilling and brief airlift recovery tests. The production data can be collected by the driller, and airlift recovery tests can be conducted by technician-level personnel. There are no additional costs associated with obtaining the airlift production data, and the airlift pumping tests can be conducted at an additional cost of about 2 hours of standby time (for the drilling contractor) per borehole.

The main point to be emphasized here is that with a very limited additional expense to the exploration drilling budget, hydrogeologic data can be obtained that eventually would have to be collected by additional boreholes drilled specifically for that purpose. This program has been successfully implemented at three major gold mines in Nevada as an integral part of the dewatering investigation.

A brief description of the procedures with which to collect hydrogeologic data during exploration drilling and the use of these data is as follows.

Airlift Production Data

Airlift production during drilling is measured after each drillpipe joint has been drilled down by either noting the time it takes the discharge from the cyclone/sample collector to fill a container of known volume or by using a calibrated weir or flume set up in the discharge ditch. The required accuracy for temperature data can be obtained with a simple armored pocket thermometer.

A typical plot of such data is shown in Figure 1. Gradual increases in the cumulative groundwater production rate with depth are associated with uniform permeability as well as the effect of submergence (i.e., the depth of the borehole below the water table). Relatively abrupt increases in the rate of production (Zone A on Figure 1) indicate a zone of relatively higher permeability. Zones of decreased production (Zone B) can also occur. They are usually the result of a clay "boot" forming on the drillpipe and cutting off the flow of water from above.

The resulting data from any large exploration drilling program should be stored in a computerized data base. This allows data from a large number of boreholes to be statistically analyzed and reproduced graphically using available software packages. In addition to plots of individual boreholes such as Figure 1, maps and cross-sections of the data can be created at the same scale as the mine plans and geologic cross-sections for areal and "hydro-stratigraphic" interpretation.

The airlift production data collected and processed as described should readily identify areas of relatively large ground-water production. In addition, temperature data could define hydrologic structures that connect the shallow ground-water system with deeper sources of potential recharge. This information is not only essential in developing the conceptual hydrogeologic model (described in a following section), but is also useful in defining sources of groundwater supply for the mine/mill and in guiding subsequent phases of the dewatering investigation by identifying potentially problematic areas.

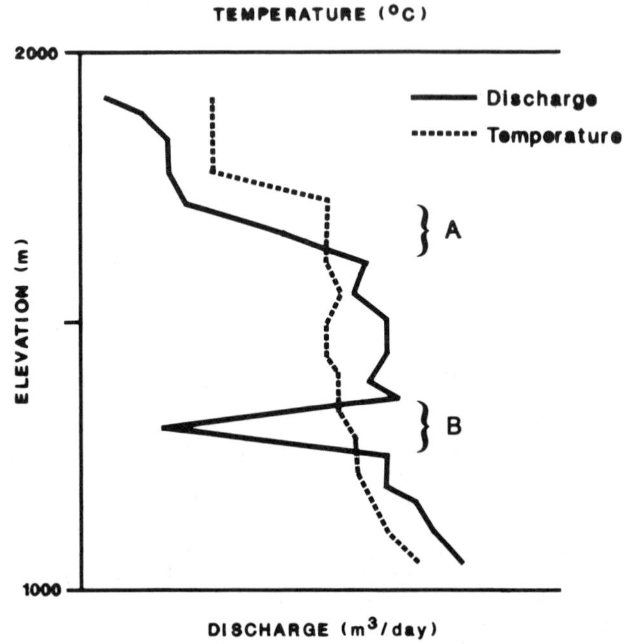

FIGURE 1. TYPICAL PLOT OF AIRLIFT PRODUCTION AND TEMPERATURE WITH DEPTH

Airlift Recovery Tests

Airlift recovery tests are performed by airlift pumping (with the drillpipe at or near the bottom of the borehole) a completed borehole for 1 to 2 hours and subsequently measuring the recovery water levels inside the drillpipe with an electric water level sounder for a period equal to the duration of pumping. The water level recovery data are then plotted vs the logarithm of t/t′ where t (min) is the time since airlift pumping began and t′ (min) is the time since pumping stopped. For example, if a borehole is pumped for 60 minutes, at 15 minutes after pumping ceases t′ = 15, t = 75, and t/t′ = 5.

An example of a typical airlift recovery plot is shown in Figure 2. Using a modification of the Jacob method (Kruseman and de Ridder, 1976), the transmissivity of the saturated interval of rock penetrated by the borehole can be estimated from

$$T = 0.183 \cdot Q/\Delta s \qquad (1)$$

where

T = transmissivity [m^2/day],
Q = average pumping rate [m^3/day], and
Δs = change in recovery water level [m] per log cycle of t/t′.

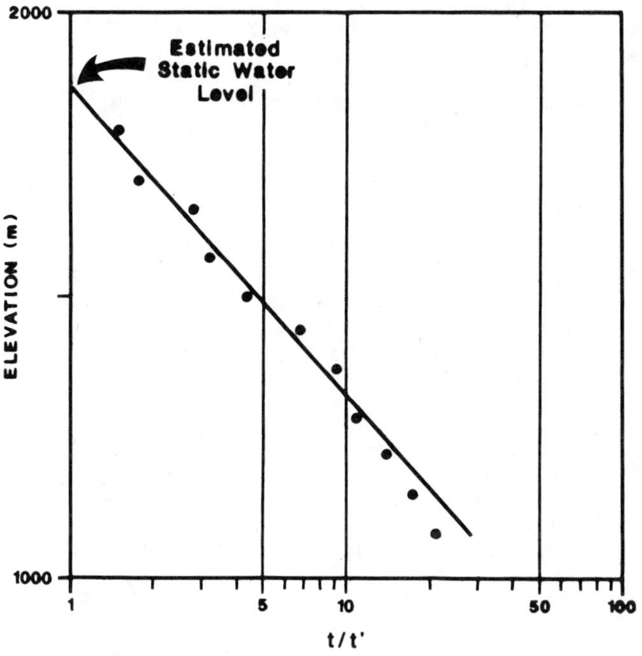

FIGURE 2. TYPICAL AIRLIFT RECOVERY PLOT

Transmissivity is a primary aquifer characteristic whose value is used in many quantitative analyses related to dewatering (several of which will be described in following sections). Another important parameter is hydraulic conductivity (frequently referred to as permeability) which can be derived from

$$K = T/b \qquad (2)$$

where

K = hydraulic conductivity [m/day], and
T = transmissivity [m^2/day], and
b = saturated thickness of aquifer penetrated by borehole [m].

Estimating Static Water Levels

Because exploration boreholes frequently collapse or are abandoned immediately after drilling, measurements of static water levels in them are often not obtainable. Fortunately, an estimate of the static water level in an exploration borehole that has been airlift pumped can be estimated from the recovery plot. In most cases, the water level at the intersection of the recovery plot extrapolated to t/t′ = 1 (which, by convention, is the left axis of a recovery plot as shown in Figure 2) provides a reasonable estimate of the static water level.

AQUIFER TESTING

The next step in a hydrogeologic investigation related to mine dewatering is, based on the preliminary data from the exploration drilling program, to design and conduct one or more aquifer tests. Aquifer tests are needed to define the hydrologic properties of the aquifer system more quantitatively and over a larger areal extent. They typically involve construction of one or more relatively large production wells and numerous, small-diameter observation wells.

A detailed discussion of the design, execution, and traditional analyses of aquifer tests is beyond the scope of this paper, but this information is well covered in the readily available literature (Kruseman and de Ridder, 1976; Lohman, 1979; Stallman, 1971; Walton, 1987; and Woodward, 1978). Instead, we will introduce a few of the less-known analytical methods more applicable to the vagaries often associated with aquifer tests in hard, fractured rocks (the host rock of many gold orebodies) and ones which we have utilized on several recent projects.

The general analysis of an aquifer test assumes an idealized aquifer that is isotropic (i.e., has the same permeability in all directions), homogeneous, of constant thickness, and of infinite areal extent. These assumptions are violated to some extent in almost all real-world systems, but nowhere are these violations more apparent than in the hydrogeologic settings of most gold deposits. Such aquifers are typically anisotropic, heterogeneous, and have multiple hydrologic boundaries.

Traditional analyses based on the classic Theis equation or any of its derivatives tend to disregard the heterogeneities and anisotropy inherent in most mine-related aquifers. The values of the aquifer characteristics, T (transmissivity, previously described) and S (storativity, a measure of the volume of water that can be extracted from the aquifer), derived using these methods might adequately reflect conditions in the immediate vicinity of the pumping well, but poorly characterize the overall aquifer properties which will most influence large scale, long-term dewatering.

Vandenberg (1977) proposed a method for analyzing heterogeneous aquifers in which the time-drawdown data from all observation wells are plotted on a single log-log graph of drawdown (s, in m) versus the parameter t/r^2 (time, in min, divided by the square of the distance, in m, of the observation well from the pumping well). The resulting band of data points is then bracketed with a pair of Theis curves from which an estimate of the range of "representative average" values for T and S can be derived (Figure 3).

Ground-water flow in many ore host rocks is principally through fractures which can impart significant anisotropy to the aquifer. A method developed by Papadopulos (1967), which utilizes data from three observation wells (or combinations of any three if more than three are available), can help define the magnitude and direction of the major and minor components of transmissivity. This information is most useful in selecting the relative placement of dewatering wells around a pit.

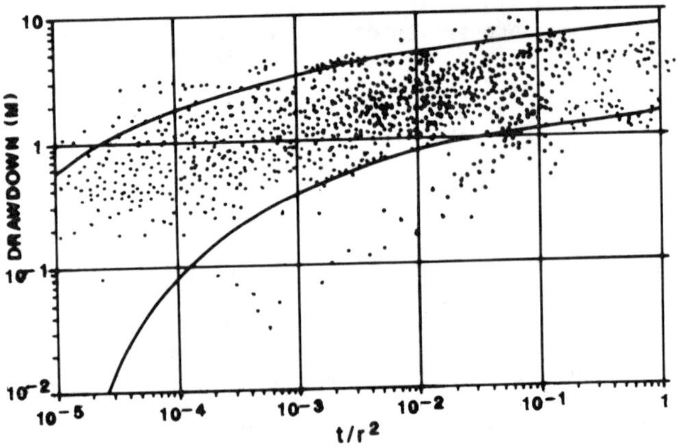

FIGURE 3. ANALYSIS OF PUMPING TEST DATA FROM A HETEROGENEOUS AQUIFER USING THE VANDENBERG METHOD

In terms of their influence on mine water problems, hydrologic boundaries are often as important, if not more important, than the average hydraulic properties of the rocks. Recognition of potential hydrologic boundaries should be possible with geologic maps prepared by the mine geologists and the results of exploration drilling. Traditionally, hydrologic boundaries such as faults and contacts with relatively less permeable rock are often treated in pumping test analyses as if they were completely impermeable. This can lead to some very erroneous conclusions regarding the source and magnitude of potential inflow to a mine. Nind (1965) developed a method to estimate the aquifer parameters on either side of a boundary with drawdown data from observation wells on opposite sides of the boundary. We have found that use of this method with an adequate network of observation wells can sometimes eliminate the need for multiple aquifer tests in a "compartmentalized" hydrogeologic system with multiple hydrologic boundaries.

DEVELOPING A CONCEPTUAL HYDROGEOLOGIC MODEL

The first major product from a mine dewatering investigation should be a detailed conceptual hydrogeologic model that incorporates all of the geologic and hydrologic components that will influence the mining operation. Without a good conceptual understanding of the hydrogeologic framework and the stresses to be put on the system, no level of effort of analytical or numerical modeling will result in conclusions that can be used with any confidence.

To produce an accurate representation of the hydrologic system, the model must be as comprehensive as possible and should include data from available literature; any surface exploration programs including geologic mapping and geophysical surveys; interpretation of topographic maps and aerial photographs; stream gauging; exploration drilling program; and aquifer tests.

A typical conceptual hydrogeologic model will incorporate the following information

- A description of the general geology including lithology, stratigraphy, and structure,
- Identification of major hydrogeologic units including formations or specific intervals within a formation, faults, fracture zones, intrusive bodies, and any discrete zones or areas of particularly high or low permeability,
- Values for the hydraulic properties of the various hydrogeologic units including hydraulic conductivity, specific storage, and specific yield, and, if thicknesses of hydrostratigraphic units are known, their bulk properties of transmissivity and storativity,
- The elevation and configuration of the water table,
- The size of drainage basins, discharge of all streams, and any gaining or losing stretches, and
- Hydrologic boundaries such as streams, drainage divides, or geologic contacts with low permeability materials.

Although not actually part of the conceptual hydrogeologic model, the "stresses" on the groundwater system that should be defined at this time in the investigation include

- The mine plan, including the vertical and lateral extent and timing of all excavation, and
- The local and regional climatology, with emphasis on precipitation and evapotranspiration.

USING SIMPLE ANALYTICAL MODELS FOR PRELIMINARY ESTIMATES OF INFLOW

Prior to the time consuming development of a numerical model (described in the following section), it is often necessary for initial mine planning purposes to use simple analytic solutions to make preliminary estimates of ground-water inflow to pits and underground openings. A suite of potentially applicable analytical solutions is presented by Singh and Atkins (1984); of these methods, a few of the most useful will be described in more detail. It should be emphasized that the estimates from these methods are "order-of-magnitude".

Inflow to a Pit

An open pit is analogous to a large-diameter well, so a preliminary estimate of ground-water inflow to a roughly circular pit can be made with a combination of a well equation to calculate the lateral component of inflow and the classic Darcy equation to estimate vertical inflow through the pit floor. The estimate of total inflow is then the sum of these two components.

The transient, lateral component of inflow is described by a slight modification to the equation for flow to a well under constant drawdown conditions (Jacob and Lohman, 1952)

$$Q = \frac{12.6 \cdot T s_a f_s}{\ln[(2.25Tt)/(r_w^2 S)]} \qquad (3)$$

where

Q = inflow [m^3/day],
T = transmissivity [m^2/day],
t = time after excavation [days],
S = storage coefficient [dimensionless],
s_a = available drawdown [m],
f_s = "skin factor" [dimensionless],and
r_w = radius of well/pit [m].

A "skin factor" is included to account for the effects of turbulent flow near the edge of the pit. This can be an important factor in limiting inflows, particularly in a fractured-rock aquifer (Dudgeon, 1988).

The available drawdown is defined as the difference in elevation between the static water level and the floor of the pit. This assumes that the full available drawdown is present when the pit reaches the depth considered. In other words, the pit is considered to be excavated "instantaneously" with previous activities having caused no drawdown. These assumptions are not very realistic, but they have limited effect on the more important later time (greater than a couple of years) values.

The vertical component of inflow through the floor of the pit can be estimated from a modified version of the steady-state Darcy equation

$$Q = \frac{3.14 K r_w^2 f_s s_a}{L} \qquad (4)$$

where

Q = inflow [m^3/day],
K = hydraulic conductivity [m/day],
L = effective length of flowlines from their source to floor of pit [m],
s_a = available drawdown [m],
f_s = "skin factor" [dimensionless],and
r_w = radius of well/pit [m].

Considerable judgment and knowledge of the site hydrogeology have to be used in selecting an appropriate value for L. Such values are typically in the range of 300 to 1000 m.

Inflow to Underground Openings

Ground-water inflow to underground workings can be estimated with a combination of Equations 3 and 4. The geometry of a horizontal or near-horizontal drift, however, precludes estimation of inflow using a radial-flow solution.

A very simplistic method used to generate an order-of-magnitude estimate of inflow to a drift was developed by Goodman et al.(1965). After the water table has declined to the level of the drift, the rate of inflow can be estimated by

$$Q = 0.707L\left(\frac{KH^3S_y}{t}\right)^{1/2} \qquad (5)$$

where

Q = inflow [m^3/day],
L = length of drift section [m],
K = hydraulic conductivity [m/day],
H = depth of drift below initial water table [m],
S_y = specific yield of aquifer [dimensionless], and
t = time since water level reached top of drift [days].

There are major uncertainties in the selection of an appropriate value for t, hut values in the range of 10 to 50 days are typically used.

USING NUMERICAL MODELS

Numerical ground-water flow modeling is a modern tool that has application to some degree on almost all mining hydrology investigations. The four primary purposes of the numerical modeling are

- To obtain more reliable estimates (relative to the preliminary order-of-magnitude estimates from the analytical solutions) of time-dependent and pit-dependent passive (i.e., with no active dewatering system in operation) inflows to pits and underground openings,

- To evaluate the most likely effects of various, alternative dewatering systems,

- To predict the local and regional environmental impacts of dewatering, and

- To predict pore pressures in highwalls relative to slope stability.

On a large scale mining project, the time and cost (perhaps in the range of US$75,000 to 125,000) to develop a fully 3-dimensional finite element model is easily justified. The model should include

- All of the key components of the conceptual hydrogeologic model,

- Appropriately located and defined boundary conditions (e.g., streams, drainage divides), and

- An accurate temporal and geometric description of the mine plan.

We have recently developed a suite of numerical models specifically for mining hydrology investigations. It includes a 3-dimensional flow code (Durbin and O'Brien, 1987) which is used at two scales to estimate inflows and a 2-dimensional flow code to aid in analysis of pit slope stability.

The first is a regional scale model (with the entire mine simulated by only 1 or 2 "constant head" nodes representing the floor of the pit at any point in time) that is used to define the effects of the regional ground-water system on water levels and primary flow directions that result from dewatering of the mine. The results of this model will identify the areal extent over which environmental impacts such as drying up of springs could occur.

The second application of the 3-dimensional code is at a scale that can incorporate the mine plan and allow specific dewatering facilities (e.g., a series of wells or a drainage gallery) to be simulated. This model uses the water levels generated by the first model as boundary conditions. The primary purpose of this model is to calculate both the active (i.e., the water intercepted by wells or galleries) and the passive (the remaining inflow which would go into sumps) portions of total inflow under various dewatering schemes. Considering factors such as locations for wells within the mine plan, the relative amounts of "clean" (from wells) and "dirty" (from sumps) water that would be produced and other logistical considerations, many simulations can be made with this model to help design the components and schedule of a dewatering system that is effective both technically and economically.

The final code in the suite is a 2-dimensional flow code that incorporates both laminar and turbulent flow. The effects of turbulent flow can be quite significant in both reducing local inflow and in creating higher than expected (relative to that resulting from laminar flow) pore pressures in a highwall in fractured rock (Dudgeon, 1988). A series of vertical slices through various highwalls are modeled with this code to examine the potential distribution of pore pressures relative to slope stability.

ALTERNATIVE METHODS OF DEWATERING

The following is a brief summary of the advantages and disadvantages of the various methods that traditionally have been used for mine dewatering. We will also introduce for discussion and debate two other methods which we feel warrant consideration at some of the gold mines that are planned or currently being developed.

Traditional Methods

There are four methods used either exclusively or in combination that comprise the majority of mine dewatering operations around the world. These are shown schematically in Figure 4 for an open pit mine, but some of them obviously have applications to other methods of mining.

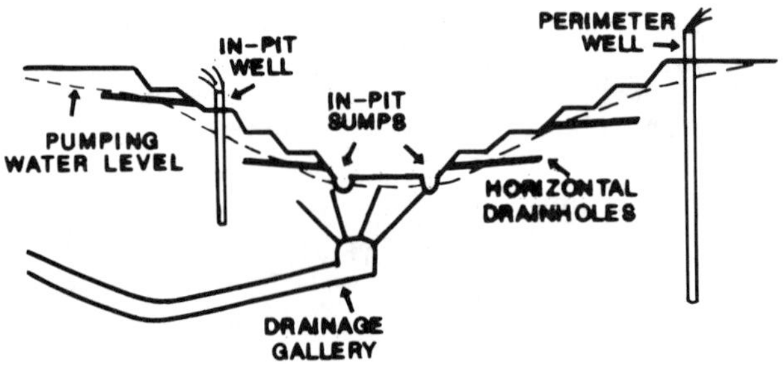

FIGURE 4. ALTERNATIVE MINE DEWATERING METHODS

Drainage Galleries. Drainage galleries are drifts driven from shafts or declines from which numerous underground drainholes are usually drilled in fan-shape patterns. The primary advantages of galleries/drainholes are that they usually are out of the way of mining and can intercept (by virtue of drilling drainholes into specific water-bearing structures or at any orientation) and handle large quantities of water. Their disadvantages are high costs and the need for large lead times in their construction.

Perimeter Wells. Perimeter wells are conventional, normally vertical wells that are drilled outside the ultimate pit boundary. Their main advantages are that they can be drilled in advance of mining, are essentially out of the way of mining operations, and are effective in intercepting inflow when the primary direction of ground-water flow in the mine is lateral. Their major disadvantages are that they are not effective in low permeability or highly compartmentalized aquifers, they can be very hydraulically inefficient if not properly constructed (to be discussed in more detail in a following section), their effectiveness can be very "hit or miss" in fractured rock aquifers, and they do not intercept a major component of inflow to a large pit in rock with significant vertical permeability.

In-Pit Wells. In-pit wells are conventional, vertical wells that are constructed within the ultimate pit boundaries either on the side of haul roads, on specially constructed wide benches or as temporary wells in areas that will eventually be mined out. Their main advantage over perimeter wells is that they can usually be shallower (and, hence, less costly) and still create more drawdown where it is needed. In addition to the same disadvantages as perimeter wells, they cannot be constructed until mining commences, are susceptible to blast and mining damage, and generally are a nuisance.

In-Pit Horizontal Drainholes. Rocks with low permeability drain slowly and thereby retain pore pressure. Consequently, such rocks often have to be depressurized with horizontal drainholes as a pit is being excavated. Other local areas either where gouge-filled faults parallel to the highwalls could form ground-water dams or where turbulent flow associated with discharge from a limited number of fractures (Dudgeon, 1988) could result in relatively high pore pressure close to the highwall must also be supplementally drained.

In-Pit Sumps. In-pit sumps are usually part of any dewatering system regardless of what other components are in the system. They are used to collect runoff from direct precipitation, discharge from in-pit horizontal drainholes, and inflow from the pit floor or lateral inflow not intercepted by wells and/or the horizontal drainholes. Although sumping is simple and relatively inexpensive, the water it produces is turbid and usually must be sent to retention ponds prior to any use or release.

Non-Traditional Methods

In consideration of the magnitude of some of world's new gold mines and the complex hydrogeologic settings in which they are situated, some non-traditional methods of managing ground water are worthy of review.

Grouting. For other than locally controlling ground water during shaft sinking or driving of declines or drifts, grouting is generally dismissed as too costly a means of managing water at a mine. However, increasing environmental costs in disposing of water in most cases and outright bans to disposal in others suggests that intercepting and handling less water should be a goal. The costs of ringing a large pit with an "impermeable" grout curtain would be prohibitive and ineffective if a major component of inflow is vertically upward, but there is no reason why relatively localized areas of high permeability could not be grouted. Significantly less costly materials other than cement and chemical grouts (e.g., slurries made from native clays) might be able to be used.

Underdrains. If it is concluded that wells will not be an effective means of dewatering a mine and advancing drainage galleries would be prohibitively expensive due to weak ground, a new method which should be considered is "underdrains." This would involve completion of horizontal drillholes cased with large (on the order of 300 to 400-mm diameter), perforated pipe. The casing would be installed by drilling a small diameter pilot hole between two underground openings and then pulling a larger diameter reamer bit (similar to the ones used in raise borings) back along with the perforated casing. Obviously this technique would require the sinking of a shaft or decline to attach the bits and perforated casing for the pullback. To the best of our knowledge, this method has never been used in a mine dewatering application. Although the drilling conditions are significantly different, the method is routinely used to install large power cables and pipelines beneath major rivers. An alternative method of installing the underdrains would be to use wells collared on the perimeter of the pit and drilling under the pit using directional drilling techniques used in the petroleum industry today.

CONSTRUCTING HYDRAULICALLY EFFICIENT WELLS

Hydraulic efficiency is desirable in most water wells, but nowhere is it more essential than in mine dewatering systems utilizing several wells (Figure 5). We will define hydraulic well efficiency quantitatively later in this section; but for the purposes of the following discussion, let us introduce it as a measure of the relative difference between the water level in a pumping well and that in the immediately adjacent aquifer.

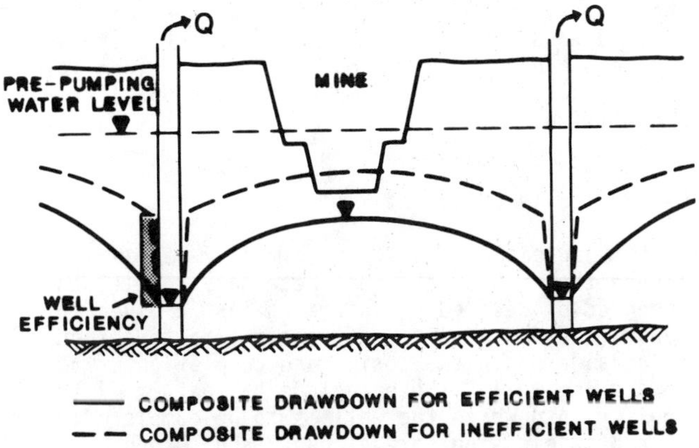

FIGURE 5. EFFECTS OF WELL EFFICIENCY IN MINE DEWATERING

Inefficient wells can significantly increase the cost of a dewatering operation by

- Increasing the number of wells required to effectively dewater a mine, and
- Creating unnecessarily high pumping lifts in the wells.

The latter factor can result in larger pumps being needed, which in turn necessitates larger wells.

In most dewatering operations, the power costs to operate pumps for the life of the operation will exceed the capital and maintenance costs of the wells and pumps. Any short-term savings in capital associated with "short-cuts" that result in poorly completed and developed wells will soon be lost after the wells are put into production. This becomes readily apparent from the equation describing the power required to pump water

$$P = Q \cdot H/(8807 \cdot E_p \cdot E_m) \quad (6)$$

where

P = power [kilowatts],
Q = discharge [m^3/day]
H = total dynamic head [m],
E_p = pump efficiency, and
E_m = motor efficiency.

Pump and motor efficiencies are typically about 0.75 and 0.85, respectively. Using for an example a dewatering operation that produces an average of 55,000 cu m per day (about 10,000 gpm) with an average lift of 200 m (about 650 ft) for 10 years and with a power cost of US$ 0.07 per kilowatt-hour, the power costs for pumping would be about US$ 12 million. As Equation 6 indicates, power and, hence, pumping costs are directly proportional to total dynamic head. A change in lift (normally the primary component of total dynamic head) of only 10 to 30 percent, which is a conservatively reasonable expectation, could reduce the long term pumping costs in this example by about US$ 1.2 to 3.6 million.

Designing, Completing, and Developing Hydraulically Efficient Wells

In most cases, a well is primarily designed around the size of the pump necessary to lift the expected or desired discharge against a given total dynamic head. A number of specific factors should be considered in the design and completion of a dewatering well. These include the nature of the aquifer into which the well is to be completed; appropriate casing, gravel pack, and screen materials; drilling methods and mud program; and well development. In any given dewatering program, there also might be other logistical considerations that dictate certain aspects of well construction.

Two key factors in assuring that a dewatering well will have maximum obtainable efficiency (within reasonable economic constraints) are

- The entrance velocity of the water into the well is kept below about 0.3 per sec, and
- The primary water-producing intervals of the wellbore are not plugged with drilling fluid or cuttings.

The first factor is easily controlled by using a proper screen. Risking the wrath of wire-wrap screen manufacturers, we will state that there are very few situations where wire-wrap screens are necessary in water wells completed into rock aquifers. We generally prefer to use mill-slotted pipe or one of the commercially available shutter or louvre screens.

There have been some significant advances in the past decade in the petroleum and mineral exploration industries related to drilling methods and fluids. Proper selection and use of drilling techniques and drilling fluid additives will greatly enhance the chances of a dewatering well being efficient. This is particularly true in fractured rock aquifers with discrete fractures or fracture zones susceptible to large drilling fluid losses. Because these same zones generally are the main water-producing intervals, it is important to use a mud system that can be completely removed from the fractures upon completion. Traditional drilling fluids for such rock have included bentonite-based muds with loss circulation materials (LCM) such as cottonseed hulls, shredded paper or tree bark, cellophane, and granular bentonite. Mud systems now exist, however, that combine poly- acrylamide-based muds with ACID WOOL (trademark of Distribution Services of Longmont, Colorado), a degradable mud and acid soluble LCM that can be removed from the well during development (Hanna et al., 1989). We believe this is an important breakthrough in dewatering well technology.

Development is critical to maximizing well efficiency by reducing any impediments to flow into the well. This is accomplished by removing mud, LCM, drill cuttings and natural fines from the screen openings, gravel pack and the aquifer immediately adjacent to the well. The most effective method of developing wells completed in fractured rock aquifers is by utilizing a specially designed tool (Lundorff, 1988) that combines jetting, surging, and airlift pumping with the use of chemicals.

Well Testing

Not to be confused with aquifer tests which are conducted to determine the hydraulic characteristics of the aquifer, well tests should also be performed. These tests, whose purpose is to determine the hydraulic efficiency and main water-producing zones of an individual well, should be conducted in any test or prototype dewatering wells and at least a few of the initial production dewatering wells. The resulting information is then utilized to design future wells to assure their maximum efficiency.

<u>Step-Drawdown Tests</u>. The most common method to assess the hydraulic efficiency of a well is to conduct a short duration (4 to 10 hours) step-drawdown test in which the well is pumped in "steps" of increasing discharge rate. The resulting discharge vs drawdown data can be analyzed by several methods. Our preferred method is one developed by Labadie and Helweg (1975) which uses non-linear regression and expresses the results in the form

$$s = BQ + CQ^n \quad (7)$$

where

s = drawdown in pumping well [m],
Q = discharge rate of well [m^3/day],
B = laminar head loss coefficient [day/m^2],
C = turbulent head loss coefficient [variable], and
n = an exponent [dimensionless].

Once the values of B, C, and n are determined, the well efficiency, WE, can be expressed by

$$WE = 100 \times BQ/(BQ + CQ^n). \qquad (10)$$

Obviously a 100 percent efficient well is one in which all of the head losses are in the aquifer and there is no turbulent flow either through the screen and gravel pack or in the aquifer. Turbulent flow in the formation in the immediate vicinity of the wellbore, however, is inevitable in most fractured rock aquifers (Atkinson and Gale, 1985). Hydraulic efficiencies in the range of 70 to 90 percent are still achievable, and any efficiencies less than this should cause reconsideration of well design and completion methods.

Production Logging. It is advisable to run production logs in wells completed during the early stages of the dewatering investigation and as part of well tests prior to ordering permanent pumping equipment. These logs, usually made with a spinner tool, yield data that can be used to define discrete water-producing intervals and correlate water production with structures and/or lithology within the well. Together with the airlift production data from the exploration program, these data define the specific targets for dewatering.

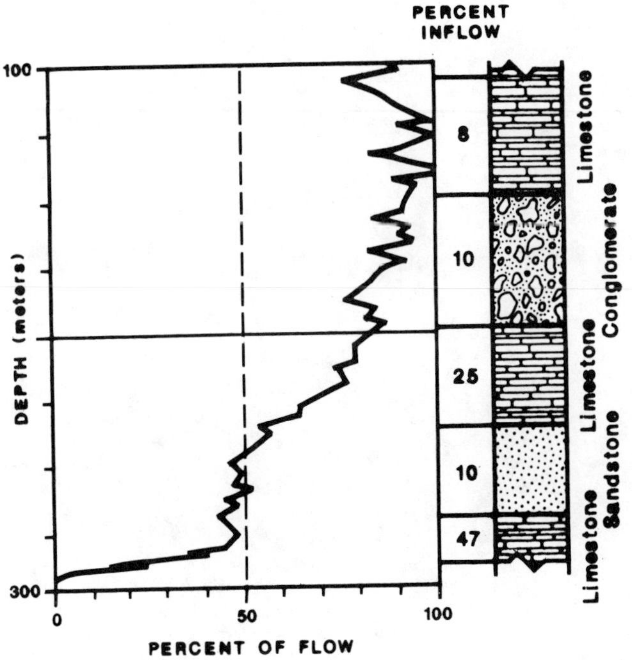

FIGURE 6. PRODUCTION LOG

For example, if production logs indicate that a limited interval is contributing most of the water (as shown in Figure 6) to a majority of the wells , it might be cost-effective to modify future well construction to produce from only that interval. Furthermore, although the concept of aquifers and aquitards breaks down in most mine settings, it might also be learned with the help of production logs that pumping from deeper intervals will not significantly effect early mining operations. Consequently, relatively shallow initial wells or modifications (e.g., plugging the bottom) to existing wells might be advantageous.

ENVIRONMENTAL AND OTHER CONSIDERATIONS

With any proposed mine dewatering operation, prior consideration must be given to the potential environmental and legal impacts of the removal, consumption and discharge of relatively large quantities of ground water. Potential impacts include alteration of existing water resources (e.g., springs, wetlands, streamflows, depths to ground water), injury to existing water rights, and changes in water quality. These issues usually require the mine to be in compliance with a plethora of governmental regulations.

In large-scale dewatering operations, one of the more problematic concerns under current environmental regulations can be the disposal of the water produced. Disposal options, which will vary depending upon regulatory considerations, land ownership, water rights, climate, and the quantity and quality of the water to be disposed, include

- Consumptive use (e.g., mill feed, dust control),
- Direct discharge into a surface-water course, and
- Re-injection or infiltration to the ground-water system.

The latter two options would have to be implemented in hydrologically isolated or sufficiently distant, down-gradient locations to prevent or limit recirculation of the water.

The quality of the water will also be a key factor in whether it can be directly discharged or returned to the ground-water regime. This could require treatment to remove suspended sediment or specific chemical constituents (e.g., arsenic) prior to disposal. As previously indicated, the selection of a dewatering system should consider the relative amounts of "clean" vs "dirty" water that will be produced and have to be managed.

Water rights can be affected by mine dewatering, but since the laws and regulations regarding water rights vary so greatly among countries and other political subdivisions, no specific procedural recommendations are appropriate here. However, in almost all cases, a thorough search of both surface and ground-water rights that could be impacted is advisable. Results of the numerical modeling will define the potentially impacted area and can also be used to help design any necessary remediation programs.

REFERENCES

Atkinson, L.C., and Gale, J.E., 1985, "Minimizing Well Losses in Dewatering Wells in Fractured Rock," Proceedings, 2nd International Congress of International Mine Water Association, Vol. 1, pp. 69-81.

Brawner, C.O., 1979, "General Report on Mine Drainage," Proceedings, 1st International Mine Drainage Symposium, pp. 33-53.

Dudgeon, C.R., 1988, "Prediction of Flow into Mines in Fractured Rock Aquifers," Reprint from Conference on Mine Dewatering, University of New South Wales, 11 pp.

Durbin, T.J., and O'Brien, B., 1987, "Documentation of a Fortran Program for Three-Dimensional Simulation of Free-Surface Aquifers by Finite Element Method," Open-File Report, U.S. Geological Survey, 154 pp.

Goodman, R.E., Moye, D.G., van Schalkwyk, A., and Javandel, I., 1965, "Ground Water Inflows During Tunnel Driving," Bulletin of Association of Engineering Geologists, Vol. 2, No. 1, pp. 39-56.

Hanna, T.M., Rothauge, F., and Pettit, P.M., 1989, "Use of ACID WOOL as a Loss Circulation Material in Fractured Rock Aquifers," manuscript in preparation.

Jacob, C.E., and Lohman, S.W., 1952, "Non-Steady Flow to a Well of Constant Drawdown in an Extensive Aquifer," Transactions, American Geophysical Union, Vol. 33, pp. 559-569.

Kruseman, G.P., and de Ridder, N.A., 1976, "Analysis and Evaluation of Pumping Test Data," International Institute for Land Reclamation and Improvement, Wageningen, Netherlands, 200 pp.

Labadie, J.W., and Helweg, O.J., 1975, "Step-Drawdown Test Analysis by Computer," Ground Water, Vol. 13, No. 5, pp. 438-444.

Lohman, S.W., 1979, "Ground-Water Hydraulics," Professional Paper 708, U.S. Geological Survey, 70 pp.

Lundorff, E.E., 1988, "Development Procedures for Well Completion," Proceedings, Well Design and Completion Seminar, California Ground Water Association.

Nind, T.E.W., 1965, "Influences of Absolute and Partial Hydrologic Barriers on Pump Test Results," Canadian Journal of Earth Sciences, Vol. 2, No. 4, pp. 309-323.

Papadopulos, I.S., 1967, "Non-Steady Flow to a Well in an Infinite Anisotropic Aquifer," Proceedings, Dubrovnik Symposium on Hydrology of Fractured Rocks, International Society of Scientific Hydrology, Vol. 1, No. 23, pp. 21-31.

Singh, R.N., and Atkins, A.S., 1984, "Application of Analytical Solutions to Simulate Some Mine Inflow Problems in Underground Coal Mining," International Journal of Mine Water, Vol. 3, No. 4, pp.1-27.

Stallman, R.W., 1971, "Aquifer-Test Design, Observation and Data Analysis," Techniques of Water-Resources Investigations, Book 3, Chapter B1, U.S. Geological Survey, 26 pp.

Vandenberg, A., 1977, "Pump Testing in Heterogeneous Aquifers," Journal of Hydrology, Vol. 34, pp. 45-62.

Walton, W.C., 1987, Groundwater Pumping Tests: Design and Analysis, Lewis Publishers, Chelsea, Michigan, 201 pp.

Woodward, D.G., 1978, "Ground Water and Aquifer Tests," Technical Paper No. 5, Minnesota Department of Natural Resources, Division of Waters, 24 pp.

Chapter 20

AUSTRALIAN NARROW-VEIN MINING - PROBLEMS AND OPPORTUNITIES

STEPHEN G. GEMELL

FELLOW, AUSTRALASIAN INSTITUTE OF MINING AND METALLURGY
MEMBER, AMERICAN INSTITUTE OF MINING, METALLURGICAL AND PETROLEUM ENGINEERS

INTRODUCTION

Gold mining has undergone a massive revival in Australia within the past decade. With very few exceptions, these operations are involved with the development of deposits which have been previously and partially worked by underground methods. In most cases, current operations have involved the open cut mining of near-surface remnants or low-grade hanging wall and footwall mineralisation. A growing number of mines are, however, progressing to the development of underground workings either as extensions of existing open cuts or as "stand alone" projects. These operations vary in scale from the Kalgoorlie operations of North Kalgurli Mines Limited and Kalgoorlie Mining Associates to the small scale "gouging" operations of individuals and partnerships. The individual production rates rarely exceed 200 000 tonnes per year, primarily because of the magnitude of reserves or the style of access available for rehabilitation.

The narrow-vein mines were initially re-opened using the traditional methods, in particular shrinkage stoping and cut-and-fill stoping. As a result of experience gained in various ground conditions, certain variants have become popular. This paper investigates the fate of the traditional methods in Australian practice, and reviews some of the recently applied variants. This involves not only a perusal of current practices, but also an assessment of the influence of recent technological development on the economics of narrow-vein mining. Some consideration is also given to the effects of previously employed stoping techniques on efforts to access remnant blocks in the re-development of old mines.

TRENDS IN CURRENT TECHNIQUES

Overhand Open Stoping

This method has been traditionally employed where orebody, hanging wall and footwall can all be described as competent. The stopes commence above two lines of stull or set timbering, or above a horizontal line of pillars separated by boxholes or millholes. The general line of face advance is up-dip. Because of the advantages in back control, each horizontal bench is worked by drilling vertical faces with machine-and-airleg rather than drilling up-holes with a stoper. Several benches can be worked simultaneously, providing a staggered face pattern. Ground support is assisted by a sequence of stull placement, with the miner working from a timber platform supported by the stulls. In later stages of the stope life, additional support can be provided in the lower sections by leaving some ore as a shrinkage fill. A general arrangement of this style of stope is shown in Figure 1.

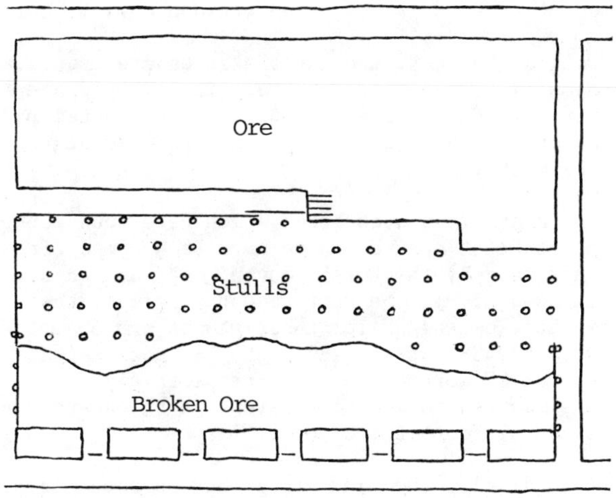

FIGURE 1. OVERHAND OPEN (STULL) STOPE

In steep-dipping workings, this method of stoping is now less popular than its alternative options. Nevertheless, it has recently been utilised in situations where either

a) the strike length of the stoping block is restricted; or

b) the irregular distribution of the mineralisation makes the retention of random pillars attractive.

In case (a), a short strike length can increase the unit costs of alternative methods. With shrinkage stopes, for example, the cost per tonne stoped of establishing rearings etc., and the difficulties associated with restriction of working face location and draw positions, are exacerbated with shortened stope length. Similarly, in case (b), shrinkage stoping is avoided because of the sometimes dangerous and always inconvenient interference to draw control imposed by random pillars.

At the Williams United Mine, near Bright in northeastern Victoria, both of these situations existed. The Williams United was discovered in 1934 and was worked intermittently until the current owners assumed control in 1983. The operation is a small one with a total operational complement of eight men, of whom five are employed underground. Mining operations have been continuous from 1983 until the mine was placed in care and maintenance in September 1988. The gold mineralisation appears in a quartz fissure vein with a strike direction which varies between 315 and 345 degrees, and with a dip varying from 57 degrees (near surface and at local "rolls") to 85 degrees to the east. The fissure width varies up to 1.7 metres (5.7 feet) and the reef material can be described as competent. The country rock comprises an interbedded mixture of sandstones, feldspathic sandstones, slates, and mudstones. This material is also competent, although some planes of weakness occur at contacts between the various rock types.

Within the reef, the payable zones are truncated physically by faults or mineralogically by grade depletion. Economic zones appear as elongated pods or shoots which plunge at approximately 70 degrees to the south.

There are four such zones which have been worked by the current owners. These have been named (from north to south) the North, South, Dab, and Anonymous Shoots. In the most recent workings, the zonal strike lengths for these shoots are 30 metres (100 feet), 38 metres (127 feet), 19 metres (63 feet) and 30 metres(100 feet) respectively. Mineralisation is most consistent in the north, and least consistent in the south.

Accordingly, the mine management selected an overhanding technique utilising random pillars located according to grade - the in-situ cut-off being 8 grams Au per tonne (0.23 oz Au per short ton) - for all stopes but the North Shoot, which is of consistently high grade throughout. Because of the relatively long stope length of the South Shoot, shrinkage stoping was initiated here. This attempt was soon abandoned due to the lack of selectivity inherent in the method.

Working faces were accessed by ladders temporarily secured to the stulls and pillars. The pillars, if physical and grade conditions permitted, were gabled to reduce ore losses from broken material retained in the stope.

Productivity ratings at Williams United were very poor, rarely exceeding 6 tonnes (6.6 tons) per man- shift underground. This resulted not primarily from the stoping techniques, however, but from the mine development and restrictions in the ore haulage facilities. The ore was transferred in 0.5 tonne (0.55 short ton) capacity trucks to the surface bins This involved hoisting the trucks sequentially in an internal inclined winze and in the main vertical shaft.

Underhand Open Stoping

The format of this technique is essentially the same as that of overhand stoping, except that the general line of face advance is down dip. As the miner is working on a solid floor, drilling can be done using either sinkers or machine-and-airleg combinations. In order to avoid extensive hand bogging, each bench is worked consecutively, with broken ore being scraped to the stope rises.

This benching method has been successfully employed in Australian Development Limited's White Devil Mine, near Tennant Creek in the Northern Territory. In this region, the method was first employed by Australian Development's contractors, Flamore Pty. Ltd., to exploit small isolated pods of gold bearing ore. These pods were individual magnetite lenses, with near-vertical dip, and with strike lengths of less than 35 metres (115 feet). The vertical dimensions of these lenses were less than 25 metres (82 feet), and the uppermost ore boundaries were from 17 metres (56 feet) to 45 metres (148 feet) below the surface.

To exploit each of these bodies, a small shaft of cross-sectional dimensions of 1.2 metres (4 feet) by 2.4 metres (8 feet) was sunk in the footwall, displaced horizontally approximately 9 metres (30 feet) from the orebody. Using a vacuum ore lift system to clean broken rock from the shaft, sinking rates of up to 8 metres (26 feet) per day were achieved. This high rate of sinking was achieved at Rising Sun West, where an existing borehole down the centreline of the shaft facilitated blasting. From the shaft, one crosscut was established to intersect below the lower extremity of the orebody whilst a second crosscut would be positioned to intersect the ore body some two to three metres (seven to ten feet) below the cap. Both cross-cuts had cross-sectional dimension of 2.0 metres (6.6 feet) height by 1.8 metres (6 feet) width. At this stage, the vacuum ore lift was removed and replaced with a conventional single-drum hoist. The lower plat was fitted with a brow box, and an Eimco 12B was then permanently stationed on that level. A rise through the ore body linked the two levels, and a sill drive was established on the upper level. The ore above the level was broken, and before being scraped to the rise, was used as a working platform from which the cap area was fully rockbolted and meshed. After scraping cleared this material, benches were established with faces approximately 2.5 metres (8.2 feet) high being worked from the central rise to the strike extremities of the ore body. Because of the short strike length of the ore body, each bench was completed prior to commencing the one beneath it. Hence, each small mine had only two working faces extant at any time.

These small mines each produced at a scheduled rate of 60 tonnes (66 short tons) per shift with a production crew of three men: miner, bogger driver, and winder driver.

Australian Development commenced underground operations at White Devil in 1986, and employed the same contractors to utilise a similar mining strategy but on an enhanced scale. A shaft was sunk 115 metres (377 feet) to exploit the Main Lode, a depth which fortuitously was just within the practical lifting limit of the twin-fan vacuum ore-lift unit available. Levels were spaced at nominal vertical intervals of 17 metres (56 feet). Stoping was scheduled from various locations on different levels simultaneously in order to optimise metallurgical blending and cash flow. All ore was scraped to the stope rises, which were used as ore passes. A rail system was installed on the lowest level for ore transfer from the rise-drawpoints to the plat loading pockets. This layout is illustrated in Figure 2.

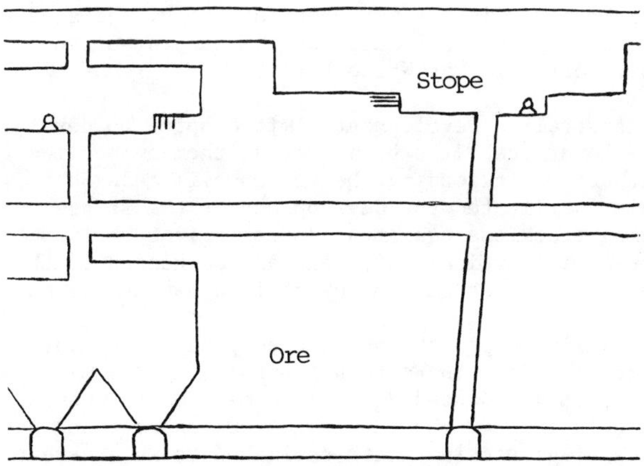

FIGURE 2. UNDERHAND OPEN STOPE

The main shaft had cross-sectional dimensions of 2.4 metres (8 feet) by 1.2 metres (4 feet). The increased depth resulted in a slower sinking rate averaging 3.5 metres (11.5 feet) per day, although individual rates of 6 metres (20 feet) per day were achieved. The shaft was equipped with a single drum hoist as the main winder, and production at a rate commensurate to 50 000 tonnes (55 000 short tons) per year was achieved. This rate was retarded slightly on the transfer of production responsibilities from the contractor to the mining company, but is expected to increase with development of the lower orebody zones (see "Shrinkage Stoping").

Shrinkage Stoping

Shrinkage stoping remains the most popular technique for working steeply dipping narrow-vein deposits in competent ground conditions. The traditional approach in the Eastern Goldfields of Western Australia has been to establish leading stopes above reef drives, then to instal timber decks above the drive to support the broken ore. Although this approach is currently employed in remnant areas, for larger stoping blocks the favoured technique is to establish a haulage drive either in the footwall or hanging wall, and to connect it by frequent crosscuts to the orebody. Rail mounted pneumatic loaders are used to load out from the drawpoints or mill holes, and discharge across the sides of the waiting ore trucks. The cross-cuts are short in order to minimise both development cost and load out time. Their tracks are linked to the haulage drive rails by small hand -operated turntables, so that the loader can be readily moved from one mill hole to the next.

The obvious alternative layout to these is to retain the reef drive and establish the stope workings from an intermediate level several metres above the drive. The intermediate is connected to the main level by a series of short rises (or "jump ups") which serve as ore passes. The rises have to be reasonably closely spaced in order to ensure even draw. This alternative is popular at North Kalgurli Mines Limited, where 40% of the underground production rate of 220 000 tonnes (242 000 short tons) per year is achieved by shrinkage stoping. Two problems arise from the use of this variant:

1) The combination of restricted dimensions on the pillars between the rises, and the incompetent ore/country rock boundaries often associated with narrow-vein deposits can create unstable working conditions. At the Williams United, this variant was used to exploit the North Shoot. On the 93 metre (305 foot) level no less than three pillars failed during the life of the stope.

2) A higher percentage of the in-situ ore is retained in pillars than in either of the other two variants. Pillar recovery operations are therefore more significant

as components of overall stoping costs, as are stope development costs.

One of the major disadvantages of shrinkage stoping is the delay in revenue caused by the retention of broken ore within the stope. This material, which forms the working platform, may not be fully drawn until the end of stoping operations. In order to overcome this, a variant known as gallery stoping has been adapted and used in several of the Western Australian gold mines.

The Paringa Mine in Kalgoorlie was amongst the first to utilise this variant in recent times. The layout, as shown in Figure 3, and operations are well described by Upton. These may be summarised as follows. The stopes are divided at the mid point horizontally and the lower half is worked by conventional shrinkage operations. At the mid-elevation line, a series of long strike pillars are left in position, and an intermediate drive is developed above this pillar line. Stoping operations commence above the pillar, whilst below the pillar line the stope can be fully drawn. The area above the pillar line is actually worked as a leading stope, and is cleaned with the assistance of scrapers. As the face of the broken ore retreats from the scraper winch position, the caps are removed from previously established rises in the pillar line. This ensures a minimised scraper operation to get the broken ore into the drawn workings below the pillar line. The final operation is the removal of both the pillar line and the series of rib pillars protecting the manways at each end of the stope.

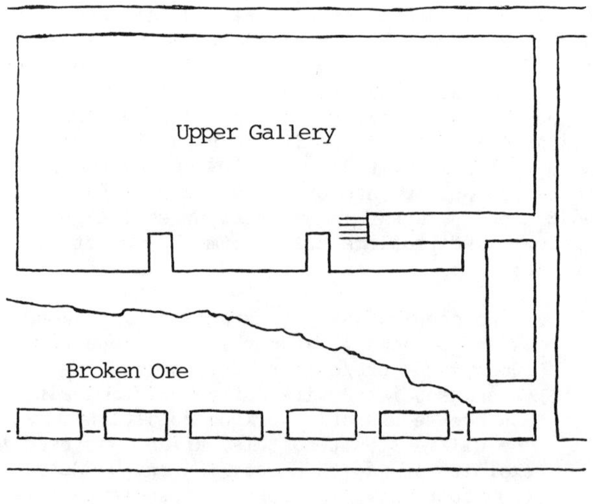

FIGURE 3. GALLERY STOPE

Although this method was adapted to replace shrinkage stoping, it was employed initially in some of the wider stoping areas i.e. at widths greater than 2 metres (6.6 feet). In mines with narrow stoping widths, it has been found convenient to limit the intermediate drive spacing, particularly where these are accessed by rises rather than being connected to a decline.

At the Timoni Mine, near Copperfield in Western Australia, the Mount Ida Joint Venture have deepened an old shaft at the southern end of the workings, using a vacuum ore-lift unit. Cross-cuts were driven on two levels to intersect abandoned drives which were then rehabilitated in order to access undeveloped reserves beyond the southern extremity of the old workings. High grade remnant blocks were also accessed. In some areas, the previous operators had completely removed the floor pillars from the reef drives. This necessitated the pinning of brackets to both footwall and hanging wall to provide support for a timber deck and the rails. A detailed description of this technique is provided later in this paper.

Initially, underhand open stoping was employed, but this was found to cause blast damage to timbers in the vicinity of drives and remnants. The gallery stoping variant then adopted used the following procedure. In each of the new stoping blocks, a rise is established in the longitudinal centre, and sub-levels at 10-metre (33-foot) vertical intervals are driven to the pay limits. Each block is then worked by leading stoping, leaving a pillar of 2 metres (6.6 feet) between the stope and the intermediate level above. All ore is scraped back to the central rise, a distance of less than 15 metres (49 feet). No pillars are recovered until all primary stoping operations in the block are completed.

The concept behind this method has been further adapted to produce yet another variant: rill shrink stoping. This method has been successfully used at the Granites in the Northern Territory, and as a result has been selected for the stoping of the lower orebodies at White Devil.

Australian Development Limited opted to develop the lower orebodies by accessing them by decline rather than extending the White Devil Main Shaft. This facilitated the development of the stopes using trackless equipment. Reef drives at 15-metre (49-foot) vertical intervals are connected to the decline. The schematic layout is shown in Figure 4.

A similar approach has been adopted at Rothsay near Perenjori in Western Australia. This mine has previously produced 323 kilograms (10 380 ounces) of gold from 40 500 tonnes (44 600 short tons), at an average stoping width estimated to be 1.4 metres (4.6 feet). The actual width of payable mineralisation is approximately 0.9 metres (3 feet). The vein is remarkably consistent with regard to strike and dip, with little structural interruption of the orebody. The angle of dip is approximately 70 degrees.

Here, however, the rill stoping design is more ambitious. The production drives are being placed at 38-metre (125-feet) vertical intervals, immediately adjacent to the reef in the hanging wall. The ore is then stripped off with minimal dilution. At the extremities of the orebody the drives are to be linked to rises serving as return ventilation airways. The length of individual stopes will be determined by grade characteristics once each level is developed. As at White Devil, the stopes will be worked as a series of en echelon faces, in this case 3 metres (10 feet) high with each face leading the face above by 2 metres (6.6 feet), retreating back towards the decline access. These face dimensions will vary in practice according to the rill angle of the broken ore. Minimum stoping width is anticipated to be 1.2 metres (4 feet).

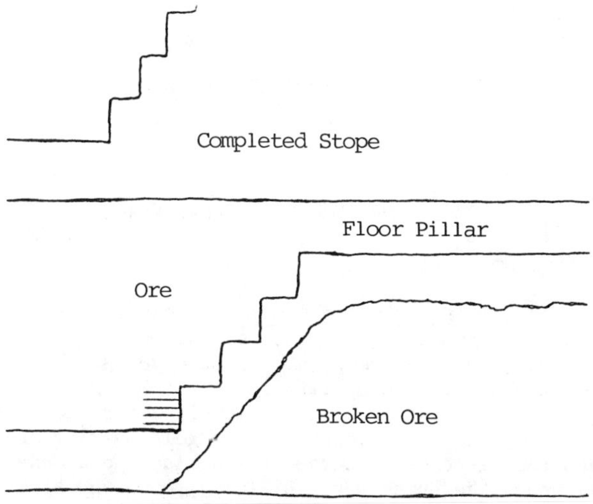

FIGURE 4. RILL SHRINK STOPE

At the time of writing, the mining company is investigating the potential of converting an Ingersoll Rand LM100 drilling rig for drilling within the stope.

THE INFLUENCE OF PREVIOUS MINING TECHNIQUES

Resue Stoping

This is one method which remains distinctly unpopular, primarily as it has to date proved difficult to mechanise. It's historic use has proved to be of considerable interest as mine owners are finding that the significant ore losses into the stope fill can convert previously neglected old workings into a source of additional revenue. As there is often no stope establishment or ore breaking cost associated with mining this resource, particularly if level rehabilitation is required to access fresh reserves beyond the old workings, the concept of bleeding these stopes becomes more attractive in inverse proportion to the efficiency of the previous operation. In Western Australia, individual stope grades of more than 3 grams per tonne (0.1 oz per short ton) are known to exist as resue stope fill.

Several mining companies in eastern Australia have undertaken some tentative and generally abortive research into the application of resue stoping in a modern productive environment. The companies most interested are those who hold mining tenements in the narrow-vein mining locales of Victoria and New South Wales. The Dick's Reward Mine, near Wellington in New South Wales, is one such site. This mine was established to work a gold-bearing quartz vein with an overall thickness varying from 0.3 metres (1 foot) to 0.9 metres (3 feet). The orebody has a dip of approximately 20 degrees to the north, but is locally displaced by folding and faulting at frequent intervals. Although pinching or swelling of the orebody - beyond the thickness limits described above - occurs at the points of displacement, the orebody has been systematically extracted over a strike length of 400 metres (1300 feet). Mining operations commenced prior to 1909, when the first of two inclined shafts were sunk on reef. From these and other subsidiary shafts, reef drives were developed to what appeared to be terminating faults. From the reef drives, stopes were resued, with the waste being packed by hand. The boundaries of the packed areas were carefully "stone walled" between stulls, with the internal packing being less consistently installed. The only pillars other than remnant blocks remaining from previous operations are the 2-metre (6.6 foot) wide pillars adjacent to the shaft. Reef drives are protected by 0.3 metre (1 foot) diameter round timbers supporting the barricades for the waste packing in the stopes above each drive. Historic records are not of sufficient accuracy to be able to reasonably estimate past production tonnage and grade. The mine currently has two types of development target: the remnant blocks, and the adjacent fault-displaced blocks not detected by earlier miners. The owners of the mine, Cluff Resources Pacific Limited, have assessed the application of a slightly revised variant of resue stoping to this situation. This variant involves the use of suspended blasting mats and scraper cleaning.

Remnant Areas

Most operators experienced in the re-establishment of mining operations in partially worked areas are well aware of the difficulties that can occur in scheduling production and cash flows, and in maintaining a safe operation. These problems include:-

a) the competence of walls and the stability of openings which have been exposed for periods of time,

b) the potential for undesired rill cascade in previously filled or incompletely drawn stopes,

c) temporary or permanent losses of broken ore during the establishment of a workable draw system in old stope areas,

d) the permanent losses of in-situ ore as pillars resulting from the irregular operations of predecessors, and

e) negotiating worked out areas in order to access reserves beyond.

As has been explained previously, the operators at Timoni preferred to extend a shaft to re-establish access on the lower levels rather than completely clear the deeper shaft to the north and rehabilitate longer sections of the reef drives. Nevertheless, in some accessed areas of these drives, it was discovered that the previous operators had completely stripped the floor pillars. The three options available to the new managers were:-

a) to abandon the attempt to access the reserves beyond the open stope and to delay stoping schedules until the block could be accessed from lower levels, or

b) to develop a drive through the footwall of the affected section thereby by-passing the open stope, or

c) to construct a deck across the stope at the level.

The third option was determined to be the most cost effective. A series of brackets suspended at 2.5-metre (8.2-foot) horizontal intervals along both footwall and hanging wall. Each bracket, or box, was secured to the wall by 2 rockbolts each 1.5 metres (5 feet) in length. Bearer sets of karri hardwood of 150mm by 150mm (6 inch by 6 inch) cross-sectional dimensions were then cut to lengths corresponding to the local horizontal stope width. These sets were then placed in the brackets and were securely wedged to prevent movement. A timber deck comprising of 50mm (2 inch) planks was overlain on the bearer sets, and sleepers of 100mm by 75mm (4 inch by 3 inch) cross-sectional dimensions were then secured. Locomotives and rolling stock were selected with the prime criteria being to obtain the smallest point loads conformable with economic practice. The weight of rail used was only 9.9kg/m (20lb/yd). A schematic section is shown in Figure 5.

This particular incident is merely one example of the effect of previously undertaken stoping and pillar recovery operations on a recent redevelopment.

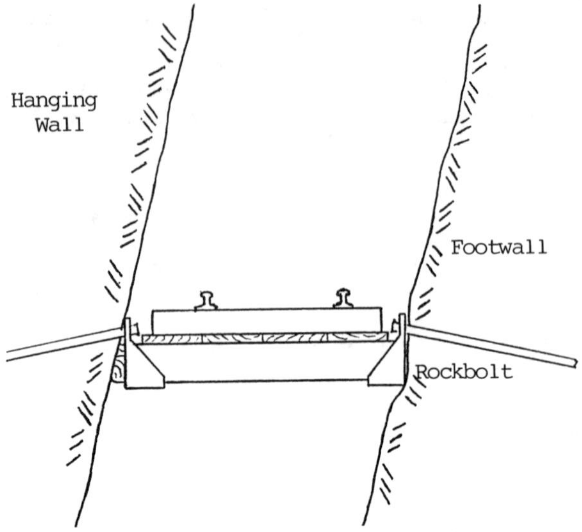

FIGURE 5. SUSPENDED LEVEL DECK

EFFECTS OF TECHNOLOGICAL ADVANCES

Vacuum Ore Lift

The use of pneumatic ore lift systems, often known as "suckers" or "blowers", has been common in soft-ground development for some years. Their development in the opal fields of Coober Pedy in South Australia has been previously described by Calbert, and their adaptation to gold-bearing ores has been further discussed by Dorricott and Jones. Although the units were initially perceived by most operators to be an extremely efficient cleaning mechanism for mud-filled shafts, they have since gained general recognition as the most efficient cleaning mechanism for shaft sinking to depths of less than 120 metres (390 feet). Some small scale operations have used the units to lift the ore from crusher stations adjacent to the shaft. Most operators prefer to use the vacuum ore lift for shaft sinking only, and change over to a conventional hoisting system for ore transport.

The application of vacuum ore lift has had the dual effect of decreasing shaft sinking time and decreasing pre-production costs. The previously described development of the magnetite pods at New Hope and Rising Sun West near White Devil illustrates the transition in economics of extraction of some of the smaller orebodies which are too far apart to use common shafts or declines for access.

The New Hope was developed with a 62-metre (203-foot) deep shaft for the extraction of 1 111 tonnes (1225 short tons) grading 38.5 gram Au per tonne (1.13 ounces Au per short ton). Rising Sun West yielded 6950 tonnes (7660 short tons) at a run-of-mine grade of 19.1 grams Au per tonne (0.56 ounces Au per short ton).

Narrow-Width LHD Units

Narrow-width LHD units have not yet gained popularity in Australia, despite their known recent history in the United States and Canada.

The Golden Crown Mine, near Cue in Western Australia, is currently utilising one LHD unit of this type.

CONCLUSION

The several described examples of narrow-vein mining practice currently adopted in Australia by no means provide an exhaustive investigation of the state-of-the-art techniques employed in working narrow-vein mines. They do, however, provide some insight into the trends of adaptation occurring. Although recent fluctuations in the gold price and the foreshadowed restructuring of the gold mining taxation system have dampened the enthusiasm of the more speculative underground developers and have resulted in the closure of some of the smaller marginal mines, the number of narrow-vein producers has continued to grow steadily. As more of the short-lifespan open cut operations exhaust their mineable reserves, these underground operations will become more and more significant participants in the production of Australian gold.

ACKNOWLEDGMENTS

The author wishes to express his appreciation to those people who provided permission to discuss aspects of their companies' mining operations. In particular, thanks are gratefully extended to Messrs R.G. Adamson (Cluff Resources Pacific Limited), S.A. Sylvester (Metana Minerals N.L.), E.P. Ashley (Poseidon Limited/Australian Development Limited), K.C. Craig (Mt. Ida Joint Venture), A.M. Norton (United and Commercial Holdings Limited), and R.G. Anderson (Flamore Pty Ltd.).

REFERENCES

Calbert, B.L. 1979 "Use of Pneumatic Conveying Systems on the Opalfields", Quarry Mine and Pit, Vol.18 No.5.

Dorricott, M. & Jones, I.O. 1984 "Vacuum Lift Systems for the Transport of Broken Rock in Shafts", Gold-Mining, Metallurgy and Geology: AusIMM Regional Conference, October 1984, pp 289 - 299.

Upton, B.S. 1985 "Gallery Stoping at Gold Resources' Paringa Operations", AusIMM Underground Operators Conference, October 1985, pp 85 - 87.

Chapter 21

MINING OPERATIONS AT THE SONS OF GWALIA GOLD MINE

J. KALNEJAIS

CHIEF GEOLOGIST

GWALIA GROUP OF COMPANIES, PERTH, WESTERN AUSTRALIA

ABSTRACT

Discovered in 1896, the Sons of Gwalia orebody was initially mined by underground methods. During the mine's underground operations workings reached a vertical depth of 1080 metres. The same orebody is now being mined by open cut methods.
The present operation mines a broad envelope of mineralisation surrounding old stopes, whereas the original underground operation concentrated on mining high grade parts of the Gwalia Lode System. Underground mining continued up to 1963 when operations ceased due to high operating costs and the low grade of ore mined. In 1984 a new company, Sons of Gwalia NL, commenced open cut operations.

INTRODUCTION

The Sons of Gwalia Mine is one of the most famous mines in Western Australia. It has had one of the longest operating periods of all the mines in WA, excluding the mines on the Kalgoorlie Golden Mile.

Originally worked as an underground mine until 1963, the mine has been re-opened and is now being mined by open cut methods.

LOCATION

The Sons of Gwalia Gold Mine is located 3.5 kilometres due south of the township of Leonora which is 640 kilometres east-north-east of Perth and 200 kilometres north of Kalgoorlie. Perth, the capital of Western Australia, is 650 kilometres by road from Kalgoorlie which is connected to the mine by a 236 kilometre sealed road. (Figure 1)

The standard gauge railroad from Leonora to Kalgoorlie traverses the lease.

Most of the terrain of central Western Australia is an elevated plain formed from an earlier plateau by the process of arid erosion. Remains of this older plateau occur as mesas, butes and more extensive tablelands. The country around Leonora is a typical example of the plateau, with broad gently sloping valleys and a few remnant hills such as Mt Leonora, which is near the Sons of Gwalia. The major drainage feature is Lake Raeside, an elongated salina trending east-south-east and passing 4.5kilometre south of the mine. Lake Raeside is transverse to the structural grain of the country and may be aligned along a major regional cross fault or lineament.

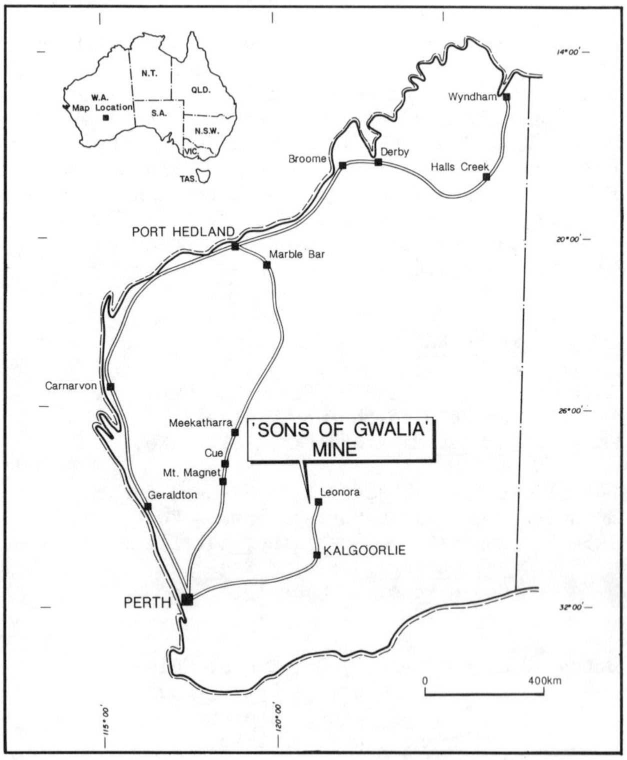

Fig 1 Location Plan

CLIMATE

The climate is arid to semi-arid with an average annual rainfall of 200mm. Usually the first half of the year is wetter, but rainfall is unreliable and the area is subject to drought.

Summers are hot and dry, the average maximum temperature in January being about 35°C with an average minimum for the same month of about 20°C. Occasionally, temperatures reach 46°C. Winters are cool to mild, the average minimum temperature in July being about 5°C and the average maximum about 18°C.

HISTORY

The Sons of Gwalia orebody was discovered in April or May 1896 by Prospectors Carlson, White and Glendinning who were part of a prospecting syndicate funded by Welshman Tommy Tobias, a Coolgardie storekeeper. The new find was only one of a number of reefs opened up in the Mt Leonora district in that year, but proved to be the most significant.

The syndicate registered mining lease 190C over the main outcrop in June 1896 and by September had commenced small scale extraction from a vertical shaft. George W. Hall, an engineer with the London and Western Australian Exploration Company Limited, obtained an option over the property and purchased it in November for £5000 cash.

In order to obtain capital for further development, Hall contacted Bewick Moreing and Company, a London based firm of mining engineers. A young mining engineer, Herbert Hoover, was sent to prepare an evaluation of the mine. Hoover was later to become the 32nd President of the United States of America in 1919.

In January 1898, Sons of Gwalia Limited was launched in London. Most of its capital of $300,000 was issued to the London and Western Australian Exploration Company to acquire the property. Bewick Moreing was engaged as managing consultant at the mine and Hoover was appointed general manager from 1st May 1898.

Hoover had inherited a mine with a vertical shaft down to about 200ft. and a small and inefficient mill. All indications were that the reefs extended to great depth and long term production must be established. In a harsh climate, seventy miles from the railhead and faced with high material and labour costs, Hoover saw that his critical task was to reduce production costs by any means at his disposal.

Hoover's cost-cutting strategy took two main directions: to convert labour to a contract basis wherever possible, and to minimise dead work such as unproductive sinking and driving in the mine. A new underlie or inclined shaft was sunk at 45° in the lode itself and a new 50hp Risdon winding engine installed.

During his time at Gwalia, Hoover was also responsible for designing a new treatment plant and layout for the mine. Hoover left the Sons of Gwalia on 21st November 1898 for a posting in China but was to return on a number of occasions as consulting engineer in later years.

By 1901, annual production had climbed to 115,380 long tons yielding 60,627 ounces.

Although impressive gold yields and generous dividends disguised the fact, ore grade was steadily declining and productivity was only maintained by ever increasing throughput. In 1900, average mine head grade was 22 dwt/ton but by 1910 had fallen to 9 dwt/ton.

In 1914, it was necessary to convert the treatment process to fine grinding and cyanidation. The Wilfley tables were abandoned and the stamp mill outflow was reground.

A disastrous fire in 1921 destroyed the powerhouse and most of the plant. Luckily, the winder, headframe and outside plant were saved. Production was reduced to almost nothing. It was not until 1923 that full production was restored.

The Sons of Gwalia Limited operated the mine continuously until 1963 during which time the mine reached a maximum vertical depth of 1.08 kilometres at the 32 level.

Mounting costs, shrinking reserves and a static gold price combined to reduce the mine's profitability in the 1950's. The Sons of Gwalia closed down a week before Christmas, 1963.

In a working life of 67 years, the mine raised over seven million tons of ore yielding over two and a half million ounces of gold and made a nett profit of £4.5 million. It paid nearly £2.5 million in dividends which is equivalent to eleven times its original capital and paid over £1 million in income tax.

The Sons of Gwalia Limited went into receivership within days of the closure. The major plant items of the mine site were bought by the State and smaller plant and stores were sold at auction in October 1965. The site was progressively stripped of useful machinery and remaining buildings fell into disrepair.

On the 5th May 1983, Sons of Gwalia NL was listed on all stock exchanges in Australia. The public issue comprised 10,000,000 shares at $0.25 for a capital raising of $2.5 million. This new company was floated with a view to initially re-treating the old mine tailings dump. Some re-treatment took place but later in 1983, the operation was suspended.

At the same time, following a successful drilling programme, an assessment was carried out of the open pittable reserves. Drilling was completed in September 1983. A feasibility study was completed by December and construction of a new treatment plant commenced in January 1984. Full scale open cut production commenced on the 22nd July 1984.

GEOLOGY

In a regional setting, the Sons of Gwalia is situated on the faulted eastern limb of the Tarmoola Anticline in a succession of volcanic and clastic sediments, interbedded with acid to mafic volcanic lavas and pyroclastics.

This layered Archean age sequence has been intruded by granite. The nearest granite to the Sons of Gwalia is half a kilometre to the west.

In the mine area, the volcanic pile is interpreted as being of intermediate to basic extrusives and pyroclastics. This original sequence has been folded, sheared and metasomatised to produce a band of chlorite, biotite quartz and actinolite schist up to 150 metres wide, which now hosts the gold mineralisation. This schist has been termed the mine schist. The mine schist strikes N16°E and dips at 40-50° to the East.

MINERALISATION

Mineralisation consists of partially silicified sericitised schists with disseminated sulfides. Sulfide bearing quartz lenses and stringers are generally present in the richer sections. This is particularly so in structurally controlled sites that have been prepared by folding and shearing.

Below the level of oxidation which extends to 45-55 metres, the main sulfides present are pyrite, chalcopyrite, pyrrhotite and arsenophyrite.

The ore bodies mined by the old Sons of Gwalia mine were high grade shoots or ore bodies from within the ore schist. These have been designated into various lode systems, namely:

1. Main Lode
2. West Lode
3. The South Gwalia Series
4. Hanging Wall Lodes
5. Southwest Branch

Main Lode which dips at 45° to the east has been mined along strike for 500 metres and from the surface right down to the 32 level. Mining width varied from 1.8 to 2.4 metres. West Lode has a maximum strike of 300 metres but has a slightly flatter dip of 38° to the east. All the lodes have a 70°S plunge.

The orebody currently mined in the open pit operation is a broad envelope of mineralisation that occurs on both sides of the old mine stopes. Mineralisation occurs on both the hanging wall and footwall sides of the old stopes. (Figure 2)

The projection of the orebody produces a horseshoe shape at the surface. The eastern limb of this orebody corresponds to what was previously mined as Main Lode, while the western limb represents the envelope around West Lode.

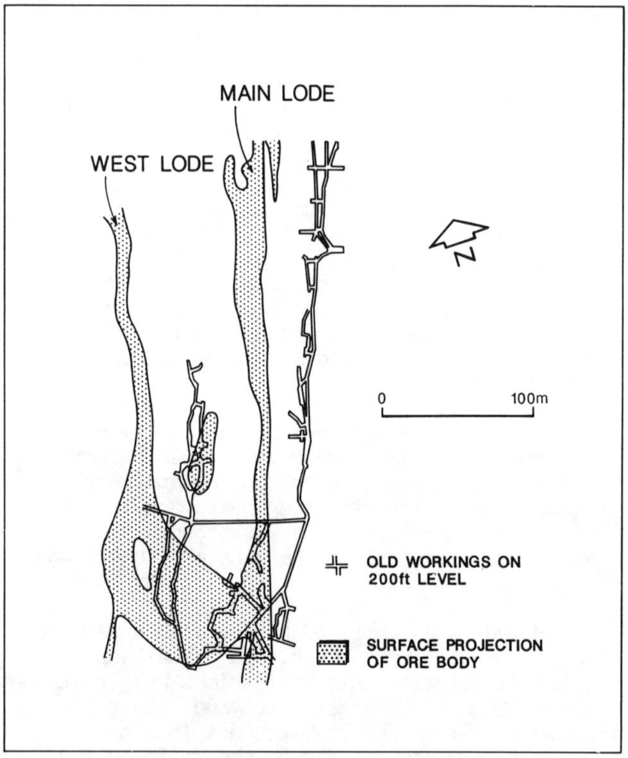

Fig 2 Plan of Orebody Projected To Surface

The orebody dips to the east at 45° with slightly flatter dips on the west limb. The whole orebody has a southerly plunge of 70°. This is identical to the plunge of the individual ore shoots mined by the old Sons of Gwalia Company. In the open pit the orebody has a relatively simple configuration in the northern end in that it parallels strike of the country.

In the southern end of the pit, there are a number of steeply dipping faults that cut the orebody. Individual displacement on these faults varies from a few centimetres to 10 metres. It is this faulting that has resulted in an apparent horseshoe shaped closure of mineralisation. It can also be suggested that this represents the nose or keel of a tight isoclinal fold.

It is considered that structure is very important in the distribution of mineralisation, particularly in that the orebody has been developed down pitch for 1750 metres while it has a strike length of only 550 metres.

UNDERGROUND MINING

The main entry to the old underground mine was by a three compartment inclined shaft with dimensions 6 metres x 2.4 metres. Level intervals varied from 31 metres to 90 metres. The deepest level served by the main shaft was the 30 level at an inclined depth of 1392 metres. (Figure 3)

Ore was hoisted in 3.5 ton self dumping skips using a steam winder.

An internal inclined shaft was sunk to serve the levels below the 30 level shaft, with dimensions 6 metres x 2.4 metres.

Open stoping was originally practiced leaving pillars for hanging wall support. By the time 25 level was reached, the restrictions on mining due to rock pressure forced the adoption of cut and fill mining methods. Later practice was to break the ore in horizontal cut and fill stopes. The width of the stopes was governed by assay results and 1.8 metre sample holes bored into both walls as a routine procedure.

Waste rock was obtained from a glory hole in the hanging wall of the lodes. Owing to the flat dip of the ore bodies, transfer levels from footwall to hanging were necessary to keep the mullock passes steep enough for the fill to run. Fill was trucked from one pass to another on the 8, 17 and 25 levels for this purpose. On working levels fill was trucked to stope winzes and distributed in stopes by scrapers.

All power requirements to the mine were supplied by suction gas engines coupled to alternators or compressors except for the main shaft winder and minor steam driven plant. Local mulga wood was used as fuel for both steam and suction gas engines. An extensive wood-cutting organisation was maintained with wood being carried up 80 kilometres distance on a 0.508 metre gauge railway.

Owing to the isolation of the property, full workshop facilities were provided at the mine capable of handling most repair jobs.

The mine was relatively dry and pumping offered no great problems. The drainage from the upper levels was effected by a pumping installation on the 7 level. Water from below this level was pumped in stages at intervals of five levels.

For ventilation two exhaust fans were installed on the 29 level and on the 30 level. The main shaft was the upcast channel and the down cast channel was at the south end of the mine isolated from the workings by doors and brattices. Auxiliary ventilation to working faces was by electric and air driven fans.

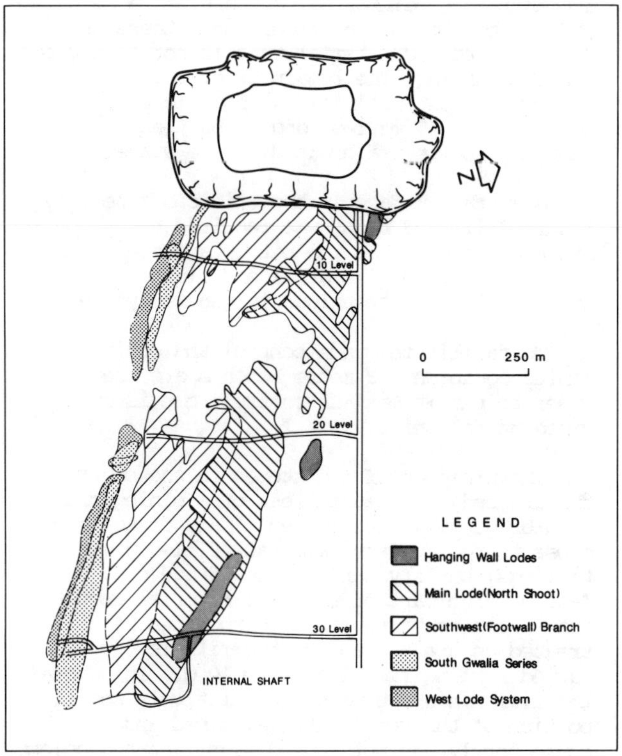

Fig 3 Plan Showing Open Cut and Stoping

OPEN CUT MINING

When open cut operations commenced in 1984 mining of both ore and waste was carried out by a contractor. Sons of Gwalia staff were responsible for ore definition, grade control, and supervision.

The initial pit plan, which provided for mining to a depth of only 60 metres, had a strip ratio of 1.8:1 waste to ore. Following successful definition of additional reserves the current pit plan takes the pit down to a depth of 150 metres at a strip ratio of 8:1.

Studies are presently in progress to determine the ultimate depth of the pit as mineralisation continues well below the base of the 150 metre pit.

Mining to date has involved a staged approach with a number of cut backs to minimise excessive waste removal at an early stage. The pit is being mined with 20 metre faces at 65° with 4 metre wide berms on the north, east and south sides.

The west wall of the pit is being mined as an unbenched wall down the western footwall of West Lode, that is, down the cleavage of the schist which is normally 45°. The upper 20 metres is not reinforced, but a 6 metre berm is left at the toe of the face. Here there is highly weathered material that is not suitable for reinforcing. The remainder of the wall is reinforced systematically with cable bolts. Each bolt is 15 metres long of 50 tonne capacity. Bolts are grouted in two stages.

Near the surface in the oxidised zone very little drill and blast was required prior to mining. However, as the pit gets deeper both ore and waste require drill and blast. A paddock blasting technique is generally used.

To facilitate grade control and optimise mining equipment, 3 metre benches are used. In order to permit maximum control, ore is only mined on day shift.

The presence of old stopes in the centre of the ore body has caused less of a mining problem than originally anticipated. Many of the stopes close to the surface were back filled with tailings, battery sands or low grade ore. To date a large part of the fill material except that which contains ash or carbon has been treated as ore. Grade of the fill is very varied ranging from 1 g up to 10 g. In general the fill mined represents a relatively small portion of the total material mined and consequently has not caused significant problems in the plant.

Mine timbers have been more of a problem, mainly in the plant. Timber is generally screened out before it reaches the crusher or alternatively material containing much timber is stock piled separately for later treatment. Generally the only time timbers are a mining problem is when drives or ore passes are encountered that are closely timber lagged. The close spaced drill pattern used to block out ore enables the position of stopes and development headings to be accurately determined, so that mining equipment can be positioned accordingly.

Late in 1988, following the definition of additional ore reserves, Sons of Gwalia NL purchased its own mobile mining fleet and ceased contractor mining. At the same time the rate of ore mining was increased from 450,000 to 750,000 tonnes per annum.

At present blast hole drilling is performed by two Atlas Copco RO81 hydraulic crawler drills. Ammonium nitrate fuel mixture is used as the blasting agent. Loading of ore and waste is carried out by two Komatsu hydraulic shovels, one of 5 cubic metres and the other of 4 cubic metres. Eighty tonne Komatsu haul trucks are used for both ore and waste cartage.

ORE RESERVES

At the commencement of open cut mining in 1984, ore reserves were:

0 - 60 metres.

PROVEN	1,468,000 tonnes @ 3.57 g/t Au
PROBABLE	254,000 tonnes @ 4.23 g/t Au
TOTAL	1,742,000 tonnes @ 3.67 g/t Au

60 - 100 metres

PROBABLE	496,000 tonnes @ 4 g

The reserves were calculated utilising the results of 93 reverse circulation percussion drill holes, a total of 4,500 metres. It was based on these reserves that the decision to commence mining was based.

The latest ore reserves for the mine dated 31st December 1988 are:

0 - 150 metres

PROVEN	3,410,000 tonnes @ 3.42g/t Au
PROBABLE	235,000 tonnes @ 3.02g/t Au
TOTAL	3,645,000 tonnes @ 3.39g/t Au

At the south end of the pit between 150 metres and 190 metres below the surface, there is an Indicated Resource of 669,000 tonnes @ 3.37g/t Au.

Since commencement of open cut mining to the end of December 1988, 1,689,598 tonnes of ore have been mined with a recovered head grade of 3.68g.

Ore reserves are based on reverse circulation percussion and diamond drill hole results. Holes are drilled on a nominal 20 x 20 metres drill pattern. With mineralisation at depth the drill pattern is opened up to 40 x 40 metres.

The reverse circulation drilling technique is preferred, as it enables relatively easy drilling through old mine workings. This is particularly important, as mineralisation occurs not only in the hanging wall but also in the foot wall of old workings.

In order to ensure the reliability of samples, reverse circulation down hole hammers are used. The water table in the mine is lowered so that no reverse circulation drilling needs to be carried out below the water table.

Currently only one water bore is required to lower the water tables ahead of any reverse circulation drilling. This dewatering bore intersects mine workings just above the old No. 7 level. The water from the mine is used in the plant, as its quality is better than that from the company's bore field which is on the edge of a salt lake.

In determining ore reserves, an estimate is made of ore previously mined by underground methods. Unfortunately the old mine records are not sufficiently accurate to use them to establish the amount of ore already mined from particular areas.

All sample and drill data is stored in a computerised data base which is used for reserve computations. Surpac mining software is used for both geological and survey data.

GRADE CONTROL

Ore mining in the open pit is based on assay cut off. Using a 1g cut off it is not possible to visually determine ore limits. It is however, possible to some degree to differentiate between high grade ore and waste. Transitional grade ore can only be delineated using assays. As there is often a halo of low grade mineralisation around higher grade ore the use of assay control for ore definition is essential.

Prior to mining, the ore body is drilled out to a 10 metres x 10 metres drill hole spacing. All grade control drilling is reverse circulation. Samples are collected at 1 metre intervals, logged and then submitted for fire assay.

Analysis is carried out by a commercial assay laboratory in Leonora that can provide a 24 hour sample turnaround. Generally the ore body is drilled out to block out 3-4 months ore production in advance. This enables the interpretation of ore boundaries and permits scheduling of production to maintain the desired head grade.

Head grade is generally maintained by varying ore sources in the pit. There is little need to blend from coarse ore stock piles to maintain the desired head grade.

To better define ore limits, channel sampling is also carried out. This is generally restricted to areas where correlation between grade control holes is such that additional sampling is warranted. In general, more channel sampling has been necessary in the lower grade parts of the ore body than in the higher grade areas. The grade control drill pattern used is designed so that channel sampling can be kept to a minimum.

In mining the ore body a combination of selective and bulk mining is used. Generally greater selectivity is required in the low grade area of the ore body to enable recovery of less material at a higher grade.

Material grading greater than 0.5g/tonne but less than 1g/tonne is consigned to the low grade stock pile for future treatment.

Ore boundaries and waste boundaries are interpreted from assay results and are marked out in the pit. Ore and waste are then removed separately. The degree of control is limited to a certain degree by the fact that both ore and waste now has to be blasted.

TREATMENT PLANT AND INFRASTRUCTURE

The treatment plant treats all ore produced from the open cut, including both oxidised and primary sulfides ore.

In July 1984, a 350,000 tonne per annum Carbon in Leach process plant with contractor owned and operated crushing section was commissioned. Two years later, in 1986, after modifications, the plant was increased to treat 450,000 tonnes per annum. At the same time a second plant involving separate grinding and leach tanks was established. This 150,000 tonnes per annum second plant was initially used to treat ore from a joint venture mining operation 30 kilometres north of Leonora at King of the Hills. On cessation of the joint venture early in 1989 this second plant has treated Gwalia ore.

In 1988, the main Gwalia Plant was converted to carbon in pulp and upgraded to 600,000 tonnes per annum with the addition of extra leach tanks. Since September 1988 the company has operated its own crushing circuit capable of 230 tonnes per hour.

The two plants give an annual treatment capacity of 750,000 tonnes per annum.

Primary crushing is by means of a single toggle jaw crusher and there is a two stage screening process with secondary and tertiary crushing performed by two 1560 Omnicone Norberg gyratory ore crushers. Cyanide is added during grinding in a 12 x 19 Morgardshammer ball mill and a 9 x 9 Morgardshammer ball mill. Classification is by cyclones. In the No. 1 plant leaching is carried out in a leach tank over a 12 hour period followed by a total of 12 hours residence in five carbon in leach pachucas. The No. 2 plant has six leach tanks mechanically agitated with nominal 24 hours residence time.

All carbon passes through an elution process and the resultant electrolyte is processed through a series of four electro winning cells. Gold stripped from the electro winning cathodes is melted and cast into bars.

Tailings from the process plant are pumped two kilometres to a tailings dam and water from the decant tower in the dam is recycled to the process plant.

Power is supplied by a 4.2 Mw diesel driven power station at the site. Water is drawn at the rate of 1,800 tonnes per day from a bore field supplemented by water from the mine dewatering bore.

At Leonora the company provides staff accommodation for 90 persons with associated messing and recreational facilities. Six houses are provided for married management staff.

HISTORICAL CONSERVATION

Sons of Gwalia NL has a commitment to retaining the heritage of the old Sons of Gwalia mining operation and the society associated with it.

The original headframe and winder from the old underground mining operation were removed in 1987 and restored and reinstated at a site some 300 metres from the old mine manager's home. A new winder shed has been built to house the historic winder. A legal entity has been incorporated representing a number of interested parties to supervise the restoration and reinstatement of the nearby museum and other homes and relics of the original Sons of Gwalia operations contained within the Gwalia townsite. The Company is contributing significantly to the preservation of the social history and environment of Gwalia and the state of Western Australia.

CONCLUSION

The Sons of Gwalia mine has been producing gold since 1896 (with a break between 1921-1923 and 1963-1984). Firstly ore was mined from underground and since 1984 from an open cut operation.

In its 67 years of operations the original Sons of Gwalia Limited made a nett profit of £4.5 million and paid nearly £2.5 million in dividends. This represents a return on original capital subscribed of 833%.

Since commencement of open cut mining to 30th June 1988, a period of 4 years, the present Company has produced 195,529 ounces of gold for a pre tax nett profit of $A75,288,385.

Since Sons of Gwalia NL was listed on the Australian Stock Exchange Limited in May 1983 an initial subscriber for a 25 cent share has received $A1.43 in dividends representing a return of 574% on the original investment in less than 4 financial years of operation.

A study is presently in progress to determine the final economic limits of open cut mining. It is expected that open cut mining will be viable to at least 190 metres below the surface in the south end of the pit.

Currently a feasibility study is also underway to determine the optimum mining method and the viability for underground mining below the level of open cut mining.

REFERENCES

Anon., 1981, The Sons of Gwalia, Open Cut Study, Bewick Moreing Consulting Geologists and Mining Engineers.

Bell, P., Connell, J. and McCarthy, J., 1985, Gwalia Conservation Study, State Heritage Branch, Department of Environment and Planning, Adelaide, S.A.

Finucane, K.J., 1965, "Geology of the Sons of Gwalia Mine", Geology of Australian Ore Deposits, McAndrew J., ed., 8th Commonwealth Min. Metall. Congress, Melbourne, pp. 95-97.

Hallberg, J.A., 1985, Geology and Mineral deposits of the Leonora-Laverton Area, Northeastern Yilgarn Block, Western Australia. Hesperian Press, Perth.

Kalnejais, J., 1987, "The Sons of Gwalia Gold Mine", Second Eastern Goldfields Geological Field Conference, Abstracts and Excursion Guide (Addendum).

Kalnejais, J., 1988, "Sons of Gwalia Gold Mine", Bicentennial Gold 88, Excursion Guidebook nos. 5-7, Western Australian Gold Deposits, Groves, D.I., et al., comp., Geology Dept. & University Extension, University of Western Australia, Publication no. 14, Perth, pp. 135-137.

Noldart, A.J. & Bock, W.M., 1960, "Notes on the geology of Mt Malcolm and Mt Margaret Goldfields", Geol. Surv. West. Aust., Annual Rep., 1959, pp. 31-34.

Reid, D. & Reid D., undated, Leonora and Gwalia, an Historical Sketchbook.

Tattam C.M., 1953, "Geology of the Sons of Gwalia Mine", Geology of Australian Ore Deposits. Edwards A.B., ed., 5th Empire Min. Metall. Congress, Melbourne, pp. 208-214.

Thom, R. & Barnes, R.C., 1971, Explanatory Notes on the Leonora 1:250,000 Geological Sheet, Western Australia. Geol. Surv. West. Aust., Perth.

7

Mining III General

D.L. Simpson
Echo Bay Mines, Ltd.
Reno, Nevada

I.A. Goddard
Highlands Gold Ltd.
Port Moresby, Papua New Guinea

Chapter 22

MINE PLANNING AT ROUND MOUNTAIN GOLD CORPORATION

Stephen J. Brower
Mine Planning Engineer

Round Mountain Gold Corporation
Round Mountain, Nevada

ABSTRACT

Located in Central Nevada, The Round Mountain Gold deposit has been continuously mined since 1977. Present proven and probable reserves are 226 Mt (250 million short tons) at 1.1 g/t (.032 oz per short ton). The current open pit mining rate of 118 Kt (130,000 short tons) per day is producing 41 Kt (45,000 short tons) per day of ore to be processed by heap leach methods. At this rate, the current reserves provide a 16 year mine life.

The high mining rate and strict ore delivery requirements mandate comprehensive and realistic mine planning. The mine planning process is described in three general steps: Deposit Modeling; Mine Design; and, Production Coordination and Support.

INTRODUCTION

The Round Mountain Gold Mine is located in the Big Smoky Valley, Nye County, Nevada. The local district has seen several episodes of mining activity. Initial interest centered on underground mining of narrow high-grade gold veins in the early 1900's. A period of relative inactivity was followed by placer mining of near surface material adjacent to the underground lode working in the 1950's. The current era of large scale open pit mining and heap leaching began in 1977 and has continued through the present. In 1988, the Round Mountain Mine produced 234,000 ounces of gold, 1989 production is expected to be 320,000 ounces. Operations at Round Mountain are conducted by Round Mountain Gold Corporation, which is a partnership of Echo Bay Mines Ltd. (50%), Homestake Mining Company (25%) and Case, Pomeroy and Company, Inc. (25%). Echo Bay Mines Ltd. is the operating partner.

Open pit mining is conducted on 10.7 m (35 ft.) benches by 136 t (150 short ton) trucks and 18 cubic meter (23 cubic yard) hydraulic excavators. The current mining rate of 118 Kt per day (130,000 stpd) produces 41 Kt per day (45 stpd) of ore which is delivered to two separate crushing and processing systems.

Ore processing begins with crushing to minus 1 cm (3/8 in.) and stacking on reusable asphalt pads. Cyanide solution is applied to the heap in two stages. Barren solution is applied to the oldest ore in circuit, yielding a lean solution. Lean solution is then applied to the fresher ore, producing the fully loaded solution for the adsorption circuit. At the end of the 100-day leach cycle, the spent ore is rinsed with fresh water and removed to final waste dumps. Final gold recovery is achieved through carbon adsorption, pressure stripping, electrowinning, and refining. Gold-silver dore' is the final product shipped.

The majority of the ore is shipped directly from the pit to the primary crusher requiring a continuous ore exposure and a sustainable ore release. The strict ore delivery schedule and high tonnage rate demand a comprehensive mine plan that provides the correct blend of ore and waste needed to assure continuous operations. The mine planning process is described in three general steps: Deposit Modeling; Mine Design; and, Production Coordination and Support.

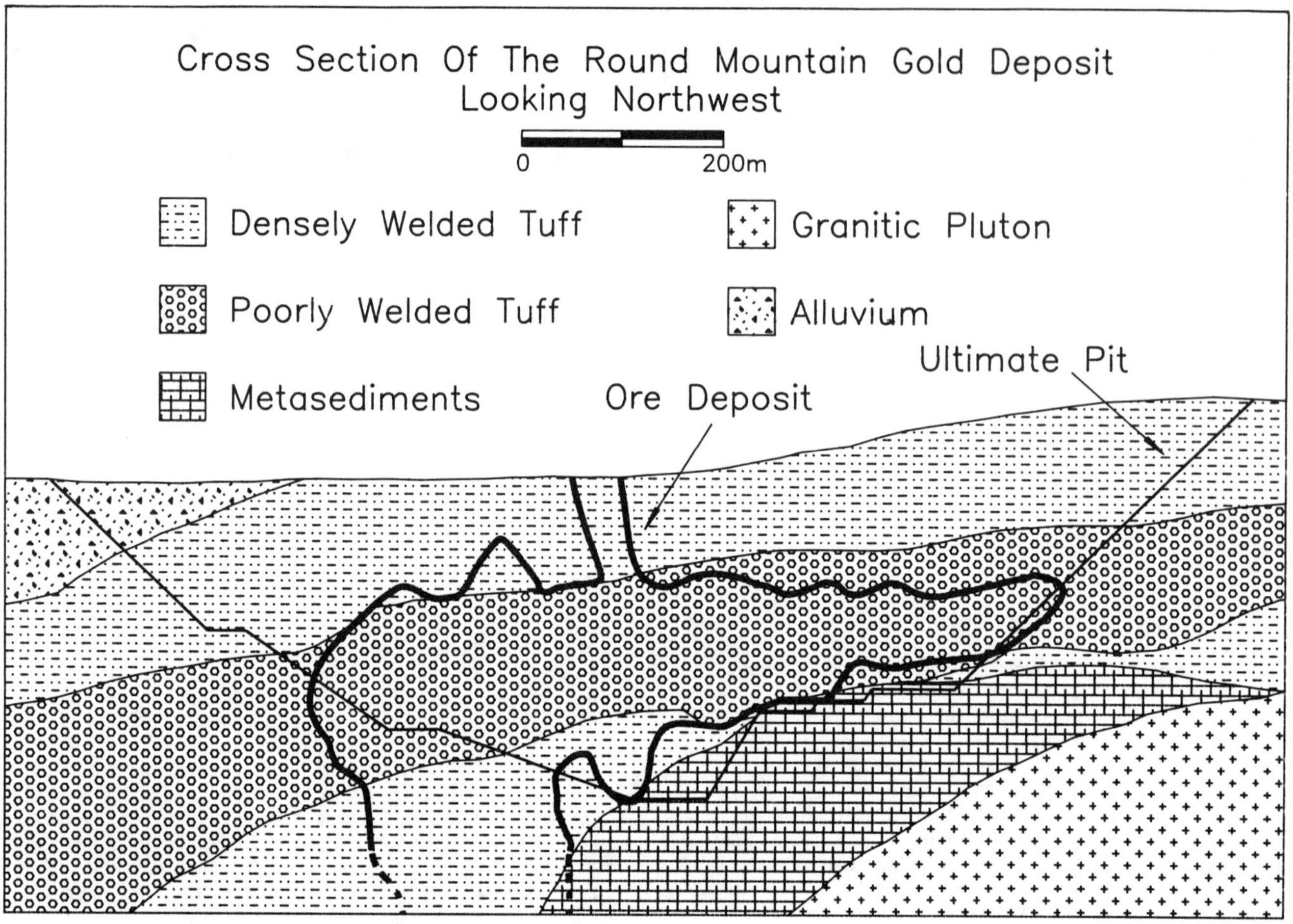

Figure 1. Idealized geologic cross-section of the Round Mountain Gold deposit. Looking northwest.

DEPOSIT MODELING

Geologic Overview

The Round Mountain Gold deposit is located in the Basin and Range province in Central Nevada. Dominant local geology consists of a thick sequence of Tertiary rhyolitic ash-flow tuffs. Underlying the volcanics are Paleozoic metasediments and a Cretaceous granitic pluton. Erosion of the volcanics has produced alluvium which overlies portions of the deposit. While all of these geologic units contain gold mineralization, the bulk of the reserve is contained in the ash-flow tuff.

The volcanic tuff sequence is subdivided into three distinct units: upper or Type 1, middle or Type 2, and lower or Type 3. Type 1 [100 to 200 meters (330 to 660 feet)] is a brittle, densely welded tuff, with gold mineralization structurally controlled primarily along joint sets. This structural control produces a more variable grade distribution than that of the remainder of the deposit. Approximately 18% of the current reserves are contained in Type 1. Type 2 [130 to 300 meters (425 to 1,000 feet)] is a softer, poorly welded tuff with gold mineralization showing more continuity than Type 1. Type 2 accounts for 70% of the current reserves. Type 3 (variable thickness) is a moderate to densely welded tuff containing mineralization characteristics that are a combination of the upper and middle flows. Type 3 contains 8% of current reserves, the remaining 4% is contained in near surface placer ore. All of the volcanic units have experienced post-mineral faulting. Figure 1 shows an idealized cross-section of the deposit looking northwest.

Modeling Methods

The deposit is modeled using the three-dimensional block modeling concept on 35-foot benches. All relevent geology is encoded into the block model. This includes different rock types (i.e., placer, tuff units, metasediments, etc.) and also the subdivision of each rock type into smaller zones. This subdivision represents unique domains primarily related to post-mineral displacement. Drill hole data is composited into bench intercepts, and the appropriate rock and zone codes are assigned. The oxidation state is encoded into the block model based on the nearest drill hole composite.

The different rock types have markedly different gold distributions and require different estimation methods to accurately model the gold grade. Where geologic evidence and statistical evaluation show grades to be continuous and evenly disseminated, Ordinary Kriging is the estimation technique used. If the gold distribution is more variable in nature, an estimation technique which limits the impact of that variability is required. This is achieved through Discriminator Kriging, which is a modified version of Indicator Kriging.

The objective of the discriminator approach is to limit the influence of high-grade composites to blocks close to the high-grade hole, so as to prevent "undue-smearing" of high-grade material. In practice, this is achieved by selecting a cutoff value that discriminates between higher and lower grade population. This value is picked from the population distribution at a point in which the data thins out, usually in the upper 5 to 10 percentile. The standard indicating process of assigning 0 (above discrimination) and 1 (below discrimination) to bench composites and estimating these indicators into blocks is applied. The gold grade in blocks passing this high grade discrimination (0.5 indicator or less) are then estimated from the full composite population using Ordinary Kriging. For the remaining block grade estimates (blocks with an indicator value greater than 0.5), all composite grades above the discriminator value are cut back to a value which approximates the general ore grade of the deposit. The same technique is also used to prevent gold estimation into areas of grouped waste composites by selecting a low end discriminator to identify blocks that will be classified as waste.

Geologic evidence and statistical evaluation determine whether to use Ordinary Kriging or the more restrictive Discriminator Kriging. Mineralization in Rock Types 1 and 3 is structurally controlled, with grade closely correlated to fracture density. Statistically, these rocks do not lend themselves to straightforward variogram modeling and display a Coefficient of Variation (standard deviation divided by the mean) of 1.6 to 2.0. Discriminator Kriging is used in Rock Types 1 and 3. The softer, more porous qualities of Rock Type 2 have promoted dissemination of gold throughout the rock mass. This is evidenced by a Coefficient of Variation of 1.0 to 1.2 and considerably better behaved variograms. Ordinary Kriging is used in Rock Type 2. Mineralization in the placer material is estimated using Ordinary Kriging with somewhat restrictive search parameters. The deeper metasediments contain gold mineralization, but are not included in the reserve at the present time.

For estimating indicator values and gold grade, the experimentally determined range and variograms are used. All composite search ellipses and variogram models have a 2 to 1 anisotropy in the northwest direction. In Rock Types 1 and 3, the horizontal search and variogram ellipse has a major axis radius of 76 to 91 m (250 to 300 feet); the vertical axis is restricted to one bench up and down. The spherical variogram has a nugget to sill ratio of 0.50. Rock Type 2 has a horizontal search and variogram ellipse radius of 91 m (300 feet) with a nugget to sill ratio of 0.3 to 0.5. For a block grade to be classified as proven, at least six drill hole composites must have been used in the grade estimate. Probable classification requires three composites. All composite searches are restricted to the structurally controlled sub-zones within each rock type.

For use in all subsequent planning exercises, a recovered gold value for each block is calculated. The recovered percentage is a function of gold grade, as shown in Figure 2. These recovery curves are a composite of information from the historical heap leach recoveries and laboratory column testing. The majority of ore processed in the past has been oxidized, providing a large amount of data for defining the oxide recovery curve. The sulfide recovery relationship is based primarily on column tests.

Reconciliation

The true proof of any deposit modeling exercise is actual mining performance. A continuous program of reconciling as-mined results with model predictions will allow fine tuning of estimation

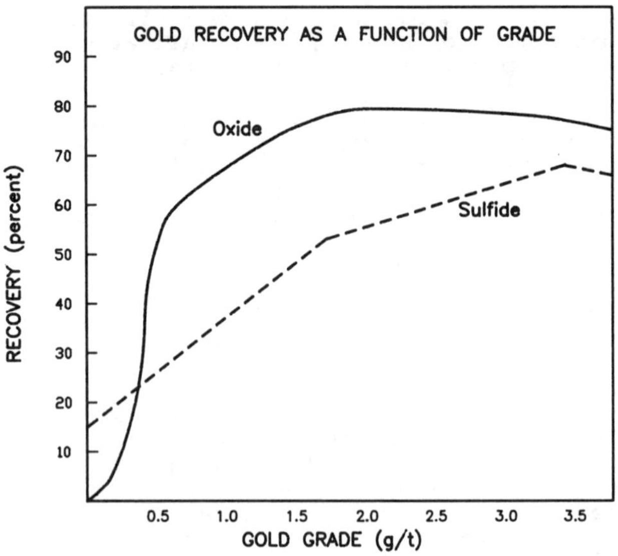

Figure 2. Gold recovery curves.

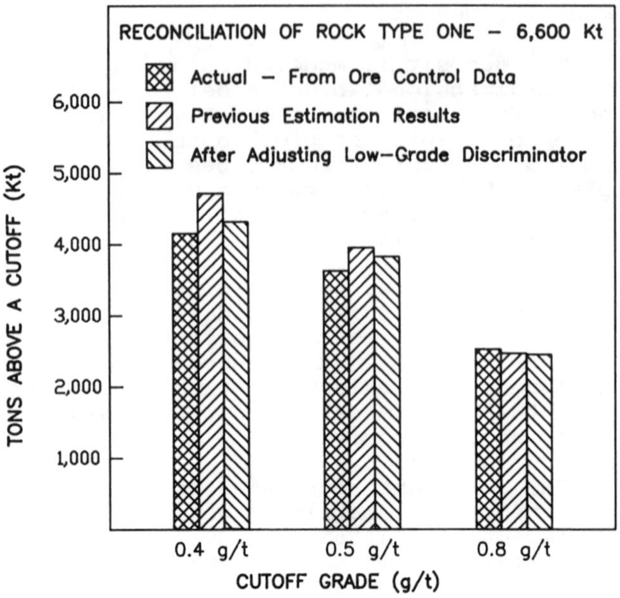

Figure 3. Reconciliation results showing tonnage above a cutoff grade. Shown are the actual tons mined, results of current estimation, and results of further restricting estimation in the lower grade range.

methods to produce an accurate model for planning purposes.

To track model performance, two reconcilation techniques are used. Each month a comparison of predicted versus actual ore tons and grade is compiled at the current cutoff value. While the one month period has proven too small a window to define the complete situation, it does provide insight into current model performance. On a yearly basis, a comprehensive reconciliation is performed. For this reconcilation, the previous year's ore control data (in our case, blasthole assays) is reevaluated to provide grade-tonnage information at a full range of cutoff values. This allows comparison of actual versus predicted information over the full range of interest. If a significant difference is indentified, corrective action can be taken to produce reliable results.

Figure 3 illustrates a reconcillation involving 6.6 Mt (7.3 million short tons) of Rock Type 1. The reconcilliation results indentified an overestimation of lower grade ore. This was corrected by slightly raising the waste discriminator to further restrict the number of blocks qualifying for gold estimation. This method of verifying the estimation technique over a full range of cutoffs, instead of at one single value, will eliminate surprises when a change to the operational cutoff grade is planned.

MINE DESIGN

Phase Sequence and Design

Due to the large size of the Round Mountain deposit, a single push-back to an ultimate pit limit is not feasible. Instead, a program of sequenced development is adopted. This mining sequence is composed of a series of push-backs or phases which are balanced to provide required ore release with a reasonably constant total mining rate. This approach to open pit development minimizes pre-stripping and smooths out equipment and manpower needs.

The actual boundary of each phase is positioned to develop the "next best" area. This "next best" location is the area that provides the highest potential profit, and generally progresses from high-grade low stripping to low-grade high stripping. To identify the area of highest potential profit, a floating cone algorithm is applied to the geologic block model. The floating cone simulates the open pit mining of all profitable material based on block grade, operating costs, recovered value, and gold price. By varying the gold price, the floating cones are manipulated to expose the highest profit portions of the deposit. When plotted on bench maps, the incremental cone results are used as a guide for actual phase layout and design.

Figure 4 displays this approach to phase layout. In this example, the starting topography is labeled Phase 4 (as this is the most advanced operating phase at the time of consideration). From this starting topography, a series of floating cones are run at $125, $175, and $225 theoretical gold prices. The cone results and a series of sequential phase boundaries are shown. For simplicity, Figure 4 does not illustrate the complete phase sequencing to the ultimate pit.

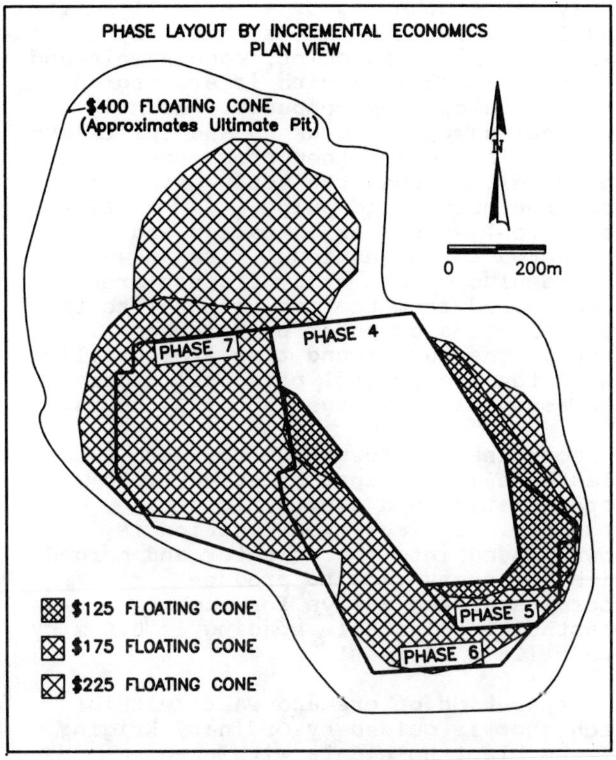

Figure 4. Showing incremental economics and resulting phase layout.

Phase geometry is designed to promote efficient equipment utilization. Design criteria follows:

Haul Roads
- 8% grade on the loaded road edge
- Minimum 30 m (100 feet) width
- Inside rock catch area where road will be used below an active mining phase
- Minimum 30 m (100 feet) inside curve radius

Inter-Ramp Pit Slopes
- 45 degrees in all but southwest sector
- 42.6 degrees in southwest sector
- 45-foot rock catch bench every sixth bench (not included in design slope angle)

Geometry
- Sized as large as possible and still achieve scheduling requirements
- 120 m (400 feet) minimum mining width
- 45 degrees minimum interface to existing pit wall

Necessary deviations from these design parameters are evaluated on a case by case basis.

Mining Schedule

Material from sequential phases is scheduled for mining at a rate that provides an appropriate amount of stripping to achieve the required ore release. The rate is constrained not only by total mine capacity, but also by individual loading machine capacity. Primary loading is handled by four hydraulic excavators. These machines are considered non-mobile for scheduling purposes, which results in a minimum unit of production (one shovel) that can be scheduled in a working phase. If a realistic mining schedule cannot be achieved, then the blend of ore and waste in a phase is manipulated by shifting the phase boundaries as appropriate. This iteration between phase design and mine scheduling is a crucial step in developing a mine plan which keeps the deposit development on schedule while still providing efficient operating faces for the equipment fleet.

The final workable schedule provides an operational calendar of events which must be met to achieve the predicted results. Information derived from the schedule includes: required equipment buildup, major shovel moves, start time of next phase in sequence, and necessary changes in cutoff grade, etc. The schedule also determines the basic grade-tonnage results for the time periods under consideration. Table 1 shows the actual ore balance obtained in the schedule. This example covers years 1990-1994.

TABLE NO. 1
EXAMPLE MINING SCHEDULE

		PHASE					
		3	4	5	6	7	Total
Year		Mt	Mt	Mt	Mt	Mt	Mt
90	Ore	11	3				14
	Total	14	21	7			42
91	Ore		12	2			14
	Total		16	26			42
92	Ore		3	11			14
	Total		4	14	24		42
93	Ore			4	10		14
	Total			4	25	13	42
94	Ore				6	8	14
	Total				7	35	42
TOTALS							
	Ore	11	18	17	16	8	70
	Total	14	41	51	56	48	210

Figure 5 is a predicted "snapshot" of the Round Mountain open pit showing the phased development. In this example, Phase 3 is the primary ore supply, Phase 4 is a secondary source of ore with substantial included waste, and Phase 5 is in early stages of stripping.

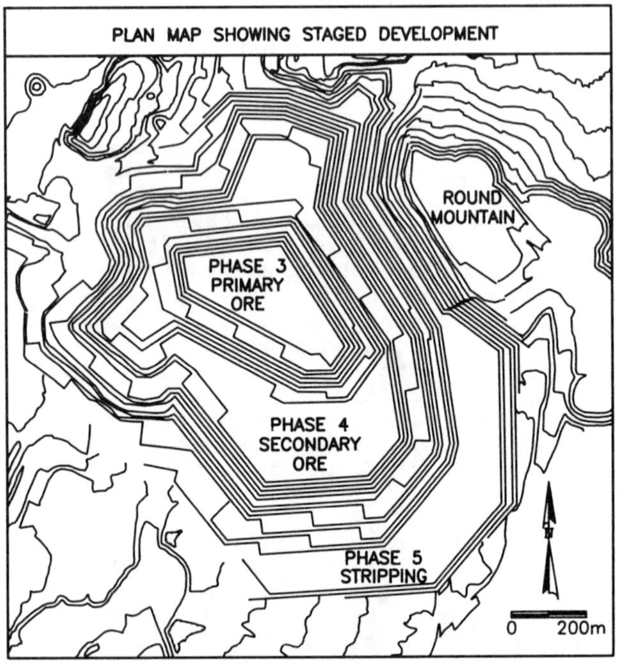

Figure 5. Plan view showing staged sequence of mining phases - lowest is in ore, highest is in stripping.

PRODUCTION COORDINATION AND SUPPORT

Production Scheduling

A realistic balance of ore and waste mining in the operating phases is achieved through production scheduling. This scheduling exercise addresses the next three months and is considerably more detailed than the phase scheduling exercise previously discussed. The primary goal is to determine an actual time schedule for the drilling, blasting, and loading unit operations which will achieve the required ore and waste balance.

In theory, the schedule is achieved by juggling the material available in the various working faces to achieve the desired results. Material information is obtained from the geologic block model and any relevant grade control data. In practice, the proper mining direction is assured through blast pattern layout. The control of blast location initiates the mining sequence in the proper direction and sets the stage for efficient and continuous mining of the correct ore to waste ratio.

Grade Control

Grade control starts at the blasthole drill when the cuttings stream is sampled by the driller using a through-the-deck sampler. Samples are delivered to the assay lab at the end of each drill shift and logged into the lab's computer system. From this point, each sample and the data associated with it are stored and manipulated by computer. Each original sample is crushed and split into two samples, which then progress separately through the assay process. From the pulverized pulp, a 146 g (five assay ton) sample is extracted and subjected to conventional fire assay. The results of the assay process are transferred from the lab computer to the mine planning computer by telephone modem. The turnaround time from drilling until the ore control personnel receives the assay data averages about 24 hours.

Coordinate information for each blasthole is obtained by survey with a Total Station and electronic data recorder. The surveyed location is down-loaded into the computer and merged with the assay data to produce coordinates and assays for each blasthole. Blasthole spacing is 5.5 m by 5.5 m (18 x 18 feet).

Separation of ore and waste within each shot is guided by Ordinary Kriging of the blast hole data within and surrounding the shot. The dual fire assays for each blast hole are averaged for use in estimation. A uniform 3 m by 3 m (10 by 10 foot) grid is overlayed onto the blasthole data, and the value at each grid point is estimated. A spherical variogram with a nugget to sill ratio of 0.3 and range of 12 m (40 feet) is used. A spherical search of 12 m (40 feet) is used, allowing blastholes from the bench above to influence the estimated grade. Color plots of the estimation are used to guide ore control personnel in blocking out mineable ore zones.

It is important to note that the kriged blasthole grades are used as a general guide only; ore boundaries are modified to conform to verifiable geologic zones and practical mining geometry. Once ore zones have been finalized and average grades tabulated, the boundary coordinates are down loaded to the electronic data recorder for field stake out. The boundaries are flagged in the field with color codes to direct the shovel operator in material separation.

CONCLUSION

The Round Mountain Gold Mine is a large open pit mine processing 41 Kt (45,000 short tons) per day of gold ore by heap leach methods. The operation produced 234,000 ounces of gold in 1988 and expects to produce 320,000 ounces of gold in 1989. A comprehensive mine plan is a key factor in achieving the targeted material movement and gold production goals. Mine planning at Round Mountain is a fully integrated process of deposit modeling, mine design, and production coordination and support.

Deposit modeling provides the foundation for all subsequent planning decisions. Relying heavily on detailed geology and innovative estimation methods, the Round Mountain geologic model has proven a reliable forecast of in-place reserves. Based on the geologic model, a phased sequence of mining push-backs are scheduled. These phases develop the "next best" one in sequence. "Next best" is defined as material providing the highest profit potential. In practice, the necessary blend of ore and waste mining is achieved through equipment scheduling and ore control. The results of accurate predictions and balanced mining schedules promote the efficient and continuous operation of the Round Mountain mine and processing system.

Chapter 23

GOLD EXPLORATION AND MINING IN VENEZUELA

A PRIVATE ENTERPRISE APPROACH

ALBERTO SARMENTERO,P.Eng

Gold Projects Org. Inc.

ABSTRACT

Venezuela, once a world's top gold producer until 1886 Witwatersrand discovery, did let fall its role then because of a serious lack of infrastructure and population, and lately (since World war II times), because of discouraging mining and taxation laws, laying upon an oil prizes bonanza.

Recent research leading to estimates of about 5.000 mt. of gold, mainly related to the greenstone belts of the Guiana Precambrian shield outcropping in Venezuela, are opening the scope for both Government and international investing and mining companies, for gold exploration and development projects in the country.

This paper overviews Venezuelan Gold mining industry, as seen from the point of view of a participating mining company.

INTRODUCTION

G.P.O., Gold Projects Org. Inc. is a mining consulting corporation registered in U.S.A. and Venezuela, with representatives in Canada, U.K., Spain, Costa Rica and Dominican Republic, that has grouped several geomining professionals whose experience in the mineral industry averages in excess of 20 Years, with 10 years as consultants. G.P.O. corporate strategy has been to concentrate its activities in precious metals and gemstone exploration, development and mining consulting in several Latin American and Caribbean countries.

This paper intends to represent a contribution to the knowledge of Venezuela's gold mining environment and their possibilities in what we feel its present-day pre-booming stage, closely related to the comparatively low operating cost of many prospecting and mining areas, despite of alternances in gold prices at Stock exchange centers.

MINING HISTORY

When Spaniard Admiral Columbus and his crew made their first landfall in the New World, they found the natives in possesion of gold nuggets (though putting little enphasis beyond its use as ornaments, sacrificial knives, etc). In 1494, mining exploration began, and in that year Alonso de Ojeda discovered the rich Cibao placers in La Hispaniola (Dominican Republic). Few months later, mining teams arrived from Spain with mining equipment and large amounts of Almaden's mercury for amalgamation. In 1949 Juan de la Cosa mapped Venezuela, and few years later North, Central and South American organized mining was initiated arround gold and other few products with enough value per ton to pay the fleet costs of those times.

Since that time of the Conquistadores (misrepresented later by Spanish Crown's competitors), almost 500 years ago, the formerly province of New Granada, where Venezuela was included, has produced more than 20 billion (of today's US) dollars in gold (60 million ounces), about 65% from placers and the remainder from underground mines. The famous El Callao mine of Venezuela (near El Dorado) recorded more than 8 mt. of gold production one single year in 1885, from a lode grading up to 215 g/mt.

THE COUNTRY

Venezuela is known as the "Gateway to South America", because of its favourable geografical situation. It is the sixth largest country of the Hispanic comunity, and it is organized as a Federal Republic with a democratic and representative form of government, centered at Caracas, the national capital.

The increasing participation of Venezuela in the world economy was first related to a fairly well managed oil production, which exports have penetrated European and U.S. markets during last six decades. Moreover, since 1960, the country steadily reduced dependence on many imported goods by encouraging local and international manufacturers to build factories in the country, in adition to further expanding its petrochemical, steel, aluminiun, and hydroelectric industries.

The economic crisis initiated in 1983, which included the devaluation of the local currency and the expansion of its foreign debt to about 30 billion US$, encouraged the new government, that started in Feb.1989, as well as legislative representants to change several laws that restricted the investment in many potential fields. One most keystoning is the mining law, that kept away the country from significant private mining activities in this highly potential geological environment.

Answering a question from the author of this paper, Mr. Carlos Andrés Perez, present-day president of Venezuela, during the First Symposium of Gold at Caracas, said that all the administrative machinery is going to be set up in his goverment to incentivate the attractive foreign investment in mineral resources in the country, and specifically on the fast growing gold mining industry, by avoiding bureaucratization and high taxation on profits from gold mines, derived from risk capital for exploration and development expenditures.

INFRASTRUCTURE

Up to few decades ago, exploration was very restricted by a tremendous lack of infrastructure in the scarcely populated jungles and savannahs of the Precambrian Guayana(*) shield outcropping South of the Orinoco river, in an extension of over 170.000 sq.miles.

(*) Guayana: Literally: "Not monkeys" (from a Guarao indians dialect, Guai = monkey, ana = not). Suposedly the name given by the natives to the spaniards when they first arrived sailing down the wide Orinoco river, hanging from the masts and ropes of their galleons and vaixells. Other acceptation is "Land of many rivers".

So, exploration and mining for gold was only feasible in the vicinity of navigable rivers, such as Cuyuní and its tributaries (e.g. La Amarilla placer gold basin), and where land transport was practicable, not very far away from Orinoco river towns, such as Ciudad Bolivar the intermediate village of Upata, and El Callao mine.

Some private gold mining manuscripts of that time (though difficult to find) described with a lot of details the ardous task of conducting exploration in those times.

Nowadays, though transportation ways as well, as other means of comunication, are fairly well developed in Northern Venezuela, the South remains mostly remote for land access, except for the road that has been open from the new industrial city of Ciudad Guayana, crossing SE Venezuela down to the Brazilian border in Santa Elena de Uairén. This magnificent penetration road, that has been improved and paved in last three years, together with the airstrips opened in the nearby villages, constitute today's axis of the development of gold mining industry. A second road was built in the other extreme of Venezuelan Guiana, once C.V.G. geologists (*) discovered in 1976 what is probably the biggest bauxite deposit on Earth, with estimated 5.8 billion tons of high grade bauxite in Los Pijiguaos, La Rabaldana, La Cerbatana and Cerro Mocho orebodies.

This road crosses a portion of South Venezuela, from the western Orinoco village of Caicara, to nearby bauxite mines and down to Puerto Ayacucho, capital of the venezuelan Amazon Territory, near the sources of Orinoco river.

GEOLOGY

Most of Venezuela's known gold deposits and exploration targets are located in the Precambrian shield of Guiana region that covers the southern half part of the country. This region contains four geological provinces, as described in detail by Sarmentero and Menendez (1986, at AIME Symposium).

(*) De Ratmiroff, and the author of this paper, together with A.Menéndez.

The oldest province is Imataca, some 3.6 billion years old, located northernmost along a strip limited from Caura river at the East and sunk under Orinoco Delta sediments at the West. This province is considered of a Granulite belt, known to be poor in precious metals, but very rich in Iron formations, being mined by Ferrominera Co. at the Iron quadrangle of Cerro Bolívar, San Isidro, and other deposits, with ore production rates of 20 million mt/y formerly mined by a U.S. Steel subsidiary and Bethlehem Iron Mines Co.

The rest of the shield geological provinces has gold potential, according with the following remarks:

1.-Pastora province.-

Guri-Santa Barbara fault system traces its northern boundaries with Imataca province. Is is the second oldest, 2.3 billion years old), and the most important at present in gold mining. It consists of metasedimentary series intruded by the sodic granites and granodiorites of Supamo complex. Gold deposits are related to Pastora Supergroup (the oldest greenstone belt of the region) and Botanamo Group, a younger one, which basal formation Caballape is the host of Botanamo gold deposits. El Callao, basal formation of the older greenstone belt, and the lateraly unconformal Cicapra formation hosts the El Callao lode underground mine, and Caratal deposit. Overlaying Yuruari formation shear zones, veins and stockworks are parcially mined in Lo Increible shafts, and at "Km.88" and Bochinche placers and lodes.

Pastora province was formerly subdivided into five traditional gold mining regions, named El Callao, El Dorado, Tumeremo, La Paragua, and El Manteco, averaging more than 14.000 sq.Km. each.

Recent grassroot exploration carried out by C.V.G. and private companies on some less known gold bearing areas of this province is leading to give increasing importance to localities such as Marwani, Bochinche, and Venamo, formerly included into Tumeremo and El Dorado regions, at the Easternmost boundaries with the so called Barama-Mazaruni Guyana Republic's greestone belts, that are continuations of the same extensive ones outcropping in Venezuela's Pastora.

Gold deposits of Pastora are the base of the past and present gold production in the country. Mines are being erected in this province to recover gold from quartz-vein lodes, stockworks, eluvia and placers.

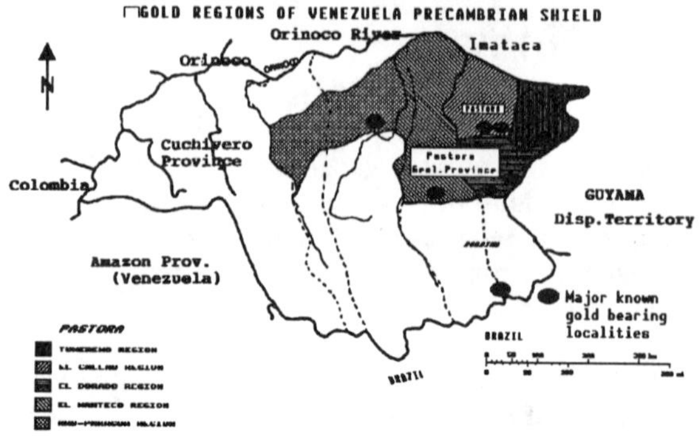

After Lopez (1981)

2.-Roraima province

It consists of late precambrian tabular sandstone series deposited over previously existing terraines and covering parcially Pastora and Cuchivero-Amazon provinces. World highest waterfalls runs down by the dissected scarps of these thick series, such as Churún Merú (literally: vertical river), known as Angel falls. From its basal conglomeratic Uairén formation are drained out the famous diamond placers of this part of Venezuela, that usually contains not negligible amounts of gold. However, Roraima is one of the least known provinces of Venezuelan Guiana, because of its remoteness. Gold is being mined in the few explored areas where rich placers derived from old (polimictic metamorphized pyritic) pebble conglomerates (Pre-Roraima), and Quartz hematite conglomerates of Uairén Formation occur, as in the gold deposits East of Santa Elena de Uairén, near the Brazilian border.

3.-Cuchivero-Amazon Province

This province came to the mining scene by the discovery of the previously mentioned plateau-type bauxite deposits developed on the Parguaza rapakivi batholit, and surrounding granites that intruded an older ignimbrite series of ryolitic tuffs. Gold nuggets have been reported by occassional panners in placers derived from overlaying conglomerates, formerly reported as Cinaruco formation (Mc Candless 1969). Th, Ce-group Rare Earths carbonatite of Cerro Impacto, the Aguamena tin deposit, and Tantalite-Columbite occurrences west of Parguaza were the aditional reported mineral potential of this province until present.

As a matter of fact, this extensive southernmost region has been only very scarcely explored, because of its difficult access until present times, thus very little is known regarding its gold potential. However, a large sequence of pebble-conglomerates, as well as quartz conglomerates has been positively tested for gold by M.E.M. and C.V.G. geologists, and some native gold panners have extracted some gold out of Yapacana and Yutagé conglomerates.

EXPLORATION

A.-Government

There are two government entities that made exploration for gold in Venezuela: The Ministry of Energy and Mines (M.E.M.), and the Guayana Development Corp. (C.V.G.), a regional development agency. Both worked separately and in a discontinuous way, and plans were depending on the politics of the government in turn. So, gathering of much previous official exploration data is sometimes very difficult for a mining company.

The M.E.M. have done or contracted in the past two decades some basic studies, including SLAR imagery and airborne E.M. maps, on specific regions, besides its supervising of mining areas for reglamentary purposes. Recently the M.E.M. founded a small company named SERVIGEOMIN, planned to be the germ of a National Geological survey depatment.

Some exploration campaigns on gold areas conducted by M.E.M. geologists, such as Botanamo are noteworthy (Benahim,l968).

C.V.G., on his side founded in 1974 his Geological Division that explored in detail the 30.000 sq.Km. region from Orinoco Delta to Bochinche in 1975, as well as other metalogenic areas, besides the discovery and development of the bauxite mines, creating Bauxiven Co. More recently, C.V.G. founded TECMIN Co., to be devoted to mineral exploration in this region.

Based on geological data, C.V.G., has got exploration permits (for 5 years) of two regions of prime exploration potential, one located in the biorder region "Arabicí-El Dorado-Yuruaní", and the other in El Callao Gold District, that are currently explored mainly by means of field geology and geochemistry. However plans are being made by the new Central Government to create the National Mining Corporation in the near future, in order to facilitate the steps for granting private concession, coordinate the scatered basic information, and negotiate agreements with international mining corporations.

B.-Private

Although all the old mining exploration has been carried out by private entrepreneurs, only few private exploration companies were active in the Venezuela Mining scene until few years ago, and their geologist barely scratched the surface, because of the reasons given above. Present day private exploration is being carried out by a still reduced number of firms, that include the following:

Riotinto (ERc)'s (Cavosa, together with local partner Vencemos) Promiven, evaluated most of the main Venezuelan reported gold prospects and orebodies during the last recent years. The company have evaluated among others, the "Km.88" and other Cuyuní river basin gold placer deposits, with average grades of about 3 g/cu m., as well as did initial exploration of large areas in Lo Increíble, El Callao, Chocó, and Bochinche.

Monarch Resources Ltd., a consortium of South African, British and some Venezuelan investors is exploring two concessions covering 10 sq.Km. arround La Camorra and Canaima, through its subsidiary Sudamericana de Mineria.

Based on preliminary sampling, the existence of a gold bearing structure has been hypothesized measuring 5 Km. in length x 2-3 Km. in width. Sampling is reported by the company to yield an average grade of 24.4 g/mt, suggesting the existence of over 100.000 oz. of gold. The area seems suitable for open pit mining, with extraction costs estimated at $10/mt vs 40$/mt for underground mining. Monarch also plans to explore some concessios arround El Callao. Other differently sized companies that carried out recently some exploration on specific orebodies include Greenwich Ltd. (Frijoles, and other concessions), China Clay Co. (Libertad), Prominsur S.A., Tranarg (Choco 1 and 2), Kaycuse S.A. (E.Sta.Elena), Tame Corp. (Chicanán), Aurisur (Callao placers), Andreyev-Donally (Union), Brewer Carías (Amazonas Territory, Triunfo), Coramin (G.Moya) (El Paují, Atlántida), Reynolds Metals Co. and several others whose exploration reports has not been made public, and hence they can not be given in this paper without their permission.

Private persons also conducted some limited exploration programs in specific areas. Examples as Mr. Michel Mir (Muliveta) and Mr.Baez (Icabarú) are remarkable.

Other international private companies are reported to be initiating their scouting in the country, such as B.P. Minerals, Total, and Noranda Corp.

METAL PRODUCTION

Venezuela is today, according to Goldfields Ltd. estimations, the 10th. Western World is gold producer with about 15 mt./year. It is the 4th.in the South America area, after Brazil, Colombia and Chilí, and it has the highest positive output growth rate of the region, though actual production has no relation with its tremendous potential, officially estimated at 5,000 mt in gold reserves and possibly up to 10,000 mt, equivalent to 10% of the World's known reserves.

The recent increases in production are due almost entirely to the higher prices now being paid by the Central Bank of Venezuela, the most solid purchaser. It has set its pricing on the Zurich and London daily fix and a free market exchange rate for the local currency. The switch by small miners to the more lucrative gold sector, together with the opening to foreign investment, has been largely responsible for the substantial increase in gold production and correspondent fall in last year's diamond output.

Table I gives an indication of the present production scenario, where some gold sources (that could include recycling) could not have been officially identified. To change this situation, the national government is organizing the so called "informal miners", from which a big deal of the reported production comes from, in specific areas under contracts, installing facilities to provide health care, and buying their production through Central Bank.

TABLE I
ESTIMATED PRODUCTION
OF GOLD MINING AREAS

Mine area	Output in t/y.
Minerven(Callao)	1.2
Venorca(Callao)	0.6
"Km.88"	4.0
ICABARU(E.S.Eln)	1.4
EL FOCO(&Surr)	0.5
Caroní-Parag.	0.4
Other Concess.	2.0
Non Concess.	3.0
Other sources.	2.0
TOTAL OUTPUT	15.0

GOLD MINING ACTIVITIES

A.Unofficial mining

About 20,000 persons are belived to earn their liviving from gold and diamond mining in Southern Venezuela, primarily from panning and usage of rudimentary sluice boxes, with recovery rates averaging less than 40%. They are belived to produce by themselves, according to local estimations up to 15 mt/y of gold. Of this amount, some 10 mt/y are sold illegally, and are not shown in table I, because of its low confidence degree.

Analysts estimate that as much as 10 mt/y of Brazil's reported gold production in 1987 was actually obtained from this unlisted miners. Until Government's tolerance to this situation started to finish early this year, (by creating the well trained and equiped mining police, deporting the illegal mining inmigrants and relocating the miners in ad hoc union's legal concessions), this adventurous and meritorious gold rushers have been mined some of the practicable part of the little known gold orebodies existing in the following areas:

1.-Ventuari-Amazon basin: Left bank of Ventuari river (Yureba, Porvenir and Marieta placers), and Yacapana hill.

2.-Aro River and left bank of Sipao creek.

3.-El Callao gold district: Small underground riskful operations at the Lo Increíble, Carotal and Carapo belt, plus Mocupia, Nacupay, Potosí, Eureka and othe placers.

4.-Caroní-Paragua: Arround 40 small mine operations in locations as Oris, Asa, Chiguao, etc.

5.-Other scattered areas: Such as Delta territory, Parapoy, Guarichito, Icabarú, Venamo, La Planada-El Palmar, Botanamo and Imataca.

B.-Private organized mining

1.-Promiven.

This well managed mining company is exploiting since early 1989 the central part of concession ORO 1, located at the La Amarilla basin of "Km.88" mining district.

As in other placer deposits of this district, gold is located in the gravel, interbedded in the placer's lower sands, as well as beneath the river sediments, in the so called "conductorite lodes" within the cristaline underlaying metavolcanic basement rocks of Pastora Supergroup. It is also present in extensive relics of disected old terraces, grading about 1 g/cu m with no overburden, which is generally considered the poorest ore of that district.

The venture has an output rate of about 10,000 oz/y of gold, and uses gravimetric separation and concentration equipment designed by a Riotinto (ERC)'s subsidiary in Spain. Operating costs are well below 150 $/oz, based on a throughput of 120,000 t/y of ore gravel, for a life span of over 5 years.

Aditionally, the company is completing detailed exploration programs, mine planning and development in adjacent secured concessions Tapaya and Libertad carried out during last months, with production starting shortly afterwards.

Promiven also signed a join venture agreement with the state entity C.V.G., founding a company named PMG. The new company will develop several large areas in the Bochinche and Chocó (El Callao gold district) mining areas, located in the northeast of Bolívar State.

2.-Monarch Resources.

This company operates through several subsidiaries, eg.: Protemin, that has got the contract to increase the Minerven's Colombia plant's efficiency in order to expand gold output to design capacity of 64,300 oz/y, and to explore and mine some concessions near El Callao, covering a leasing area of 30 sq Km.

Remevin, another Monarch subsidiary, is going to mine the Mocupia Gorge tailings, grading in average 1.35 g/mt that have been deposited in the Yuruari river since 1928, and now extend over some 5 Km at depths of up to 11 m. Plans are to build a 830 mt/d carbon-in pulp processing plant. Recovery tests indicate an 85% recovery rate. Direct operating costs are estimated at U.S.$ 126/oz of gold, at a treatment rate of 300,000 mt/y of tailings for a life span of 15 years.

3.-Other private mines

Legal mining activities are under way in the some of the 87 in force granted concessions, widespread in Venezuelan Guayana territory. This number of private concessions as well as the area covered by them is negligible and meaningless when compared to the extension of the reportedly gold potential regions.

Mining procedures both in underground or opencast mines in such concessions are frecuently very primitive, and recovery rates are thus usually low. Many claimers have been taking advantage of the frecuent high grade ores, and the former absence of enviromental regulations, specially in the placer areas, where the usage of large amounts of free mercury thrown directly in the so called "waterholes" (pits open with jets) to catch the gold by amalgamation in the suction pipes or in primitive sluice tanks is a comon practice and a standard recovery process. The process is highly dangerous, since mercury contaminates the miners and their environment.

Although there are few good examples of efficiently operated small gold mines and concentration plants, such as Molitec and Recuper's experimental heap leaching plant, near El Callao town, and several other developments, the lack of production-scale related modern mining experience of most concessionaires -and the pressures of the government to set up rational mining procedures, (as legally signed when the claims were granted), which usually require more than negligible front end investments- is often leading them now to look for leasing, or (in a lesser extent) join venture agreements with established large international mining companies to increase the production and profits. Negotiations to arrive to such an agreements on a convenient riskless royalty basis, normally require honest and experienced consultancy advice.

C.Government mining.

Government's gold output represents a 9% of the total mined in Venezuela. Despite of the very profitable results, the aim of government has primarily been to serve as a pilot example, and to train the formerly scarcely technified personnel at El Callao gold district.

1.-Minerven.-

This state owned company whose partners are 75% FMO (A subsidiary of CVG) and 25% FIV (Financial entity of Venezuelan government), was formed in 1970 to develop and exploit the El Callao concession area N 7.

A full geological study of the concession resulted in the construction of a processing plant and the sinking of a new mine shaft in 1976.

Drilling indicated ore reserves, to a depth of 250 m, of 1.55 million mt with an average grade of 9.2 g/mt.

Ore is found in two vein structures , the America and the Colombia, with a known length of 800 m and a variable thikness of from 2 to 10 m.

Minerven is producing 80% Au ingots from Colombia mine and other mineral sources. In situ mineral is blasted, then crushed, grinded, leached, precipitated with Zinc, filtered and then smelted in 5 Kg. bullions for sale.

Production peaked in 1988 at 140 Kg./mo. Minerven leases part of their concessions to third parties for exploration and exploitation purposes.

2.-Venorca.

It is an old recovery plant, with a current processing capacity of some 200 mt/day, from ores brought by small private mine operators of the El Callao-Lo Increíble vicinity, according to a commitment with a local gold mining enterpreneur's association. Improvements of this plant are under way to reach this year a 600 mt/day processing capacity.

D.- Current plans

A booming investment market is expected in Venezuela during the next decade for the gold mining industry, as long as long-term commitments are being made to provide the adequate incentives to the foreign mining corporations. Official goal is to develop through private investment the capacity to reach an output of 5 million oz/y by 1997 and 8 million oz/y early next century, placing the country into the 5 largest forecasted gold producers of the world.

That could conceivably solve, together with the oil, aluminium and other nations star industries, its present foreign exchange and balance of payments problems in less than a decade. Gold would have an even more important effect on Venezuela's international credit rating.

CONCLUSION

- According to the limitations of this paper, the exploratory and mining framework of Venezuela's gold scenario have been only barely squetched here.
- Exploration carried out by both government and particulars during last years, on variable scales: from the grassroot investigation to the scale of prospect areas, resulting in the finding of several large unclaimed high grade placers, (with high operational flexibility, and low capital investment required) and some surprisingly unexploited low operating cost hard rock deposits, are a powerful magnet for any mining company.
- Venezuela's governments in turn during last five decades have discouraged gold exploration and mining investment in the country by mine taxation rules, issued in 1945, that did not provide any adequate incentives to potential investors to recoup their required outlays.
- Although most of legal, financial, institutional, and economical gold mining related considerations are out of the reach of this paper, the general situation has been changing, and the author considers advisable for gold mining enterpreneurs and investor to take steps into scouting this potential country, where, according with some analyst's opinion, "actual opportunity rate for gold mining investment ranks top in Venezuela".

ACKNOWLEDGEMENTS

The author wishes to thank the professional staff of Promiven, C.V.G., and M.E.M.,for their continuous support.
He is also grateful to:
Ms.A.Ortiz, Mr.R.Relayze, Mr.A.Monge,
Mr.J.H.Noguerol, Ms.S.Armengol,
Mr.J.Urquidi, Mr.G.De Ratmiroff,
Mr.M.Uso, Mr.J.M.Gomez, Mr.A.Menendez,
Mr.B.Nicoli, Mr.E.Del Monte,
Mr.V.Mendoza, Mr.A.Pazo, Mr.A.Baez,
Mr.C.Estevez, Mr.V.Guerrero,
Mr.M.Mir and other colleagues who provided stimulating discussions about this topic.

REFERENCES

Benahim,N.et al., 1968.
"Revision de la estratigrafia de la provincia de Pastora" .Bol Geol MEM.Caracas.

G.P.O., 1989.
"Gold in Venezuela. Investment opportunities". GPO Consulting report. Grbro.N.C.(On press). 400 pp.

Mc.Candless, 1969.
"Geology of the Northern part of Venezuelan Guayana". Bol.Geol.MEM.Caracas.

Mendoza,V., 1985.
"Prospecciones geologicas en las zonas fronterizas de Guayana". Cong.Geol.Venez.,T.IX.Caracas.

Menendez,A., 1972.
"Geologia de la región de Guasipati, Guayana Venezolana". MEM, IV Cong.Geol.Venez.,V.IV.Caracas.Editorial Sucre.

Sarmentero,A. and Menendez,A., 1986.
"Geology of the Los Pijiguaos Bauxite Deposits.Venezuela". AIME Baux.Symposium.Chapter 17.USA.

Sarmentero et al, 1988.
"Treatment of LPJB in digestion plant". AIME Meet.Proc.New Orleans.

Uso, M. et al., 1987.
"Informe geologico del area de Bochinche". CVG Tecmin report.Cd.Guayana.Venezuela.

Chapter 24

ECONOMICS OF CONTRACT MINING
or
Should I Use a Contract Miner?

by
Donald D. Haas

Assistant Director - Mine Engineering
Morrison-Knudsen Company

Abstract. This presentation will discuss a mine owner's evaluation to utilize a contract miner in the development and production of gold properties. I will address both the quantitative and subjective aspects of the evaluation process. The quantitative aspects include cost evaluation, capital preservation, acquisition and disposal of new and used equipment, administrative costs, recruiting and training, and organizing a staff and craft labor force. The subjective areas of this discussion include corporate philosophy, administrative concerns, contractor competence, loss of owner control, operating flexibility, financing assistance, recruiting and training program utilization and risk reduction. The evaluation of alternatives provides a system for comparing, on a cost basis, an owner/operator system vs. a system using a contract miner. The supplement provides a concept on how a contract miner may be the cost effective system to be utilized.

INTRODUCTION

Contract mining has a definite place in the gold mining industry. Contract mining is a potential operating method in the development of a project anywhere in the world. I will discuss owner considerations as he evaluates various alternatives available in the development of a gold project. Factors to be considered may be quantitative or subjective, while some are both. I will discuss evaluation techniques that may be used to determine the use of a contract miner. I will supplement this talk with the reasons why contract mining can, in some cases, be a cost effective method of mine operation.

This presentation correlates with a paper presented at the Northwest Mining Association conference in December, 1986 on the "Contract Mining Bidding Process", and a workshop at the Northwest Mining Association conference in December, 1987 on "How to Contract Your Surface or Underground Mine". This information is a portion of the chapter on "Project Operating Strategy" in the yet to be published 2nd edition of the "SME Mining Engineers Handbook".

Definition.

"Contract mining" as I define it for surface mining is:

> At a minimum, the removal of waste or overburden, extraction of ore and delivery of that ore to a processing facility. Contract mining may include many other activities. For example, it may include development of infrastructure for the operation, construction of access to the site, clearing and grubbing the site, removal, stockpiling and replacement of topsoil, water control around the mine area, crushing and other preprocessing activities, construction of the processing facilities and management of the ore processing plant.

I will first discuss the quantitative factors, then the subjective factors to be considered. The subjective factors must consider the many components unique to the owner. I will follow that discussion with a cost evaluation technique for comparing costs only. I will conclude with a supplement on the reasoning a contract miner can be cost competitive with the owner/operator system.

QUANTITATIVE FACTORS

Quantitative factors are those factors having an actual or projected cost and can be estimated and used in the evaluation process. The primary quantitative factors are 1) operating costs, 2) capital preservation, 3) acquisition and disposal of equipment, 4) direct payroll and administrative costs, 5) recruiting and training costs, and 6) employment and termination of staff and hourly labor force. These factors are discussed in the following paragraphs.

1) Operating costs.

The "direct operating cost" for an owner operated mine will be less than the price charged by a contract miner because the contract miner has ownership costs for his capital plus a return on his investment in the contract price. In comparing costs of an owner/operator system vs. a contract miner, the total cost of operating a mine must be incorporated in the analysis, including the cost of capital. The analysis should be performed using a discounted cash flow analysis, insuring that proper recognition is given to the time value of money. A sample evaluation illustrating the time value of money is explained in the "Evaluation of Alternatives" later in this paper.

2) Capital Preservation.

Contract mining reduces the magnitude of the initial capital investment required to bring the property into production, thereby preserving the owner's capital, especially in the early stages of development and operation. The owner may have invested considerable capital resources in exploration and/or acquisition, engineering, and permitting. Capital outlay for equipment represents a substantial financial load on company resources and may restrict project development or prohibit project development. The owner's capital funding can be preserved for construction of other facilities, equipment, or projects by employing contract mining.

3) Acquisition and disposal of equipment.

Acquisition and disposal of equipment can be costly, not only for the capital cost required, but also for the time consumed by the owner's management personnel. Owners seldom have the personnel who are familiar with used equipment, either for acquisition or disposal. This cost is especially magnified for a short term project.

4) Direct payroll and administrative costs.

Contract mining reduces direct payroll and administrative costs. Items such as staff and hourly direct payroll, record keeping for labor, regulatory reports, administrative costs, and supervision are performed by the contract miner. In addition, the responsibility of health care, benefits, insurance, equipment record keeping, and all costs and planning required to maintain a qualified work force are the responsibility of the contract miner.

5) Recruiting and Training Costs.

Recruiting and training of the labor force is expensive. Frequently, a trained labor force is not readily available, especially in remote areas where much of the gold mine developments occur. Major contract mining firms have training programs in place to train local labor to create a skilled labor force to operate a mine. In some parts of the world, training is critical due to the lack of skilled labor.

6) Employment and Termination of Staff and Hourly Personnel.

Major costs may be incurred during employment and termination of staff and hourly labor personnel. These costs include the mobilization of management and higher level staff personnel, time required to develop a well coordinated staff, severance pay for terminated personnel, and the demobilization (relocation) of management and higher level staff personnel. Most of these costs can be quantified with some certainty, but the development of a well coordinated staff is difficult to quantify with accuracy. The staff, especially in remote areas of the world, must work together in a cooperative and coordinated mode in order to have an efficient operation. The experienced contract miner can help eliminate this problem by including these problems in his contract.

SUBJECTIVE FACTORS

Subjective factors are difficult to quantify but do represent significant factors to be considered in the evaluation. Factors to consider include 1) corporate philosophy, 2) risks associated with operating costs, 3) financing the project, 4) operating control, 5) operating flexibility, 6) recruiting and training of personnel, 7) employment and termination, 8) administrative concerns, and 9) owner/contractor relationship.

1) Corporate Philosophy.

The strategy of a corporation including its goals on mining and the philosophy of a corporation toward operating its own properties and contract mining may have a major impact on the consideration of contract mining. For example, corporate strategy may require reducing capital investment, therefore favoring contract mining. However, corporate strategy may dictate becoming an operating company which would disfavor contract mining. Corporate philosophy may be to maximize return on investment therefore dictating a discounted cash flow analysis determined by the cost analysis (example illustrated later in this report). Another corporate philosophy might be to operate all its own properties "because we are expert miners and no one can operate better than we can", therefore precluding contract mining. This philosophy may be detrimental to the corporation because it eliminates an alternative that could be beneficial. Another philosophy considered may be that the corporation may not want to obtain and/or currently does not have the expertise necessary to do its own mining. Therefore, a contract miner is necessary. Corporate strategies and philosophies may be short term or long range. The examples given above are only a few of the policies that may influence or dictate the use of contract miners.

2) Risks Associated with Operating Costs.

All mining ventures have risks involved with the development and operation of a mine. These risks can be minimized by using qualified technical personnel to perform appropriate tests, studies, evaluations, and proper decision making. However, risks can not be totally eliminated. A risk that may be minimized is the mine operating cost. The owner may contract out the mining operation, thereby transferring the risk to the contract miner. However, this transfer of risk is a double edged sword. The risk of higher operating costs is transferred by contract mining. In addition, the benefit of lower operating costs through higher operating efficiencies is also transferred to the contract miner.

3) Financing the Project.

Fixed operating expenses through a contract miner may facilitate financing for the project. If a "reputable and qualified" contractor is selected and operating costs are contractually defined, financial institutions often recognize that the project has a better chance of meeting projected costs and production levels. The financial institutions consider that the contractor has assumed some of the risks inherent in the project both at start-up and the continuing operation and the operating costs are fixed. This, of course, is largely dependent on a properly engineered mine plan.

4) Operating Control.

The owner relinquishes some control of the project through the use of a contract miner. The owner must work closely with the contract miner to ensure that ore control is performed properly and the production goals are achieved. Selection of a reputable and capable contract miner is beneficial to an efficient operation and reduces the owner's work load to control his project.

5) Operating Flexibility.

The owner/operator has more flexibility operating his own equipment but is limited to the equipment on-hand and the ability of that equipment to perform the required tasks. The owner has greater flexibility with a contract miner when setting his production requirements and varying them to meet the needs of the operation. In addition, the owner has greater flexibility in changing production goals to meet the changing market conditions for his product (in this case, gold). Most contract miners have several operations and can change the equipment to meet the changing production goals while minimizing the cost to the owner.

6) Recruiting and Training of Personnel

Recruiting and training of personnel is both a quantitative and subjective factor in the analysis. The quantitative costs were discussed under the quantitative factors. The subjective aspects are related to the dilution of efforts of the owner's management personnel, establishment of a cohesive staff and development of a trained labor force. The use of a contract miner reduces the required management effort to establish a cohesive staff. A cohesive, qualified staff is important to the optimum operation of the mine. The contract miner generally assembles his staff from other operations, thereby reducing his difficulties in achieving a cohesive, qualified staff. The owner is no longer concerned directly with the establishment of a trained labor force. The owner is indirectly concerned in insuring that the work by the contract miner is being performed in a manner beneficial for the owner. Frequently, the contract miner has well trained and qualified specialty hourly labor personnel available to reduce his difficulty in the mine operation. All this effort, if undertaken by the owner, dilutes his effort toward the optimum operation of the mine.

7) Employment and Termination

Most mining companies are concerned with long term employment of their personnel. Generally, mining companies are not equipped to employ and particularly not, the termination of large groups of both staff and hourly personnel. Contract miners perform these tasks easily as they are experienced at initiating and terminating the work at a project.

8) Administrative Concerns

Utilizing a contract miner reduces the administration of accounting, engineering, personnel, inventory, equipment, etc. for the mine operation. However, it adds for the owner the administration of the agreement with the contract miner. The establishment and administration of accounting for mining costs, establishment of engineering procedures for the operation, administration of personnel policy, establishment and tracking of parts and supplies, the purchase, operation and maintenance of an equipment fleet, etc. can be handled by the contract miner rather than the owner. However, the owner must administer the contract to include monitoring the contract miner to insure that the goals of the owner are achieved.

9) Owner/Contractor Relationship

One of the most important aspects in utilizing a contract miner is the owner/contractor relationship. This relationship has a great impact on the successful operation. An adversarial relationship will create additional work for the owner and will impact the costs on the project. A spirit of cooperation will have a positive impact on the morale of both the owner and the contract miner with an attitude that "there is no problem that can't be overcome as we work together to solve it". The relationship is a two-way street. Both parties are attempting to achieve their respective goals and this relationship is critical to the attainment of those goals.

EVALUATION OF ALTERNATIVES

Introduction

As a means to discuss the evaluation of alternative methods, I have used a hypothetical mining project as an example. The evaluation discussed is strictly an economic evaluation based on the expected costs of each of the alternative ways of developing and operating the property. Factors other than economics may enter into the decision making process, such as corporate philosophy, risks, ability to finance the project, etc. Costs given below are typical of an open pit, heap leach gold-mining operation in Nevada, but do not represent a particular operation. However, the procedure shown below can be used in any mine operation, whether open pit or underground, precious metal or base metal, metallic or industrial.

Base Criteria

The basic criteria for the property and the mine operation is shown in Table 1.

Table 1
Base Criteria

Minable Reserves	10,000,000 tons
Grade	0.06 oz. per ton gold (all oxide)
Strip Ratio	2.5 tons waste per ton ore
Production Rate	2,000,000 tons per year

Capital and Operating Costs

The capital and operating costs shown for the owner/operator case and the contract miner case are illustrated from the perspective of the owner of the property. The operating and capital costs are the owner's costs as determined by an estimate prepared by his engineering department. The contract mining costs assume the owner decided to use a contract miner to remove the overburden, mine and crush the ore and place the crushed ore on the heap leach pad.

Owner/Operator Case

The analysis has been simplified to illustrate the evaluation process. All the complexities of all possible ventures could not be addressed in one example.

The owner has determined the mine will be operated on the production schedule shown on Table 2.

Table 2
Production Schedule
(000 tons)

Year	Ore	Waste	Total
-1	-0-	2,000	2,000
1	2,000	5,000	7,000
2	2,000	5,000	7,000
3	2,000	5,000	7,000
4	2,000	5,000	7,000
5	2,000	3,000	5,000
Total	10,000	25,000	35,000

The owner has performed a cost estimate and has determined expected capital and operating costs are as shown on Table 3.

Table 3
Owner/Operator Case
Capital and Operating Cost

Capital Costs	
Mobile Equipment	$7,750,000
Crushing and Agglomeration	$6,250,000
Leach Pad and Plant	$7,000,000
Ancillary Facilities	$4,000,000
Infrastructure	$2,000,000
Total	$27,000,000
Operating Costs	
Ore Mining	$0.70/ton
Waste Removal	$0.65/ton
Crushing and Agglomerating	$0.80/ton
Heap Construction	$0.30/ton
Leach Pad/Plant Operation	$2.85/ton
Overhead	$2,200,000/yr

Capital costs include mobilization of equipment (freight and erection) and the working capital for the project. The operating costs for the tasks are labor and supplies only. Supervision, property taxes, insurance, etc. are in the overhead cost.

The owner spent $6,000,000 in exploration and development drilling and in permitting and other developmental costs and has capitalized those costs under exploration/development.

Preproduction and development years are designated as year -3, -2, and -1. These exploration, development, and permitting costs were incurred as follows: year -3, $2,000,000; year -2, $3,000,000; year -1, $1,000,000. These costs have been capitalized and amortized over the ore production years. The annual amortization cost is $1,200,000.

The annual operating costs for the project are summarized in Table 4.

Table 4
Owner/Operator Case
Annual Operating Costs
($000)

Year	Ore	Waste	Crush/ Agglom.	Heap Constr.
-1	-0-	$1,300	-0-	-0-
1	$1,400	$3,250	$1,600	$600
2	$1,400	$3,250	$1,600	$600
3	$1,400	$3,250	$1,600	$600
4	$1,400	$3,250	$1,600	$600
5	$1,400	$1,950	$1,600	$600
Total	$7,000	$16,250	$8,000	$3,000

Year	Leach & Plant	Overhead	Total
-1	-0-	$2,200	$3,500
1	$5,700	$2,200	$14,750
2	$5,700	$2,200	$14,750
3	$5,700	$2,200	$14,750
4	$5,700	$2,200	$14,750
5	$5,700	$2,200	$13,450
Total	$28,500	$13,200	$75,950

For simplification, the depreciation for all the capital equipment and facilities is used as 10 years, straight-line with zero salvage value. The depreciation for the project is $2,700,000 per year. Salvage value is assumed to be the book value and is taken in the last year of the project. The salvage value reflects the recoupment of working capital and the demobilization of equipment.

Revenue for the project is calculated on an average ore assay of 0.06 ounces of gold per ton with an average recovery of 70% and an average gold price of $400 per ounce. The annual revenue for this project is $33,600,000 with a total project revenue of $168,000,000.

A simplified discounted cash flow analysis is shown in Table 5. No royalty exists on the mining claims. Taxes, other than income tax, are included in the overhead. (Nevada does not have a corporate income tax.) The federal tax rate is assumed to be 34%. The tax calculation assumes consolidated returns with other opera-

Table 5
Owner/Operator Case
Discounted Cash Flow Analysis
($000)

Years	-3	-2	-1
Revenue	-0-	-0-	-0-
Operating Costs (-)	-0-	-0-	$3,500
Operating Income	-0-	-0-	($3,500)
Depreciation (-)	-0-	-0-	$2,700
Amortization (-)	-0-	-0-	-0-
Depletion (-)	-0-	-0-	-0-
Taxable Income	-0-	-0-	($6,200)
Income Tax (-)	-0-	-0-	($2,108)
After Tax Income	-0-	-0-	($4,092)
Depreciation (+)	-0-	-0-	$2,700
Amortization (+)	-0-	-0-	-0-
Depletion (+)	-0-	-0-	-0-
Explor/Develop (-)	$2,000	$3,000	$1,000
Capital (-)	-0-	-0-	$27,000
Salvage (+)	-0-	-0-	-0-
Cash Flow	($2,000)	($3,000)	($29,392)

Years	1	2	3
Revenue	$33,600	$33,600	$33,600
Operating Costs (-)	$14,750	$14,750	$14,750
Operating Income	$18,850	$18,850	$18,850
Depreciation (-)	$2,700	$2,700	$2,700
Amortization (-)	$1,200	$1,200	$1,200
Depletion (-)	$5,040	$5,040	$5,040
Taxable Income	$9,910	$9,910	$9,910
Income Tax (-)	$3,369	$3,369	$3,369
After Tax Income	$6,541	$6,541	$6,541
Depreciation (+)	$2,700	$2,700	$2,700
Amortization (+)	$1,200	$1,200	$1,200
Depletion (+)	$5,040	$5,040	$5,040
Explor/Develop (-)	-0-	-0-	-0-
Capital (-)	-0-	-0-	-0-
Salvage (+)	-0-	-0-	-0-
Cash Flow	$15,481	$15,481	$15,481

Years	4	5	Total
Revenue	$33,600	$33,600	$168,000
Operating Costs (-)	$14,750	$13,450	$75,950
Operating Income	$18,850	$20,150	$92,050
Depreciation (-)	$2,700	$2,700	$16,200
Amortization (-)	$1,200	$1,200	$6,000
Depletion (-)	$5,040	$5,040	$25,200
Taxable Income	$9,910	$11,210	$44,650
Income Tax (-)	$3,369	$3,811	$15,179
After Tax Income	$6,541	$7,399	$29,471
Depreciation (+)	$2,700	$2,700	$16,200
Amortization (+)	$1,200	$1,200	$6,000
Depletion (+)	$5,040	$5,040	$25,200
Explor/Develop (-)	-0-	-0-	$6,000
Capital (-)	-0-	-0-	$27,000
Salvage (+)	-0-	$10,800	$10,800
Cash Flow	$15,481	$27,139	$54,671

tions so that tax losses can be taken in the year in which they occur.

The cash flows developed in Table 5 were discounted at 15% to determine the net present value of the cash flows, as shown in Table 6.

Table 6
Owner/Operator Case
Present Value
($000)

Year	Cash Flow	Discount Factors	Net Present Value
-3	(2,000)	1.000000	(2,000)
-2	(3,000)	0.869565	(2,609)
-1	(29,392)	0.756144	(22,225)
1	15,481	0.657516	10,179
2	15,481	0.571753	8,851
3	15,481	0.497177	7,697
4	15,481	0.432328	6,693
5	27,139	0.375937	10,203
Total	54,671		16,789

Contract Miner Case

The same operation example is used in this case as was used in the owner/operator case. The only difference is the consideration of a contract miner to do the waste removal, ore mining, crushing and agglomerating, and the placement of the crushed and agglomerated ore on the leach pad. The owner has developed a bidders list of contractors and transmitted a request for proposal (RFP) to the contractors on the bidders list. The owner has received proposals from the contractors and has selected the most qualified contractor based on price, qualifications of the contractor and his personnel, availability of equipment and personnel, ability to perform, contractual terms and the type of owner/contractor relationship he expects. The contract miner's bid that was selected is shown on Table 7.

Table 7
Contract Miner's Bid

Operating Costs	
Ore Mining	$1.15/ton of ore
Waste Removal	$1.10/ton of waste
Crushing and Agglomeration	$1.50/ton of ore
Heap Construction	$0.50/ton of ore

The owner's expected cost for the remaining operations are shown in Table 8.

During exploration and development, the owner has spent $6,000,000 in exploration, development drilling, permitting and other developmental costs. These costs were incurred as follows: year -3, $2,000,000; year -2, $3,000,000; year -1, $1,000,000. These costs have been capitalized and amortized over the ore production years. The annual amortization cost is $1,200,000.

Table 8
Contract Miner's Case
Capital and Operating Costs

Capital Costs	
Leach Pad and Plant	$7,000,000
Infrastructure	$2,000,000
Total	$9,000,000
Operating Costs	
Leach Pad/Plant Operation	$2.85/ton
Overhead	$1,000,000/yr

The annual operating costs for this operation are shown in Table 9.

Table 9
Contract Miner's Case
Annual Operating Costs
($000)

Year	Ore	Waste	Crush/ Agglom.	Heap Constr.
-1	-0-	$2,200	-0-	-0-
1	$2,300	$5,500	$3,000	$1,000
2	$2,300	$5,500	$3,000	$1,000
3	$2,300	$5,500	$3,000	$1,000
4	$2,300	$5,500	$3,000	$1,000
5	$2,300	$3,300	$3,000	$1,000
Total	$11,500	$27,500	$15,000	$5,000

Year	Leach & Plant	Overhead	Total
-1	-0-	$1,000	$3,200
1	$5,700	$1,000	$18,500
2	$5,700	$1,000	$18,500
3	$5,700	$1,000	$18,500
4	$5,700	$1,000	$18,500
5	$5,700	$1,000	$16,300
Total	$28,500	$6,000	$93,500

For simplification, the depreciation period for all capital equipment and facilities is taken as 10 years and assumed to be straight line with no salvage. Depreciation for the project is $900,000 per year. Salvage value on the project is assumed to be book value and is taken in the last year of the project. The salvage value reflects the recoupment of working capital and the demobilization of the equipment.

Revenue for the project is calculated on an average ore assay of 0.06 ounces of gold per ton with an average recovery of 70% and an average gold price of $400 per ounce. The annual revenue for this project is $33,600,000 with a total project revenue of $168,000,000

A simplified discounted cash flow analysis is shown in Table 10. No royalty exists on the mining claims. Taxes, other than income tax, are included in the overhead. (Nevada does not have a corporate income tax.) The federal tax rate is assumed to be 34%. The tax calculation assumes consolidated returns with other operations so that tax losses can be taken in the year in which they occur.

The cash flows developed in Table 10 were discounted at 15% to determine the net present value of the cash flows, as shown in Table 11.

Table 11
Contract Miner's Case
Present Value
($000)

Year	Cash Flow	Discount Factors	Net Present Value
-3	(2,000)	1.000000	(2,000)
-2	(3,000)	0.869565	(2,609)
-1	(11,806)	0.756144	(8,927)
1	12,394	0.657516	8,149
2	12,394	0.571753	7,086
3	12,394	0.497177	6,162
4	12,394	0.432328	5,358
5	17,446	0.375937	6,559
Total	50,216		19,778

Evaluation Summary

Results of the evaluation of the two alternatives using the examples stated for the owner/-operator case and the contract miner case are shown on Table 12.

Table 12
Net Present Value Comparison

Alternative	Net Present Value
Owner/Operator	$16,789,000
Contract Miner	$19,778,000

The numbers used here are simulated and should not be construed as an endorsement of contract mining as the proper choice from an economic standpoint in all circumstances. The choice in this case, using the net present value only, is the use of the contract miner. The second choice would be the owner/operator. However, other factors may enter into the decision making. Corporate philosophy may dictate that the company desires to become an owner/operator and is unwilling to use a contract miner. The accuracy of the cost number used in the evaluation must be considered when making the decision.

Table 10
Contract Miner's Case
Discounted Cash Flow Analysis
($000)

Years	-3	-2	-1
Revenue	-0-	-0-	-0-
Operating Costs (-)	-0-	-0-	$3,200
Operating Income	-0-	-0-	($3,200)
Depreciation (-)	-0-	-0-	$900
Amortization (-)	-0-	-0-	-0-
Depletion (-)	-0-	-0-	-0-
Taxable Income	-0-	-0-	($4,100)
Income Tax (-)	-0-	-0-	($1,394)
After Tax Income	-0-	-0-	($2,706)
Depreciation (+)	-0-	-0-	$900
Amortization (+)	-0-	-0-	-0-
Depletion (+)	-0-	-0-	-0-
Explor/Develop (-)	$2,000	$3,000	$1,000
Capital (-)	-0-	-0-	$9,000
Salvage (+)	-0-	-0-	-0-
Cash Flow	($2,000)	($3,000)	($11,806)

Years	1	2	3
Revenue	$33,600	$33,600	$33,600
Operating Costs (-)	$18,500	$18,500	$18,500
Operating Income	$15,100	$15,100	$15,100
Depreciation (-)	$900	$900	$900
Amortization (-)	$1,200	$1,200	$1,200
Depletion (-)	$5,040	$5,040	$5,040
Taxable Income	$7,960	$7,960	$7,960
Income Tax (-)	$2,706	$2,706	$2,706
After Tax Income	$5,254	$5,254	$5,254
Depreciation (+)	$900	$900	$900
Amortization (+)	$1,200	$1,200	$1,200
Depletion (+)	$5,040	$5,040	$5,040
Explor/Develop (-)	-0-	-0-	-0-
Capital (-)	-0-	-0-	-0-
Salvage (+)	-0-	-0-	-0-
Cash Flow	$12,394	$12,394	$12,394

Years	4	5	Total
Revenue	$33,600	$33,600	$168,000
Operating Costs (-)	$18,500	$16,300	$93,500
Operating Income	$15,100	$17,300	$74,500
Depreciation (-)	$900	$900	$5,400
Amortization (-)	$1,200	$1,200	$6,000
Depletion (-)	$5,040	$5,040	$25,200
Taxable Income	$7,960	$10,160	$37,900
Income Tax (-)	$2,706	$3,454	$12,884
After Tax Income	$5,254	$6,706	$25,016
Depreciation (+)	$900	$900	$5,400
Amortization (+)	$1,200	$1,200	$6,000
Depletion (+)	$5,040	$5,040	$25,200
Explor/Develop (-)	-0-	-0-	$6,000
Capital (-)	-0-	-0-	$9,000
Salvage (+)	-0-	$3,600	$3,600
Cash Flow	$12,394	$17,446	$50,216

Possibly, an upside potential exists for lowering the costs so that the owner/operator is the preferred alternative. Another overriding consideration that may dictate the use of a contract miner, may be the ability to raise capital to invest in this or other operations. The decision is seldom black or white, but one must consider many factors when making the decision.

SUPPLEMENT

Mine owners and operators ask how can a contract miner operate at a lower total cost. The major direct cost advantages of a contract miner are discussed in the "quantitative factors" section of this report (page 1). The cost advantages involve personnel, equipment, and experience. The personnel costs involve recruiting, training, employment, and termination of personnel. The equipment costs include the buying and selling of equipment, current ownership of the equipment, utilization of equipment at completion of the project at a future or another project. The experience costs involve both personnel and equipment and the ability to perform the tasks mentioned in both areas.

However,the greatest cost advantage of the contract miner is motivation. "Motivation" or "incentive" or "inducement" are the ingredients that make people perform at a higher level. A contract miner has an incentive to keep costs down because he has a set price to perform the work. Changes in the price of the commodity (gold) or the grade of the ore will not affect or change his return on his investment. The contract miner can only improve his profitability by becoming more efficient in his work. An owner/operator can have his profitability affected significantly with a change in the price of the commodity (gold) or a change in the grade of the ore. In most operating mines, if the project is making a good profit, the operation tends to become less efficient and costs are incurred that do not benefit the operation. An example of the "lack of motivation" was the Bingham Canyon operation prior to 1980. G. F. Joklik stated in his address at the Utah AIME Student Night on February 18, 1988 that "Enjoying easy profitability year after year, the Bingham Canyon operation had become big, fat, and inefficient by 1980." As he stated, this was true not only of the Bingham Canyon operation, but the entire copper industry. The Bingham Canyon operation began the decade losing money which provided the incentive to improve the operation. Productivity increased at the operation from 31 tons of copper per man year in 1980 to 110 tons of copper per man-year in 1988, nearly 3-1/2 times the level of 1980. This example illustrates the efficiencies that can be developed when people are motivated to perform.

Contract mining is one of the potential operating methods that may be used in the development of a project anywhere in the world. It is not the panacea to all operating problems and isn't always the most cost effective method for development, but it should be considered as you, the owner, considers the development and/or operation of a gold project.

REFERENCES

Haas, D. D., 1986, "The Contract Mining Bidding Process," Northwest Mining Association annual meeting, Spokane, WA, December, 10pp.

Haas, D. D., 1987, "How to Contract for Your Surface or Underground Mine," Short Course at Northwest Mining Association annual meeting, Spokane, WA, December, 89pp.

Haas, D. D., 1992 (projected), "Project Operating Strategy," SME Mining Engineering Handbook, H. L. Hartman, sr. ed., AIME, New York.

Joklik, G. F., 1988, "BP Minerals America," Utah AIME Student Night, Salt Lake City, Utah, 12pp.

Chapter 25

DOMINION MINING LIMITED - COSMO HOWLEY GOLD MINE

M COLLINGS[1], D LENARTOWICZ[2], K R ALEXANDER[3]

[1]Group Mining Engineer - Dominion Mining Limited
[2]Pit Superintendent and [3]Senior Geologist - Dominion Gold Operations Pty Ltd

ABSTRACT

Cosmo Howley is a gold deposit in the Northern Territory of Australia. The geology is complex and the input of geological data is an important part of the grade control process. As the geology of the deposit has become better understood, grade control and mining techniques have evolved to meet the changing circumstances. Present grade control depends on the use of angled blast holes, 2½ m benches and selective mining. Data is processed on site using 2 386 PC's running Surpac software.

An environmental protection programme is employed, including soil stripping and stockpiling, sulphide waste containment and water quality monitoring.

INTRODUCTION

The Cosmo Howley Gold Mine is located in the Northern Territory of Australia approximately 160 km southeast of Darwin and 12 km north of Hayes Creek near the main north-south Stuart Highway (Figure 1). The mine is owned and operated by Dominion Gold Operations Pty Ltd, a wholly owned subsidiary of Dominion Mining Limited. Current production is from two open pits, Cosmo and Phantom, the lower grade oxide ore being processed by heap leaching and the higher grade oxide material and all sulphide ore being fed to the recently commissioned CIL plant.

Current production rates are 900,000 t/annum of heap leach feed and 1,000,000 t/annum of CIL feed, with the latter expanding to 1,500,000 t/annum as a second ball mill is commissioned. It is expected that the heap leach stacking operation will shortly terminate.

Recoverable reserves in the two pits presently stand at 6.1 Mt at 2.56 g/t. To January 1989, 1.34 Mt of oxide ore had been stacked at an average grade of 1.64 g/t. Gold recovery from the heaps to date is 61% and is projected to achieve 70% at the end of the final leach cycle.

The climate is sub tropical with mild dry winters. In the summer months (November-April) monsoonal conditions can prevail, an average of three cyclones a year affecting the northern areas. Some of these move inland and can cause very heavy rain to fall in the mine area and can cause a number of operational problems.

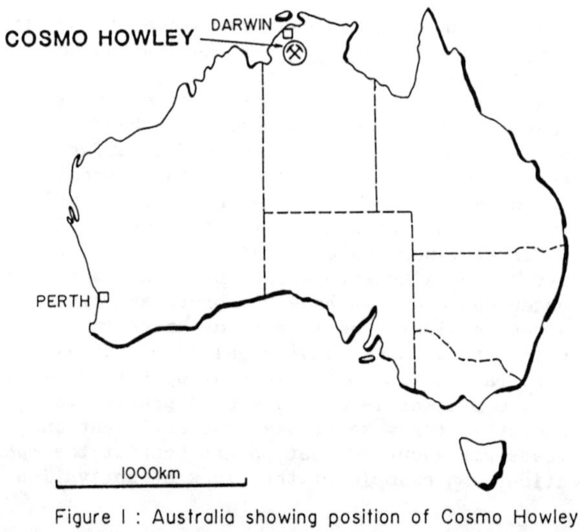

Figure 1 : Australia showing position of Cosmo Howley

PROJECT HISTORY

Gold was discovered at Cosmo Howley in 1873 during the construction of the Adelaide to Darwin overland telegraph. The mine was worked under tribute by the Chinese from 1892 to 1903 with a recorded production of 34,000 oz making it one of the more significant deposits in the Pine Creek field.

Further evaluations of the deposit were undertaken in the 1930's and 1950's but did not lead to any new gold production. The most recent phase of exploration began in the late 1970's and involved a succession of companies including Homestake, BHP, Geopeko and Anaconda.

A deep drilling program completed by Geopeko-Anaconda in 1982 intersected significant ore grade mineralization at vertical depths between 300 and 600 metres below surface. In 1985 Dominion acquired 50% of the property and in late 1986 became the sole owner and took the project through to production.

REGIONAL GEOLOGY

Cosmo Howley occurs in the Early Proterozoic sediments of the Pine Creek Basin.

From about 2,400 Myr to 2,100 Myr sediments were deposited onto a crystalline Archean basement. After a period of uplift and erosion, sedimentation continued until approximately 1,800 Myr when a major phase of deformation accompanied by greenschist to amphibolite facies metamorphism took place. This deformation produced tight non-cylindrical folds, considerable brittle fracture and a widespread axial plane slatey cleavage in pelitic rocks. Major faulting also occurred at this time.

Granite intrusion at 1,760 Myr was widespread and resulted in open, irregular refolding of the earlier structures and a rare crenulation cleavage.

The gold deposits of the Pine Creek Basin occur in various lithologies with all exhibiting strong structural control both regionally and at the mine scale. They appear to be broadly synchronomous with the 1800 m/yr phase of major deformation.

As well as being an important gold producing area the Pine Creek Basin is a uranium province of worldwide significance.

MINE GEOLOGY

Stratigraphy

Cosmo Howley is hosted by sediments of the Koolpin Formation of the South Alligator Group.

The Koolpin Formation is divided into Lower, Middle and Upper members. The Lower Member is up to 250 metres thick and consists of carbonaceous mudstones, slates and phyllites. The Middle Member consists mainly of silicate facies of iron formation (10-15% Fe including iron sulphides), carbonaceous mudstone and phyllites and attains a thickness of up to 100 metres. The Upper Member consists of carbonaceous mudstone and varies in thickness from 50 to 150 metres.

Several 5-20 metre thick dolerite sills intrude the mine sequence.

Economic mineralization at Cosmo Howley is found within the whole range of lithologies of the Middle Koolpin but is preferentially located in the uppermost silicate iron formations.

Structure

Folding Styles Cosmo Howley is located on a major regional structure, the Howley anticline. The plunge of the fold axis of the structure is quite shallow and in the mine area varies from 20° grid North through horizontal to 20° grid south (Figure 2).

The anticline is asymmetrical with the west limb dipping moderately west while the east limb dips steeply east at the surface and steepens through vertical to overturned at depth. The closure of this major structure is rarely observed due to intense faulting of the axial region. At the mesoscopic scale, folds associated with this structure are similarly uncommon but when observed are tight to isoclinal. A sub-vertical to west-dipping axial plane cleavage developed during the formation of the anticline.

The major anticline has been modified by two later folding events. The first and more important of these two events produced steeply plunging open to close folds with amplitudes up to several metres. This phase of folding crenulated the early cleavage and was locally intense enough to generate a new cleavage.

Faulting Intense faulting was associated with the late stages of development of the Howley anticline. The faulting is of a brittle-ductile nature and occurs across the whole structure but is best developed in the axial zone.

Fault zones are up to ten metres wide and contain a variety of sheared and brecciated fragments. Quartz is found in or near fault zones as a fracture infill and as lithologically conformable veins and saddle reefs.

MINERALIZATION

Gold mineralization is closely associated with sulphides. The main sulphide minerals are pyrite and arsenopyrite with traces of sphalerite and chalcopyrite. Pyrrhotite occurs at depth (+300 m below surface). In the weathered zone the sulphide have been altered to limonite, goethite and hematite.

Two generations of pyrite occur. The first generation of pyrite is aligned within the earlier cleavage and has been flatted during the original folding event and may be of sedimentary or diagenetic origin. It is disseminated throughout the ironstones, phyllites and carbonaceous slates. A second phase of subhedral pyrite occurs within and adjacent to shear zones, quartz zones and the hinge zones of the later folds.

Arsenopyrite occurs as thin layers (2-3 mm) of fine grained prismatic crystals within the iron formations. Large euhedra (1-2 cm) of arsenopyrite also occur and are associated with

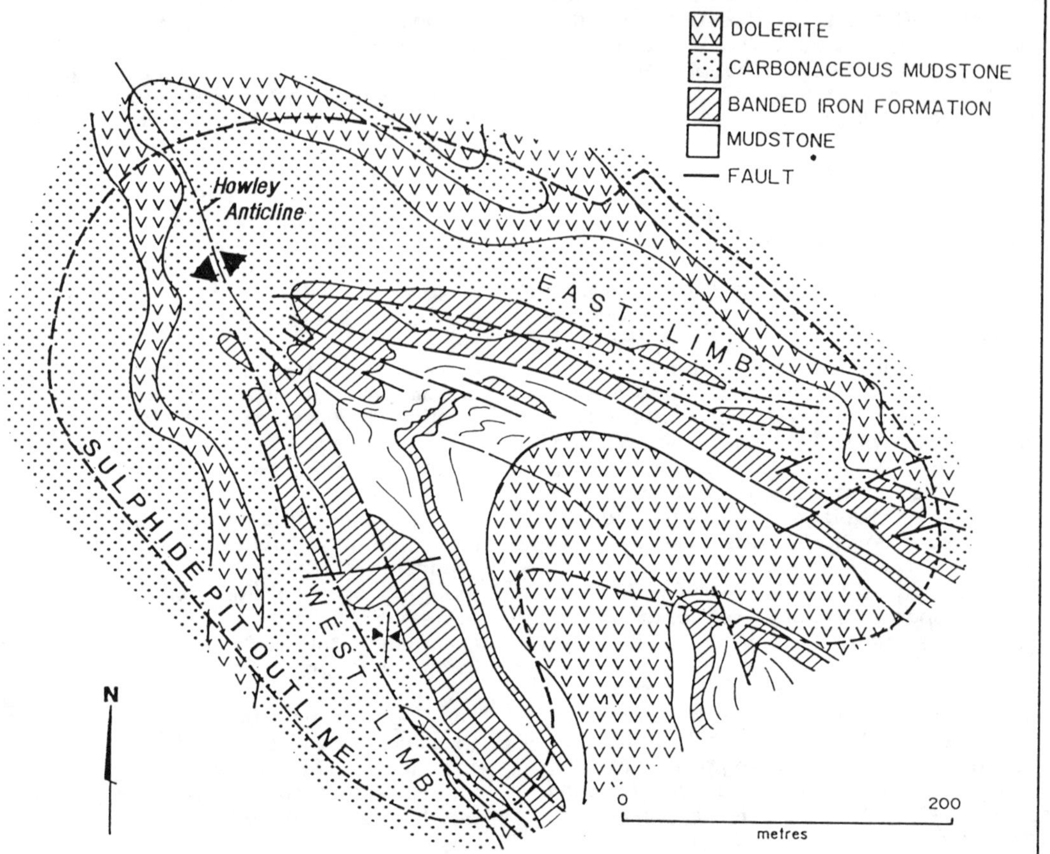

Figure 2 : Cosmo Howley Mine Open Pit Geology - 1157m Level.

the second phase of pyrite. The two arsenopyrite types may be of the same generation.

Gold is very fine grained and rarely seen. The specific relationship of the gold to the sulphide types is not known.

Controls on Mineralization

There is a strong lithological control to mineralization, with most gold occurring in the iron formations although all lithologies host some mineable ore.

At bench scale most ore blocks are associated with the slightly discordant breccias, shears and quartz veins that accompanied the intense faulting and folding of the major anticlinal structure.

Higher grade mineralization is common where the axial zone of the major structure has been modified by well developed steeply plunging later folds, and in these places the ore lenses plunge in a similar steep manner.

GRADE CONTROL AND SELECTIVE MINING

Introduction

Most grade control practices in highly selective gold mining operations rely partly or wholly on visual control during mining. At Cosmo Howley direct visual identification of lithologies and structural features such as folds and shears is only a guide to location of ore, with ore waste delineation being ultimately dependent on sampling and assay.

Initial Mining and Grade Control

To minimise capital expenditure the initial mining at Cosmo Howley was based on the concept of a preliminary 'oxide pit' which would remove oxidised ore down to RL 1140 m, about 15 m below the level of the plain (the crest of the hill which would be removed was at RL 1172 m). This was to be followed by a 'sulphide pit'.

In keeping with this philosophy the excavation, crushing and stacking was contracted out, and following a tender and contractor selection process Roche Bros commenced excavation in October 1986.

Equipment used was industry standard and comprised one and subsequently two Caterpillar 245 back hoe type excavators, 50 t trucks and associated plant.

The conceptual plan was to remove and stockpile the soil from the hillside for later rehabilitation, allowing access to the ore and then to remove the hill in horizontal 2.5 metre slices. During this phase of the operation, grade control was carried out by the sampling of channel lines that had been ripped at right angles across the ore body by a bulldozer tyne.

Samples were taken on 2 metre spacings on lines 5 metres apart with sample reliability dependent upon precise survey control and the collection of uncontaminated samples.

Based on these samples the ore was marked out using survey flagging and spray paint.

Before a new bench could be rip sampled it had to be cleaned of contaminants, chiefly spillage from trucks, by a road grader. The rip line was graded several times until, upon inspection by the geologist, it was clear that bedrock had been exposed. The traverse line was then ripped by a D-7 bulldozer in one pass using a single tyne with the operation being supervised by a geologist to ensure an even depth of ripping was achieved and that smearing, by dragging material in front of the tyne, was kept to a minimum. At Cosmo Howley, unlike many West Australian mines, it was found that oxide ore was too hard to be ripped by a tyne attached to a road grader whilst any bulldozer larger than a D-7 produced uneven ripping and too much smearing.

Approximately 5 kg of material was collected over a continuous 2 m interval by placing fragments into a calico sample bag. Maximum sizes of fragments were around 5 cm with large fragments being broken up before sampling. All sample material was transported daily to Australian Assay Laboratories in Pine Creek approximately 70 km away with a turnaround of between 24 and 48 hours. Assay grades were determined by fire assay.

After ripping, the geology of the channel was mapped at 1:250 scale. 3-dimensional exposure was not good enough for detailed structural mapping but was adequate for identification of the major lithologies and fault zones. This geology was then used to complement and interpret bench assay plots.

Location of the sample lines was accurately controlled from known survey pegs. On occasions scheduling requirements necessitated truck haulage over the sampling grid and the ore was later marked out by precise resurvey of the sampling grid. An example of part of a grade control plan produced from rip line assays is shown in Figure 3.

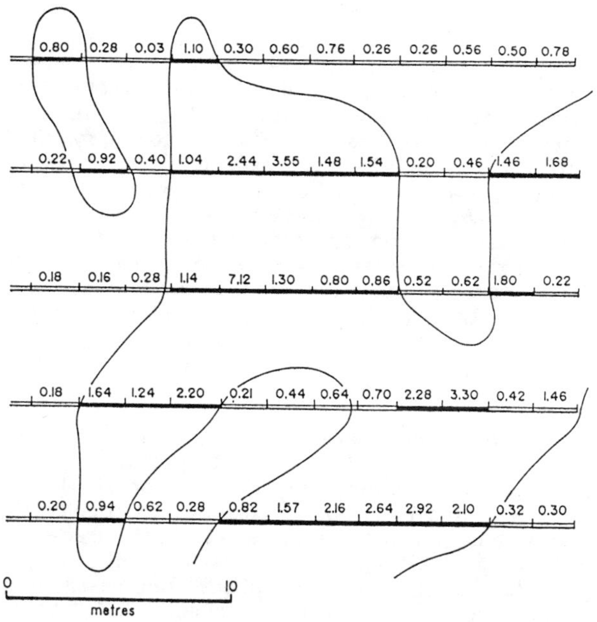

Figure 3 : Grade Control from rip lines.

Introduction of Paddock Blast and Grade Control Drilling

After several months of mining it was evident that ground conditions were too hard for acceptable production rates to be achieved by ripping so a drill and blast operation was initiated. Shortly after June 1987 the sampling of channel lines was discontinued and grade control was then undertaken by sampling of the blast holes.

Under the terms of the mining contract there had always been a provision for the sampling of blast holes. Sample drilling costs were, therefore, largely included within the "drill and blast ore" contract rate with only minimal extra cost being incurred to compensate for the slightly lower production drilling rates obtained when the cuttings were sampled.

In order to maintain continuity of information in section, holes were drilled on the same lines as used for ripping, i.e. 5 m apart. Initially holes only 4 m apart along the line were required for blasting purposes and where additional assay information was needed, usually in areas of complex geology, an infill pattern of 5 m x 2 m was produced. As the ground became harder a pattern of 2.5 m x 4 m was used, i.e. holes 4 m apart along lines 2.5 m apart with the pattern staggered to give a 3.2 m x 3.2 m rhombic pattern for blasting. This represented a compromise between the needs of the geologists for sampling and the optimum blasting pattern. Holes are angled at 70° depending upon which side of the anticlinal axis they are sited and drilled to 6.4 m allowing for 1 m subdrill. Two samples are taken per hole, i.e. one per bench.

Holes are drilled by a Tamrock DHA-800 hydraulic trackdrill with a 300 CFM compressor on board. Initially a 140 mm hole diameter was drilled but it was found that the flushing velocity obtained was insufficient to carry drill cuttings out of the hole especially in areas of broken ground where flushing air was being lost into cracks. A hole diameter of 102 mm is now being used. The contract miner would prefer the larger diameter holes as a given powder factor can be attained with less drilling overall, however the smaller drilling diameter is being maintained in the interests of sample recovery.

Ore Block Delineation

During the period of the rippable ore operation a geological interpretation was produced by mapping the channels or rip lines. Ore was blocked out according to geological trend and assays using a cut off grade of 0.8 g/t Au. At one stage material in the 0.5 to 0.8 g/t range was stockpiled for resampling while occasional sampling checks of waste material were also undertaken. All processing of assay and geological data was done manually without specialist drafting assistance.

At the beginning of the drill hole phase of grade control the geological staff was increased and a geologist was employed to map the 2.5 m high bench faces. The detailed geological interpretation produced from the mapping is used as an underlay to all bench grade plans before any ore-waste boundaries are drawn.

Initially ore block delineation from blast hole data was also carried out manually. Data for each hole was plotted in section and line intercepts of ore were produced for each section. This is now being accomplished using computer software developed by Surpac Mining Systems Pty Ltd and run on Compaq 386 Personal Computers. A file containing survey data for drill holes is created with an additional file containing sample assay data. These two files form the basis of the grade control and can be manipulated to produce sections and plans of various parts of the orebody. At Cosmo Howley a plan with the assays for each drill hole projected to its mid-bench position, i.e. 1.25 m down hole, is used for daily grade control activities (Figure 4).

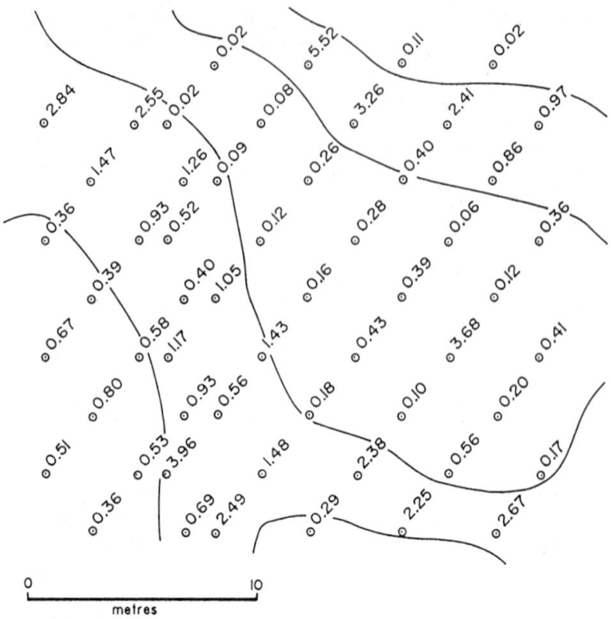

Figure 4 : Mid-bench down hole projections of blast holes

Selective Mining

The mine operates 2 x 11 hour shifts, 5½ days per week. As far as possible the ore is scheduled to be removed during the day shift but on occasions it has been successfully mined at night.

The original 20-30 m deep oxide pit was commenced in September 1986 and in November 1987 pre stripping of the deeper (130 m) sulphide pit commenced. At this time, the mining fleet consisted of the following:

2 Caterpillar 245 Hydraulic Excavators
6 Caterpillar 773 (50 t) Haul Trucks
1 Caterpillar D7 Dozer
1 Caterpillar 14G Grader
3 Tamrock DHA 800 Hydraulic Drills

and ancilliary equipment to produce an average of 280,000 BCM per four week period.

The 245 excavators proved to be very reliable, working well in the local conditions and achieving an average production rate of approximately 390 BCM/hr.

The powder factor on the staggered 4 m x 2.5 m pattern was between 0.21 and 0.25 using ANFO explosives.

In September 1988, following a successful exploration programme, additional ore reserves were delineated. A new mining schedule required production rates of 335,000 BCM per four week period.

Following discussions with the mining contractor the mining fleet was upgraded to:

1 Komatsu PC 1000 Hydraulic Excavator
1 Komatsu PC 650 Hydraulic Excavator
4 Caterpillar 773 (50 t) Haul Trucks
2 Caterpillar 777 (85 t) Haul Trucks
1 Caterpillar D8 Dozer
1 Caterpillar 14G Grader
1 Explosives Slurry Truck
3 Tamrock DH 800 Hydraulic Drills

The upgrading of the production fleet did not make an immediate impact on the production as the initial availability of the Komatsu excavators was poor. At the same time the contractor found it difficult to provide suitable quantities of broken material using the aging Tamrock drill rigs. The net result of these two factors was that the budgeted production rate was rarely attained.

Further discussions with the contractor resulted in replacement of the Tamrock drills with two new new Montabert hydraulic drill rigs mounted on Komatsu PC220 excavator chassis. These rigs are achieving an average of 48 m/hour and combined with the increasing availability of the excavators budgeted production rates are being sustained.

The powder factor currently varies between 0.25 and 0.35 but this is being reduced by a recent change of drill pattern which widened the drill hole spacing from 2.5 m to 3 m. This will effectively give a burden of 3.9 m diagonally and result in more broken ground for the same drill metres.

An investigation is also taking place to see if there are any deleterious effects of a decision to increase the bench height to 3.3 m, which is more suitable for the larger equipment now being used.

As the water table has been approached it has been necessary to change from ANFO to water resistant slurry explosives. Initially Dupont SOL was tried and later substituted by ICI EP Gold 500 which had a consistency better suited to the slurry truck used by the mining contractor on site.

Currently Dominion staff directly concerned with mining are as follows:

1 Mine Superintendent
1 Mining Engineer
3 Geologists
1 Pit Foreman
1 Surveyor
8 Sampling/Grade Control Crew
2 Dewatering Crew

MINE PLANNING AND SCHEDULING

Mine planning is done in three distinct stages.

1. Long term planning and scheduling. The bulk of the work is normally done at the head office in Perth with some input from the mine staff. During this phase the basic design and long term strategies are established.

2. Medium term planning and scheduling consists of more detailed breakdown of the long term plans normally on Period by Period basis (there are 13 four week periods per Year). This is done on site using the mining reserves generated in Stage 1.

3. Short term scheduling consists of preparing a plan and a conceptual strategy on a week by week and day by day basis. The whole mine has been divided into nominal blocks 20 m x 20 m and these are used as the basis for mine planning. As drilling and sampling are completed, tonnes and grade for each block are calculated using the SURPAC software. This information is used by the planning engineer who prepares a map showing the conceptual progress of mining for the following week. It is also used to calculate the tonnes and grade of various ore types that are expected to be mined during the week. The prepared mining schedule is discussed with the contractor and where necessary alterations are made.

MONSOONAL RAIN AND GROUNDWATER

During the wet season from December to March all aspects of the mining operations can be hampered by water. For example, during the last 88/89 wet season about 1 m of rain fell, including 50 cm in one fortnight.

In order to minimise the impact of flooding, the geometry of the pit benches is modified in the 2-3 months prior to the wet season so that there is the maximum amount of ore available above the bottom bench which forms a large sump. Nuisance water is dealt with by small portable pumps.

Pit drainage of groundwater has not been a problem to date due to the relatively impermeable wall rocks. It has been handled by a series of 8" (20 cm) boreholes within the pit sited on known favourable structures. It is intended to maintain the drilling of the bore holes as the pit develops.

Holes will be drilled in the pit walls to relieve pore pressure if necessary.

REHABILITATION AND ENVIRONMENTAL CONCERNS

The environmental aspects of any mining operation are very important today. The position of Cosmo Howley, relatively close to the Kakadu National Park, has made Dominion even more aware of these constraints. Consultants have been retained to advise the Company on all aspects of rehabilitation and environmental concerns as the mine progresses.

Waste Dumps

During the life of the mine approximately 20 million BCM of waste rock will be removed from the pits. This comprises essentially two types:

(a) rock containing sulphides and capable of producing acidic run off,
(b) inert non sulphidic rock, either dolerite or weathered rock from the near surface.

The problem of acidic run off is compounded by the high monsoonal rainfall.

In order to minimise the environmental impact of the dumps the following has been agreed with the authorities and is in progress:

(a) The waste dump is sited in a valley with a single, northerly drainage. Topsoil has been removed and stockpiled for use in later revegetation.
(b) The sulphide waste will be enclosed in a layer of inert rock.
(c) The water storage and retaining dam to the north will trap any water run off. This is then fed to the CIL plant.
(d) The final dump batters will be reduced to 15° and the whole dump revegetated with natural flora. As far as possible this will be an ongoing exercise as the dump progresses.

Tailings Dam

The tailings are being discharged into a valley to the north of the plant. Topsoil has been removed from the valley and is stockpiled.

As each section of the dam is completed and drained, topsoil will be replaced and the surfaces revegetated.

ORE RESERVE RECONCILIATION

As with many highly selective gold mining operations there is a significant difference between indicated and measured resources and recoverable reserves at Cosmo Howley. This is of particular importance as a new preliminary total resource (oxide and sulphide) of 6.1 Mt at 2.56 g/t has recently been announced and will form the basis of the 1.5 Mt per year sulphide ore CIL operation.

During the life of the oxide operation comparisons of the actual production obtained versus the in situ reserves are continually being made. The results of a recent exercise are given below.

Cosmo Pit - Production Pds 1 to 3

	Tonnes	Grade	Contained Gold
Geological Resource	247,344	2.94	727,191
Grade Control Resource	315,928	2.35	742,431
Actual Mined	386,895	1.96	758,314

These figures indicate a 56% increase in ore tonneage compared with the indicated and measured resource and a 4% increase in the total contained gold.

The reasons for these differences are various and cannot be discussed at length in this paper. Suffice to say that an increase in tonnes of this magnitude (which should not necessarily be considered 'dilution') is not unexpected given the complex morphology of the Cosmo Howley ore bodies.

The recoverable reserves derived from the indicated and measured resource and on which all budgeting is based take into account increases in ore tonneages of this order.

SUMMARY AND CONCLUSIONS

The Cosmo Howley gold mine has produced over 43,000 ounces of gold since heap leaching operations began in early 1987 and 15,000 ounces from the recently commissioned CIL plant.

The geology of the deposit is complex but is becoming better understood as the programme of face mapping proceeds. The use of geological data is an important part of the grade control process while the mining of ore is supervised by geologists, especially in the very complex areas of the ore body where major dilution is most likely to occur.

Heap leaching has proved to be a very cost effective way of treating the low grade oxide ore at Cosmo Howley. A major CIL plant treating sulphide material of significantly higher in situ grade was commissioned in November 1988 and is currently processing 200 tonnes per hour (1.5 million tonnes per year) of sulphide and high grade oxide ore.

Cosmo Howley was the first gold mine to be developed by Dominion Mining Limited. With its complex geology and relatively low grade it is difficult to mine selectively at high rates. The grade control and mining practices have been developed on site to meet the specific requirements of the deposit and have involved considerable experimentation. The willingness of the company to try new approaches has been an important factor in the success of the operation so far and augurs well for its continued success.

ACKNOWLEDGMENTS

The authors would like to express their thanks to fellow members of staff of Dominion Gold Operations Pty Limited for their assistance and to the management of Dominion Mining Limited for permission to present this paper.

REFERENCES

ALEXANDER, K.R., 1988. The Cosmo Howley Gold Mine, NT. Geology, Grade Control, Selective Mining and Heap Leaching (unpub).

8

Processing I
Heap, Dump, and In Situ Leaching

J.B. Hiskey
University of Arizona
Tucson, Arizona

R.C. Dunne
Western Australia School of Mines
Kalgoorlie, W. Aus., Australia

Chapter 26

STATUS OF HEAP, DUMP, AND IN SITU LEACHING OF GOLD AND SILVER

Paul D. Chamberlin

Chamberlin & Associates
Littleton, Colorado

ABSTRACT

The evolution of heap and dump leaching has followed the development of agglomeration, materials of construction for leach pads and ponds, solution distribution systems, testing techniques, and state regulations. They are now mature processes. Possible future directions and developments are discussed.

In situ leaching of precious metals is waiting to be commercialized. Regulators are reluctant to allow the use of cyanide because of the potential for excursions and operators are reluctant because of the potential for lawsuits. Developments in alternative lixiviants are needed before in situ leaching is accepted.

INTRODUCTION

Heap and dump leaching of gold and silver ores have matured over the past 10 years. They are no longer emerging technologies. Enough has been written and discussed about them to give prudent operators a good chance of success on the first effort. This was not the case 10 years ago when the failure rates were very high (Kappes, 1979).

Heap and dump leaching operations can be quickly put into operation. They have always had this reputation. It is common to have an operation online within a year of AFE approval and even four months has been reported. The environmental setting plays a large role in the schedule and it has become obvious to the industry that early and continual contact with the regulatory agencies is essential to a successful project.

Heap and dump leaching also have a reputation as low cost operations. As a result, low grade ores can be treated. The lowest grade, stand-alone heap leach operation that has been reported is AMAX Gold's Wind Mountain project which has an average ore grade of 0.023 oz gold/tonne. Other low grade operations are Coeur d'Alene's Rochester project which processes an ore containing 0.012 oz gold/tonne plus 1.65 oz silver/tonne and Pegasus Florida Canyon operation which processes 0.026 oz gold/tonne. A profitable low grade operation will have one or more of the following characteristics: >5000 tonnes/day throughput, >80-85% recovery of values, no requirement for crushing run-of-mine ore, and stripping ratios that are <1:1. Some dump and heap leach operations treat only the low grade portion of an ore body. In these cases, their operating costs are often only incremental costs because another process that is used to treat the high grade ore is carrying the bulk of the costs. In these cases, material as low as 0.01 oz gold/tonne has been treat profitably.

In situ leaching, on the other hand, doesn't have a single commercial experience to its credit, yet. This technology was widely used in the uranium industry and acceptance is gaining in the copper industry but it has not found acceptance in the precious metals business.

This paper discusses how heap and dump leaching technology has matured and discusses some of the reasons that projects have failed in the past. It then suggests future directions that heap and dump leaching may take. The paper also takes a look at current thinking for the in situ leaching of precious metals and gives prospects for the future.

HEAP AND DUMP LEACHING - PROGRESS TO DATE

The heap and dump leaching industry has made much progress in its understanding of what causes project failures. Some of the reasons for past failures are discussed in the following paragraphs along with improvements that have made heap and dump leaching the successes they are.

Agglomeration

The merits of agglomeration were not widely recognized until the US Bureau of Mines published their work (Heinen, et al., 1979). Since the application of agglomeration technology, the failure rate of new heap leach projects has decreased dramatically. Agglomeration has not only provided great technological advantage but it has also given heap leaching an air of respectability (Tibbals, 1987). Prior to the Bureau of Mines' work, some operators in the copper vat leaching business, such as Yerington and Inspiration, practiced agglomeration by wetting finely crushed ore before loading it into vats.

This simple wetting gave better permeability than if dry ore was loaded into the vats. The precious metals industry and the USBM built on these early copper experiences.

Agglomeration does not imply that binders must be used. In the precious metals business, it simply means that deleterious fines are bonded to coarser particles long enough to prevent segregation of coarse and fine particles as the heap is being constructed. In some cases where the amount of fines is minimal, this "agglomeration" can be achieved by simply wetting the ore with water. The surface tension of the water is enough to bind fines to coarser particles.

In most cases, however, a binder is needed, particularly if there are many fines, which is loosely defined as being >10%-100 mesh. The standard binder is Portland Type II cement which has good sulfate resistance. All cements hydrate down to temperatures as low as 15°F, although the rate slows considerably as the temperature drops. For air temperatures below 40°F, the use of Type I or Type III cement is helpful because their higher tricalcium aluminate content makes them more reactive. Type III Portland cement develops early high strength which is a benefit for agglomeration in cold weather. However, it has 25% to 50% more surface area per unit of weight than Types I and II and will have a greater tendency to agglomerate onto itself than coarser cements do. Therefore, it must be well blended into the ore before moisture is added to get full benefit of all the cement. Otherwise the cement may ball up and agglomerate on itself and be wasted.

Other binders such as lime, fly ash, or flue dust have been used because they are less costly than Portland cement. But, they are generally not as uniform as cement, either chemically or physically, and they have not performed quite as well. Relatively few operators use these other binders.

Empirical test procedures have been developed to quantify the amount of binder and moisture needed for agglomeration (McClelland, 1988). The results of these tests require careful interpretation. If used blindly, they could result in excessive use of binder.

Several machines for agglomerating have been tried but the most commonly used are drum agglomerators, pug mills, and conveyor belts (Chamberlin, 1986). Generally, the pug mills are best on ores with many fines; conveyor belts are best on ores with few fines; and drums are in between. There are some notable exceptions to this observation, however, such as Cominco's Buckhorn operation.

Heap Leach Testing Procedures

Testing ores for their amenability to heap or dump leaching sounds deceivingly simple, but many former operators can testify to the pitfalls of not performing a rigorous test program. Lists of things to do, and not do, have been compiled over the years and reported (Chamberlin, 1981, 1988; McClelland, 1988). A quarterly newsletter called *Heap & Dump Leaching* is published to inform the industry about techniques and procedures. And countless articles have been published about all aspects of the business. Prudent operators today have a wealth of information that can be tapped to avoid pitfalls.

In the past, it was common to get metallurgical data on recovery of precious metals and then go build an operation. That simplistic approach is no longer applicable because environmental concerns require much more. Process development work is done not only on the recovery of gold and silver from ore, but also on the detoxification of solutions and solids and the attenuation of trace elements that may leak from the circuit. Procedures have been established for attenuation testing (Rouse, 1988) and these are becoming more important as state regulators ask, "what if?" questions.

Leach Pad Construction

Ten year ago, the primary leach pad construction materials were natural soils and clays, sometimes mixed with bentonite, soil cements, or sodium salts to reduce permeability of the compacted mass. More often than not, these pads leaked badly because it was difficult to maintain uniform quality during construction. The labor and equipment necessary to do a good job of construction made the pads quite expensive, also. Then geomembranes came on the scene: PVC, Hypalon, and the present favorite, HDPE. State regulators started implementing requirements that were more strict, such as leak detection systems and redundancy in the leach pad itself. The use of synthetics blossomed into geogrids, geotextiles, and perforated pipe. Leach pads, which started as one-liner structures more than a decade ago, became two-liner structures and now, on rare occasions, even three. Sometimes the leach pad is a composite of liner materials, some synthetic and some natural. The complexity has raised costs from <\$4.64/m^2 to more than \$23.23/m^2 and designing the leach pads is a major part of any project (Strachan and van Zyl, 1988; Richins, 1988).

There are two basic types of leach pads, permanent and reusable. The permanent pad has a heap built on it which will remain forever. Permanent leach pads are much more common because most ores take months if not years to leach completely. The reusable leach pad has a pile of ore leached on it for a finite period of time and then it is removed and replaced by another pile of fresh ore. Only ores that leach within days or a couple weeks are leached on reusable pads. The reusable pads are generally a multiple layered structure incorporating natural materials, synthetics, and asphalts.

Geotechnical Data

The design of leach pads, heaps, and the solution collection system all require an assessment of geotechnical characteristics of materials. These characteristics include permeability of the ore and construction materials, settlement of the ore and construction materials, and maximum compressive loading on agglomerated ore. The slope stability of heaps has become very important, requiring an understanding of the friction angles of the materials of leach pad construction, and the ore itself. Years ago, the industry didn't recognize the importance of such data, but today it is commonplace and extensively used in designing facilities (van Zyl, 1987; Strachan and Dorey, 1988).

Water Balance

The industry has been embarrassed many times by operators who failed to adequately account for surface water control, precipitation, and snowmelt when designing their facilities. The resultant spills or unintended discharges have brought retaliation by state regulators, and the entire industry suffers.

The knowledge to design adequate leach pads, heaps, and ponds is available (Hutchison, 1988), but the biggest problem is the adequacy of the meteorological data base that is available (or not available). Some operators feel they have been subjected to 100-year storms back to back. They probably haven't, but when pond levels are rising and mud has shut down the mining operations, it is natural to wish that Mother Nature would comply with the historical records used in the process design and be more kind and gentle.

Heap Building

A major reason for heap and dump failure has been poor heap building techniques. Improper heap building methods can create areas of compacted ore, cause extreme segregation of particle sizes, and cause lower and slower gold-silvery recovery (Chamberlin, 1981; Muhtadi, 1988). For example, (1) driving trucks all over the surface of a heap is sure to compact the ore, and (2) building cones of unagglomerated ore with a stacker is sure to create segregation. Many methods of heap building have been tried. Stackers are gaining popularity and are a good way to build heaps provided they move continuously and lay down narrow windrows of ore. Truck and dozer systems are also common but they require careful control of truck traffic and deep ripping between lifts.

Solution Distribution

Leach solution has been applied to the surface of heaps and dumps by several devices over the years. The two most favored devices today are Wobbler sprinklers and drip irrigation systems (Dixon, 1988; Chamberlin, 1988; Muhtadi, 1988).

Typically the leach solution is percolated through the heap or dump once, is collected, and pumped to the process plant to recover the gold and silver. As a result, the grade of the solution is relatively low, flow rates are relatively high, and the cost of the recovery system is higher than it needs to be. Improvements in these areas are becoming more accepted, as described later in the section, "Efficient Use of Leach Solution."

Scale Control

The chemistry for controlling the buildup of scale on pipelines and equipment has been known for a long time, but it took heap leach operators a long time to recognize that scaling problems must be controlled from the very first day of operation. Chemical companies have been good about servicing the industry and severe scaling problems are largely a thing of the past.

Detoxification of Heaps and Dumps

The cyanide solution that remains in permanent heaps must be removed according to criteria specified by the states when leaching operations cease. Removing the solution could be delayed many years until active leaching of the ore is completed. The heaps on reusable leach pads must be similarly cleaned, i.e., detoxified, before the leach residue can be removed from the leach pad. A summary of the recent state cyanide regulations has been compiled (Thatcher, et al., 1988; Smith and Struhsacker, 1988). ASTM's standard D 2036-82 defines standard analytical procedures for cyanides in waters.

Most cyanide species will decompose with time or they can be flushed out of a heap with cyanide free aqueous solution. Both of these processes take time. If the removal/destruction of cyanide must be speeded up, the addition of oxidizing chemicals is the most likely alternative. The chemical most used in the past for destroying residual cyanide has been calcium hypochlorite but it is gradually being replaced by hydrogen peroxide (Griffiths, 1988) or Inco's SO_2-air system (Devuyst, 1988). Other approaches to destroying cyanide have been developed. The simple use of ferrous sulfate is reportedly enough to minimize bird kills at solution ponds. At the opposite extreme, Homestake developed a bacterial system to meet their effluent requirements. The CNC process by Radian Corp. combines both chemical and biological unit operations to give clean effluents.

Actual detoxification experiences in Canada have been reported by Zaidi (1987).

HEAP AND DUMP LEACHING - THE FUTURE

The dump and heap leaching industry has matured but there is still room for improvement. Some of the improvements that are imminent or may occur in the future are described next.

Computer Simulation

Performing all the testwork that is required for designing a heap or dump leach operation takes much time and money. The testwork should be done, of course, but computer simulation of heap leaching will give management a preview of what can be expected and it can be used to help design the testwork. The simulation can be used to forecast the consequences of changes in ores or design parameters (Prosser, 1989).

Today's mineral processing models are fairly well developed and simulation packages have been described in the literature, but they tend to be academic and research-oriented; therefore, operators and engineering companies don't use them routinely. Popularizing the process models and simulation packages and making them user friendly will be necessary before they are widely used (Sastry, 1989).

Use of Oxygen

The use of oxygen to speed up leaching rates has reportedly become accepted practice in South African CIP and CIL circuits (Stephens, 1988). Kamyr, Inc. has been promoting its use in this country (Elmore, 1988). The leach rates are nearly proportional to the dissolved oxygen content of the leach solution. Although the use of oxygen under heaps or dumps has not been reported, it seems logical that it will be tried, especially in deep heaps and those that contain sulfides. Oxygen flow rates would be slow to allow maximum adsorption in the leach solution and to minimize losses to the atmosphere. The benefits may be increased recovery, passivation of preg robbing constituents, and faster leach rates.

The use of air rather than oxygen may offer cost advantages, i.e., power cost for blowers vs. cost of oxygen. And, if the heap has sulfides that generate heat and the resultant "chimney effect" within the heap, then air may be naturally aspirated through pipes at the base of the heap and the power cost for blowers may be reduced.

Air, however, will not raise the disssolved oxygen content of the leach solution as much as oxygen. Testing of the two systems needs to be done in column tests.

Leaching in Cold Weather

The prevailing thinking years ago was to heap and dump leach only during warmer weather so that solution spray systems did not freeze or cause a thick ice cap on the heaps. More recently, however, especially with the introduction of drip irrigation systems, year-round leaching in severe winter areas is gaining acceptance. Burying the drip irrigation lines in the surface of the heap prevents them from freezing and eliminates the build up of large amounts of ice on top of the heap. Deep heaps and dumps work better for this than shallow heaps because of the reservoir of heat they contain. Careful attention to engineering and construction details will make wintertime heap and dump leaching very acceptable.

If the ore contains sulfides, wintertime heap and dump leaching may be quite desirable because the heat generated within the pile will create an upward air flow or "chimney effect" which will naturally draw in fresh air through the sides of the pile. The chemical reactions are enhanced by the fresh source of oxygen and leaching rates will remain high. In fact, leaching rates in the winter could be faster than in the summer because of the chimney effect.

Crushing

Roller presses, recently introduced by KHD Humboldt Wedag Ag, produce a product that appears to have more microfractures than more crushed products (Esna-Ashari, 1988). The presses also appear to consume less energy than other crushers. For a heap leaching operation, the microfractures are advantages that will lead to higher recovery, faster leaching rates, and lower costs. Roller presses may also reduce the need for binder in agglomeration because the press forms flakes of aggregated ore that are fairly stable when exposed to a trickling cyanide solution. The gap between the rollers is typically 10 to 30 mm and the reduction ratio is about 3:1.

Better Binders

The binders that are used for agglomeration are alkalis, such as Portland cement. This makes it costly to use neutral or acidic leaching solutions because the pH is always rising. What is needed is a binder that will work in acidic media. Nalco has developed such a polymeric chemical and is testing it on a copper heap leach in Arizona. If cost effective, this should open the door to alternative lixiviants such as the halogen chemicals without ruining the permeability of the heaps and dumps.

Better Leach Pad Construction Materials

Geomembranes typically have slippery surfaces and so manufacturers have been building in a texture which increases the friction angle and therefore increases leach pad and heap stability. The geomembranes are relatively delicate, however, and puncturing them is easy to do. They also require field assembly of the relatively small shippable panels into the large monolithic sheets that must cover the entire leach pad. The field assembly is not always perfect and leaks that are difficult to locate can result. Methods or materials are needed to reduce the ease with which leaks can be made and the difficulty of repairing them.

One such material may be to use geocomposites such as Claymax. It is a sandwich of treated bentonite clay between polypropylene and polyester geotextiles. Field fabrication is not needed; the material self-heals if punctured; and reported permeability is lower than state regulations require.

Efficient Use of Leach Solution

Heap and dump leaching are slow processes that require large volumes of leach solution. The usual procedure is to percolate the solution through the heap once then pump it to a carbon or zinc recovery system to recover the gold and silver. This requires an expensive recovery system because large volumes of low grade solution are processed.

There is a gradual awakening to the fact that countercurrent heap leaching should be practiced just as countercurrent agitation leaching and countercurrent decantation are practiced. Passing the leach solution through the heap three times (low grade ore, middle grade ore, and fresh high grade ore) before pumping it to the recovery system will not only leach three times the tonnage with the same amount of solution but also give a higher grade preg solution.

Countercurrent heap leaching could also be combined with intermittent spraying to achieve even greater use of the leach solution before it is pumped to the recovery system. Intermittent spraying means applying leach solutions to an area of the heap only 8 to 12 hr per day. During the remainder of the day, it is "resting" and the leach solution is being used to spray another area of the heap. Many column leach tests have shown that intermittent spraying does not usually reduce the rate of leaching compared to full-time spraying. If the rest period was many days or weeks, the rate of leaching would decrease. But the rate is not decreased if the rest period is only 12 to 16 hr. By using this concept, two to three times as much ore can be kept actively leaching with the same volume of leach solution that is required when leaching a single area full time. The primary advantage of this concept is that the size of the recovery system can be reduced and capital dollars can be saved.

Bacterial Heap Leaching

Bacterial oxidation of pyrite and other sulfides plays a very important role in biohydrometallurgy. It provides sulfuric acid and ferric ions which are needed in the heap and dump leaching processes as leaching agent and oxidant (Torma, 1989). Chemical lixiviants will not be replaced by bacterial leaching of heaps, but bacterial leaching will be used to supplement chemical leaching when acidic systems are used on sulfidic ores. This type of heap leaching will probably be used on the low grade portion of an ore body and the high grade ore will be treated by agitation leaching. Success will not come easily, but the interdisciplinary approach that operators have been forced to learn for heap leaching is a major step in the right direction (Holmes, 1988).

Alternative Lixiviants

Cyanide is the lixiviant of choice today. Thiourea has not been applied very much and thiosulfate even less so. However, companies are working on proprietary reagents containing bromine, or iodine, or chlorine. Halogen compounds have great promise but have not been widely accepted yet (Sergent, 1988). Perhaps that is because most of the ore deposits that have been developed so far are relatively clean and easy to treat. When companies are forced to consider deposits containing cyanicides, in spite of the high cyanide consumption, then alternative lixiviants will become more common. Pressures will also be brought to bear on operators to try them. The pressures will be imposed by the increasingly difficult chemical and physical nature of available ore deposits and by state regulators responding to public environmental demands.

Counteracting the use of alternative lixiviants will be the use of cyanide regeneration systems to keep cyanide costs under control. All of the systems use variations of a common theme: the cyanide solution is acidified; hydrogen cyanide gas is evolved; and the gas is reabsorbed for reuse. Not many of these systems are in commercial use today, but they have been proposed (Kitney, 1988; Witteck, 1986; Resource Technology, 1986). If the consumption of NaCN is >1.0 kg/ton of ore, the use of cyanide regeneration may be economic.

Another development that will maintain cyanide's position as the premier lixiviant for gold and silver is the low cost generation of cyanide systems at the mine site. Cyano Systems, Inc. claims that the direct production cost of cyanide from their modular system is about \$0.53/kg. Amortization, maintenance, and fees will probably double that cost to ∿\$1.00/kg, which is still a very attractive price. The system is claimed to be safe and requires little operator or maintenance attention.

Use of Surfactants

The use of surfactants to increase recovery or leaching rate, or both, has had occasional successes. On the other hand, they have occasionally fouled carbon circuits. More definitive work needs to be done to determine which of the thousands of surfactants are helpful, and why. The prospect of an additional 5% extraction is the incentive.

Percolation Rate vs. Heap Height

A device is needed to determine the percolation rate through ores that are under pressure -- pressure that simulates heaps up to 200 ft deep. One of the most important economic questions every heap designer faces is, "how high can I build the heap and still get adequate percolation?". Column tests at 6, 10, 20, and 30 ft give useful information, but it would be risky to extrapolate this data to 100 or 200 ft. The "percolation under pressure" tests should not be performed quickly, i.e., within less than one day because the strength of agglomerates and even the mineralogical character of the ore itself changes with time within the heap leach environment (Kurtz, et al., 1987).

IN SITU LEACHING - PROGRESS TO DATE

The concept of in situ leaching is quite simple. In essence, two or more holes are drilled into an ore deposit. A leach fluid is injected into the injection holes and pregnant liquor is withdrawn from the production holes. The ore body is not disturbed.

In situ leaching enjoys certain advantages over more commonly accepted mining and milling methods. Primary among the economic advantages is the ability to recover values from deposits that are uneconomic for conventional mining methods due to depth, low tonnage, water saturation, ground support problems, or low grade. In addition, the capital and operating costs are generally lower; start-up times are faster; and the leaching operation usually has less impact on the environment. It is a way of processing small high grade or large low grade deposits that are too deep for open pitting but too small for underground mining. It is not generally applicable to surface deposits, but it could be used to extract extra values from deep dumps, pit walls, underground stopes, and subsidence zones (Hiskey, 1986).

In situ leaching has been used for the recovery of uranium, copper, salt, potash, and sulfur but not for gold and silver. In spite of the fact that recovering uranium and copper with acids mobilizes more ions than leaching gold ores with cyanide does, state regulators are more attuned to the fears of cyanide. Operators understand the resulting politics of cyanide and are hesitant to be the first to try cyanide in situ leaching for gold and silver.

However, cyanide is not the only lixiviant for in situ leaching gold and silver ores. For example, Jacobson and Murphy (1988) have presented a case for the in situ mining of argentite with a lixiviant of ammonium thiosulfate and ammonium carbonate. Halogen compounds also show promise for in situ leaching operations not only because they are less toxic than cyanide but also because they are effective oxidizing and complexing reagents for gold and silver. Jacobson and Murphy (1988) developed an iodine/iodide system for leaching deep gold placers. An electrowinning diaphragm cell regenerates the iodine as it recovers the gold and well field cleanup is accomplished by capturing residual iodine/iodide on a strong base anionic resin.

A common fear that in situ leaching operators must contend with is the escape of leach solutions, i.e., an excursion. If a leach fluid were injected into the deposit with no attempt to produce the pregnant liquor, then the fear of excursions and environmental damage would be great. But, if the volume of fluid recovered from production wells is slightly greater than the amount injected, which is always the case, there is a driving force imposed on the fluid in the ore zone and its direction of flow can be controlled. In situ operators have developed several lines of defense against fluid excursions beyond the boundaries of the ore zone. These include not only the driving force on fluids but also (1) guard wells into which water can be injected to drive leach solutions back to the ore zone, (2) monitor wells to warn of impending excursions, and (3) grout curtains around the perimeter of the ore zone to physically prevent excursions. Which of these defensive measures is needed is dictated by the stratigraphy, size, shape, depth, permeability, porosity, and mineralogy of the ore zone and by the chemistry of the system (Huff, 1988).

It is generally believed that in situ leaching must be done below the water table. This is desirable but not a necessity. In fact, most of the gold and silver ore bodies that have been mined are above the water table. Ore bodies above the water table can be operated within an artificial water table that is created by injecting water into barrier wells around the perimeter of the ore body. The use of polymers to form a grout curtain around the ore body is another technique for encapsulating the deposit so it can be leached within an artificial water table. Grout curtains have been used in the oil industry, but the spacing of polymer injection points has been <2 m. Tests need to be done at wider spacings, i.e., 8 m, to make grout curtains economically attractive for in situ leaching.

It is also believed that in situ leaching must be done in rubblized rock in order to have adequate permeability. The uranium sandstone deposits were an exception because their permeability was naturally quite high. But there is a growing belief, even for deposit permeabilities as low as 2 to 5 millidarcies, that rubblizing is not the best approach to take. Rather, the operator should use the same flow channels that Mother Nature used when forming the deposit. It is along these flow channels that the gold and silver mineralization was deposited and the most efficient use of lixiviants is to pump them only through these same channels (Wargo, 1988). Why break the rock and expose barren rock to the lixiviant? Keep the lixiviant only in the natural flow channels and the grade of the pregnant solution can be increased while minimizing the volume. Admittedly, this may require high injection pressures. The chemistry of the system must also be clearly understood and controlled to prevent the formation of unwanted precipitates that could plug the ore body. Overall, the advantages of leaching unrubblized rock will probably outweigh the disadvantages.

IN SITU LEACHING - THE FUTURE

The near-term outlook for in situ leaching of precious metals with cyanide is poor because of the perceptions held by the public and state regulators. It seems that they generally believe that anything injected into the ground is bad and that anyone who does so is guilty of pollution until proven innocent. It is the "proving of innocence" that deters most companies from using in situ leaching. They recognize that they will probably end up in court trying to prove their innocence and they fear the liability that may ensue.

On the other hand, in situ leaching with a noncyanide lixiviant may be tried by some innovative company that has an ore body which is not economic with either open pit or underground mining. The economics of in situ leaching are very attractive on many ore bodies, even with two or three levels of redundancy to prevent excursions, and some company will try it for the money that can be earned. Success will be dependent upon a good working relationship with state regulators, a good ore body, and proper testing and implementation of all technical aspects of the project. The technical know-how is available today. The missing ingredient is a reputable company willing to give it a try.

CONCLUSION

Heap and dump leaching have reached the stage of mature technologies and they have done so within the past 10 years. They have been widely used because they are low cost technologies capable of processing low grade ores and because they can be put into production quickly. There have been many failures along the way because of insufficient attention was given to items such as heap building procedures, agglomeration of the ore, adequate testing, leach pad construction, geotechnical assessment of materials, and water balances. But the lessons have been learned, written about, and distributed to the industry through various media. Today the chances of success for a new venture are very high.

Those involved with heap and dump leaching are continuing to make improvements. The rate of leaching will be increased via the use of oxygen, operations in cold weather environments will be more common, new binders will be developed to allow the use of lixiviants other than cyanide in acidic systems, more efficient use of leach solution will allow reductions in capital cost, and computer modeling and simulation will allow a faster and more complete evaluation of new projects.

In situ leaching of precious metals has not been commercialized yet. The technical knowledge is available from operations in the uranium and copper industries, but the environmental concerns about using cyanide underground are pervasive. In situ leaching has the potential for being a very low cost technology, but it may sit on the shelf until lixiviants other than cyanide are accepted.

REFERENCES

Chamberlin, P.D., 1981, "Heap Leaching and Pilot Testing of Gold and Silver Ores," *Mining Congress Journal*, April, pp. 47-52.

Chamberlin, P.D., 1986, "Agglomeration: Cheap Insurance for Good Recovery When Heap Leaching Gold and Silver Ores," *Mining Engineering*, December, pp. 1105-1109.

Chamberlin, P.D., 1988, "Heap Leach Testing," *Randol Gold Forum 88*, January, pp. 71-74.

Chamberlin, P.D., 1988, "Applying Leach Solutions to Heaps," *Randol Gold Forum 88*, January, p. 70.

DeVuyst, E.A., Conrad, R.B., and Robbins, G., 1988 "Commercial Performance of INCO's SO_2-Air Cyanide Removal Process," Randol Perth International Gold Conference, pp. 87-89.

Dixon, S.N., 1988, "Benefits of Drip Irrigation versus Spray Irrigation," Randol Perth International Gold Conference, pp. 226-228.

Elmore, C.L., Brison, R.J., and Kenney, C.W., 1988, *Randol Gold Forum 88*, January, pp. 197-201.

Esna-Ashari, I.M., and Kellerwessel, H., 1988, "Roller Press Comminution Improves Heap Leach Recovery," Randol Perth International Gold Conference, pp. 50-53.

Griffiths, S.A., 1988, "Detoxification of Gold Mill Waste Water with Hydrogen Peroxide: Three Case Studies," SME preprint 88-27, SME Annual Meeting, Phoenix, AZ, January.

Heinen, H.J., McClelland, G.E., and Lindstrom, R.E., 1979, "Enhancing Percolation Rates in Heap Leaching of Gold-Silver Ores," RI 8388, US Bureau of Mines.

Hiskey, J.B., 1986, "The Renaissance of Copper Solution Mining," *Heap & Dump Leaching*, June, p. 1.

Holmes, D.S., 1988, "Biotechnology in the Mineral and Metal Processing Industries: Challenges and Opportunities," *Minerals & Metallurgical Processing*, May, pp. 49-56.

Huff, R.V., 1988, Ray V. Huff & Associates, Inc., Golden, CO.

Hutchison, I., 1988, "Surface Water Control - Water Balance," *Introduction to Evaluation, Design and Operation of Precious Metal Heap Leaching Projects*, D. van Zyl, I.A. Hutchison, and J. Kiel, eds., SME, Littleton, CO, pp. 203-253.

Jacobson, R.H., and Murphy, J.W., 1989, "Gold Solution Mining," *Precious and Rare Metal Technologies*, A.E. Torma, ed., Elsevier Science Publishers, Amsterdam, pp. 157-174.

Jacobson, R.H., and Murphy, J.W., 1988, "Insitu Mining of Silver," *Proceedings*, Silver - Exploration, Mining and Treatment, Institution of Mining & Metallurgy, London, pp. 279-288.

Kappes, D.W., 1979, "Precious Metals Heap Leaching; Simple - Why Not Successful?," Northwest Mining Association, December 7.

Kitney, M.J., 1988, "Cyanide Regeneration from Gold Tailings; Golconda's Beaconsfield Experience," Randol Perth International Gold Conf., pp. 89-96.

Kurtz, J.P., Hauff, P.L., and Chamberlin, P.D., 1987, *Small Mines Development in Precious Metals*, T.M. Li and T.M. Plouf, eds., SME, Littleton, CO, pp. 161-167.

McClelland, G.E., 1988, "Testing of Ore," *Introduction to Evaluation, Design and Operation of Precious Metal Heap Leaching Projects*, D. van Zyl, I.A. Hutchison, and J. Kiel, eds., SME, Littleton, CO, pp. 61-67.

Muhtadi, O.A., 1988, "Heap Construction and Solution Application," *Introduction to Evaluation, Design and Operation of Precious Metal Heap Leaching Projects*, D. van Zyl, I.A. Hutchison, and J. Kiel, eds., SME, Littleton, CO, pp. 92-106.

Prosser, A.P., 1989, "Simulation of Gold Heap Leaching as an Aid to Process Development," *Precious Metals '89*, Jha and Hill, eds., TMS, Warrendale, PA, pp. 121-135.

Resource Technology Associates, 1986, Boulder, CO.

Richins, R.T., 1988, "Emerging Regulatory Developments in the USA and Their Implications for Pond/Pad Liner Selection," *Randol Gold Forum 88*, January, pp. 7-17.

Rouse, J.V., and Pyrih, R.Z., 1988, "Natural Geochemical Attenuation of Trace Elements in Migrating Precious Metal Process Solutions," Randol Perth International Gold Conf., pp. 77-81.

Sastry, K.V.S., and Lofftus, K.D., 1989, "Challenges and Opportunities in Modeling and Simulation of Mineral Processing Systems," *Challenges in Mineral Processing*, K.V.S. Sastry and M.C. Fuerstenau, eds., SME, Littleton, CO, pp. 369-380.

Sergent, R.H., 1988, "The Use of Brominated Hydantoins for Gold Leaching," *Randol Gold Forum 88*, January, pp. 167-169.

Smith, A., and Struhsacker, D.W., 1988, "Cyanide Geochemistry and Detoxification Regulations," *Introduction to Evaluation, Design and Operation of Precious Metal Heap Leaching Projects*, D. van Zyl, I.A. Hutchison, and J. Kiel, eds., SME, Littleton, CO, pp. 275-292.

Stephens, T., and Arnold, K.M., 1988, "Pure Oxygen for Gold Leaching in South African and Australian Gold Mines," Randol Perth International Gold Conference, pp. 189-194.

Strachan, C., and Dorey, R., "Geotechnical Studies," *Introduction to Evaluation, Design and Operation of Precious Metal Heap Leaching Projects*, D. van Zyl, I.A. Hutchison, and J. Kiel, eds., SME, Littleton, CO, pp. 152-175.

Strachan, C., and van Zyl, D., 1988, "Leach Pads and Liners," *Introduction to Evaluation, Design and Operation of Precious Metal Heap Leaching Projects*, D. van Zyl, I.A. Hutchison, and J. Kiel, eds., SME, Littleton, CO, pp. 176-202.

Taylor, P.R., 1986, "Small Heap Leach Operations, Problems and Promise," *Heap & Dump Leaching Newsletter*, March, P. 1.

Thatcher, J., "Regulatory Aspects and Permitting Requirements for Precious Metal Heap Leach Operations," *Introduction to Evaluation, Design and Operation of Precious Metal Heap Leaching Projects*, D. van Zyl, I.A. Hutchison, and J. Kiel, eds., SME, Littleton, CO, pp. 40-58.

Tibbals, R.L., 1987, "Agglomeration Practice in the Treatment of Precious Metal Ores," *Proceedings*, International Symposium on Gold Metallurgy, Salter, et al., eds., Pergamon Press, New York, pp. 77-86.

Torma, A.E., Guay, R., and Ghosh, J., 1989, "Kinetics of Microbiological Production of Ferric Ion for Heap and Dump Leaching," *Biotechnology in Minerals and Metal Processing*, B.J. Scheiner, S. Kawatra, and F.M. Doyle, eds., SME, Littleton, CO, pp. 95-106.

van Zyl, D., ed., 1987, *Geotechnical Aspects of Heap Leach Design*, SME, Littleton, CO.

Wargo, J.G., 1988, "In Situ Leaching of Disseminated Gold Deposits -- Geological Factors," *Mining Engineering*, October, p. 973.

Witteck Development, Inc., 1986, Mississauga, Ont., Canada.

Zaidi, A., et al., 1987, "The Art and Science of Treating Wastewaters from Gold Mines," Canadian Mineral Processors Annual Conference, January.

Chapter 27

NEW GOLD PROCESSING TECHNOLOGIES: AN ENGINEER'S PERSPECTIVE

Kevin A. Foo, Murray D. Bath and Arnaldo Ismay
Minproc Engineers Inc. 5600 S. Quebec St, Englewood, CO 80111, United States of America

John H. Canterford
Minproc Technology Pty Ltd, 364 Albert St, East Melbourne Victoria 3002, Australia

ABSTRACT

There is no doubt that the dominant future source of gold will be the so-called refractory ores, especially those with gold in a pyrite/arsenopyrite matrix. Minproc Engineers is taking an active role in establishing detailed design criteria for a number of apparently competing processes. This involves generating a detailed understanding of the metallurgy involved, systematic pilot plant testing, and the application of advanced engineering technologies. This presentation outlines the approach adopted by Minproc as well as the design criteria for several current projects.

INTRODUCTION

Interest in the treatment of gold ores in general, and refractory ores in particular, has greatly increased over the last 5 years. Numerous papers and books have been written, reviews compiled, and conferences devoted to the subject. Some good reviews have been published recently [Brown, 1984; Jha, 1989; Lakshmanan and Bosch, 1984; Wells, 1989] and most professional societies now devote considerable space and time to the subject in their publications and meetings. The definitive review of the treatment of refractory gold ores has yet to be written, but the Randol reports [Randol, 1987] provide the most comprehensive coverage of the technology.

This paper is not intended as a review, but rather a discussion of how one engineering company has viewed and participated in the development of a number of processes for treating refractory ores. This experience has resulted in significant progress, as well as the highlighting of some major deficiencies in how ores are assessed for possible commercial treatment.

While it seems obvious, the need to assess an ore adequately in terms of mineralogy, geology, geography, environment and laboratory testwork (and pilot plant testing if needed), is often overlooked in the excitement of developing a new resource. Shortages of sample, funds or time are often cited as justification for limiting, or ignoring the gathering of vital metallurgical data. This decision invariably returns at a later stage of development to highlight deficiencies, to the embarrassment of all concerned. The "fix-up costs" are usually many times the cost of doing the work properly in the first place.

Knowing when "enough" assessment has been done requires considerable experience and maturity on the part of the developer and his engineer. The willingness of a resource developer and his engineer to take a calculated risk often results in some great innovations and provides a stimulating project environment. However, risk also has its downside!

Minproc has had considerable experience in the development of some modern refractory gold treatment processes, from laboratory testing and flowsheet development to plant design and construction. These are:

* Bacterial Oxidation
* Nitric Acid Oxidation
* Artech/Cashman Process
* Sirosmelt

Features of these processes and their current status are discussed and compared with existing technology. Some guidelines are provided to the steps required in assessing ores, together with comparative processing costs for the first two of the above processes.

ORE TYPES

A free-milling gold ore is one which requires no treatment before cyanidation, other than crushing and grinding. A refractory ore, on the other hand, is not amenable to cyanidation without some form of pre-treatment. Pre-treatment is generally aimed at oxidising sulphides to release encapsulated gold, but may also have other objectives, for example, de-activation of preg-robbing carbon.

Except for roasting, most refractory ore treatment processes have been commercialised only recently, or are just emerging from the pilot stage. It is beyond the scope of this presentation to discuss in detail the potential or current processes for treating the various forms of refractory gold ore. The reader is directed towards the references cited for each ore classification.

* Carbonaceous Ores [Guay and Peterson, 1973; Nice, 1971]
* Pyrrhotite Ores [Davie, 1989; Royston, Spencer and Winborne, 1984]
* Pyrite/Arsenopyrite Ores [Anon, 1988; Archibald, 1949; Hansen and Laschinger, 1967; Lawrence and Bruynesteyn, 1983; Marchant et al., 1989; Van Aswegen and Haines, 1988; Wright, 1961]
* Telluride Ores [Cornwell and Hisshion, 1976; Markham, 1960]

METALLURGICAL ASSESSMENT

The evaluation of metallurgical processes for the treatment of an ore seldom proceeds in a logical and systematic manner. It is not unusual for work to be duplicated, or for promising approaches to remain unexplored because of the lack of a consistent, coherent program. Results of later testing may contradict those obtained earlier in the program, and various investigators may draw different conclusions from the data. As a result, the "wrong" flowsheet may be selected, or the project abandoned because of apparently unfavourable process economics. There are good reasons for this:

* The property may pass through the hands of several owners, each of whom may use different consultants and laboratories. Often only enough work is done to take the property to the next stage of development, before it changes hands again.

* It is unusual for adequate metallurgical samples to be available early in the life of a project. Frequently, the first samples are oxidised material from surface outcrops or shallow exploration trenches. Metallurgical tests are often conducted on these, but the results may be of very limited applicability. Even when drilling gets underway, core samples provide only limited material for treatment investigations, and do not represent the orebody as it will finally be delineated.

* Those in charge of the investigation may have fixed ideas about those processes which may or may not be applicable, based on their personal experience or because of a lack of knowledge of new developments. Potentially promising approaches therefore may receive only cursory attention and be dropped before enough work has been done.

There is no universal remedy for this problem. It is the nature of mining projects that development proceeds in stages. Projects may be started, then put on hold when economic or market conditions change, then restarted. Concepts change when ore reserves are found to be greater (or less) than anticipated, or the easily treated surface oxide ore is found to overlie a far greater tonnage of refractory sulphides. The course of metallurgical investigations is inevitably affected by these changes in direction. Despite these practical realities, it is worth trying to set down some guidelines for the planning and conduct of metallurgical test programs. Figures I, II and III show one possible approach to assessment of a gold ore. It is not intended to be rigidly followed, but does provide a basis for those planning testwork.

Figure I covers the initial phase during which it will be determined whether the ore is refractory or not. If processing is found to be relatively straightforward, efforts can be concentrated on checking the amenability of samples of as many ore types as are available, and on optimising grind and other treatment conditions.
If the extraction by direct cyanidation is found to be less than some arbitrary target, typically 80%, then the ore is classified as refractory, and the steps outlined in Figure II are followed. The purpose here is to identify whether a fairly simple remedy for the refractoriness can be found, for example, alkaline pre-aeration, or the use of CIL to combat mild preg-robbing.

If none of the foregoing is successful in achieving the target gold recovery, then more severe oxidation treatments must be tried (Figure III). This stage of the investigation may be protracted and

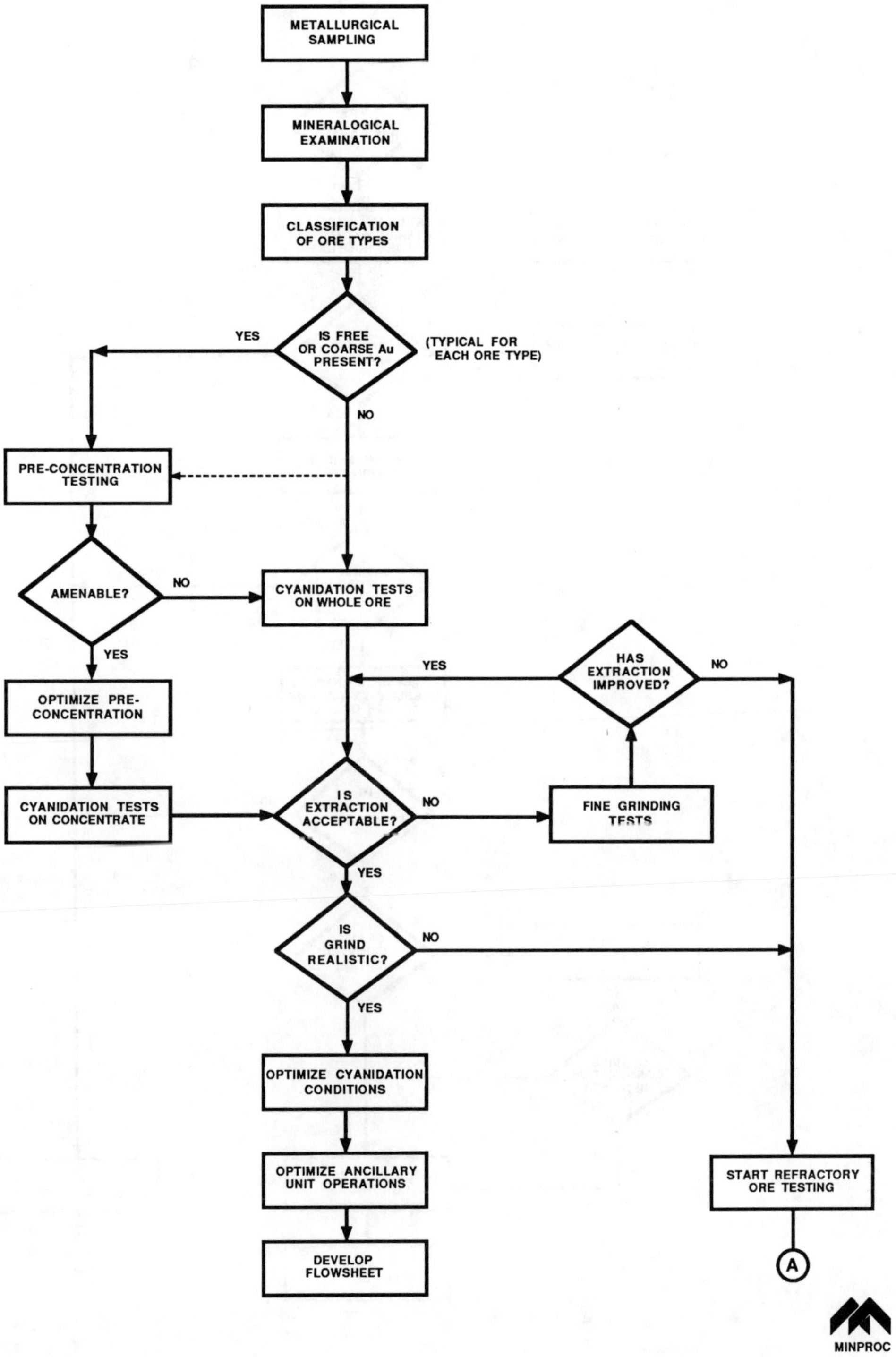

Figure I. Metallurgical testing of gold ores - Phase I.

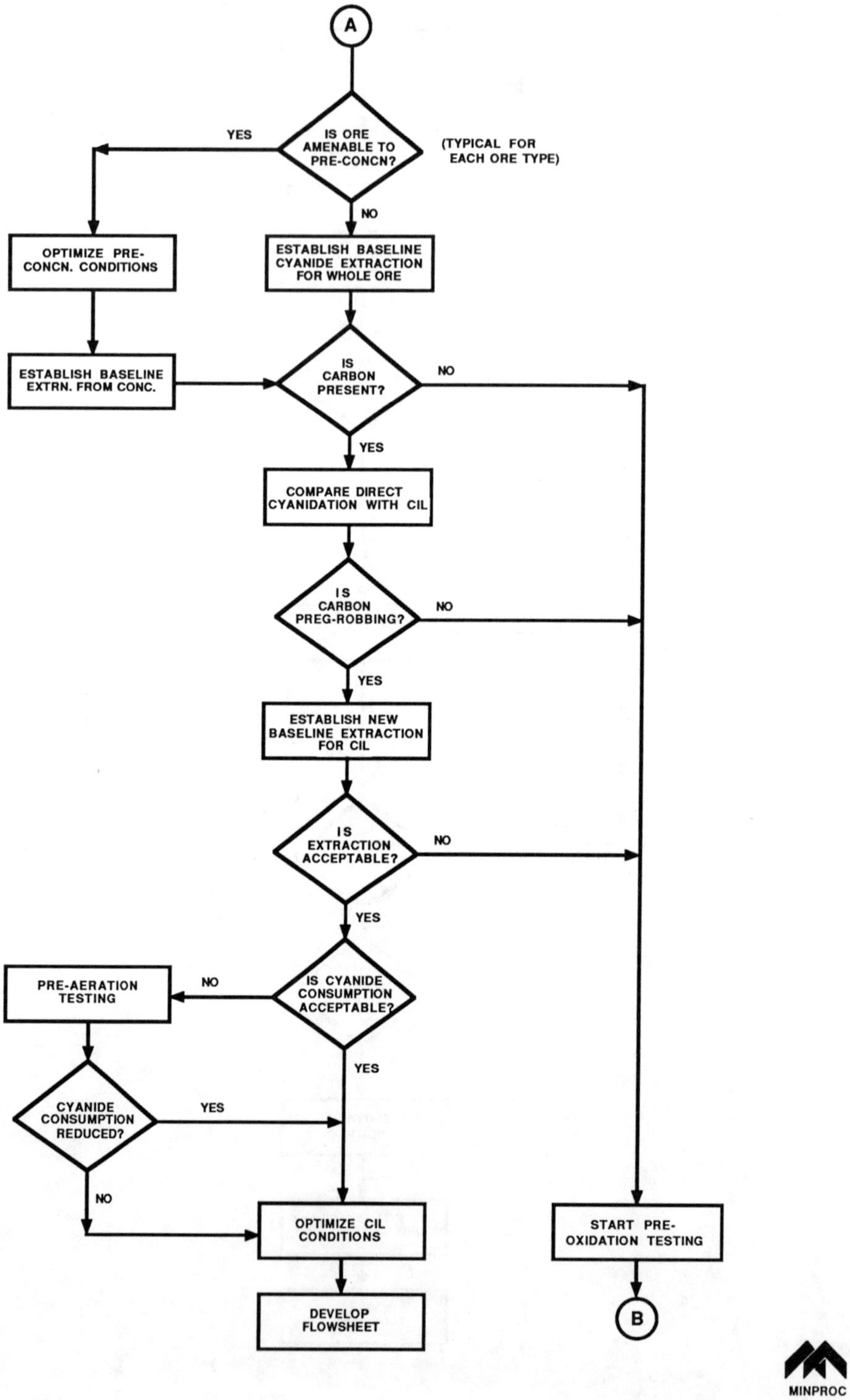

Figure II. Metallurgical testing of gold ores - Phase II.

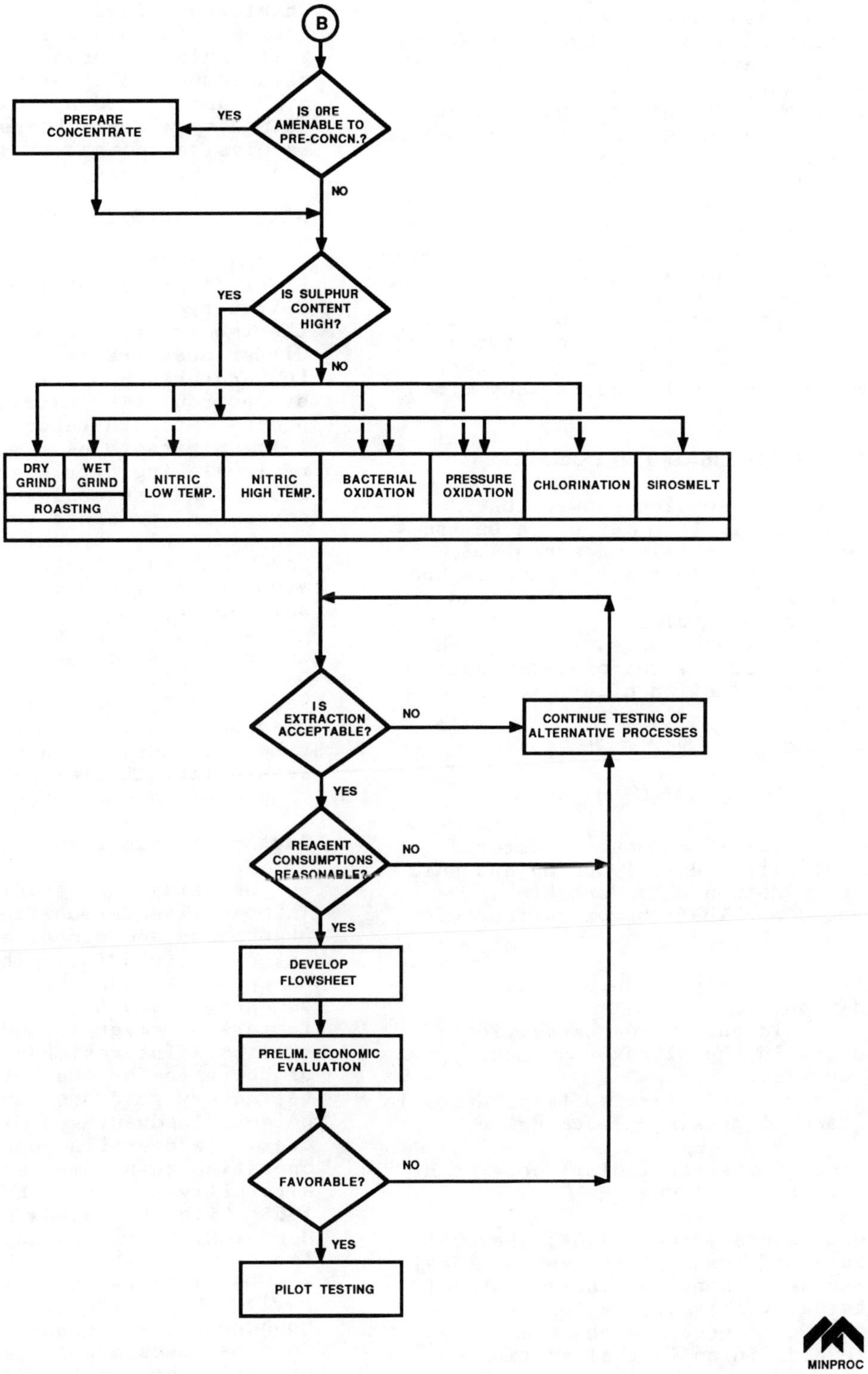

Figure III. Metallurgical testing of gold ores - Phase III.

expensive. It is not unusual to find that several processes will yield acceptable recoveries, and at this stage preliminary economic evaluations will be useful. These must, of course, take into account the specific conditions of the site in determining reagent costs, environmental constraints, power costs and tailings disposal options and also the likely scale of operations. Finally, one or more of the successful process flowsheets may be piloted. The need for this will depend on whether one of the mature technologies is being considered, or whether a less-established process such as bio-oxidation or nitric acid oxidation is contemplated. For a large project and a new process, piloting will repay its costs many times over.

SOME PROCESSING ALTERNATIVES

The previous section showed that oxidation of sulphides may often be the only solution to obtaining acceptable recoveries from refractory gold ores and concentrates. While the hardware and chemistry of the various hydrometallurgical processes discussed below differ widely, the overall chemical reaction for oxidation of pyrite is essentially

$$4FeS_2 + 15O_2 + 2H_2O =$$
$$2H_2SO_4 + 2Fe_2(SO_4)_3$$

The reaction is common to bacterial oxidation, nitric acid leaching and acid pressure oxidation. The notable differences are that the oxidation is greatly accelerated by

* bacterial catalysts in biological oxidation,
* nitric acid and its decomposition products in the Nitrox and Redox process, or
* pure oxygen and elevated temperature in the case of pressure leaching.

The choice of the particular "accelerant" is really the key issue.

Recent papers [Brown, 1984; Jha, 1989; Lakshmanan and Bosch, 1984; Wells, 1989] have provided a range of assessments of the alternatives available for treating refractory gold ores and these are valuable aids in an initial sorting of processes. However, such comparisons are of limited value because circumstances vary widely for each project.

Environmental factors are now a dominant influence on process selection. Roasting, for instance, was simply not an option for the Cinola project, currently under development by City Resources (Canada) Limited on Graham Island in the pristine environment of the Queen Charlotte Islands of British Columbia. Economic factors will always rule and while sulphide pressure leaching is now used commercially at several mines, the high capital and operating costs have helped spawn many competing processes. It is considered that this technology is in the mature stage.

Nitric Acid Oxidation

This technology is represented by the Redox (formerly Arseno) process and the Nitrox process. The overall chemistry of the two is similar but distinctive differences are in the operating temperatures and pressures and the means of recirculating nitric acid. The Nitrox process, for instance, uses soluble calcium nitrate as a mechanism for recirculating nitric acid. The Redox process has several variations that use oxygen and pressure vessels on high sulphide, high arsenic feeds, or air regeneration of nitric acid and atmospheric leach vessels on lower sulphide feeds. Developments in the Redox and Nitrox processes have recently been described [Beattie and Raudsepp, 1988; Davie, 1989; Fair, Schneider and Van Weert, 1987]. The Redox process is to be used on the Cinola orebody and the Snow Lake tailings in Manitoba. Bankable feasibility studies have recently been completed on both properties.

Bacterial Oxidation

The oxidation of sulphides by Thiobacillus ferrooxidans is a natural phenomenon and can be observed in most sulphide orebodies. The bacteria behave as catalysts and under ambient conditions can accelerate the oxidation reaction by factors of several hundred thousand to one million. Interest has turned to the use of bacteria for the oxidation of refractory gold ore, driven by the potential advantages of lower capital cost, low operating cost and simplicity of operation when compared to the alternatives [Gibbs, Errington and Pooley, 1985; Gilbert, Bounds and Ice, 1988; Hardwick, Errington and Miller, 1988].

Observations that bacteria attack pyrite and arsenopyrite along grain boundaries, where gold often resides, has in some cases meant that only partial oxidation of the feed is required for high gold recoveries to be achieved [Southwood and Southwood, 1986].

Gencor, at their Fairview Mine in South Africa, have been pioneers in this field and they operate a 12 t/d commercial plant on a pyrite-arsenopyrite concentrate [Van Aswegen and Haines, 1988]. A similar size plant is operating in Zimbabwe. Gencor has found that as long as the natural system is maintained within reasonable operating bounds of pH, residence time, temperature and solids concentration and sufficient nutrient is added to the system, it is relatively robust. An interesting finding in their pilot plant operation was that after two years, residence time was reduced from 10 days to 4 days. This was believed due to the mutation of a bacteria strain that had gradually adapted to conditions, allowing the increased treatment rate. Hopefully this internal "expansion capacity" is a standard feature of bacterial oxidation plants!

U.S. Gold Corp. is constructing the world's largest bacterial oxidation plant at its Tonkin Springs, Nevada, mine to treat 1500 t/d of low sulphide ore. The project is discussed more fully in a later section of this paper. Giant Bay Resources is building a demonstration plant at the Congress Mine in British Columbia to batch leach 600 t of concentrate.

Artech/Cashman Process

This process is currently under development to treat arsenical base metal flue dusts. By choice of appropriate operating parameters, the process can also be applied to gold-arsenopyrite ores and concentrates. A suitable calcium salt is added to the reaction system to provide sufficient cations to precipitate the arsenic as a mixture of calcium and ferric arsenates, as well as to control the acidity of precipitation. Typical reaction temperature is 110^{0}C and operating pH is 1-2, with reaction time 30 minutes to 2 hours.

A particular advantage of the process is that the arsenic-containing residue produced appears to be very stable. Gold recoveries on a variety of ores tested have exceeded 90% in all cases.

Sirosmelt

This is a high-intensity smelting process developed in Australia by CSIRO and a number of mining companies. It is used on a limited commercial scale to smelt lead and copper concentrates, and to fume zinc and antimony from concentrates and slag. Small-scale commercial use is made of Sirosmelt to smelt gold-containing sulphide concentrates, refinery drosses and electronic scrap. The technology has good potential for treating high sulphide copper-precious metal or base metal-precious metal concentrates [Lanyon and Floyd, 1987]. The technology has been proposed for the Laloki copper-gold ore in Papua New Guinea [Floyd and Lightfoot, 1984].

CONSIDERATIONS IN ASSESSMENT OF ALTERNATIVES

An earlier section outlined guidelines for metallurgical testing of refractory ores. It did not include decisions associated with other project influences that are critical to the ultimate process choice. Some of these are:

* Environmental issues
* Process chemistry
* Sulphur disposal
* Arsenic disposal
* Water quality and disposal options
* Gold recovery
* Flowsheet complexity and workability
* Materials of construction
* Process control
* Safety and hygiene issues
* Quality of labour required
* Scale of operations currently in use
* Further development required
* Capital costs
* Operating costs
* Ability of company to manage new technology
* Risks versus rewards

Any one of these factors could be sufficiently important enough to delay or stop development of a project.

Attention is often focused on the chemistry of new processes and their promise to extract gold more completely, rapidly, cheaply or easily than other methods. A point needs to be brought out here relating to new technologies and their sponsoring laboratories. There are not many truly independent testing laboratories. Some have particular expertise in a variety of processes, and some have proprietary knowledge in a particular process. None has first class expertise in all processes.

Opportunities can also be missed if the developer or laboratory is too impatient for results. Refractory ores are also refractory in the laboratory and a rational assessment of a few of the logical alternatives, considering the factors above, usually pays dividends. Attempting to "pick winners" on flimsy evidence or gut feeling is foolish.

Aspects of the process chemistry and flowsheet development and selection deserve special mention. These are:

* Sulphur formation during leaching
* Sulphur disposal
* Nitrate disposal
* Neutralisation techniques
* Heat balance
* Ancillary equipment
* Oxygen utilisation

Sulphur Formation

In pressure leaching, bacterial leaching and nitric acid leaching processes, the net chemistry is similar for pyrite decomposition (see previous chemical equation). However, for some ores and concentrates, elemental sulphur does form to varying degrees, especially when base metals, arsenopyrite or non-stoichiometric pyrite are present. The presence of sulphur has a number of potential effects.

* Gold is physically entrained or coated with sulphur, making its extraction by cyanide very difficult.
* Thiocyanates are formed, consuming cyanide.
* Sulphur can react with lime during neutralisation at high temperature to form soluble polysulphides, however it does not form polysulphides at lower neutralisation temperatures.

Suggested solutions to the sulphur formation problem have included operating at temperatures above the sulphur formation temperature, flotation of sulphur from the leached residue, burning the sulphur or recycling slurries to aid in sulphur dispersion. All of these solutions add to processing costs.

Sulphur Disposal

Stack discharge of sulphur dioxide from roasting or smelting of refractory ores has been the traditional method of sulphur disposal. The opportunities to continue this option are diminishing, and perhaps rightly so. The problem is compounded if arsenic is present in the stack gases. Production of sulphuric acid is the only practical means of disposal of sulphur dioxide, and the practice of marrying low-cost acid plants with roasting or smelting plants may increase. An example of this is the Salsigne plant in France where a low-cost Fattinger plant [Lesoille, 1987] treats roaster gases to produce 70% sulphuric acid. For coastal sites, sea water scrubbing of roaster gases may be an option.

In pressure leaching, bacterial oxidation and nitric acid leaching, sulphide sulphur is substantially converted to soluble sulphate, which must be neutralised with limestone and lime. There is rarely any way of avoiding this and if alkali costs are high, the economics of a project can be seriously damaged. For high sulphide feeds, neutralisation costs can dwarf all others. The formation of gypsum scale is a problem not often encountered in laboratory tests but it can be a difficult operating problem in full-scale plants.

Nitrate Disposal

This is a special consideration associated with the Redox and Nitrox oxidation processes. Loss of nitrate from either process is undesirable because make-up acid is costly and the environmental effects of discharged nitrate are unacceptable. The key lies in close process control. For instance, in the Redox process, operating excursions that lead to excess nitrate in solution can be reduced by adding small amounts of pyrite or transferring reacting slurry to the leach.

Neutralisation Techniques

Neutralisation technique is an important factor in all modern hydrometallurgical processes. Some acid recirculation may be desirable to minimise acid consumption, or to help acidify occasionally alkaline ores. Prolonged storage of oxidised slurry prior to neutralisation and cyanidation may dramatically reduce gold recovery. Consequently, optimisation of the neutralisation process should be an integral and preferably continuous step in the pilot test program.

Heat Balance

Rarely will an ore or concentrate contain the optimum amount of sulphur to satisfy the energy requirements of the oxidation process. High levels will dictate cooling, and insufficient sulphur will necessitate the addition of heat. For processes conducted at temperatures near ambient, for example bio-oxidation, seasonal temperature changes can alter the heat balance to the extent that cooling may be required in summer and heating in winter.

The use of dry grinding for roaster feed preparation is receiving interest because of the potential for overall energy savings. Bone-dry ore containing as little as 8-10% sulphide sulphur will roast autogenously, whereas a typical flotation concentrate slurry must contain 20-30% sulphur, depending on the water content. Since many refractory ores cannot successfully be floated in any case, dry grinding is a logical choice.

Pressure leaching and nitric acid oxidation of concentrates or moderately high sulphide ores require tight temperature control to prevent overheating, or in extreme cases, explosive reaction of the feed. Heat exchangers, or control of the feed by dilution are the usual solutions. Both of these methods can add considerable cost and complexity to the processes.

Bacterial oxidation is not spared this problem either. Thiobacillus ferrooxidans, the bacterium most commonly used, requires fairly tightly controlled temperature to function properly as a catalyst for the oxidation reactions. Experience has shown that a temperature between 30 and 40°C is usually required. Outside of this range, the bacterial activity slows down. In temperature excursions above 45°C, considerable bacteria kill can be expected. Consequently, temperature control by heat exchange, dilution, air-slurry mixing or refrigeration is necessary. For very low sulphide feeds (say less than 1% S) in cold climates, slurry heating may be necessary. These peripheral requirements can add considerably to the cost of a bacterial oxidation plant.

Ancillary Equipment

All refractory ore treatment processes require ancillary equipment to support or supplement the basic oxidation process. Flowsheets may include:

* Specially designed pumps
* Feed storage tanks
* Thickeners
* Filters
* Heat exchangers
* Oxygen plants
* Refrigeration units
* Baghouses
* Boilers
* Pressure letdown tanks
* Nitric acid regeneration plants
* Reagent preparation systems
* Gas scrubbers

Process requirements often justify the need for this hardware, but the installed cost of any piece of equipment is usually three times its purchase price. Simplification of the design of the plants to treat refractory gold ores takes considerable skill and time. Many dollars can be saved by asking (and answering) three simple question:

Why do it?
Why do it this way?
Why do it now?

Oxygen Utilisation

Recent papers have demonstrated the effect of sulphide content on process economics [Carter and Litz, 1988; Wells, 1989]. In summary, a high sulphide feed to any refractory gold treatment process means

* More oxygen is required to oxidise the feed.
* Greater amounts of acid wastes or sulphur dioxide are produced requiring more lime or limestone for neutralisation.
* More heat is generated requiring dissipation or recovery.

Without discussing the relative merits of direct ore treatment against concentrate treatment, it follows that maximum oxygen utilisation is important.

The oxidation of one ton of sulphur requires about 2 tons of oxygen. Oxygen enrichment is an option in roasting, some smelting processes, pressure oxidation and some versions of the Redox and Cashman processes. The use of pure oxygen has apparently not been seriously considered in bacterial oxidation or the Nitrox process.

Processes using oxygen have several advantages over those using air:

* Smaller reaction vessels can be used
* Reaction times are shorter
* Utilisation of oxygen is generally high (>80%)
* Gas volumes are reduced and sulphur dioxide concentrations are higher

The main disadvantage is cost, because air is essentially "free" and the cost of producing or buying oxygen is about $25 - $50/t. It follows that processes using air have an inherent advantage <u>provided that</u> the efficiency of oxygen utilisation is acceptable.

The Redox and Nitrox processes both use air to regenerate nitric acid which in turn oxidises the sulphides. Oxygen utilisation is therefore quite high, about 95%.

Factors influencing the design of bacterial oxidation plants include:

* Limited solubility of oxygen - about 8 mg/l at sea level
* Usually long residence times for slurries are required -rarely less than 60 hours
* Sometimes high sulphide contents
* Maximum solids concentration of about 30%
* Poor oxygen mass transfer coefficients
* Mixing and suspension of slurries

Optimising all of these interrelated factors presents complex engineering problems. Ensuring an adequate supply and transfer rate of oxygen into the slurry requires considerable energy. Compromises have to be made between low efficiency air dispersion and high shear rates required for good slurry suspension. A recent review [Hardwick, Errington and Miller, 1988], suggested a submerged diffuser for air with mechanical pulp circulation as the best solution. A proprietary mixing system, which is a radical departure from conventional methods, has been selected by U.S. Gold Corporation for its Tonkin Springs, Nevada, plant. This system offers considerably higher oxygen transfer efficiencies and low unit energy consumption.

Other Influences

It hardly needs to be said that developing a mine is hard work. Developing a refractory gold mine is exceptionally hard work and that is one of the reasons why so few have so far been brought into production, other than those using roasting. Some of the companies using or presently installing refractory ore process plants are listed below.

Company	Operation	Country	Process
Gencor	Fairview	SA	Bio-Oxidation
Homestake	McLaughlin	USA	Pressure Oxidation
Gencor	Sao Bento	Brazil	Pressure Oxidation
Barrick	Mercur	USA	Alkaline Pressure Oxidation
Barrick	Goldstrike	USA	Pressure Oxidation
FMC	Getchell	USA	Pressure Oxidation

These great advances were not made without considerable research, technical and engineering skills; the willingness of the developers to take a risk was also a vital influence. The interplay of some of these factors is shown in Figure IV. The willingness of emerging gold producers such as U.S. Gold and City Resources to take calculated risks in backing new technology is an encouraging trend.

EXAMPLES OF RECENT DEVELOPMENTS

A selection of mines currently treating refractory ores was shown in the previous section. Some examples of recent projects in which Minproc has been involved are discussed, and Tables 1 and 2 summarise key project data and design criteria.

Cinola Gold Project

This is currently under development by City Resources (Canada) Limited, and project go-ahead is expected before mid 1989. The Cinola orebody, discovered in 1970, is situated in the centre of Graham Island, in the Queen Charlottes of British Columbia. Because the mine is located in a sensitive environment, substantial environmental safeguards have been required. Many unsuccessful attempts were made by previous companies to treat this ore. It is highly siliceous with fine-grained, disseminated and vein sulphides containing fine gold.

All of the standard methods were evaluated to establish the best treatment route and a development of the original Arseno process was finally selected and piloted. The attraction of the process for Cinola is the low capital and operating costs of the oxidation step, and the high gold recoveries achieved. A simplified process flowsheet is shown in Figure V.

Snow Lake Gold Project

This project, under development by Sikaman Resources will treat arsenical tailings from a property near Snow Lake, Manitoba. Plant startup is expected by the end of 1989. The tailings are very fine and are composed primarily of arsenopyrite (50%), pyrrhotite (10-15%) and silicates (34%). Direct cyanidation generally achieved less than 15% gold recovery. The low-temperature, atmospheric pressure version of the Redox process was tested but gold occlusion by elemental sulphur complicated the process and this was changed in favour of the high-temperature nitric acid leach. Testwork conducted at higher temperatures showed that complete oxidation was possible in 2-5 minutes at 380^{0}F (195^{0}C).

A pipe reactor has been used at pilot plant scale and is proposed for the commercial plant. Nitric acid regeneration will be "in-situ" with the aid of direct-injected oxygen. Gold recoveries are expected to exceed 91%. A simplified process flowsheet is shown in Figure VI.

Tonkin Springs Gold Project

U.S. Gold has operated an oxide heap leach at this property since 1985 and is now constructing a 1500 t/d bioleach plant for treatment of sulphide ore. The mine is located at Tonkin Springs, 60 miles south of Carlin in Nevada. The ore is highly siliceous with fine-grained sulphides containing fine to ultra-fine gold. Variable amount of carbonates and shales are also present.

Conventional cyanidation with fine grinding could at times achieve 60% recovery but this was unacceptable and an extensive metallurgical campaign was completed. Flotation, concentrate and

TABLE 1

KEY PROJECT DATA

PROJECT DATA	CINOLA	SNOW LAKE	TONKIN SPRINGS
Oxidation Process	Nitric Acid(Redox)	Redox (Pressure O_2)	Bacterial
Developer	City Resources (Canada) Limited	Sikaman Resources	U.S. Gold Corp.
Location	Graham Island, Queen Charlottes, British Columbia	Snow Lake, Manitoba	Tonkin Springs, Nevada
Annual Throughput, t	2,310,000	140,000	550,000
Design Throughput, t/h	289	17.6	69
Gold Grade, oz/t	0.08 (2.45 g/t)	0.377 (11.7 g/t)	0.125 (3.9 g/t)
Silver Grade, oz/t	0.096 (3 g/t)	0.086 (2.7 g/t)	0.020 (6.2 g/t)
Recovery: % Au	92	91.5	90
% Ag	50	25	-
Assays:			
% S (sulphide)	1.70	14	1.34
% Fe	2.25	24	1.76
% As	0.03	22	0.25
% SiO_2	86.0	40	-
% $CaCO_3$	-	-	3.3
% C (org)	0.1	-	0.39
ppm Hg	5	-	20

ore roasting, alkaline leaching, nitric acid leaching and bacterial oxidation were tested. Bacterial oxidation was selected because of acceptable recoveries, simplicity of flowsheet and low capital and operating costs. The basic flowsheet is shown in Figure VII. The development by U.S. Gold of Tonkin Springs using bioleaching of whole ore at such a large scale is an industry first and considerable interest has developed in the project. Plant startup is expected in the third quarter of 1989.

CONCLUSIONS

This paper has attempted to show that although several mature technologies exist for the extraction of gold from refractory ores, new approaches are being actively developed and about to be implemented on an industrial scale. The development of these processes has required the commitment and the willingness of the owners to take risks and much hard work and foresight by the process developers and engineers. It is certain that the industry is keenly awaiting the startup of new plants incorporating bio-oxidation and nitric acid oxidation processes in their flowsheets.

TABLE 2

KEY PROJECT PROCESS DESIGN CRITERIA

DESIGN CRITERIA	CINOLA	SNOW LAKE	TONKIN SPRINGS
Bond Work Index, KWh/t	20.0	N.A.	20
Grind Size, P_{80}, Mesh	240 (60 um)	400 (38 um)	325 (44 um)
Oxidation Conditions:			
Pressure, psi	Atmospheric	400	Atmospheric
Operating Temp. F^0	185	380	95
% Solids	50	17	30
Residence Time, Hours	2	5 minutes	60
Stages	4	1	3
% S Oxidation	95	98	90
pH	<1.2	1-2	1-2
Eh, mV	600-650	N.A.	500-600
Design Oxygen Efficiency, %	97 (As HNO_3)	80	N.A.
Pre-thickening	Yes	No	Yes
Post-thickening	Yes	Yes	Yes
Filtration	No	Yes	No
Power Consumption kWh/t (oxidation only)	7.5	26	N.A.
Neutralisation Steps	Two Stage	Two Stage	Two Stage
Limestone Consumption, lb/t	74	923	-
Lime Consumption, lb/t	14	25	60
CIL/CIP			
Stages	2 + 7	6	7
Residence Time, hr.	18	8	24
Cyanide Consumption lb/t	3.9	1.6	5
Capital Costs: (US $ million)			
Total Project	96.0	13.5	23
Plant	44.0	10.5	19
Oxidation Section	8.0	6.8	N.A.
Operating Costs: ($ US/t)			
Total Project	14.5	29.0	20
Plant	8.9	26.2	19
Oxidation Section	2.0	12.0	6

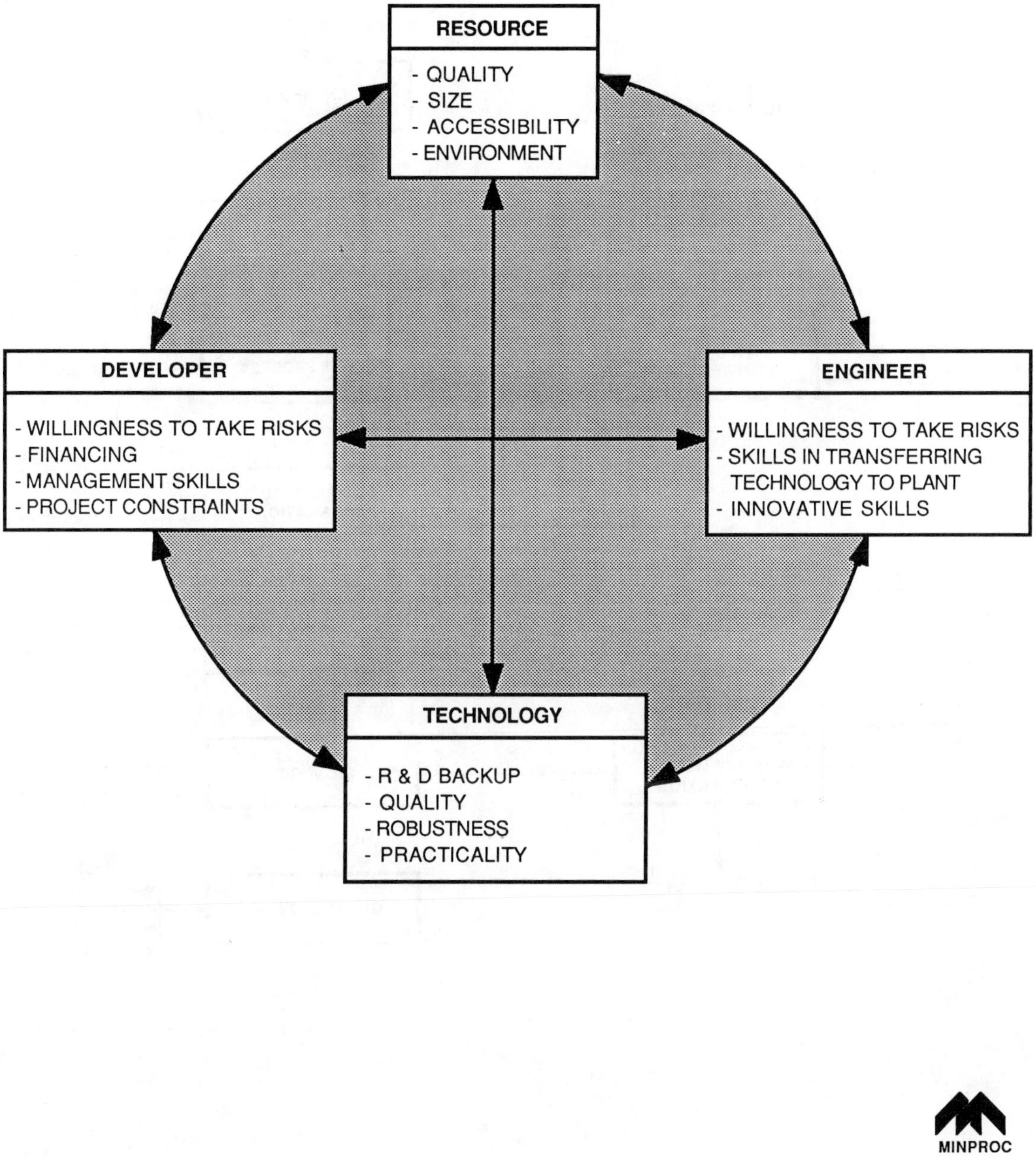

Figure IV. Interactions in project development.

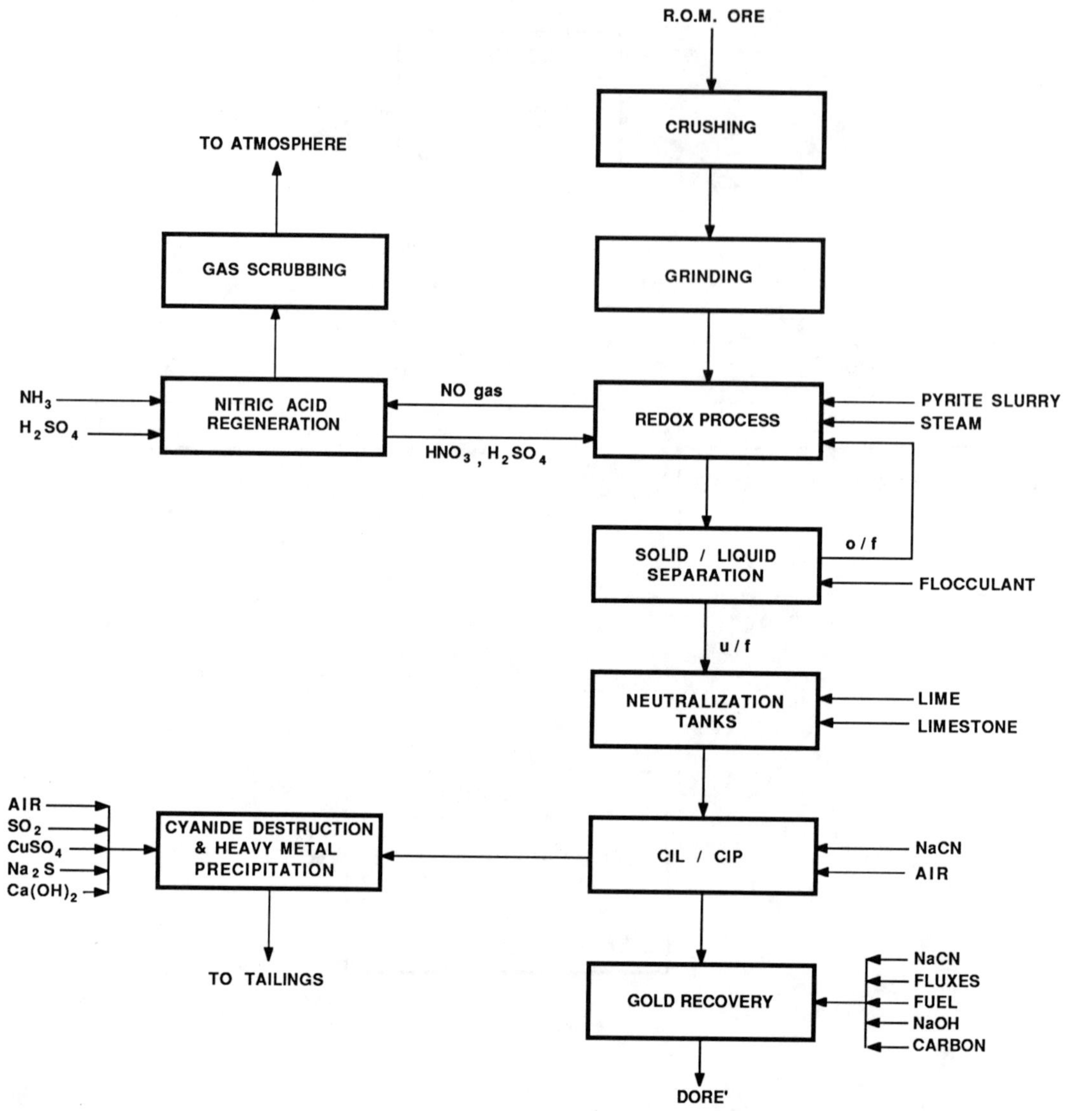

MINPROC

Figure V. Cinola gold project - Simplified process flowsheet.

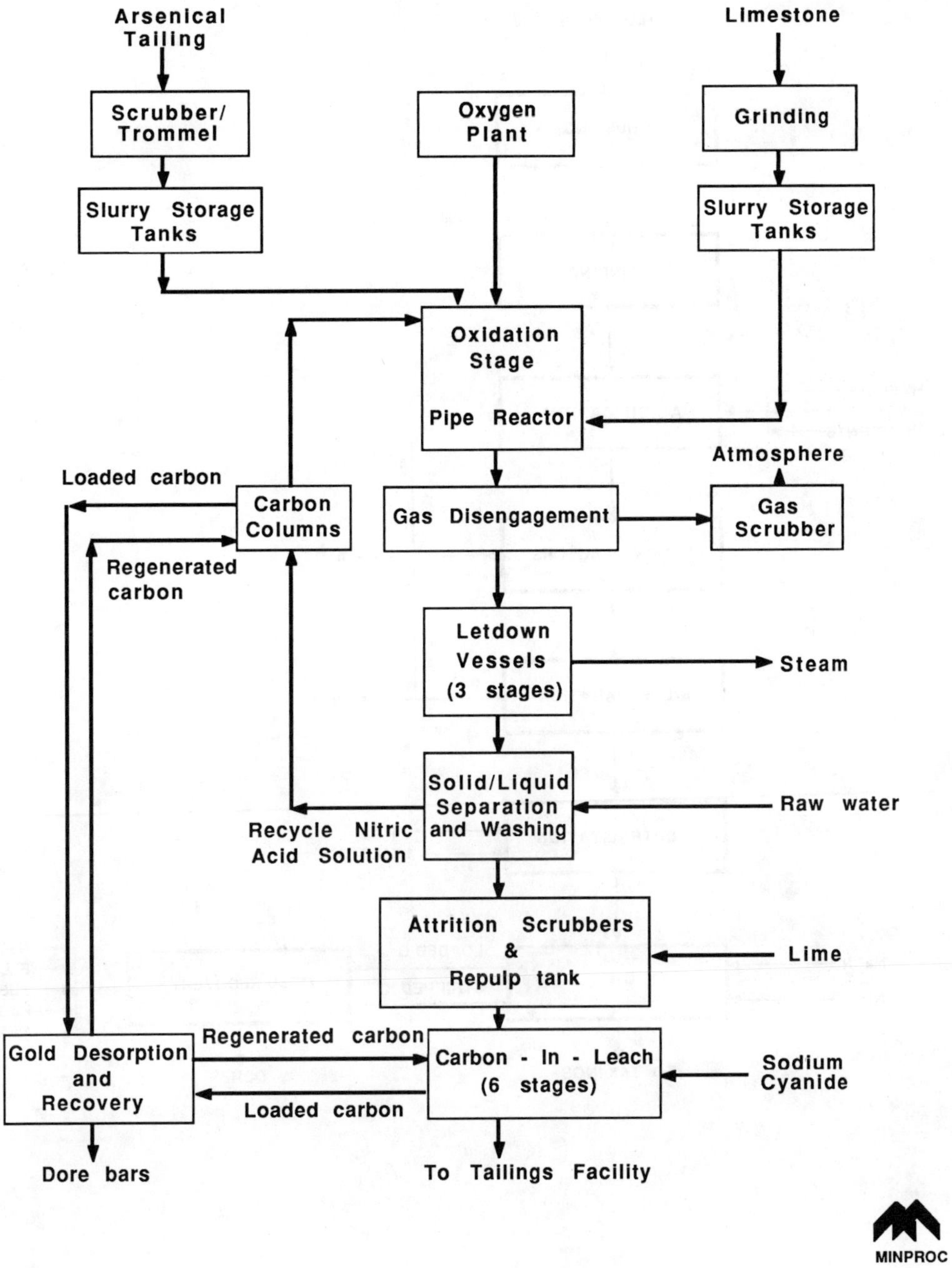

Figure VI. Snow Lake gold project - Simplified process flowsheet.

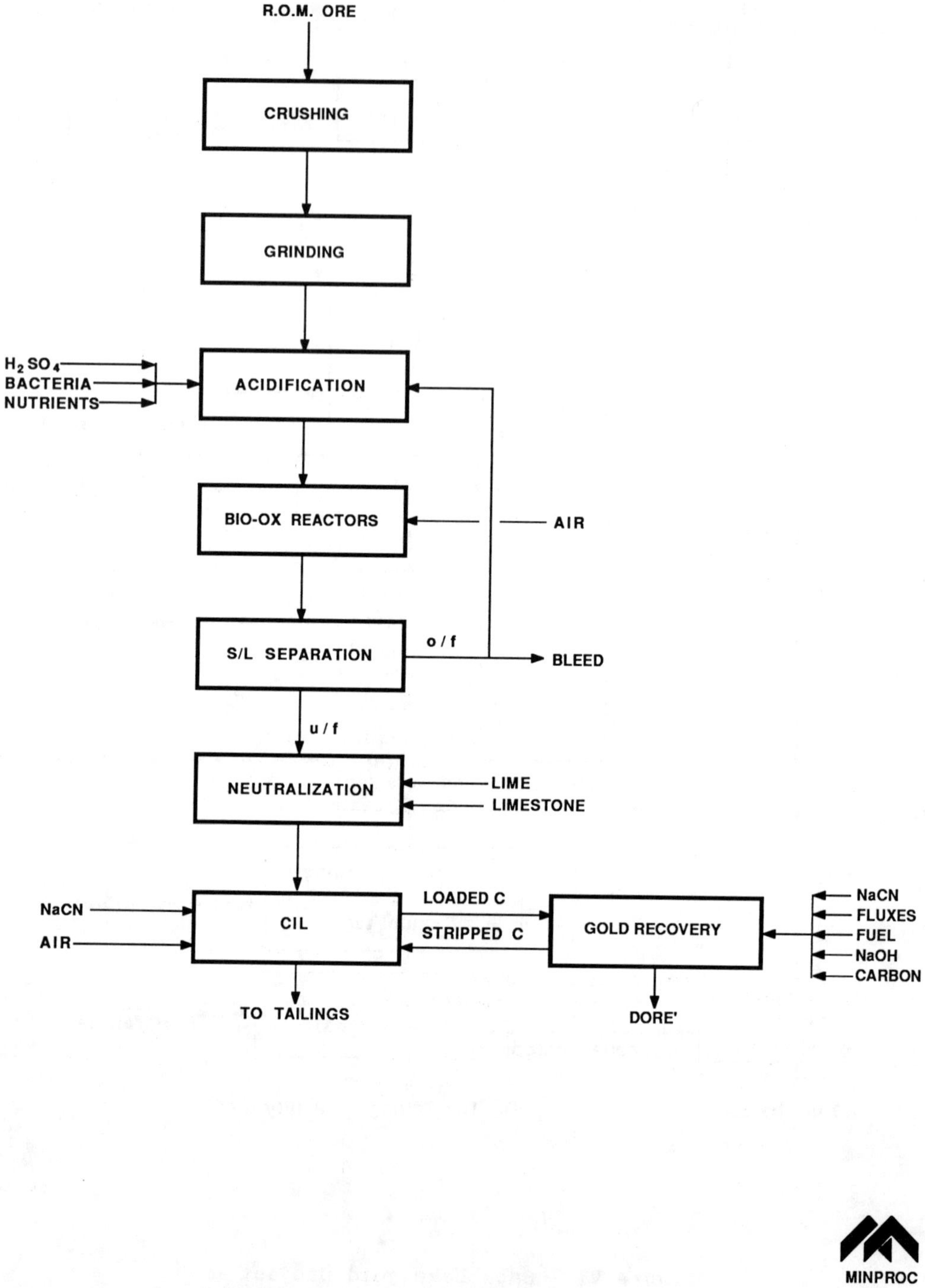

Figure VII. Tonkin Springs gold project - Simplified process flowsheet.

REFERENCES

Anon.,1988, "Mercur Starts up its New Alkaline Pressure-Oxidation Autoclave Plant", Engineering Mining J., Vol.189, No. 6, pp. 26-31.

Archibald, F.R.,1949, "Roasting Arsenical Gold Ores and Concentrates", Canadian Institute Mining Metallurgy Bulletin, Vol.42, pp. 129-139.

Beattie, M.J.V., and Raudsepp, R., 1988, "The Arseno Process - An Update", Annual CIM General Meeting, Edmonton.

Brown, A.J.,1984, "Economics of Recovering Gold from Refractory Sulphide Concentrates", First International Symposium on Precious Metals Recovery, Reno.

Carter, R.W., and Litz, J.E.,1988, "Comparative Economics of Refractory Gold Ore Treatment Processes", Randol Gold Conference, Perth.

Cornwell, W.G., and Hisshion, R.J., 1976, "Leaching of Telluride Concentrates for Gold, Silver and Tellurium - Emperor Process", Trans. AIME, Vol.260, pp. 108-112.

Davie, C.J.A.,1989, "The Cinola Project, Process and Environmental", Annual Canadian Mineral Processors Meeting, Ottawa.

Fair, K.J., Schneider, J.C., and Van Weert, A.,1987, "Options in the Nitrox Process", Annual CIM Conference Metallurgists, Winnipeg.

Floyd, J.M., and Lightfoot, B.W., 1984, "Direct Smelting of a Complex Copper Ore Using Sirosmelt Technology", Annual CIM Conference Metallurgists.

Gibbs, H.E., Errington, M., and Pooley, F.D.,1985, "Economics of Bacterial Leaching", Canadian Metallurgical Quarterly, Vol.24, No. 2, pp. 121-125.

Gilbert, S.R., Bounds, C.O., and Ice, R.R.,1988, "Comparative Economics of Bacterial Oxidation and Roasting as a Pre-Treatment Step for Gold Recovery from an Auriferous Pyrite Concentrate", Canadian Institute Mining Metallurgy Bulletin, Vol.81, No. 910, pp. 89-94.

Guay, W.J., and Peterson, D.G., 1973, "Recovery of Gold from Carbonaceous Ores at Carlin, Nevada," Trans. AIME, Vol.254, pp. 102-104.

Hansen, K.R.N., and Laschinger, J.E., 1967, "The Roasting of Refractory Gold Ores and Concentrates" National Institute Metallurgy NIM 85.

Hardwick, W.E., Errington, M.T., and Miller, P.C.,1988, "A Brief Techno-Economic Assessment of the Provision of Oxygen to Bio-Oxidation Reaction Systems for Refractory Sulphide Ores", Randol Gold Forum, Randol International Ltd, Golden, pp. 205-208.

Jha, M., "Refractoriness of Certain Gold Ores to Cyanidation: Probable Causes and Possible Solutions",1989, Annual SME/AIME, Las Vegas.

Lakshmanan,V.I.,and Bosch,D., 1984, "Treatment of Refractory Gold Ores", Annual CIM Hydrometallurgical Meeting, Timmins.

Lanyon, M., and Floyd, J.M.,1987, "Recovery of Gold From Refractory Ores and Concentrates Using the Sirosmelt Reactor" Research and Development in Extractive Metallurgy, AusIMM, Melbourne, pp. 45-55.

Lawrence, R.W., and Bruynesteyn, A., 1983, "Biological Pre-oxidation to Enhance Gold and Silver Recovery from Refractory Pyritic Ores and Concentrates", Canadian Institute Mining Metallurgy bulletin, Vol.76, No. 857, pp. 107-110.

Lesoille, M.,1987, "Schema de Traitement de Concentres Arsenies Auriferes a l'usine Metallurgique de Salsigne", Industrie Minerale, August/September.

Marchant, P.B., Lawrence, R.W., Chapman, J.T., Brooks, W., and Kuipers, L.,1989, "Austin Gold Venture Bioleaching: Project Summary", Annual Canadian Mineral Processors Meeting, Ottawa.

Markham, N.L.,1960, "Synthetic and Natural Phases in the System Au-Ag-Te", Economic Geology, Vol.55, pp. 1148-1178.

Nice, R.W.,1971, "Recovery of Gold from Active Carbonaceous Ores at McIntyre", Canadian Mining J., Vol.92, No. 6, pp. 41-49.

Randol, "Gold and Silver Recovery Innovations, Phase III", 1987, Randol International Ltd, Golden.

Royston, D., Spencer, P.A., and Winborne, D.A.,1984, "Alkaline Oxidation of Pyrite in Gold Recovery", Extractive Metallurgy Symposium, AusIMM, Melbourne, pp. 61-67.

Southwood, M.J., and Southwood, A.J., 1986, "Mineralogical Observations on the Bacterial Leaching of Auriferous Pyrite: A New Mathematical Model and Implications for the Release of Gold", Fundamental and Applied Biohydrometallurgy, Elsevier, New York, pp. 98-113.

Van Aswegen, P.C., and Haines, A.K., 1988, "Design and Operation of a Commercial Bacterial Oxidation Plant at Fairview", Randol Gold Conference, Perth.

Wells, J.A.,1989, "A Generic Study of the Capital and Operating Costs for the Recovery of Precious Metals from Refractory Ore", Annual Canadian Mineral Processors Meeting, Ottawa.

Wright, K.P.,1961, "Fluid Bed Roasting Practice in the Red Lake Camp", Canadian Institute Mining Metallurgy Bulletin, Vol. 54, pp. 629-634.

Chapter 28

GOLD RECOVERY AT GOLD FIELDS - CHIMNEY CREEK

by

JAMES R. ARNOLD, P. E.

Plant Superintendent

Gold Fields Operating Co. - Chimney Creek

Gold Fields Mining Corporation started processing ore from it's Chimney Creek operation in northern Nevada in November, 1987. Since that time production has averaged about 200,000 ounces per year from the milling and dump leaching circuits combined.

The events that led up to commissioning of the plant started in January of 1985 when the Gold Fields exploration team drilled their first hole in an area now known as Discovery Hill. The major orebody was located later in the year in what is currently the major focus of activity, the North Pit. Subsequent drilling since plant startup has found another 14 million tons of ore two miles south of the original orebody. This South Pit orebody is currently being stripped and metallurgical testing is underway.

Engineering

Engineering on the Chimney Creek plant started in earnest in May, 1986. Davy McKee Corporation from San Ramon, California, did both the engineering and the construction management. Before engineering began the entire senior staff of the Plant Department was hired and assigned to the project. Senior staff included the Plant Superintendent, Plant General Foreman, Maintenance General Foreman and Chief Metallurgist. Gold Fields does not employ an internal central engineering staff to design their plants. We feel that the people who design the plant should be the ones who will be required to run it once it has been commissioned. This philosophy affords a continuous accountability from start of design through commissioning and operations. All four of the senior staff spent the most critical phase of the engineering stage at Davy McKee, so the plant is truly a reflection of all four disciplines.

In bringing on the Maintenance General Foreman from the start, there was a focus placed on ease of maintaining the plant. Another philosophy that was adopted from the start was that the plant be easy to work in and maintain for both operators and maintenance personnel. Numerous design criteria were given to Davy McKee to ensure this. For example, we required that every piece of equipment could be reached easily with a crane or fork lift. All conveyors were required to have walkways on both sides for ease of maintenance and a minimum of 30 inches of clearance under them for ease of cleanup. Visibility for operators was considered very important and the control rooms and general plant layout were designed with this in mind. A list of about 100 general rules were given to Davy McKee during the early layout meetings and these were held as the general design criteria throughout engineering and construction.

While engineering was progressing in San Ramon, California, dump leach test work was being done at the site. Three 6 foot diameter fiberglass columns were constructed and filled to a 20 foot height with low grade ore. Leaching characteristics were determined and the dump leach design was based on information collected from these columns.

Construction

The major contractors on the project were The Industrial Company for the bulk of the piping and structural, EMKO on the electrical/ instrumentation and Hunewill for most of the earth work and leach pad construction.

All plant operating and maintenance personnel were employed eight weeks before startup for intensive training. Instructors were Gold Fields personnel, including foremen, and Davy McKee engineers. On occasion, selected suppliers representatives were brought in to discuss major pieces of equipment.

Project completion was targeted for January, 1988 at a projected cost of 78.9 million dollars. Actual startup was about 1 1/2 months ahead of schedule and slightly under budget. Full design capacity was achieved the first full month of operation.

Crushing

Run of mine ore is delivered to the crusher and dumped into a 90 ton hopper which is topped with a 30" X 30" opening stationary grizzly. Ore is fed to a 50" X 60" Telsmith jaw crusher with a 5' wide apron feeder. The closed size setting of the crusher is set at eight inches. Crusher feed rate is about 800 tons per hour and is run for 10 hours, four days per week. The crushing facility was sized to crush heap leach material from either the North Pit or the South Pit in the future as requirements mandate. Experimental work to date has shown that the incremental costs of crushing dump leach ore can not be borne by the increased recovery. This situation may change as ore comes from deeper areas of the pit or as new orebodies are processed.

Crushed ore is weighed with a belt scale then placed on a 35,000 ton stockpile with a stacking conveyor capable of slewing 180 degrees and luffing. By using the stacking conveyor in conjunction with the loader, mill feed can be thoroughly blended or selectively segregated depending on mill requirements. Flexibility was a major concern in designing the entire plant complex. Ore to the mill has averaged about 0.20 oz/ton, but the grade will drop as mining gets deeper into the pit.

Grinding

A front end loader feeds ore from the stockpile to a 50 ton dump hopper. A 48" wide apron feeder loads the mill feed conveyor. Feeding with the front end loader avoided the problem of rat holing or stove piping of clayey ores that sometimes can be a problem in Nevada. The grinding circuit consists of a SAG mill, SAG discharge screen, two recycle conveyors, a bank of cyclones and a ball mill. The entire plant circuit is shown in Figure I.

The 900 Hp DC variable speed Marcy SAG mill (18' X 6.5') grinds the ore to pass a 5/8" internal grate with a charge of about 8% four inch balls. SAG discharge passes over one of two 5' X 14' Derrick vibrating screens. Screen oversize is conveyed back to the mill feed conveyor and undersize is pumped to the cyclone bank from the feed sump. Both the mill feed conveyor and the recycle conveyor are fitted with weightometers and samples are taken for moisture once per day and smoothed with a three day moving average. A second SAG discharge screen sits on a rail trolley attached to the first, so that one screen can be maintained out of the circuit, then slid into service rapidly allowing maintenance on the other. Due to the excellent wear characteristics of the screen panels, one screen has been fitted to make a 14 mesh cut and the other at 28 mesh. This way the screens can be switched rapidly and the load distributed efficiently between the SAG mill and the ball mill, depending on changing ore characteristics.

The undersize is fed to a bank of four Krebs 20" cyclones fitted with 3 1/4" apexes. The undersize passes through a 1150 Hp Marcy 11-1/2' X 17' ball mill charged with a mixture of 1 1/2" and 2" balls. The cyclone feed pump is an Allis Chalmers SRL 8"x 8" that has performed exceptionally well. The volute of the pump currently in service has over 8000 hours and shows almost no discernible wear. The cyclone overflow density and grind are monitored with an Autometrics particle size monitor, PSM 400. Grinding is done in cyanide as both cyanide and pebble lime are introduced into the SAG mill from the start. By the time the ore reaches the cyclone overflow, over 60% of the available gold is dissolved (Table I).

Process

Ground ore from the cyclone overflow is diluted to 15% solids, passes through a 3 square meter Delkor screen to remove all wood fiber then goes to the double thickener circuit. In the double thickener circuit the ore is thickened in a 70 foot diameter thickener. Thickener underflow takes 15 hours to pass through three in-series leach tanks, then is diluted with mill solution and thickened again in a second thickener which is identical to the first. Air for the leach tanks is supplied down the shaft of Lightenin' A310 agitators. The second thickener underflow goes to CIP (carbon in pulp). The second thickener overflow makes up the dilution water for the first thickener feed (see Figure I). The first thickener overflow goes through the CIC (carbon in columns) circuit consisting of five carbon columns.

The purpose of the double thickener configuration is to leach as much gold as possible out of the ore before the pulp reaches CIP, and to transfer this dissolved gold to the CIC section of the plant, rendering the CIP section effectively a scavenger circuit. The major advantages of the double thickener circuit are:

1) Lower capital costs. Since most of the gold is collected in the CIC section, the carbon loading isotherm shows much higher loadings than would have been the case if most of the gold were loaded in CIP. This allows a reduction in the size of the carbon handling circuit.

2) Lower operating costs. The higher carbon loadings require less carbon to be stripped and handled.

3) Increased gold recovery.

4) Added flexibility. The plant is designed such that a thickener can be bypassed and the circuit operated conventionally if a thickener is down for any reason.

5) Increased silver recovery in CIP.

6) Circuit stability. The overall circuit is very insensitive to changes in feed tons and feed grade. These fluctuations are absorbed much better than by a conventional single thickener circuit, which not only results in better gold recovery in stable conditions, but also much better recovery during upset conditions.

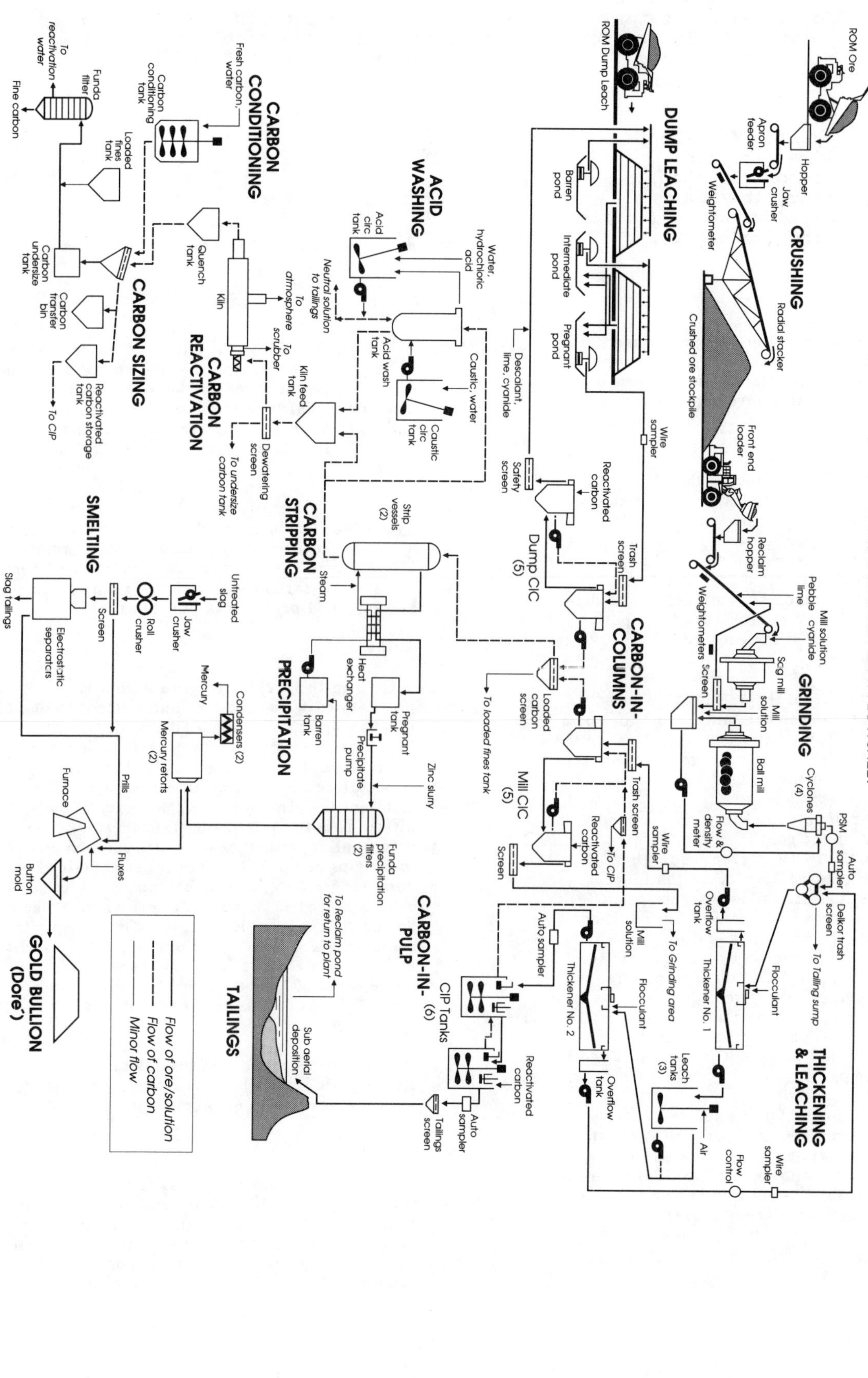

Figure 1: GOLD FIELDS OPERATING CO.-CHIMNEY CREEK PLANT FLOWSHEET

Table I
FY 1989 Operating Data (10 months)

Grinding dissolution %	61	% Au to CIC	84
Solid tails (oz/ton)	0.0081	Power consumption (kW/T)	15
Solution tails (oz/ton)	0.0004	Grind (% -200 mesh)	78
Mill availability %	95	Carbon density CIP (g/l)	15

Reagents	Mill	Dump Leach
cyanide (lb./ton)	0.42	0.10
lime (lb./ton)	1.59	0.19
carbon (lb./ton)	0.10	
floc (lb./ton)	0.07	
grinding media (lb./ton)	2.57	
zinc (oz/oz Au)	1.3	

7) Improved metallurgical performance. Several smaller benefits have become apparent since startup in the areas of carbon kinetics, inventory control, trace metal management, carbon attrition and circuit stability.

The CIP circuit takes the #2 thickener underflow and passes it through six CIP tanks which collect the last of the available gold. Each tank has a retention time of one hour and contains 15 grams of carbon per liter of pulp. Kambalda screens are used for interstage screening. Each screen has an independent drive and uses woven wire screen decks. The pulp discharge from CIP is screened and pumped to the tailings dam.

The overflow from the #1 thickener passes through five columns containing 2 to 4 tons of 6 X 12 activated carbon (the same carbon that is used in the CIP circuit). The discharge from the last column is screened then reused in the process.

Because the double thickener circuit removes 80 to 90 percent of the gold from the ore before it reaches the CIP circuit, carbon loading in CIP stays very low; in the 30 to 50 oz/ton range. Since the loadings are so low, carbon from CIP is transferred to the #3 CIC column for additional loading rather than to the strip circuit. This also helps clean the carbon by washing the silica out prior to stripping. Carbon is transferred out of the #1 CIC tank every other day and sent to one of two strip vessels. Carbon advance in the CIP circuit is done with Sala recessed impeller vertical pumps and in the CIC circuit with Wemco Hidrostal pumps.

Loaded carbon, containing 150 to 200 oz/T of gold, has the fines screened out then is sent to one of two strip vessels where it is stripped with a caustic cyanide solution (5 g/l NaCN) at 65 psi and 300 degrees F. Steam is used as a heat source for the strip solution. The pregnant strip solution is stored for further processing. Stripped carbon typically assays less than 0.5 oz/ton after a 5 to 6 hour strip.

Loaded fine carbon is filtered on a dedicated Funda filter then stripped with cyanide solution on the filter. Stripped carbon fines are then discarded. This averts sending out loaded carbon and paying a processing fee for gold removal. Stripped fine carbon assays from 0.5 to 1.0 oz/ton.

Stripped carbon is acid washed in dilute (2%) hydrochloric acid, and reactivated in a 500 lb/hr 40" X 34' horizontal Lochhead Haggerty rotary kiln. Kilned carbon is screened then stored for reuse. New carbon is attritioned for two hours and screened before it is put into use.

Bullion

Gold in the pregnant strip solution is precipitated with zinc dust and filtered in one of two 75 square foot Funda filters. It is interesting to note that no deaeration is required for precipitation because of the elevated temperature of the pregnant solution and the resultant low dissolved oxygen content. Lead nitrate is not required for efficient precipitation, either. Each filter can accept 4 to 6 strips or up to 7000 ounces of contained dore' before they have to be dumped. The filtered cake is moved to a vacuum retort where a small amount of mercury is collected. The precipitate, containing about 70% precious metal, is fluxed and melted in an electric induction furnace then poured into 600 to 900 ounce buttons. Dore' typically runs 780 to 850 fine gold and 100 to 150 fine silver.

Slag is reprocessed by crushing and electrostatic separation. The product is remelted with the next batch of precipitate.

Tailings Disposal

Tailings are pumped to a subaerial dam that is split by a center dike. The south half of the dam is generally used in the winter which gives the north side time to completely drain and dry out. The process is reversed, south to north in the summer. This drying of the tailings under subaerial conditions creates a negative pore pressure that autocompacts the tailings and affords three major advantages;

1) More tailings can be placed into a given volume which allows construction of a smaller dam,

2) The increased stability of the tailings makes possible upstream dam construction (at least in the case of Chimney Creek),

3) The negative pore pressure compaction lessens the problems associated with wind and dust on the dam.

On the negative side, subaerial deposition requires considerably more manpower and maintenance than a conventional dam. Migratory birds, especially puddle ducks and shore birds are more attracted to a subaerial dam because of the larger wetted area and the apparent increased shallow areas. Also, because of the smaller pond size, the concentration of cyanide remains higher than in a pond that has a larger retention. The smaller pond size also increases the risk of pumping muddy water back to the plant due to the smaller settling volume.

The actual pond area at Chimney Creek covers about three acres. The decanted solution is pumped back to the plant with floating raft pumps. Return solution usually assays about 75 ppm of cyanide as NaCN. A drainage blanket runs under the entire tailings disposal area, except the pond, and drains into a seepage pond below the dam. Seepage solution is pumped automatically back to the pond from a steel tank which sits in a plastic lined overflow pond. The overflow pond is only for emergencies and is kept dry.

Control System

The control system is a Modicon PLC based system which performs virtually all analog and discrete functions in the plant. The general architecture of the system can be seen in Figures 2 and 3. Basically, the plant is controlled by a Modicon 984B PLC with a second identical unit to provide redundant backup. The primary PLC does a self diagnostic scan every 90 milliseconds and reports to the backup. If there are any failures reported, the backup takes over and the operator is alarmed in the control room. The PLCs report through a modem and modbus multiplexer to the operator and engineering interfaces in the control room. The interfaces are industrially hardened IBM AT's running a commercial software package called "The Fix" made by Intellution. "The Fix" gives the operator and engineer a credible graphic display that not only facilitates control of the plant, but collects data, writes reports, tunes loops, performs trending and records all alarms (Figure 5). A more detailed description of the control system is available in the literature.

Power

Chimney Creek generates it's own power using natural gas tapped from a main Southwest Gas header about 23 miles from the plant. Four 1.5 megawatt units supply the normal 3.4 megawatt requirements for the entire facility. On day shift three normally run on line while one is on standby. At night, when the crusher is down, two will supply site requirements, at least until 1991 when the South Pit operations come on line. Each generator is driven by a 16 piston, 8 cylinder engine that operates at 900 rpm. Maintenance is low. One mechanic spends about six hours per day in the power plant checking the system and there are checks on a regular basis at night and on weekends. Power costs are considerably less than through the public utility and reliability is much higher.

Dump Leach

Dump leach tonnages to the pads are dependent on the amount of low grade material that must be moved to supply the mill with ore. Dump leach tonnage varies from one to three million tons per year. Ore grade runs between 0.015 to 0.060 ounces/ton and averages about 0.03.

Run of mine ore is delivered to the dump leach pads and loaded from the top in 40 foot lifts. The liner is constructed of a layer of 20 mil pvc sandwiched between 6 inches of silty soil below and 12 inches of silty soil above. On top of the liner is placed 12 inches of crushed gravel to act as a fluid conduit and a cushion to protect the liner from large run-of-mine rock . Ore is ripped after a pad has been constructed and drip irrigation is used to deliver 0.004 gpm/ft^2 of solution to the ore. Solution assays about 135 ppm of cyanide as NaCN at a pH of 11. Originally, solution was delivered with wobblers. Drip irrigation replaced the wobblers because of better leaching in the winter and improved flow rates through the pad. However the drip irrigation has been found to be more difficult to place on the pad and is more susceptible to scaling.

Barren solution is used for leaching for the first month on each pad section. The leach pad discharge can be routed into either the pregnant or intermediate solution ponds. When the solution tenor drops the discharge is routed from the pregnant pond to the intermediate pond. Intermediate solution is then pumped back on the pad and the cycle is repeated until the gold tenor is high enough to route the solution to the pregnant pond. The tenor cutoff is not fixed, but is dependent on the flow of solution off the pads and the tonnage of ore being leached at any point in time. In the final stages, just before solution flow is stopped, barren solution is again used to give a final washing. Leach cycles are highly variable, depending on ore type, but range from 100 to 300 days.

Figure 2: CONTROL SYSTEM ARCHITECTURE OVERVIEW

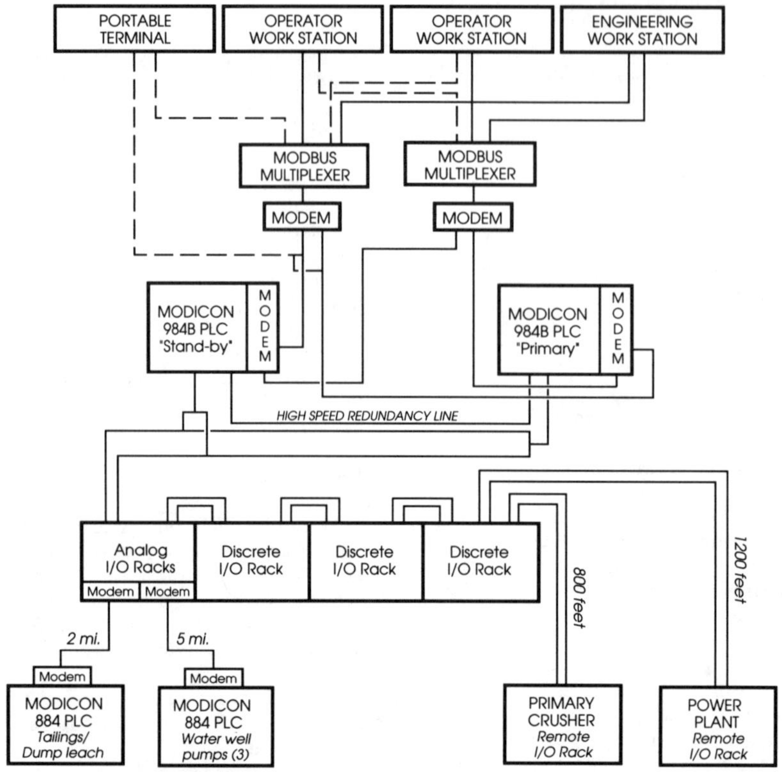

Figure 3: SUPERVISORY SYSTEM ARCHITECTURE

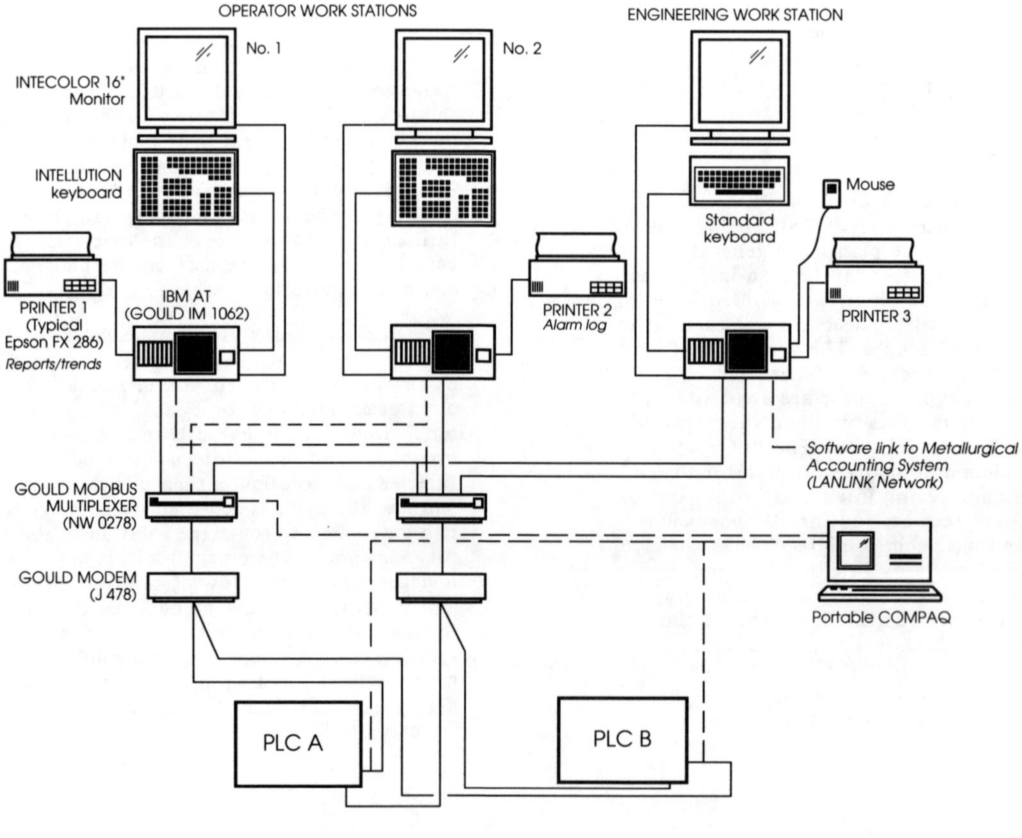

Figure 4: "THE FIX" - SOFTWARE ELEMENTS

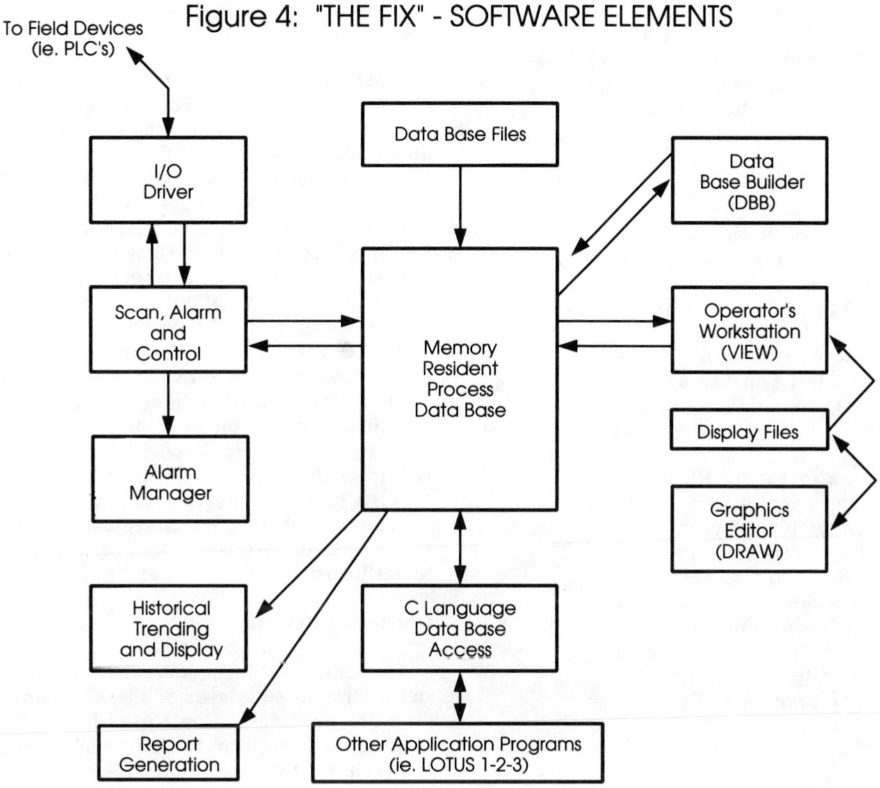

Pregnant solution is pumped to the mill building where it flows through a set of five carbon columns that is an exact mirror image of the mill carbon columns. The solution is kept separate from the mill solution and is returned to the barren pond. Tailings reclaim solution and fresh water are used as make-up water. Dump leach carbon is stripped and reactivated in an identical manner to the mill carbon.

Reagent consumption data are listed in Table I. Lime is delivered to the barren pond with a Denver lime slaking package and cyanide is pumped from the plant mix area. Both cyanide and bag lime can be added to the intermediate pond in the collection ditches, which are concrete with Ceilcote epoxy liners.

Environmental Controls

Migratory birds are a problem with most Nevada mines. At Chimney Creek numerous measures have been taken to protect wildlife. All solution ponds have been fenced to prevent access from terrestrial wildlife. Avian wildlife is hazed from solution ponds by a combination of propane cannons, owl decoys and flagging spaced on 20 foot centers. Plans are underway to net the solution ponds in the near future.

The tailings pond is not a threat to wildlife because the cyanide concentration is kept below lethal doses. Shortly after startup, when cyanide concentrations in the tailings were over 250 ppm, there were a few mortalities. Some success was achieved using the hazing techniques utilized on the solution ponds, and patrolling the area. Eventually the problem took care of itself when the concentration of cyanide in the tailings was lowered to below 50 ppm of cyanide as CN.

All solution pipes on the property lie in plastic lined ditches to contain potential spills from pipe breakage. Monitoring wells keep watch on the ground water, vadose zones, drainage blankets and solution pressures in the tailings and dump leach pads. Dust is controlled by using surfactants and monitored with opacity readings done by trained personnel on the site.

Ancillary Facilities

Other plant facilities include a reagent mix area where sodium cyanide and sodium hydroxide are mixed and pumped to their various addition points in the plant. A boiler-compressed air station is situated near the reagent mix area. Maintenance is handled out of the maintenance shop which provides office space, drafting facilities, electrical shop and the mechanical shop for major repairs and fabrication.

A fully staffed laboratory provides analytical and metallurgical services for mine oil analyses, gold assaying for mine, mill and exploration, metallurgical balances and development engineering and research.

A small lab was installed right in the mill building and all operators are trained to assay for pH, cyanide by titration and gold with an AA. This training has allowed the operators to check their solutions during precipitation and strip and to check solution tenors at any point in the plant. As an added advantage, because of the ability to generate quick answers, operators can test their own "theories" in the plant and receive rapid answers. This encourages them to learn about the systems and base reactions in the plant, and at times, we find operators teaching metallurgy to the metallurgists.

Staffing consists of three departments reporting to the Plant Superintendent. Maintenance includes 6 Electricians with a Foreman and 9 Mechanics with a Foreman. Mechanics and Electricians work a 8 hour day with skeleton coverage on the weekends. A Maintenance Clerk and a Maintenance Planner also report to the Maintenance General Foreman. Operations has five Foremen, one on each rotating crew in the mill and one who takes care of tailings, dump leach, reagents, bullion and the crusher. The mill crews have a Control Room Operator, 3 Mill Operators, a Shift Mechanic and a Loader Operator, who all work 12 hour shifts. Twelve hour shifts have been run for about a year and have been both popular and successful. All five foremen and the Mill Clerk report to the Mill General Foreman. The lab consists of ten Assayers who cover two shifts, a Chief Chemist, two Metallurgical Technicians and three Metallurgists all reporting to the Chief Metallurgist.

Operating Results

The Chimney Creek startup was a tremendous success by all standards of measurement. The first ore was crushed on November 9, 1987, almost two months ahead of schedule and 22 months after the first hole was drilled in the orebody; the first ore was milled on November 14; and the first gold was poured on November 23. Figure 5 shows the targeted gold production for the first six months of planned operation versus the actual production. It should be noted that June is a 36 day month because it ends our fiscal year, so June figures are slightly misleading.

Original plant design targeted production at 700,000 tons per year in the mill. Numerous minor improvements have been made since startup that focused on working out bottlenecks in the circuit and current production is 900,000 tons per year. Mill life-to-date recovery is 95.7% versus the original lab indicated target of 92%; the major factor in the improvement being that original metallurgical work was done on a CIP circuit rather than a double thickener circuit, which gives better recovery.

Dump leach recovery is estimated at 65% but is difficult to accurately determine because we have not yet sampled nor drilled spent pads. Leaching of the second 40 foot lift will start in late 1989.

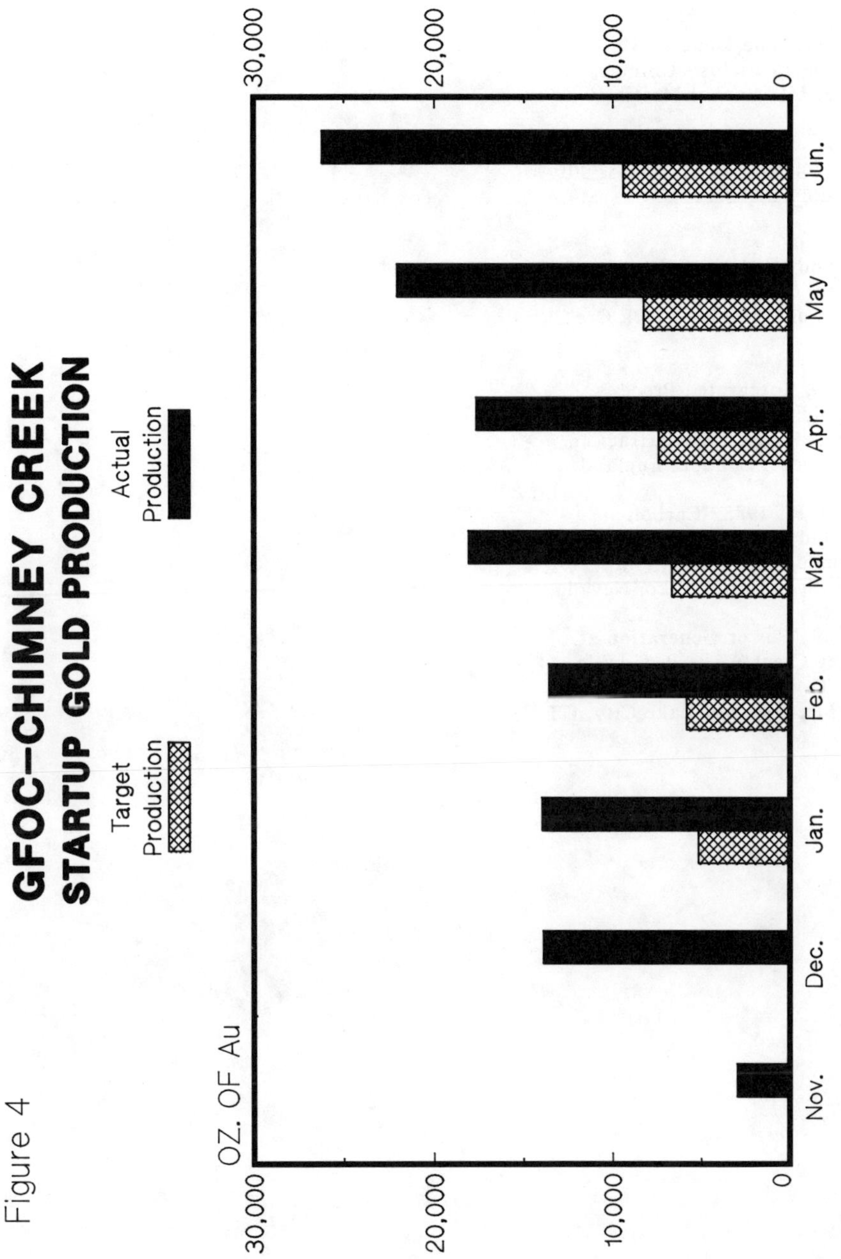
GFOC—CHIMNEY CREEK
STARTUP GOLD PRODUCTION
Target Production
Actual Production
OZ. OF Au
30,000
20,000
10,000
0
Nov.
Dec.
Jan.
Feb.
Mar.
Apr.
May
Jun.

Figure 4

Acknowledgments

The author wishes to thank the management of Gold Fields Mining and Chimney Creek for permission to print this paper. The contributions of **John Marsden, Mike Gleason** and **John Mansanti** were indispensable and greatly appreciated.

References

Mansanti, J. G., et al., 1988, "The Double Thickener Circuit at Gold Fields - Chimney Creek," Paper, SME-AIME Annual Meeting, Phoenix, Arizona.

McQuiston, F. W. Jr., Shoemaker, R. S., 1980, Gold and Silver Cyanidation Plant Practice, Volume II.

McClelland, G., Heinen and Lindstrom Consultants, 1986, "Carbon in Pulp, Carbon in Column Testwork for Samples of Chimney Creek Ore," HLC Job No. 1139.

Marsden, J. O., et al., 1989, "Integrated Process Control System at Gold Fields Operating Co. - Chimney Creek Mine," Paper, IMM Extraction Metallurgy '89 Conference, London, England.

Mansanti, J. G., Arnold, J. R., 1988, "Carbon Fines Handling at Gold Fields - Chimney Creek," Paper, Intermountain Mining and Processing Operators Symposium, Elko, Nevada.

Gleason, M. F., et al., 1989, "Power Generation at Gold Fields -Chimney Creek, The Profitable Alternative," Paper, To be presented at the SME-AIME Annual Meeting, Salt Lake City, UT.

Chapter 29

In Situ Leaching of Western Australian Palaeochannel Gold

Whincup, P.R.[1], McDonald, D.C.[1], Duke, H.[2]

[1]BHP-Utah Minerals International, Asia Pacific Division, Melbourne, Australia
[2]BHP Engineering, Sydney, Australia

ABSTRACT

Due to grade, overburden depth and stability and the presence of groundwater it was considered impractical to recover gold contained in unconsolidated palaeochannel sediments by conventional mining, dredging or borehole mining techniques.

A series of field experiments was conducted to determine the feasibility of recovery of gold by in situ leaching from sediments located at a depth of about 80 metres.

The test wellfield site and plant are described together with field test procedures, results and wellfield rehabilitation procedures. Choice of sodium cyanide as leachant and laboratory testing are discussed in relation to groundwater and sediment chemistry and problems encountered in field tests.

INTRODUCTION

Placer gold deposits within Tertiary palaeochannel systems of the Yilgarn Block in Western Australia collectively represent a mineral resource which although low in gold concentration could provide the basis for a substantial new gold mining venture in that State. The channels, described in detail by Smyth and Button, (1989), reach between 60 and 100 metres below the surface and represent old river valleys which have filled with unconsolidated gravel, sand, clay and groundwater of high salinity.

Due to depth, stability of overburden and the presence of groundwater at a shallow depth, techniques including open cut mining, borehole mining and dredging would be both impractical and uneconomic. Desktop studies were conducted which showed that in situ leaching offered the only possibility for technical and commercial success in exploitation of this resource.

In situ leaching has been applied for a number of years to extraction of uranium and copper and although there have been proposals for in situ leaching of gold ores dating back to 1896 [Mineev and Shutov, (1979)], little work to develop the concept is known.

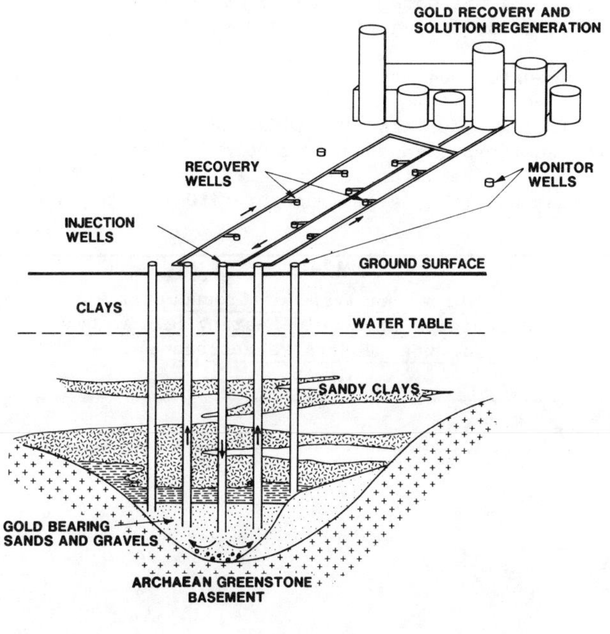

Figure 1
Conceptual Diagram for In Situ Leach Operation

Figure 1 shows a conceptual in situ leaching system for extracting gold from auriferous alluvial deposits. The commonly postulated requirements of such a deposit for application of in situ leaching are:

- The size of the deposit and its recoverable gold content are such that treatment is economic.
- For all practical purposes the aquifer containing the gold bearing sediments should be static or slow flowing.
- Ideally, impermeable zones should exist above and below the gold bearing sediments.
- Gold particles must be accessible to the leach solution.

- The leach solution should not irreversibly damage the aquifer environment.

- The sediments should be low in minerals other than gold which consume or otherwise adversely affect the leach reagent.

- The gold bearing sediments should be permeable.

In 1985 BHP-Utah Minerals International commenced a program of laboratory and field testwork to determine the feasibility of in situ leaching of gold from Western Australian palaeochannels involving auriferous sediments of the Rebecca palaeochannel.

LEACH SYSTEMS

Development of a leach system suitable for in situ leaching of Tertiary Channel sediments was a process involving consideration of channel groundwater quality, initial laboratory lixiviant screening testwork and the solution of problems which arose during field testwork.

Groundwater Quality

Typical analysis of groundwater contained in the auriferous sands and gravels is shown in Table 1. The water is saturated with respect to sodium chloride, gypsum and brucite $[Mg(OH)_2]$.

Table 1

Typical Analysis of Groundwater Associated with Gold Bearing Sediments in the Rebecca Palaeochannel

Component	g/1
Na^+	83
K^+	1.12
Ca^{2+}	0.61
Mg^{2+}	8.5
Cl^-	143
SO_4^{2-}	16.6
Total Dissolved Solids	307
pH	6
Density	1.165g/cm^2
Temperature	22°C (on site - September 1985)

The buffering capacity of the magnesium prevents the pH rising above 8.7 unless lime is added in excess of 35 kg/tonne to precipitate magnesium ions.

With groundwater of this quality the choice of leachant chemistry must be such that significant increase in calcium, sulfate or hydroxyl ions does not occur. It was believed raising either the calcium or sulfate concentration would lead to formation of gypsum, which would probably coat particle surfaces, thus limiting lixiviant access to gold particles. Similarly an increase in pH was shown to restrict solution flow through sediments as a result of precipition of gelatinous brucite.

Leach Reagent Selection

Laboratory testwork was conducted to choose a suitable lixiviant for in situ leaching. Solutions of the lixiviants cyanide, bromine, chlorine, thiourea and thiosulfate in test site groundwater were trialled using gold strip. Consideration was also given to the interaction of each reagent with other minerals occurring in the sediments, the groundwater and its environmental acceptability.

Bromine and chlorine were excluded because oxidation of sulfides present in the sediments would increase sulfate concentration and produce gypsum which had been observed to coat gold particles.

Thiourea required acid to reduce groundwater pH and reagent costs would be high. The carcinogenic nature of thiourea also provided grounds for rejection of this reagent.

Thiosulfate showed low gold dissolution rates in the laboratory tests and a reasonably high concentration of reagent would be required to stabilise the gold complex in solution. Breakdown of thiosulfate could also lead to gypsum formation.

The foregoing considerations resulted in the cyanide/oxygen system being chosen as the preferred lixiviant. In addition, the chemistry of gold cyanide complex formation is relatively well understood; breakdown of cyanide was unlikely to irreversibly damage the aquifer environment; and, if necessary, cyanide retained in the aquifer could be destroyed by mild chlorination.

However, there were problems associated with the choice of cyanide/oxygen as the lixiviant including:

- Toxicity of cyanide.
- An elevated pH is usually necessary to maintain cyanide in solution.
- Low solubility of oxygen in the high salinity groundwater.
- The tendency of cyanides to form hydrogen cyanide (HCN) in high salinity groundwater with subsequent loss of HCN to atmosphere.

When sodium cyanide was added to the groundwater the pH was found to increase to 8.7 at which point brucite precipitated, buffering the solution. Laboratory experiments showed that a maximum of 200ppm sodium cyanide could be dissolved in the groundwater before brucite precipitation commenced and therefore this level was suggested as optimum for field tests from the point of view of avoiding the possibility of pore and wellscreen blockage.

Nevertheless, concentrations of sodium cyanide higher than 200ppm were used in the initial field tests as it was assumed fine diffuse precipitates would not block pores. However at the highest cyanide concentration used, 1,500ppm (0.15% w/v) sodium cyanide, nucleation and growth of brucite quickly plugged the well screens demonstrating the reality of the anticipated problem.

In groundwater of the composition shown in Table 1, solubility of pure oxygen was found to be about 3ppm maximum at 1 atmosphere which is only about 30% of that of well aerated carbon-in-leach circuits. Thus the diffusion rate of oxygen to gold particles would be slower in the groundwater than for aerated water of low total dissolved solids. Consequently some field tests incorporating oxygen injection were undertaken.

The high ionic strength of the groundwater and pH buffering both contributed to enhanced chemical stability of HCN and high partial pressure of HCN caused cyanide losses on exposure of solutions to air. It was suggested that any cyanide system using high salinity groundwater would have to be totally enclosed to limit cyanide loss.

Use of calcium cyanide or pH adjustment with lime was considered inappropriate as these would promote formation of gypsum.

Laboratory column leach tests using groundwater and auriferous sands taken from exploration bores in the Rebecca palaeochannel were undertaken in 100mm NB x 1 metre high PVC columns. Various concentration levels were used in these tests. An additional column, packed with clean quartz sand with a gold plate of known surface area, was used as a control and to assess gold leach rates.

From this testwork a possible extraction of about 12% of the gold from channel sands was estimated for a 14 day contact period using a solution containing 200ppm sodium cyanide solution with air as oxidant. The rate at which gold leached using the 200ppm sodium cyanide solution was estimated to be 0.3 mg/day/cm^2.

TEST SITE

A field test site was established over a section of the Rebecca palaeochannel on sparsely vegetated arid pastoral land located some 135 kilometres east north east of the gold mining centre of Kalgoorlie, Western Australia.

Gold had been located within sands and gravels immediately above impermeable Archaean saprolitic greenstone bedrock at a depth of about 80 metres. The channel contained groundwater, of composition shown in Table 1, with the water table approximately 7 metres below ground level.

WELLFIELD CONSTRUCTION

A wellfield comprising 6 wells designed for small scale push-pull and circulation testing was established in a pattern designed for testwork using flows over varying distances, both along and across the channel.

Wells were constructed using conventional water well techniques. The wells were rotary mud drilled to a point approximately 5 metres below the unconformity at the base of the Tertiary alluvial sequence. Each hole was then cleared of drilling debris and gamma logged. The gamma logs provided accurate identification of the unconformity and were used to locate the position for placing well screens. Each hole was reamed, fitted with 155 mm NB Class 9 PVC casing and a 1 metre long x 155 mm NB x 0.5 mm aperture, wedgewire screen located immediately above the unconformity.

Test wells incorporated a neoprene gasket immediately above the screen and gravel packing which allowed the annulus above the screen to be sealed with cement to prevent direct leakage between the target horizon adjacent to the screen and those horizons higher in the sequence.

HYDROLOGICAL CONSIDERATIONS

The objective of hydrological testing was to determine the necessary parameters to be able to model the extent of lixiviant flow within the sediments.

From regional flow studies of the Rebecca palaeochannel area, intrinsic groundwater velocity was established as about 1 metre per year. A very low groundwater velocity was confirmed by tracer test and leaching test results. This very low velocity reduced the complexity of modelling and monitoring of field testwork.

Field tests were performed to allow an assessment to be made of the hydraulic conductivity of the gold bearing sediment by three different methods; flooding bores at constant and falling heads; by using drawdown data and a two dimensional computer model; and by use of a three dimensional model developed by In-Situ Inc., of Laramie, Wyoming. Values determined varied significantly with method used. However, modelling results provided by In-Situ Inc. were considered to present the most accurate representation of the aquifer under test. A mean hydraulic conductivity of 4.1 m/day over a section of thickness 15.1 m was accepted. Table 2 shows that the value determined is highly dependent on the model chosen to represent the groundwater flow regime.

The nature of deposition of cross bedded river channels in general leads to anisotropy of permeability and hydraulic conductivity parameters. The possibility of channelling of reagents through the coarse sediments (for example in the thalweg of the river bed) was cause for concern about the variations in hydraulic conductivity within the sediments. Such variations would facilitate leachant excursions and hinder solution access to the finer grained material.

Table 2

Hydraulic Conductivites of the Aquifer at the Test Site

	Base Depth m	Thickness m	Hydraulic Conductivity m/day
Australian Groundwater Consultants	80.5	0.6	47.8* 22.3**
	81.4	0.6	0.5* 0.2**
	82.3	1.0	4.0* 2.0**
	83.5	1.0	7.2* 8.1**
BHP Engineering 2 Dimensional Model	82.0	2.0	20.0 Average+
In-Situ Inc. 3 Dimensional Model	82.0	15.1	4.1 Average+

* falling head tests
** constant head tests
\+ determined as best fit of drawdown and pump test data

FIELD TESTWORK

Field testwork undertaken comprised push-pull tests conducted with both high and low strength lixiviant, two-well circulation tests and an extended four-well circulation test. Push-pull and two-well circulation tests using a lithium tracer were also conducted.

Push-Pull Tests

Push-pull tests were undertaken to establish the methodology for testwork and to determine if recovery of gold from palaeochannel sediments using in situ cyanide leaching was possible.

Performance of push-pull tests and interpretation of results are relatively simple as these tests involve small volumes under static conditions and use of a single well. For these reasons push-pull tests were chosen as the starting point for field testwork.

Push-pull tests comprised injection of a volume of about 25 m^3 of leach solution into the aquifer and allowing it to remain for several days before withdrawal of liquor from the same well. Samples of recovered liquor were analysed to determine gold recovery, lixiviant consumption and identify chemical reactions which took place between leach solution and aquifer sediments.

The circuit shown in Figure 2 was used for push-pull tests.

Prior to commencement of a push-pull test groundwater was pumped from the wellfield into

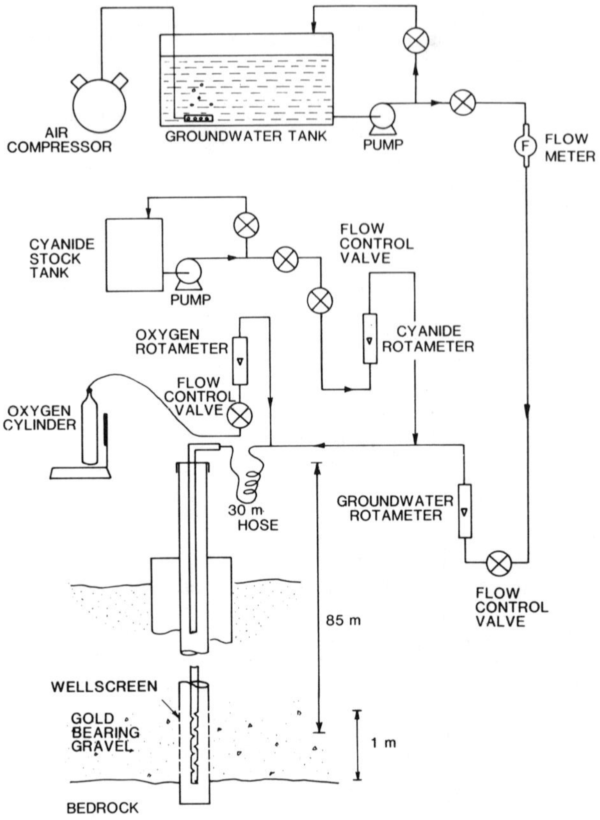

Figure 2
Equipment Configuration for Push-Pull Testwork

one or more holding tanks, each of about 40 m^3 capacity. A nominal 5% w/v sodium cyanide stock solution was mixed in a separate tank using fresh water.

Leach solution was introduced into the well at a rate of approximately 40 l/min. using a 25 mm diameter plastic pipe positioned at the well screen. Cyanide stock solution was added to groundwater stock immediately prior to introduction into the well.

Rotameters, flow control valves and in-line flow meters were used to monitor and adjust flow rates but measurements of volumes of liquor used were made by measurement of liquid levels in respective tanks.

For early tests only aerated groundwater was used in the leach solutions. However, in later tests gaseous oxygen was introduced into the leach solution immediately prior to injection into the well. In some tests a quantity of oxygenated groundwater was introduced into the aquifer ahead of the leach solution as a buffer to maintain a high aquifer oxygen level in the lixiviant pulse.

Solution was generally allowed to remain in the aquifer for about 6-7 days before recovery. In some tests the leach solution was moved or 'pulsed' daily by drawing out about 2 m^3 and replacing it with fresh lixiviant or by simply injecting an additional 2 m^3 of fresh lixiviant.

Solution was recovered by lowering a well pump to a point approximately 40 m above the well screen and pumping at the same rate at which the solution had been introduced, that is nominally 40 l/min. Samples were taken at 15 minute intervals for the first two hours of recovery and half hourly thereafter. Each sample was immediately subject to cyanide, Eh, pH and oxygen activity determination and a portion set aside for gold, iron and copper analysis.

Oxygen activity was determined using a TPS PTI-401 dissolved oxygen meter. Owing to the high ionic strength of the solution, determinations made in this way could not be considered a quantitative measure of the dissolved oxygen but simply a qualitative measure of relative oxygen activity.

Recovered liquor was stored in surface tanks for later disposal into a surface evaporation pond after passing through activated carbon columns for gold recovery.

Tracer Test

At the commencement of the program a push-pull test was conducted using lithium chloride tracer. The purpose of this test was to demonstrate that, in conducting a push-pull leach test, a high recovery of injected liquor could be achieved. Lithium chloride was chosen as the tracer as lithium was not intrinsic to the groundwater. Also, lithium ions did not react appreciably with the sediments and accurate analysis at low lithium concentrations could be performed on site using atomic absorption spectrophotometry.

Other tracers such as tritium and bromide were considered but rejected due to the extended time and extra equipment required for appropriate analysis of these ions.

An accurately measured volume of solution containing a known quantity of lithium chloride was introduced into the test bore at a nominal rate of 40 l/min. over a 9 hour period.

The injected liquor was allowed to remain underground for 7 days and recovered at a nominal rate of 40 l/min. over a period of 37 hours during which time periodic measurements of volume recovered were made and samples for lithium analysis taken.

The recovery of the lithium tracer as a function of volume of liquor extracted is shown in Figure 3. Approximately 83% of the tracer was recovered in the first pore volume extracted with a total of 94% in two pore volumes and 96.8% after almost four pore volumes. These results demonstrated that for a push-pull test a high recovery of liquor injected could be achieved.

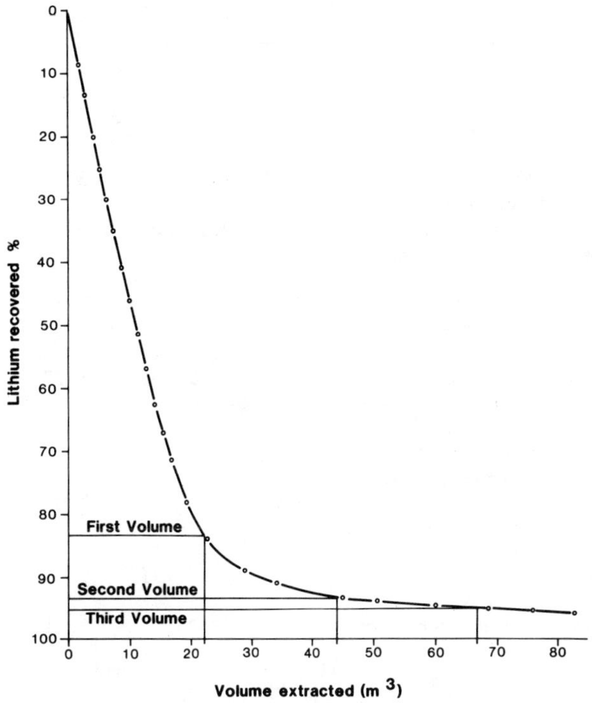

Figure 3
Lithium Tracer Remaining Underground During Recovery Pumping in a Push-Pull Test

Results

A total of 16 push-pull cyanide leach tests was conducted. Initial tests were conducted at a nominal lixiviant strength of 0.15% w/v sodium cyanide. However, following problems associated with well screen plugging attributed to brucite formation in these push-pull tests and in early two-well circulation tests, all further testwork was undertaken using nominal sodium cyanide solution concentrations in the range of 0.01-0.02% w/v.

Test conditions and results of four of the more significant push-pull tests are summarised in Table 3.

Figure 4 shows gold concentration in recovered liquor as a function of volume recovered. Sustained concentration of gold occurred in only one case.

Results from these tests showed:

- Recovery of gold from palaeochannel sediments using in situ cyanide leaching was possible.
- Oxygen activity played a significant role in the dissolution process.

Table 3

Summary of Conditions and Results of Cyanide In Situ Leach Push-Pull Tests

Test No.	B5	B6	B9	D21
Volume injected - m^3	22.2	24.6	29.0	25.0
Rate of injection - l/min.	41.1	36.4	44.1	31.4
Lixiviant strength - % w/v sodium cyanide	0.135	0.133	0.155	0.021
Oxygen added to lixiviant	No	Yes	Yes	Yes
Oxygen buffer	No	No	Yes	Yes
Residence time - hrs	166	166	120	226
Pulsed	No	No	No	No
Volume recovered - m^3	86.1	37.9	66.9	58.7
Recovery rate - l/min.	37.4	43.4	40.5	37.4
Average sodium cyanide concentration of first pore volume recovered % w/v sodium cyanide	0.062	0.061	0.073	0.008
Relative oxygen activity of first pore volume recovered	5	2	21	5
Average pH of first pore volume recovered	8.59	8.67	8.89	8.21
Calculated gold recovered - mg	723	913	2610	1355
Calculated iron recovered - g	181.2	113.3	46.8	26.9
Maximum solution gold tenor - ppm	0.03	0.06	0.08	0.08
Calculated cyanide recovery - %	59.3	49.6	54.3	51.1

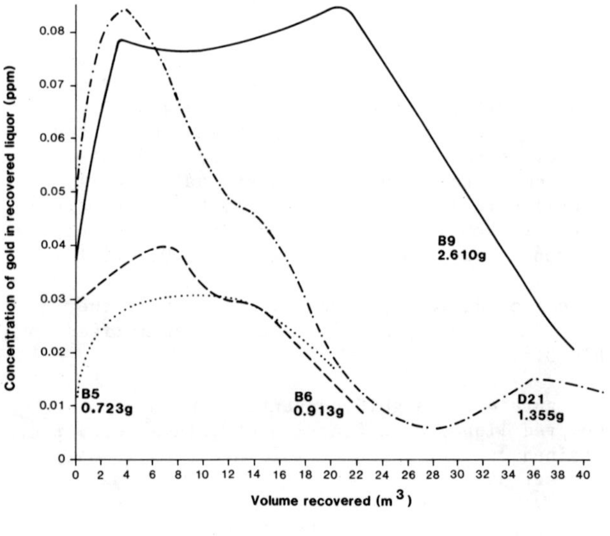

Figure 4
Gold Recovery for Selected Push-Pull Tests as a Function of Solution Volume Recovered

- At least 50% of injected cyanide was recovered. (It is believed significant HCN loss occurred from the high salinity, oxygenated groundwater during the sample taking process or during oxygenation.)
- A 0.02% w/v sodium cyanide solution will dissolve gold from the sediments but, as expected, the dissolution rate appears lower than that for a 0.15% solution.

Two-Well Circulation Tests

Following several push-pull tests conducted at a lixiviant strength of 0.15% w/v sodium cyanide, a program of two-well circulation leach testing was undertaken to establish the methodology for two-well circulation testing and to demonstrate that lixiviant flow between wells could be established.

The circuit shown in Figure 5 was set up and operated intermittently over a period of 14 days. Numerous delays to operation in the circuit were caused by mechanical failure of equipment. Periods of restoration pumping aimed at limiting excursion of lixiviant from the injection well were also undertaken. In all, the circuit was operated in the circulation mode for some 82 hours with an additional 75 hours total of restoration pumping.

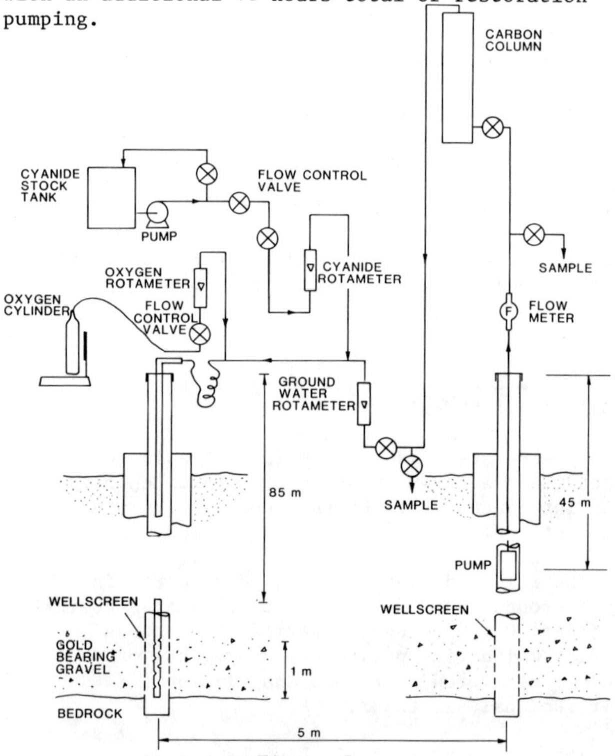

Figure 5
Equipment Configuration for Circulation Leach Tests

During operation of the circuit, significant well screen scaling and plugging of carbon columns occurred due to brucite formation. This resulted in a steadily increasing circuit pressure and decreasing solution flow.

Tracer Test

In order to establish that a high recovery of injected liquor was possible during circulation and restoration pumping phases of a two-well circulation test, a test using a lithium chloride tracer was undertaken. In this test an accurately measured weight of lithium chloride was dissolved in a known quantity of groundwater and introduced into the injection well at 35 l/min.

Extraction of liquor took place at a well 5 metres distant at a rate of about 50 l/min. and extracted liquor was pumped directly to the waste water disposal area. Circulation was undertaken over 3 pump/rest cycles, each cycle comprising 12 hours pumping and 36 hours rest. Following circulation, extraction continued until the level of lithium in extracted liquor fell below the limit of detection. Results showed that 93% of injected lithium was recovered (see Figure 3).

Results

Test conditions and results for two-well circulation testing are summarised in Table 4.

Table 4

Summary of Two-Well Circulation Test Conditions and Results

Total volume circulated - m^3	153.7
Average injection rate - l/min.	31.2
Injected lixiviant strength- % w/v sodium cyanide	0.15
Oxygen addition rate - g/m^3	132
Circulation time - hrs	82
Elapsed time - hrs	336
Volume extracted in restoration pumping - m^3	212
Calculated gold recovered during circulation - mg	4,415
Calculated gold recovered during restoration pumping - mg	8,381
Calculated cyanide recovery - %	31.3
Relative oxygen activity of extracted liquor	25
Maximum solution gold tenor - ppm	0.13

This work established the ability to pump lixiviant between two wells while maintaining oxidising conditions and confirmed the ability of a cyanide solution to recover gold from the sediments.

During circulation testwork using lixiviant containing 0.15% w/v sodium cyanide, significant operational problems were encountered due to brucite formation. This resulted in carbon column, well screen and possible sediment plugging leading to difficulty in maintaining solution flow in the circuit. This led to the conclusion that an in situ leaching operation in water of quality found in the test site using cyanide concentrations of the order of 0.15% w/v sodium cyanide would not be possible.

Four-Well Circulation Test

The final test undertaken in the program was a four-well circulation test, the primary purpose of which was to establish the methodology for future multiple well tests.

A circuit comprising one extraction and three injection wells was operated for a period of 30 days on daily cycles of 12 hours pumping and 12 hours rest. During the period 590 m^3 was circulated while maintaining an injected lixiviant strength of 0.01% w/v sodium cyanide. During circulation a calculated 9.479 g of gold was recovered.

Following completion of this test, restoration of the total test area was undertaken during which a further 12.570 g of gold was recovered.

Overall recovery of cyanide for the four-well circulation test was calculated to be 50.4%.

REHABILITATION OF THE WELLFIELD

On completion of leaching tests, wells likely to contain residual cyanide were pumped for over one month.

The extracted groundwater, totalling 893 m^3 was regularly tested for cyanide. When cyanide analysis values reduced to a consistently low level, the wellfield was allowed a rest period of 1.5 months to detect any equilibration of cyanide species between the aquifer material and the groundwater.

Following the rest period, withdrawal of groundwater (3,213 m^3) was continued for over one month, and free cyanide analysis by a modified colorimetric technique was used in determining the period of extraction required.

Finally the groundwater was sampled five times over a day's extraction, in the presence of an independent organisation's representative. A similar five samples were obtained from a well with no history of cyanide contamination. All samples taken were preserved with sodium hydroxide in clear plastic bottles, chilled and forwarded to three independent analytical laboratories for free cyanide, total cyanide and potassium analysis (the potassium concentration was used to maintain a check on sample dilution).

All values for free cyanide and total cyanide were well below 0.5 ppm as sodium cyanide. Comparative cyanide analysis figures are provided in Table 5.

All facilities were dismantled and removed from site and only the capped wells have been left. The area was left to regenerate naturally and on recent site visits regrowth of small ephemeral plants was observed. Most of the desert vegetation present in the area should re-establish over several years.

Table 5

Cyanide Analysis of Restored Wellfield Water and Virgin Wellfield Water

	Free NaCN (ppm)		Total NaCN (ppm)	
	Used Wells	Virgin Wells	Used Wells	Virgin Wells
BHP CENTRAL RESEARCH LABORATORIES				
High	0.33	0.22	0.25	0.12
Average	0.28	0.10	0.19	0.09
Low	0.17	0.05	0.14	0.05
THE CHEMISTRY CENTRE OF WESTERN AUSTRALIA				
High	<0.10	<0.10	0.30	<0.10
Average	<0.10	<0.10	0.30	<0.10
Low	<0.10	<0.10	0.30	<0.10
AMDEL				
High	0.42	0.17	0.20	0.40
Average	0.15	0.08	0.13	0.30
Low	0.07	0.04	0.09	0.12

DISCUSSION

The successful extraction of gold from Tertiary channel auriferous sediments by cyanide in situ leaching was encouraging.

Push-pull tests indicated the presence of recoverable gold in the sediments of the wellfield aquifer. However, it is believed the time the lixiviant was resident in the aquifer was too short for a significant solution tenor to develop. Attempts to increase the liquor tenor were made by 'pulsing' in which lixiviant was moved short distances through the aquifer and, while the flow of solution is laminar, some mixing and transfer of reagents was achieved at a higher rate than by diffusion alone. However, no significant effect was detected. Liquor tenor and total gold recovered were improved by increasing both the quantity of available oxidant and the cyanide concentration.

Push-pull tests with lithium tracer demonstrated that recovery of liquor was possible and thus established the ability to recover injected cyanide solution. Tracer tests also confirmed no significant groundwater flow was present in the channel. The high recovery of tracer indicated cyanide not recovered had been consumed by decompositon, by loss to atmosphere on recovery or by adsorption on the sediment surfaces.

While decomposition or atmospheric cyanide losses would be unlikely to constitute an environmental hazard, adsorption or storage of cyanide in the aquifer could become an environmental issue. For this reason stage restoration pumping with lengthy rest periods between stages was undertaken to determine if equilibration between cyanide species and sediments had occurred. It was, however, found that no significant recharge of groundwater with cyanide species occurred during the rest periods indicating such equilibration had not taken place.

Wellfield restoration, while time consuming, proved clean up of the aquifer was possible. Analysis values for free cyanide remaining in the test wellfield after clean up were found to be sufficiently close to those of virgin groundwater indicating restored groundwater to be essentially free of cyanide species. Had this water originally been potable, it could again be used for animal or human consumption. This standard of groundwater restoration was pursued and achieved to demonstrate that it is possible for an aquifer having potential as a potable water source to be restored after gold recovery by cyanide in situ leaching.

To measure low cyanide values in the highly saline groundwater with confidence, it was necessary to develop suitable cyanide analysis techniques. The standard silver nitrate/potassium iodide titration was adopted for high cyanide concentrations. At levels of cyanide below 20ppm use of rhodanine indicator was found to be preferable.

For very low cyanide concentrations a modified pyridine/barbituric acid colorimetric analysis developed by Doolan, (1989), at BHP Central Research Laboratories was used. After appropriate sample preparation the method was used to detect down to 0.01ppm free cyanide. A standard distillation was used to determine total cyanide.

The values in Table 5 show background measurements (for the virgin groundwater) as high as 0.4ppm sodium cyanide, demonstrating the problems associated with cyanide measurement below 0.5ppm. The groundwater salinity forced a tenfold sample dilution during analysis reducing the sensitivity of cyanide determinations.

Table 5 also shows free cyanide values higher than total cyanide were measured, indicating some aberration yet to be rectified in the determination of free cyanide in saline groundwater.

Murphy, McGrew and Jacobson, (1984), suggested that recovery of gold for a given sediment thickness may be determined by the following equation:

$$\text{Recovered Gold} = [1 - \exp(-k \times \text{time})] \times \text{Accessible Gold Grade} \times \text{Area Accessed}$$

where k is the extraction rate constant for the particular sediments and prevailing conditions.

For push-pull tests, values of k in the range of 0.001 to 0.003 per day were determined while for circulation testing a value of 0.003 per day was determined. These values were of the same order of magnitude as that determined in laboratory column tests (k = 0.009 per day) indicating reasonable agreement between laboratory and field data.

Calculations using a k value of 0.003 per day and the indicated grade and thickness of Rebecca palaeochannel sediments, together with the order of magnitude estimates of capital and operating costs, showed in situ leaching of this deposit would be uneconomic. These calculations also showed that for the process to be economic a rate constant or leachable gold grade greater by an order of magnitude would be required.

Since the testwork showed the rate constant to be a function of leach solution cyanide strength, attention is now being given to deposits in water of lower salinity and in particular lower in magnesium content.

CONCLUSIONS

Field testwork has shown that technically it is possible to extract gold from W.A. Tertiary palaeochannel sediments by in situ leaching using a cyanide/oxygen lixiviant. The apparent rate of gold dissolution was, as expected, dependent on the quantity of available oxidant, gold, and solution cyanide concentration. Improvement in the leach rate by increasing cyanide concentration is limited by formation of brucite in the highly saline groundwater.

An economically viable gold in situ leaching operation would be dependent on attaining a much higher leach rate than that achieved during the field tests. The possibility of improvement in leach rate rests in either increasing the leach rate constant by using a higher cyanide strength, different lixiviant, or by discovery of higher grade deposits.

In addition to the leach rate, the cost of locating the deposit and the cost and number of wells which need to be established are of critical importance in establishing the viability of any gold in situ leaching operation.

Restoration of the aquifer environment tested with respect to cyanide contamination is possible by pumping for extended periods.

ACKNOWLEDGEMENTS

The authors wish to acknowledge the contribution made by Uranerz Australia to exploration of the Rebecca palaeochannel and initial field testwork.

Thanks are due to BHP Central Research Laboratories for conducting lixiviant research and development of necessary analytical techniques, and to BHP Engineering and In-Situ Inc. for their work in modelling groundwater hydrology.

In-Situ Inc. is also acknowledged for their expert assistance in economic evaluation of the process and three dimensional modelling of the wellfield.

Thanks are extended to Miss H. Cruttenden for manuscript typing and Mr. T. Roberts for editorial comments.

Finally the authors wish to thank BHP-Utah Minerals International, Asia Pacific Division and BHP Gold Mines Limited for support of the project and permission to publish this paper.

REFERENCES

Doolan, K.J., 1989, Private Communication.

Mineev, G.G. and Shutov, A.M., 1979, "Feasibility of Mining Gold by Underground Placer Leaching", Soviet Mining Science (Engl. Trans.), Vol. 15, No. 4, July-Aug., 1979.

Murphy, J.W., McGrew, K.J. and Jacobson, R.H., 1984, "Application of Laboratory Stream Tube Testing to Economic Evaluation of Solution Mining", Minerals and Metallurgical Processing, Vol. 1, No. 1, May 1984, pp. 49-56.

Smyth, E.L. and Button, A., 1989, "Gold Exploration in the Tertiary Palaeodrainage Systems of Western Australia", Gold Forum on Technology and Practices - World Gold '89, Reno, Nevada, Nov. 1989.

Chapter 30

DETOXIFICATION OF SPENT HEAPS WITH HYDROGEN PEROXIDE

M. QUAMRUL AHSAN, PH.D. and ANDREW GRIFFITHS, PH.D.
DEGUSSA CORPORATION

EDWARD P. JUCEVIC
CONSULTANT FOR ACH-DAYTON PROJECT

Keywords

Hydrogen peroxide
Heap Leach
Detoxification
Cyanide
ACH-Dayton
Copper hydroxide
Rinse solution
Pumping and dosing unit

Abstract

Heap leaching of precious metal ores using the cyanide process is extensively used in the Western U.S.A. for extracting gold. Consequently, the impact of cyanide in the spent heap is closely examined by environmental agencies and permitting authorities. Currently, many states require that all decommissioned heaps be detoxified before abandonment. This paper describes a case study for detoxification of spent heaps at the ACH-Dayton Project using hydrogen peroxide. The chemistry of cyanide detoxification by hydrogen peroxide, together with simplicity of a detoxification operation, is discussed.

Introduction

The use of hydrogen peroxide to detoxify gold mill wastewaters containing cyanide and other toxic substances is an established technology. Degussa has many years experience in detoxifying barren bleeds, tailing pond overflow solutions and CIP tailings. In recent years, heap leach projects are becoming increasingly common in the United States and, accordingly, the regulatory agencies are also taking a closer look at the short and long-term environmental consequences of this recovery technology. As things stand now, it appears that the permitting authorities in many states will require detoxification of spent heaps before these sites are abandoned. The regulatory limits/permit conditions for abandonment may be accomplished in one of several ways:

Passive abandonment
Rinse with water and make use of natural degradation of cyanide
Rinse with water and detoxify chemically.

In this paper, the hydrogen peroxide detoxification of a decommissioned heap at ACH-Dayton Tailings Reclamation Project in Lyon County, Nevada, is described. The objective of this paper is to point out the advantages of neutralizing waste heaps by chemical treatment, compared with passive abandonment or slow rinsing with water. This paper will also discuss the simplicity of the hydrogen peroxide detoxification operation.

Background of ACH-Dayton Reclamation Project

In the early 1980s, Hughes Development Corporation acquired a 36-acre property containing about 200,000 tons of tailings. These were discharged at this site by many mills which in the 1900s used the Washoe (Pan amalgamation) process for gold extraction. These tailings contained 0.047 oz Au/T and 2.15 oz Ag/T.

These tailings were processed by heap leaching by ACH-Dayton Joint Ventures. At the end of the operation, three heaps, each containing approximately 75,000 tons of spent ore, needed to be detoxified before closure. The original state permit for closure called for rinsing the heap with fresh water until the return from beneath the heap showed a pH of 8.5 for three consecutive readings spaced 24 hours apart.

Since the material was agglomerated with Portland cement and lime, the rinse solution pH criteria would be virtually unattainable in any reasonable time frame. The Nevada Division of Environmental Protection subsequently repermitted the Dayton project to meet the alternative closure criterion: "for 3 consecutive days, the rinsate from beneath the heap must show that the free cyanide is less than 1.0 mg/l, and that total zinc and total copper concentration gain is less than 50%

increase over their background concentrations in the source process water. Alternatively, the agglomerated heap can be chemically or physically treated until the mean WAD cyanide is less than 10 mg per kilogram of dried spent ore."

Chemical Background

The ability of hydrogen peroxide to oxidize cyanide to non-toxic products is well-known. Hydrogen peroxide (H_2O_2) oxidizes free and weakly-complexed cyanide in a single step, as shown in equations 1 and 2. Metals such as copper are precipitated simultaneously as hydroxides.

$$CN^- + H_2O_2 \text{ ---> } OCN^- + H_2O \quad (1)$$

$$2\ Cu(CN)_3^{2-} + 7H_2O_2 + 2OH- \text{ ---> }$$

$$6\ OCN^- + 2\ Cu\ (OH)_2 + 6H_2O \quad (2)$$

Any excess hydrogen peroxide decomposes quickly to give water and oxygen, according to equation 4.

$$2\ H_2O_2 \text{ ---> } H_2O + O_2 \quad (3)$$

Heap detoxification described in this paper will only be concerned with the removal of free and weakly complexed cyanides (CN_{WAD}).

Detoxification Testwork By ACH

ACH decided to investigate several chemical methods available for detoxification in order to determine the most appropriate and cost-effective method for the elimination of cyanide in the spent heaps.

Processes tested were:

Hypochlorite treatment
Ferrous sulfate treatment
H_2O_2 treatment

From the results of testwork done by independent laboratories, ACH decided that H_2O_2 would be the most convenient and cost-effective process for detoxification of the heap at the Dayton project.

Degussa was requested to perform testwork to confirm the results of the independent laboratories. This laboratory testwork showed that 6 lbs of 50% H_2O_2 would be required to destroy 1 lb of cyanide contained in the heap. Calculated on the basis of 60 mg of CN/kg of ore, the total consumption of H_2O_2 (50%) for detoxification of heaps 2 and 3 would be approximately 127,000 lbs. After considering these results, ACH decided to detoxify the heaps using H_2O_2, with Degussa's assistance.

Design of the Detoxification Unit

A flow sheet for the detoxification is shown in Figure 1.

Heap rinse solution is piped into pond #1, from which the solution is pumped into hydrogen peroxide reaction tank #1. H_2O_2 and lime is added in-line with the incoming solution. The reacted solution flows from reaction tank #1 into reaction tank #2, and finally exits into pond #2. The green precipitate of $Cu(OH)_2$ settles to the bottom of the pond and the supernatant liquid from the top is pumped into a third pond, which allows time for further settlement of the precipitate. The clear solution from the top of this pond #3 is recycled back to the heap sprinkler system. In order to ensure that no $Cu(OH)_2$ was returned to the heap, the intake to the pump was mounted on a skimmer, designed with tiny holes in the bottom. This essentially eliminated pumping any floating $Cu(OH)_2$ back onto the heaps.

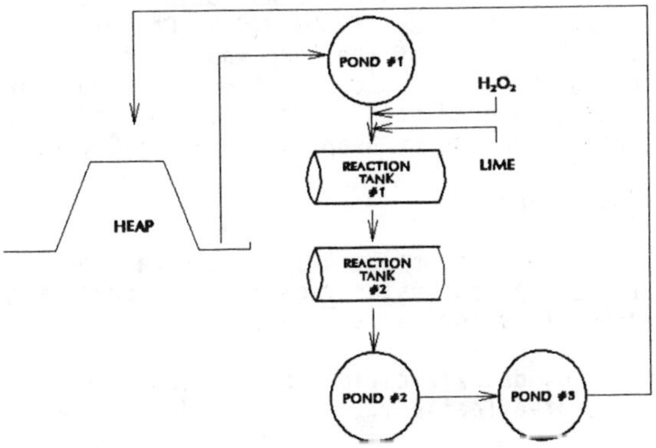

FIG. 1 Flowsheet for Heap Detoxification

The two reactor tanks were well-rinsed diesel tanks of 10,000 gallons capacity each. These reaction tanks were connected in series to give sufficient reaction time at a maximum effluent flow rate of 200 gpm. The peroxide was dosed by a pumping and a dosing unit designed and supplied by Degussa. The dosing rate of H_2O_2 could be varied between 5-50 gallons per hour. Solid lime was added to the reaction tank from a hopper bin via a screw-feeder system. The addition of lime was essential for two purposes: to reduce the foaming during reaction and also to attain a pH of 10.5 at which the detoxification reaction was found to be most effective.

Results of detoxification

The data presented here refer to heap #2, only, since data for heap #1 are incomplete. The content of cyanide in heap #2 is shown in Table 1 and Figure 2. At the start of application of H_2O_2, the average cyanide content of heaps #2 was 60 mg/kg of ore. The concentration of WAD-cyanide before and after treatment are shown in Figure 2, as CN_W influent and CN_W pond #2.

The CN_{WAD} of the rinse solution decreases continuously as the cyanide content of the heap is reduced. It can be observed that the efficiency of rinsing is decreased as the cyanide content in the heap is reduced, e.g. at 60 mg CN_{WAD}/kg of ore, the concentration of rinse solution is 300 mg/l, while at 10.0 mg CN_{WAD}/kg of the ore, the concentration of rinse solution is only 45 mg/l.

Figure 3 shows the residual CN_W concentration in the solids vs. the cumulative volume of water passed through heap #2. At the end of the treatment period, a total of 7,000,000 gallons of water had passed through heap #2 which contained 75,000 tons of ore, thereby reducing the CN content from 60 mg/kg ore to 10 mg/kg of ore. 8,250 lbs of CN were rinsed out by 7 million gallons of water. As expected, the efficiency of rinsing falls as the concentration of CN_W in the solids drops. During the 6 weeks of treatment, 7,000,000 gallons of cyanide-containing water was detoxified with about 53,000 lbs. of 50% H_2O_2. Consumption for detoxification of the heap was about 53,000 lbs. of 50% H_2O_2. This consumption was about 20% less than predicted. The total cost of H_2O_2 for detoxification of heap #2 was about $18,000.00

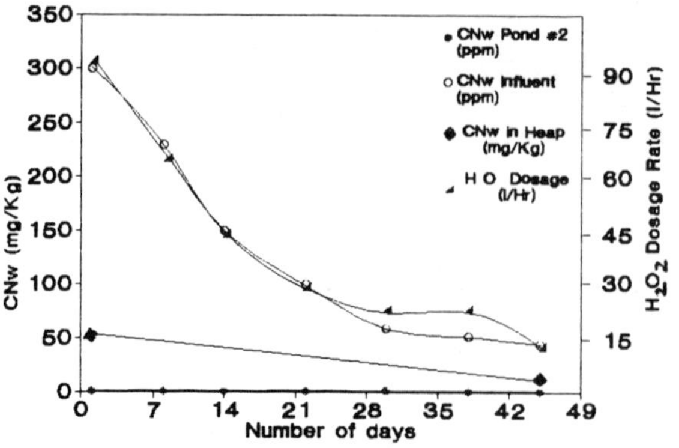

Fig. 2 Detoxification of ACH Heap #2

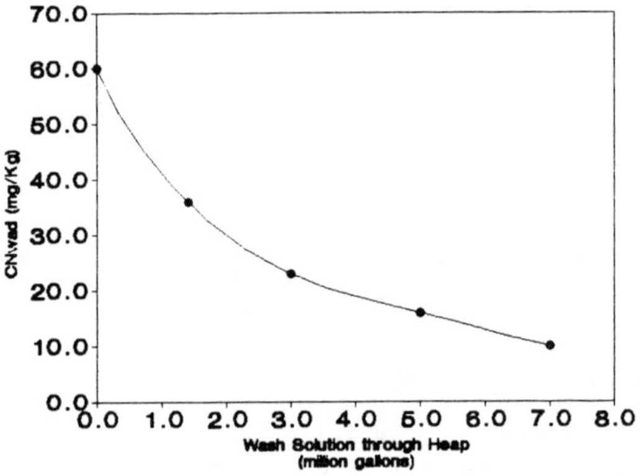

Fig. 3 Heap #2 Washing Curve

Conclusion

In the case study described here, the following conclusion regarding hydrogen peroxide detoxification can be made:

* H_2O_2 detoxification is efficient and cost-effective.

* Design and operation is extremely simple. The major item required by the operator is two reaction tanks of suitable size. Minimal amount of supervision is required.

* The detoxification of heaps can be completed in a very reasonable time. This is, of course, dependent of the cyanide content of the heap, the rinse rate, and the capacity of the reaction tanks.

* The permanent closure and final abandonment of heaps can be accomplished in a much shorter time, than would have been possible using natural degradation alone, thereby saving substantially in costs.

	Number of days	CNw in Heap mg/Kg	CNw Influent ppm	CNw Pond #2 ppm	Influent Flowrate gpm	H2O2 Dosage Rate l/Hr
1	1	53.0	300.0	1.5	200.0	92.0
2	8	-	230.0	1.7	200.0	70.0
3	14	-	150.0	1.3	200.0	45.0
4	22	-	100.0	1.3	200.0	30.0
5	30	-	60.0	1.9	180.0	20.0
6	38	-	53.0	1.2	150.0	20.0
7	45	10.0	45.0	1.3	150.0	12.0

Table 1. Results of Detoxification of ACH Heap #2

9

Processing II Agitation Leaching/CIP

L. Harris
Newmont Mining Corp.
Denver, Colorado

J. Avraamides
Mineral Processing Laboratory
Perth, W. Aus., Australia

Chapter 31

CONVERSION OF NEWMONT GOLD CO.'S PLANT NO. 1 TO A CIL CIRCUIT

Tom L. Enos

NEWMONT GOLD COMPANY

Abstract. Increased production through the plant was limited due to thickener circuit constraints. To overcome these constraints and maximize plant throughput to match grinding circuit capability, two options were considered:

A. Adding 2000 TPD CIL circuit to operate in parallel with the existing CCD circuit.

B. Adding a CIL circuit to treat the entire 4600 TPD while retaining some thickeners for tailings thickening.

Option B was selected based on simplified plant operations, reduced operating costs, minimal capital requirements, reduced solution losses, and bullion grade improvement.

INTRODUCTION

On May 27, 1989, Newmont Gold Company's Mill No. 1 entered its 25th year of operation. The plant was originally designed to treat 2,000 tons per day of oxide ore. Over the past 20 years, the grinding capacity has been increased to 4,600 tpd. In 1985, a nominal 3,200 tpd was being processed, due primarily to downstream constraints in the countercurrent decantation (CCD) circuit.

In late 1985, a program was begun to identify the constraints within the plant, with evaluation of nearly every piece of installed equipment in both the crushing and milling complex. Through this evaluation, it was found that the largest constraint was in the 5-stage CCD circuit.
In late 1986, the expansion of the through-put capacity at Mill No. 1 was planned to make maximum possible use of existing equipment within limits of practicality. Two flowsheets were developed:

A. Addition of a 2,000 tpd Carbon-in-Leach (CIL) circuit to operate in parallel with the existing CCD circuit which would treat 2,600 tons per day.

B. Processing the entire 4,600 tpd through one CIL circuit, but retaining four of the existing CCD thickeners for conventional tailing thickening.

Option B was selected for implementation based on consideration of simplified plant operations and re reduced operating costs with minimal capital requirements. This option also offered other potential advantages; among them, reduction of solution losses and improvement in bullion grade.

In March of 1987, Kilborn Engineering of Vancouver, B.C., was awarded a contract for the conversion. The CIL plant was brought into operation in October of 1987.

SUMMARY OF PROCESS

Mill No. 1 was originally designed as a 2,000 tpd oxide circuit. The crushing circuit consisted of a Pioneer 42 in. x 48 in. jaw crusher and a Symons 5-1/2 ft secondary crusher in closed circuit with a Tyler 6 x 12 ft vibrating screen.

The ore was crushed to minus 3/4 inch and stored in one of three 1,800 ton fine ore bins. Next, the material was ground in a conventional ball mill circuit by a Marcy 10 x 10 grate discharge ball mill and the pulp was agitated in the leach circuit for 15 hours. The agitated pulp was then further treated in a counter-current washing and decantation system utilizing five 125-ft diameter thickeners. Gold-bearing solution from the decantation system entered a precipitation system in which the gold was removed from the solution by deposition onto zinc dust. Periodically, the gold-bearing precipitate was removed from the presses, retorted, and melted into bullion in the furnace facilities located at the plant.

In 1969, a 600 tpd ore grinding circuit and chlorination complex was added to treat the carbonaceous ore being mined simultaneously with the oxide ore from the Carlin Mine. The only modification to the downstream circuit was the conversion of one oxide leach tank to a chlorinated carbonaceous ore leach tank.

In 1983, another 10 x 10 Marcy grate discharge ball mill was added to the grinding circuit to accommodate the harder ores that were being encountered at the Bluestar and Maggie Creek mines. The average daily tonnage had dropped to between 2,100 and 2,000 tpd for the oxide and carbonaceous ore circuits combined.

In 1985, ore from the Genesis mine was processed at Mill No. 1, and a nominal 3,200 tpd was treated.

OPTIMIZATION OF EXISTING EQUIPMENT

Before any capital expenditures could be made, the plant had to be optimized. Maximum utilization of the installed equipment was the major consideration.

Crushing. Evaluation of the crushing circuit revealed that the Symons cone crusher and the Tyler screen were worn out, so they were replaced with new equipment of the same kind.

Historically, mill throughput dropped off significantly in winter months due to the inability of the crushing plant to supply feed to the ball mills, and when ore was supplied, it was crushed coarser. The installation of soft rubber screen panels provided a cure to the winter crushing problems. The rubber screens were flexible and did not blind as a normal wire screen panel did.

Grinding. Grinding circuit optimization had been an ongoing program. In 1985 when Mill No. 2 started up at Gold Quarry, one-half of the experienced operators and shift foremen were transferred from Mill No. 1. A program of operator training was initiated with emphasis on the cyclone classification circuit.

The ball charge in all of the grinding mills was increased from 40% of mill volume to 50% mill volume, to maintain maximum power draw.

As the crushing rates improved due to higher mechanical availability, the crusher discharge product was made finer by installing 1/2 in. screen panels with no loss of efficiency. At this point, the standard ball ration of 3 in. and 4 in. balls was changed to 2 in. and 3 in.

Double wave liners were installed in both Marcy mills when the single wave liners wore out.

CCD Circuit

The installation of the second Marcy mill increased the plant grinding capacity beyond the designed capacity of the thickeners. Efforts in this area were targeted at maintaining very dilute thickener feed solids, in the 15-17% range. Plant flocculation tests were conducted on a regular basis. Thickener problems were largely those of high torques on the drive units, and resulted in extensive double-lining at times. Thickener repairs were being done on an unpredicted frequency due to overloading.

Capital Costs. Consideration for continuing to operate the existing CCD circuit at 2,600 tpd, and constructing a new 2,000 tpd CIL facility as opposed to construction a new CIL facility to treat 4,600 tpd was studied by Kilborn Engineering. Capital cost estimates for the 2,000 tpd CIL were $842,000 less than the 4,600 tpd option. The 4,600 tpd option was accepted and in March of 1987, engineering was started.

Basic metallurgical design data was derived from the long operating experience at Mill No. 1, and was used in arriving at the process design for the expansion.

CIL PROCESS DESIGN

Mill No. 1 processes two distinct types of ore which are classified as carbonaceous and oxide.

The carbonaceous ore is ground in fresh water to 60% minus 200 mesh. Cyclone overflow is trash screened by the use of a 3 ft X 6 ft high frequency, 5 degree, 10 degree, 15 degree inclined Derrick screen installed with .5 mm urethane screen panels. Screen underflow is pumped to the chlorination circuit. Screen over-size is returned to the mill. Discharge of the chlorination circuit is pumped to No. 1 and No. 2 Leach Tanks where the pH is elevated and cyanide is added. Retention time for leaching is approximately 24 hours.

Oxide ore is ground to 60% minus 200 mesh. Cyclone overflows from both oxide mills are discharged to a common sump, diluted and pumped to the oxide trash screens. The trash screens consist of four Derrick 4 ft X 8 ft "flowline" type screens installed with .5mm urethan panels. Oversize from the trash screens was designed to be returned to the grinding circuit. This was changed shortly after bringing the CIL circuit on line. Trash screen oversize is now directed to No. 2 thickener. Oxide trash screen under-size is discharged to the CIL interstage sump where it joins the treated carbonaceous ore stream and is pumped to a new leach tank in the CIL tank farm.

The one new leach tank and six CIL tanks are nested together, and each tank is of equal volume. The heights of the new leach tank and CIL tanks range from 35 ft x 45 ft 8 in. The bottom of each tank, with the exception of CIL tank No. 6, was raised by the appropriate amount by the addition of coarse granular fill overlain by concrete to obtain an equal volume in each tank.

The leach and CIL tanks are equipped with 50 hp double propeller Lightning agitators. Leach air is provided to the leach tank and CIL tank No.'s 1 and 2 by a 400 cfm compressor.

The pulp flows directly from the leach tank into the train of six CIL tanks. Pulp flows tank to tank by gravity on a 45 degree pattern. Drop from tank to tank is 2 feet.

Carbon retention and periodic transfer is accomplished by the use of Derrick interstage screens. Mill No. 1 was the first mill to use this unique system. The screens are a standard 4 ft wide by 8 ft long high-frequency, which has a 12 degree reverse slope to the discharge. The screens are set in the tank close to the pulp level.

The pulp flows over a weir onto the screen. Using a 35 mesh polyurethane screen deck, the pulp passes into the underpan and reports to the next CIL tank. Carbon, which is coarser, reports as oversize back to the same tank.

The carbon is advanced on a continuous basis counter-current of the pulp flow by a SALA vertical recessed impeller pump suspended in the pulp. The pumps deliver a small stream of 135 gpm of pulp to the prior tank containing activated carbon at the average concentration.

Loaded carbon is recovered from No. 1 CIL tank with a SALA vertical recessed impeller pump which discharges over a disengagement screen. The carbon which reports as oversize flows by gravity to a loaded carbon bin. Loaded carbon is accumulated until there is a load of suitable size for transport to the Central Carbon Handling facility at Gold Quarry. The screen undersize is returned to No. 1 CIL tank.

Reactivated carbon, returned by truck from the Central Carbon Handling facility, is unloaded and pumped to a reactivated bin. The carbon is then pumped into No. 6 CIL tank.

The CIL tailing, which passes over two 4 ft x 8 ft, 12 degree safety screens, is pumped to two of the existing 125 ft diameter CCD thickeners prior to disposal in the tailings impoundment.

Now that the downstream constraint of the CCD circuit has been removed, maximum tonnage throughput is our objective. Due to the variability of the hardness of the ore being delivered to Mill No. 1, this is being accomplished by a coarser grind in the oxide circuit.

Chapter 32

FUNDAMENTAL ASPECTS OF GOLD CYANIDE ADSORPTION ON ACTIVATED CARBON

W.G. Jones, C. Klauber and H.G. Linge

CSIRO Division of Mineral Products
Bentley, Western Australia 6102

ABSTRACT

The carbon-in-pulp process for gold processing depends on the adsorption of dissolved $Au(CN)_2^-$ from solution onto activated carbon. Our work shows that coconut and peat based carbons have the same mechanism and kinetics of adsorption. There is no contribution to the bonding from carbon-oxygen functionalities on the carbon surface, as has been invariably supposed in all previous theories for the mechanism of the reaction. It is shown that $Au(CN)_2^-$ is reversibly adsorbed intact on, and parallel to, the graphitic planes of the carbon. The physiochemical basis of the measurement of carbon reactivity for gold adsorption is discussed.

INTRODUCTION

The Carbon-in-Pulp (CIP) gold processing technology has enabled a rapid, world-wide growth in gold production. As a result, gold is one of the few winners for the mining industry in the 1980's. The key step in the CIP technology is the adsorption of the gold(I) dicyanide ion, $Au(CN)_2^-$, onto coarse carbon granules. This adsorption mechanism has been studied since at least 1909 and a number of quite different theories for this step have evolved (Bansal et al., Chapter 6.8, 1988) supposing one of the following essential steps in the process: 1) reduction of $Au(CN)_2^-$ to Au, 2) adsorption of neutral metal cation-gold(I) dicyanide anion pairs, 3) decomposition of $Au(CN)_2^-$ to AuCN, 4) adsorption of $Au(CN)_2^-$ onto charged carbon-sites with partial reduction of $Au(CN)_2^-$ to cluster-type species; or 5) involving a redox-mechanism in which specific sites on the carbon are oxidized to form a positive site for $Au(CN)_2^-$ adsorption with a concomitant reduction of quinone sites on the carbon surface (Hughes et al., 1984).

These gold adsorption theories invariably assume that carbon-oxygen surface structures play a direct role in the mechanism. Oxygen is located on the edges of the graphitic planes located randomly on the pore-surfaces of activated carbon and forms specific C–O functionalities such as quinone/hydroquinone, carboxylic acid, chromene etc., although there has not been full characterization of these structures to date.

Conclusive proof of each proposed mechanism has hitherto been lacking but one interesting result is the demonstrated effect of the presence of oxygen on carbon on the gold loading of Norit R2020, which is a peat-based extruded CIP carbon. This result was interpreted as an oxidative reaction in which AuCN was formed, but the applicability of this mechanism to coconut shell carbon was unclear (Tsuchida and Muir, 1986). Similarly, there is no consensus on the factors which determine the carbon-gold adsorption reactivity in a CIP plant and the selection of carbons for use in CIP plants has no broad scientific basis.

We have therefore studied the fundamental aspects of gold adsorption on activated carbon. In this paper we describe $Au(CN)_2^-$ adsorption/desorption in alkaline solution. Direct X-ray photoelectron spectroscopy (XPS) was used to confirm the state of adsorbed gold on typical coconut shell and extruded peat-based CIP carbon and we discuss the physiochemical basis of the measurement of carbon reactivity for gold adsorption.

EXPERIMENTAL

The procedure for gold adsorption, desorption and kinetic experiments was the same as previously described (Jones et al., 1988).

Norit R2520 and RO3515 are extruded peat-based carbons and samples were ultrasonically water washed of fines to obtain reproducible results for the specific gold loading rate. Calgon GRC22 is a granular coconut shell carbon and was prewashed in 1 M HF - 1 M HCl, followed by extensive ultrasonic water washing and storage in vacuo; but for this carbon, these preconditioning steps had no influence on the specific gold adsorption rate.

XP spectra were obtained using a VG ESCALAB Mk II with MgK_α radiation. The state of adsorbed gold cyanide was inferred from the measured Au 4f and N 1s photoelectron peaks recorded and calibrated as described elsewhere (Klauber, 1988).

RESULTS AND DISCUSSION

Solution Adsorption/Desorption Studies

In a 24 hr. interval, adsorption capacity measurements correspond to an adsorption equilibrium with carbon surfaces accessible via macropores. A key observation is that $Au(CN)_2^-$ adsorption on carbon is thermodynamically reversible in the absence of CN^- ions in solution, as demonstrated over 7 orders of magnitude for $[Au(CN)_2^-]$ (Jones et al., 1988). The same pattern was found when gold adsorbed at 30°C was eluted from the carbon by soxhlett extraction at 95°C. In this test, the carbon was presoaked in cyanide-free alkaline solution and soxhlett extracted over 24 hr. At 95°C, gold desorbs because the adsorption step is exothermic. Subsequently, the test was repeated after presoaking the carbons in 5% NaCN/10% NaOH. The elution efficiency is shown in Table 1 and proves that cyanide ion has no effect on the gold recovery from peat or coconut derived carbon (the residual gold on the carbons in these tests at 95°C would represent the equilibrium isotherm value).

Table 1
% Gold elution efficiency after soxhlett extraction at 95°C from carbons loaded to 20 wt% after presoaking in basic electrolytes.

Electrolyte	Calgon GRC22	Norit R2520
Cyanide-free basic solution*	87	87
5% NaCN/10% NaOH	83	88

*either pH 10, ionic strength 0.1M borate buffer or 10% NaOH.

The absence of any effect of CN^- on the reversibility of adsorption/desorption and on the high elution yields is compelling evidence that no Au-CN bond is broken in the adsorption step. If such a bond was broken, for example, by the proposed reaction on the carbon surface (Tsuchida and Muir, 1986):

$$Au(CN)_2^- \rightarrow AuCN + CN^-$$

then surface CN^- would either desorb from the carbon or be destroyed by surface oxidation, ultimately to NH_3 and CO_3^{2-}. Therefore, since AuCN is insoluble in cyanide-free solution (Anon, 1976) reversible desorption/elution would only be possible in cyanide-containing electrolytes. Similar arguments rule out the supposed presence of Au on the carbon surface. Our experimental facts therefore demonstrate that $Au(CN)_2^-$ does not undergo any change in structure in the adsorption step.

Gold adsorption on activated carbon appears to be ultimately limited by the solution solubility of $KAu(CN)_2$ rather than the supply of active carbon sites. Gold loadings of several mmol/g can be readily achieved (Jones et al., 1988) with an extrapolated value of 3 mmol/g (60 wt%) for coconut carbon Calgon GRC22 in 1M $KAu(CN)_2$ solution (30°C, pH 10). These high values are at odds with the theoretical requirement (Bansal, et al. Chapter 6.8, 1988) that the C–O functionalities on the carbon surface are involved in the adsorption mechanism. The theories all require that the C–O site has an equimolar participation with each mole of adsorbed gold and that it is capable of interaction with H^+ and OH^-. The reported total C–O concentrations in typical CIP carbons measured by pH-titration only give total C–O concentrations in the range 0.05-1 mmol/g (Parentich and Kinsella, 1984). These values are clearly too low not only for the present gold adsorption results but also for the value of ~0.3 mmol/g reported by Tsuchida and Muir (1986) for as-received peat carbon Norit R2020.

XPS Studies

Because much higher carbon-site concentrations are required than available from C–O functionalities on CIP carbons suggests that the graphitic planes, which make up the bulk of the carbon surface, are involved in the gold adsorption mechanism. XPS was used to further investigate this mechanism and Figure 1 shows the observed N 1s and Au 4f peaks for gold adsorbed on coconut and peat-based carbons. These spectra demonstrate conclusively that the state of gold is the same for each type of carbon, in agreement with the solution adsorption/desorption studies described above. The stoichiometry of the adsorbed species was confirmed as Au:CN = 1:2 using the relative areas of N 1s and Au 4f peaks. The spectra show that coconut and peat derived carbons have the same surface geometry for the adsorbed gold species.

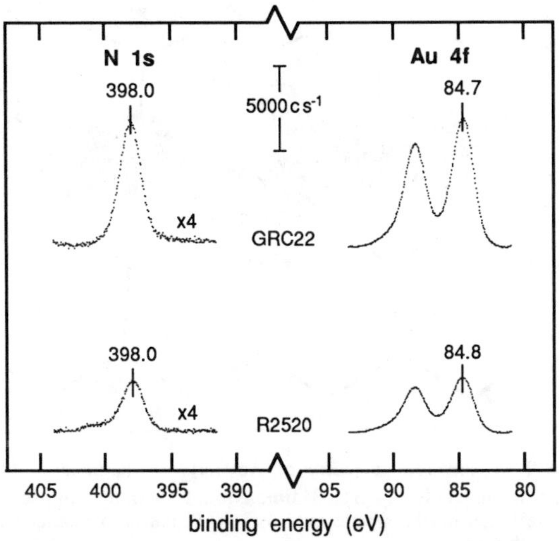

Figure 1. N 1s and Au 4f photoelectron peaks for Calgon GRC22 (gold content 40 wt%) and Norit R2520 (gold content 20 wt%) for gold adsorption from cyanide-free pH10 borate/KCl buffer at 30°C.

The N 1s state is sensitive to the local site environment on the carbon as previously demonstrated when the effect of H^+ on the structure of the adsorbed gold species on carbon was explored (Klauber, 1988). The fact that the N 1s spectrum consists of a single, sharp peak with the same electron binding energy for both carbon types shows that the terminal N atoms in the linear ionic structure $(N{\equiv}C{-}Au{-}C{\equiv}N)^-$ experience the same chemical environment on the surface of each carbon. This confirms directly that the geometry of the adsorbed $Au(CN)_2^-$ species therefore cannot involve C–O centres, because this would result in an asymmetric surface binding state for N. The only alternative adsorption geometry consistent with the symmetry of the N state found for both carbon-types is an interaction between the central gold atom and the graphitic structure, as shown in Figure 2. The geometry of the graphitic plane provides an excellent match to the structure of the $Au(CN)_2^-$ ion and there is a weak carbon to gold π-donor bond which holds the ion onto the carbon surface. The cations in the system (Na^+, K^+) coadsorb non-specifically at the carbon-solution interface and do not form identifiable ion pairs $[M^{n+}][Au(CN)_2^-]_n$ on the carbon surface as suggested by McDougall (1988), since this geometry would also remove the observed symmetry of the N interaction.

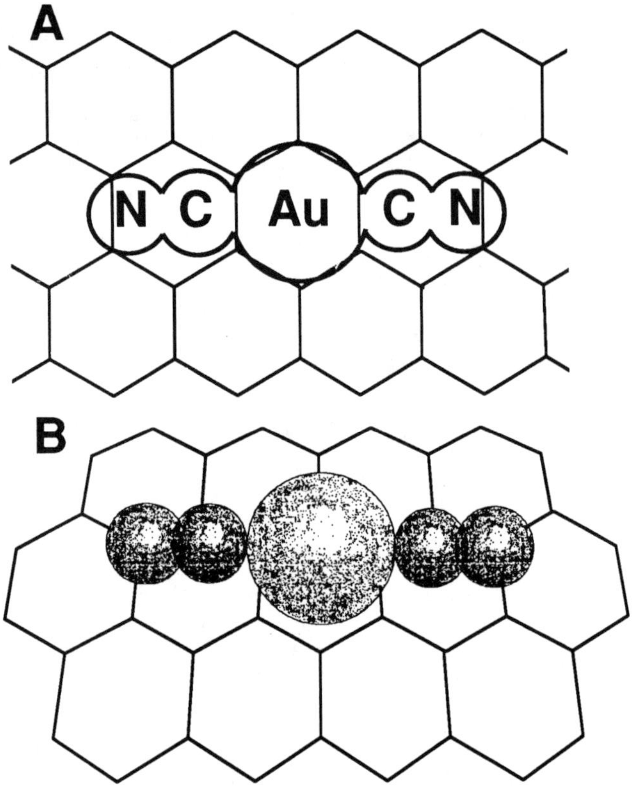

Figure 2. Adsorption geometry of $Au(CN)_2^-$ on a graphitic plane: A) Geometric match; B) 3-dimensional artistic impression. Sizes are based on the atomic radii for Au, C and N in relation to the graphitic ring size.

The present work on the adsorption mechanism of gold on both coconut and peat-based carbon identified that the most relevant characteristic of activated carbon for gold adsorption in a CIP operation is a high concentration of accessible graphitic planes. The previously reported 50% reduction in the gold adsorption capacity of deoxygenated peat-based carbon (Tsuchida and Muir, 1986) is evidently due to a decreased wetting of the carbon pores since deoxygenation increases the hydrophobicity of the carbon surface. C–O functionalities are not involved in the adsorption mechanism of $Au(CN)_2^-$ and there is no intrinsic degeneration of the carbon surface as a result of gold adsorption on the carbon surface. Thus, there is no fundamental reason why carbon has to be regenerated in a clean CIP plant, as has been supposed (Muir, 1987).

Carbon Gold Loading Reactivity Measurement

Experimentally, carbon gold loading reactivity is best obtained from a measurement of the initial reaction rate in a baffled batch reactor. The solution should be stirred rapidly to eliminate rate-control by mass-transport of $Au(CN)_2^-$ to the external surface of the highly reactive activated CIP carbons. In this work, stirring of at least 2000 rpm was required to eliminate diffusion-control in the solution film surrounding the carbon particle. For these conditions, the specific carbon reactivity can be obtained from a first-order expression describing the heterogeneous kinetics of the surface reaction:

$$R = k[Au(CN)_2^-] \quad \ldots(1)$$

where R = gold adsorption rate (mol $g^{-1}s^{-1}$) and k (L $g^{-1}s^{-1}$) a measurement of the carbon gold loading reactivity. This relationship is demonstrated in Figure 3 showing the initial rate of gold adsorption over at least two orders of magnitude of $[Au(CN)_2^-]$ for Calgon GRC22.

Equation (1) also holds for batch adsorption on fresh Calgon and Norit carbon for a carbon concentration ~1 gL^{-1} from CIP-grade gold solutions (e.g. 5 mgL^{-1} concentration) and at least 80% of the gold can be stripped from the solution in such a test without any deviation of the data from equation (1). Lower gold solution concentrations result if the batch-reaction is followed beyond about 1 hr., but for this condition the apparent carbon reactivity decreases continuously because the equilibrium position in the more dilute solution is rapidly approached at such reaction times. To treat this variation requires an equation of the form (Jones et al., 1988):

$$k = k_0[\Gamma_{(eq)} - \Gamma] \quad \ldots(2)$$

where Γ (mol g^{-1}) is the gold loading on the carbon and $\Gamma_{(eq)}$ the isotherm value ultimately attained in the experiment. k_0 is a true kinetic measurement of the carbon gold adsorption. However, k_0 is reduced by about one magnitude in the presence of pulp particles (Jones et al., 1988). Although the mechanism of this effect has been identified (Jones and Linge, 1987) there is no available mathematical relationship for the effect and prediction of gold loading rates in a CIP plant requires experimental measurements in the presence of pulp in the reactor.

Previously, Davidson et al. (1982) also recommended that carbon reactivity be tested in a baffled reactor but the authors did not stress that the solution stirring rate dependence of gold adsorption needs to be eliminated from the measurement to ensure that solution mass transport did not control the rate. The authors proposed evaluation of the carbon reactivity from a linear plot of t/Γ versus t (t=time). The full fundamental basis of this equation is unknown and in this work plots of data in this form have always produced curves.

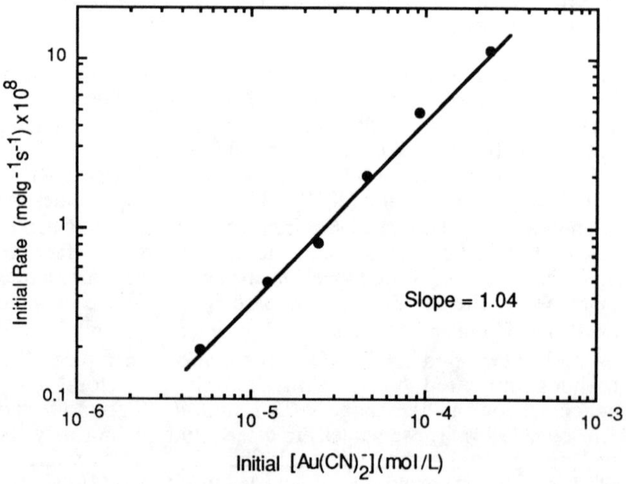

Figure 3. Initial rate of gold adsorption on Calgon GRC22 versus initial gold solution concentration in a baffled reactor for solution stirring of 2000 rpm at 30°C in pH10 borate/KCl buffer. (Note conversion factors: 10^{-11} $molg^{-1}s^{-1}$ = 7 $gt^{-1}hr^{-1}$; 10^{-5} $molL^{-1}$ = 2 mgL^{-1})

The Mintek-equation involves a plot of log Γ versus log t and was initially proposed for the analysis of the adsorption rate when the gold solution concentration was constant in the test; but again, these plots curve unless a restricted time interval is chosen for the analysis (Jones et al., 1988).

In Australia, the Murdoch equation (Muir, 1987) has been vigorously promoted for assessing the gold loading reactivity of carbons (Muir 1987, Avraamides 1988a). The Murdoch equation has the same form as the Mintek equation but the gold solution concentration is treated as a variable. Thus, for batch adsorption tests, plots of $\log(\Gamma/[Au(CN)_2^-])$ versus log t are tested but we have not been able to produce linear plots from our data using this test (Jones et al., 1988).

The Murdoch equation has usually been used for analysing gold adsorption on CIP carbons in rolling bottles (La Brooy and Bax, 1985; Avraamides, 1988a). It has been shown (Hughes and Linge, 1988) that rolling bottles do not provide sufficiently rapid solution mass transport of $Au(CN)_2^-$ to CIP carbon and that the measurement is therefore unable to distinguish the reactivity difference of different reactive carbons. Bottle rolling tests are thus unsuitable to evaluate the reactivation efficiency of kilns, as has been presumed (Avraamides, 1988b). For mass transport controlled gold adsorption tests, the main response difference of different carbons is the specific external carbon surface area, and in the bottle rolling tests the reactivity response of different carbons therefore varies inversely with the carbon particle diameter, as has been observed (Muir, figure 5, 1987).

The Murdoch equation returns apparent first-order rate constants which do not fit the measured half-life (ln 2/k) of the reaction. For example, for bottle rolled Calgon GRC22 the observed half life in the test is about 1.4 hr. whereas the Murdoch equation rate constant is quoted as 1100 $hr.^{-1}$ (La Brooy and Bax, figure 2, 1985) and this inconsistency is confusing.

Users of the Murdoch equation fail to recognize that in a batch test gold adsorption stops when the equilibrium isotherm conditions prevail and this equation is thus fundamentally unsound as a basis of a carbon gold adsorption reactivity assessment.

ACKNOWLEDGEMENT

We thank C. Langan for helping prepare this manuscript.

REFERENCES

Anon, 1976, Entry 4358, The Merck Index, M. Windholz editor, Merck and Co. Inc. Rahway N.J. USA. 9th edition, p 586.

Avraamides, J., 1988a. Selection and testing of activated carbon for gold recovery, Randol Perth International Gold Conference, pp. 261-267.

Avraamides, J., 1988b. Principles and practice of thermal carbon regeneration, Randol Perth International Gold Conference, pp 294-296.

Bansal, R.C., Donnet, J.B. and Stoeckli, F., 1988. Active Carbon, (Marcel Dekker, Inc. New York.

Davidson, R.J., Douglas W.D. and Tumilty, J.A., 1982. The selection of granular activated carbon for use in a carbon-in-pulp operation, In: Carbon in Pulp Technology for the Extraction of Gold, Aus.I.M.M. Parkville, Vic. Australia, pp 199-218.

Hughes H.C., Muir D.M., Tsuchida N., and Dalton R., 1984. Oxidation reduction potential of activated carbon during anion loading, In: Gold Mining, Metallurgy and Geology, Aus.I.M.M. Regional Conference, Kalgoorlie, Paper 7, Metallurgy Section.

Hughes, H.C. and Linge H.G., 1988. The kinetics of gold loading form gold(III) chloride solution onto fresh activated coconut carbon, Hydrometallurgy (in press.).

Jones, W.G., and Linge, H.G., 1987. Effect of ore pulp on the adsorption rate of gold cyanide on activated carbon, Aus.I.M.M. Regional Conference, Kalgoorlie and Hydrometallurgy (in press).

Jones, W.G.., Klauber, C. and Linge, 1988. The adsorption of $Au(CN)_2^-$ onto activated carbon, Randol Perth International Gold Conference, pp 243-248.

Klauber, C., 1988. Acid induced oligomerization of aurocyanide adsorbed on carbon, Surface Science, 203:118-1298.

La Brooy, S.R., and Bax, A.R., 1985. The fouling of activated carbon by organic reagents, In: Innovation in the Process and Resource Industry, Proc 13th Aust. Chemical Engineering Conference, Perth, E.A. Books, St Leonards, NSW, pp 187 -191.

McDougall, G.J., 1988. The mechanism of gold cyanide onto activated carbon, Randol Perth International Gold Conference, pp 249-252.

Muir, D.M., 1987. Recent developments in gold metallurgy from an Australian perspective, Bull Proc. Aus.I.M.M., 292:87-97.

Parentich, A. and Kinsella, B., 1984. Titrimetric studies on activated carbon, In: Gold Mining, Metallurgy and Geology, Aus.I.M.M. Regional Conference, Kalgoorlie, Paper 6, Metallurgy section.

Tsuchida, N. and Muir, D.M., 1986. Studies on role of oxygen in the adsorption of $Au(CN)_2^-$ and $Ag(CN)_2^-$ onto activated carbon, Metallurgical Transactions B, 17B:529-533.

Chapter 33

THE PULP RHEOLOGY OF SOME AUSTRALIAN GOLD ORES

Janine L. Beazley and James H. Kyle

Western Australian School of Mines
Kalgoorlie, Western Australia.

ABSTRACT

The effects of pH, pH modifier (lime or caustic soda) and cyanide on the pulp rheology of a number of ore types ranging from highly siliceous to highly kaolinitic has been investigated. All experiments were performed in fresh water at solids concentrations of 25 and 45% solids and in the pH range 6 to 11. The kaolinitic ores exhibited greater changes in rheology with pH and calcium ion concentration than did the siliceous ores.

INTRODUCTION

In recent years there has been increasing interest in the subject of pulp rheology. Viscous pulps can cause mixing problems in the Carbon-in-Pulp process for the extraction of gold which, in turn, may lead to low recoveries. This is because good mixing is essential for the effective leaching and loading of gold onto activated carbon. Viscous pulps can seriously hinder this process. In addition, low recoveries caused by pumping, grinding and screening difficulties have been attributed to pulp viscosity (Snow et al., 1987; Osan et al., 1988).

In general terms the viscosity of a slurry depends on the following properties of the ore:

(i) Mineralogy
(ii) Particle size
(iii) Solids concentration in the pulp
(iv) Chemical environment (e.g. pH, ionic strength)

This paper discusses the rheology of several different gold ore pulps and compares the effects that the use of caustic soda, lime and cyanide have upon the pulp viscosity as a function of both pH and solids concentration.

Rheology

The flow behaviour of a fluid is governed by its viscosity. Fluid flow properties may be classified as Newtonian or Non-Newtonian. For a Newtonian fluid the shear stress (force/unit area) is directly proportional to the rate of shear (velocity gradient), the proportionality constant being the viscosity:

$$s = u.dv/dy \quad \ldots\ldots\ldots \quad (1)$$

where s = shear stress (Pa)
u = viscosity (Pa.s)
dv/dy = rate of shear (s^{-1})

However, for most mineral slurries the relationship between shear stress and shear rate is more complex. Fluids of this type are classified as Non-Newtonian.

A variety of equations and models have been employed to describe flow behaviour of non-Newtonian fluids. These include the Mooney equation (Dinsdale & Moore, 1962) and the power law (Wilkinson, 1960). Although these models are useful for illustrating flow behaviour, they are not usually used in describing plant performance. The simpler term "apparent viscosity" (u_a) defined by equation (2) has found wider use (Osan et al., 1988).

$$u_a = s/(dv/dy) \quad \ldots\ldots\ldots \quad (2)$$

Non-Newtonian fluids can be divided into two groups: time-dependent and time-independent.

There are three main types of time-independent fluids - pseudoplastic, dilatant and Bingham plastic. The general relationship between shear stress and shear rate for these fluids is shown in Figure 1.

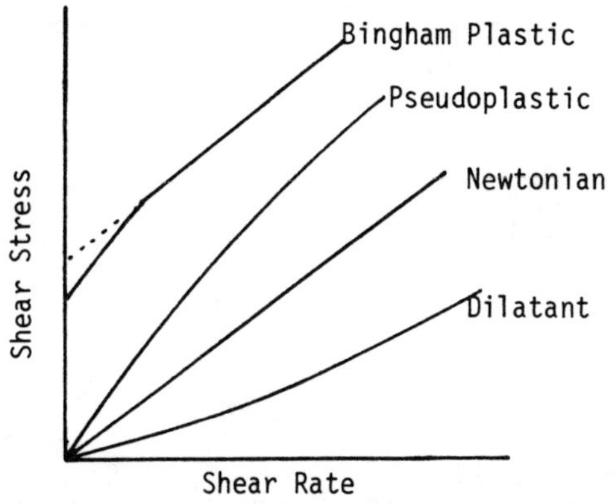

FIGURE 1. Types of Time-independent Fluid Behaviour

Many Non-Newtonian fluids obey a Power Law (equation 3) over certain ranges of shear rate:

$$s = k.(dv/dy)^n \quad . . . . (3)$$

with the apparent viscosity u_a (as defined in equation 2) being given by:

$$u_a = k.(dv/dy)^{n-1} \quad . . . . . . (4)$$

For such a fluid, a plot of log s versus log (dv/dy) will give a straight line of slope n. For pseudoplastic fluids $n<1$, and for dilatant fluids $n>1$.

For a pseudoplastic, the apparent viscosity decreases with increasing shear rate whereas for a dilatant fluid the opposite is true. Bingham plastics, however, require the application of a certain minimum shear stress before any flow will occur. This minimum shear stress is called the yield stress (Y).

For time-dependent non-Newtonian fluids, two types of behaviour are evident - thixotropic and rheopectic. In the former, the apparent viscosity decreases with increasing time of shear at a constant shear rate. This type of behaviour may be due to a breakdown in the structure of the fluid with time or with increasing rate of shear. In rheopectic behaviour, a gradual formation of structure in the fluid is established with time. This leads to an increase in shear stress at a constant shear rate.

In this paper we will present the results of a rheological study of four different ore types. All ore slurries exhibited non-Newtonian behaviour and some required a yield stress for flow to occur. These latter ores are described as "yield pseudoplastic" (Osan et al., 1988).

EXPERIMENTAL

Pretreatment of Ores

The ores examined were from Central Norseman Gold Mines (siliceous), Kidston Gold Mine (siliceous/muscovitic) and Paddington and Hannan's South Gold Mines (kaolinitic). The chemical and crystal composition of these ores as determined by X-ray Fluorescence and Diffraction are shown in Tables 1 and 2.

For the testwork, a bulk sample of each ore was crushed to <2mm, riffle split into 1 kg samples, and then ground in a rod mill to D_{80} < 75um immediately prior to pulp viscosity measurement. To ensure against changing their rheological characteristics, the pulps were not allowed to age and the kaolinitic ores were not dried prior to grinding.

Table No. 1.

MAJOR ORE ELEMENTS AS DETERMINED BY XRF.

ORE	Fe_2O_3	SiO_2	Al_2O_3	CaO	MgO	LOI
Norseman	5.94	74.2	5.01	4.27	4.92	3.30
Kidston	5.72	68.7	13.90	0.28	0.75	4.15
Paddington	11.67	58.6	13.48	0.67	2.76	8.68
Hannan Sth	16.80	51.1	16.27	1.27	0.69	9.31

Table No. 2.

MAJOR ORE MINERALS AS DETERMINED BY XRD.

Mineral	Norseman	Kidston	Paddington	Hannan Sth
Quartz	53	67	35	29
Goethite	-	-	20	18
Kaolinite	-	-	29	36
Amphibole	27	-	-	-
Muscovite	-	17	11	-
Chlorite	14	5	-	-
Pyrite	-	6	-	-
Anorthite	-	5	-	-
Biotite	2	3*	-	3
Microcline	-	-	-	4
Illite	-	-	-	3*

*Identification uncertain.

Viscosity Measurement

The following conditions were employed for all pulp viscosity measurements:

(i) Percent solids: 25 and 45%
(ii) pH: natural, 9.0 and 11.0.
(iii) pH modifier: hydrated lime ($Ca(OH)_2.2H_2O$) or caustic soda (50% solution NaOH)
(iv) Sodium cyanide (0.05% solution NaCN) at pH 11.
(v) Temperature: 25°C

Bohlin Visco 88BV Viscometer

All data were obtained using the Bohlin Visco 88BV Viscometer. The available measuring systems are summarised in Table 3. In this study, only the narrow and wide gap configurations were used employing a stationary cup and a rotating bob. The pulp was suspended using a vibromixer prior to measurements being taken.

Table 3.

MEASURING SYSTEMS FOR BOHLIN VISCO 88BV

Spindle Diameter (mm)	Cup Diameter (mm)	Height of bob (mm)	Measuring System
14	15.4	21.0	Narrow gap
25	27.5	37.5	Narrow gap
30	33.0	45.0	Narrow gap
14	33.0	21.0	Wide gap
25	33.0	37.5	Wide gap

The torque on the bob was recorded at various speeds (20 to 100 rpm) and the shear stress and shear rates calculated as follows:

$$\text{Shear stress} = T/2\pi R_b^2 h$$

where T = Torque (Nm)
R_b = radius of the bob (m)
h = height of the bob (m)

$$\text{Shear rate} = 2\pi R_c^2 w/(R_c^2 - R_b^2)$$

where R_c = radius of the cup (m)
w = angular speed (s^{-1})

The apparent viscosity was then calculated from equation (2). Thixotropic behaviour, if it occurred, was established by observing the change in torque with time at constant bob speed.

RESULTS AND DISCUSSION

Rheology of Norseman Ore

The rheograms for Norseman ore pulps at solids concentrations of 25 and 45% are shown in Figure 2. The pulp exhibits dilatant behaviour in that the apparent viscosity increases with increasing rate of shear. However, the apparent viscosities for this pulp are comparatively low, particularly at low shear rates where apparent viscosities are similar to that of water (1 centipoise, cps).

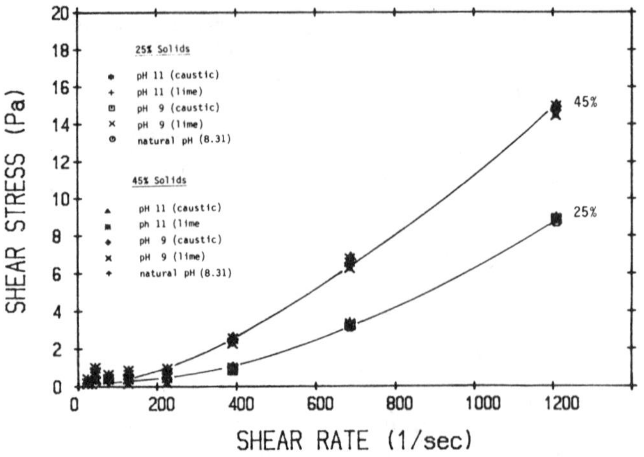

FIGURE 2. Rheograms of Norseman Ore Pulps at 25% and 45% solids.

The viscosity of the pulp at 45% solids was significantly higher than at 25%, but only at high shear rates with the difference increasing with increasing shear rate.

The effects of pH, lime, caustic soda and cyanide addition were all minimal. This behaviour is presumably due to the low clay mineral content of the ore, although the precise surface chemistry of this behaviour is unknown and is at present the subject of investigation (Meagher et al., 1988).

Kidston Ore

The rheograms for Kidston ore pulps at 25 and 45% solids are shown in Figures 3a and 3b. At 25% solids this pulp also shows dilatant behaviour, but at 45% solids the rheology is more complex. At low shear rates it behaves as a yield pseudoplastic but at higher rates of shear the rheogram shows the characteristics of a dilatant fluid.

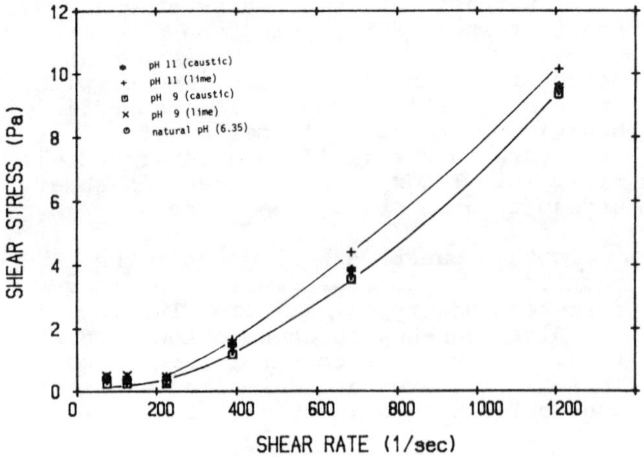

FIGURE 3a. Rheograms for Kidston Ore Pulps at 25% Solids.

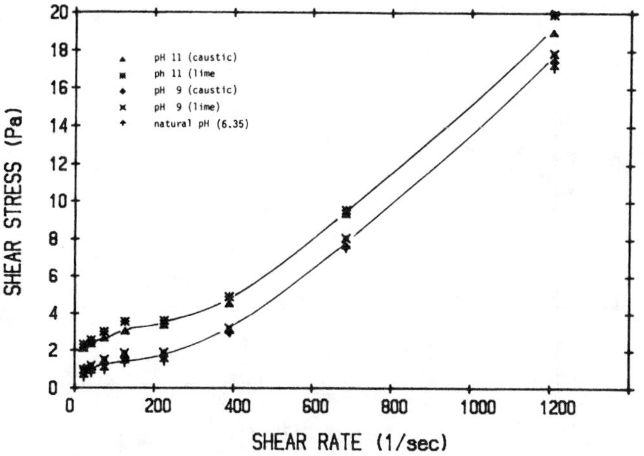

FIGURE 3b. Rheograms for Kidston Ore Pulps at 45% Solids.

This change in behaviour is demonstrated by a Power Law plot, log s vs. log dv/dy (Figure 4) which shows a distinctive change in slope at a shear rate of about 300 s^{-1}. Below this shear rate n<1 (in equation 3) which is indicative of pseudoplastic behaviour whilst above this shear rate n>1 which corresponds to dilatant behaviour.

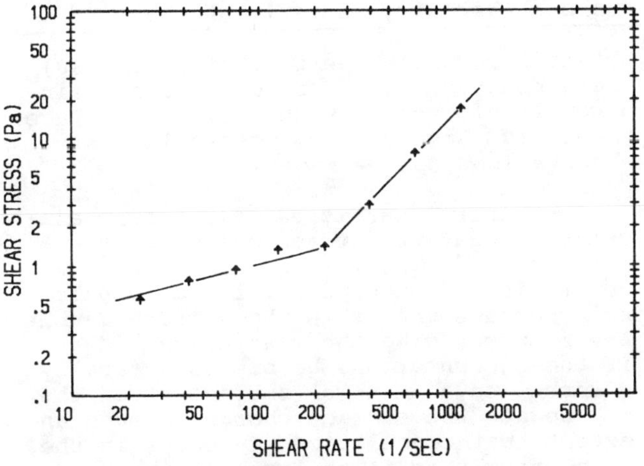

FIGURE 4. Power Law Plot for Kidston Ore Pulp at 45% Solids and Natural Ore pH (6.35).

The apparent viscosity of this ore, unlike Norseman, did increase with increasing pH. This effect was more noticeable at the higher solids concentration of 45%. There was also a measurable difference between the effect of caustic soda and lime on the pulp viscosity. Both these effects were most significant at high pH and high percent solids.

At 45% solids and pH 11, there was also a measurable yield stress which was not evident at lower pH or at 25% solids.

The differences between this ore and Norseman may be related to the presence of a significant amount of muscovite in the ore.

Paddington Ore

At 25% solids (Figure 5a), the rheograms for this ore, like both siliceous ores above, showed dilatant behaviour but with a more noticeable pH dependence. At 45% solids however, the rheology changes (Figure 5b).

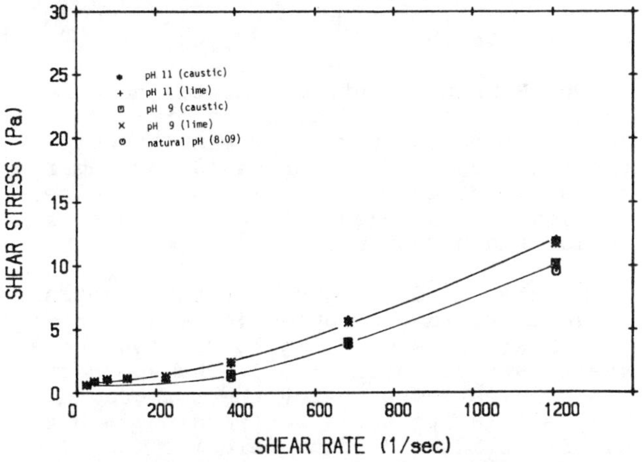

FIGURE 5a. Rheograms for Paddington Ore Pulps at 25% Solids.

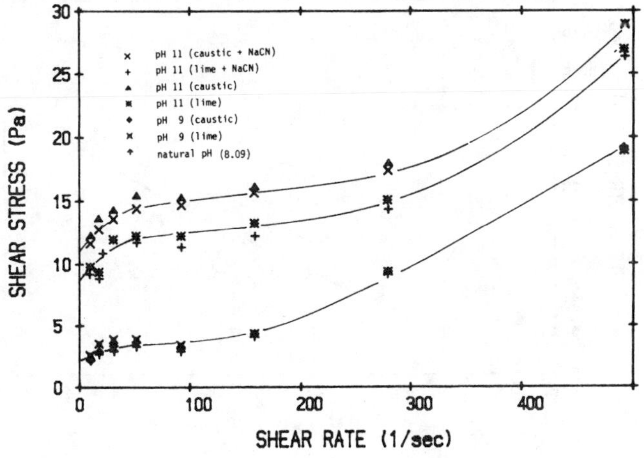

FIGURE 5b. Rheograms for Paddington Ore Pulps at 45% Solids.

At pH 9 and below, the behaviour is similar to Kidston ore - at low shear rates behaving like a yield pseudoplastic but changing to dilatant behaviour at higher shear rates. The changeover shear rate is, however, much lower - about 75 s^{-1} instead of 300 s^{-1}. The use of caustic soda produced slightly lower viscosities than lime at this pH.

At pH 11, the rheology is similar but with greatly increased yield stress and apparent viscosity. In addition, there is now a marked difference in apparent viscosity between the pulps employing lime or caustic soda as the pH modifier, with lime now producing the lower viscosity pulp, and a small but noticeable decrease caused by the addition of cyanide to the pulp.

The rheology of this ore pulp is greatly influenced by the kaolinite present in the ore.

Hannan's South

The Hannan's South ore contained more kaolinite than the Paddington ore and hence was expected to show similar pulp rheology but with higher yield and shear stresses. However, this was not the case, particularly at higher concentrations of solids and higher pH.

At 25% solids (Figure 6a) the rheograms indicate yield pseudoplastic behaviour at low shear rates changing to dilatant at a shear rate of about 75 s^{-1}. The apparent viscosities and shape of the rheogram are very similar to those for Paddington ore at 45% solids (pH 9 and below). The apparent viscosities increase slightly with pH but are independent of the modifier used.

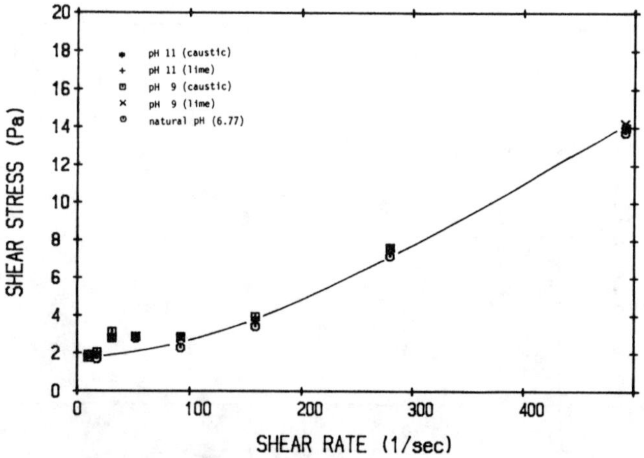

FIGURE 6a. Rheograms for Hannan's South Ore Pulps at 25% Solids.

The rheograms at 45% solids and pH 9 (Figure 6b) illustrate yield pseudoplastic behaviour. There is no evidence of dilatant behaviour at the highest measured shear rates. The yield stresses and the apparent viscosities are higher than those obtained for Paddington ore (45% solids).

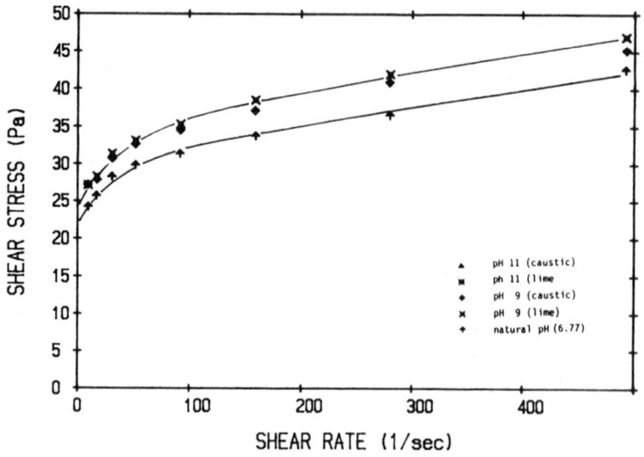

FIGURE 6b. Rheograms for Hannan's South Ore Pulps at 45% Solids.

Initial investigations of the rheograms for Hannan's South ore pulps at pH 11 indicate that the apparent viscosity is extremely high in comparison to the other ores studied in this work. In addition the apparent viscosity changed with time indicating thixotropic behaviour. No further data on this system is available as yet.

Surface Chemistry and Pulp Rheology

Until recently (Meagher et al., 1988), there has been little attempt to explain mineral pulp rheology in terms of surface chemistry. However, some general observations can be made.

The mineral particles will nearly all carry a negative surface charge at the pH's studied. The role of dissolved counterions in the pulp will be to balance this surface charge in the diffuse double-layer surrounding the mineral particles. If these counterions become specifically adsorbed to a mineral surface, then they may negate the surface charge to such an extent that coagulation may occur in the pulp, resulting in an increase in the apparent pulp viscosity.

The ion most likely to specifically adsorb on a mineral surface in the present system is Ca^{2+} or at pH >10, $Ca(OH)^{+}$. The latter ion is known to specifically adsorb onto quartz surfaces at pH 11 and above (Fuerstenau et al., 1985). It is therefore possible that specific adsorption of this ion onto clay minerals (kaolinite, muscovite) will occur to a significant extent at pH 9 and above in mineral systems. This would result in some coagulation of these minerals in the pulp resulting in a change in pulp rheology and increased apparent viscosities. This was observed for all ores containing either muscovite or kaolinite, with the one notable exception of Paddington ore, but only at 45% solids.

SUMMARY

The rheograms presented for the four gold ores studied show extremely variable pulp rheology - the only real similarity being that they all exhibited Non-Newtonian behaviour.

For the siliceous ore (Norseman), the pulps generally exhibited dilatant behaviour with minimal effect of pH, lime, caustic or cyanide addition on apparent viscosity.

The presence of muscovite in the Kidston ore pulps caused an unusual rheogram in which a change from pseudoplastic to dilatant behaviour occurred at a certain shear rate. This behaviour was also observed for the kaolinitic ores (Paddington and Hannan's South) under certain conditions. Such behaviour may be an indication of the higher shear rates causing flocculation in the pulps, although why this might occur is unknown at this stage of the investigation.

The presence of kaolinite in the ores obviously has a major effect on the apparent pulp viscosity. The highest yield stresses and shear stresses were found in the Hannan's South ore pulps, which also contained the most kaolinite. The Paddington ore , which also contained kaolinite but in lesser amounts, produced similar effects to the Hannan's South ore but showed smaller yield stresses and shear stresses.

The use of caustic soda instead of lime as a pH modifier generally lowered the apparent viscosity with one notable exception - Paddington ore pulps at 45% solids in which case the opposite was true.

The ore pulp with the highest concentration of kaolinite, Hannan's South, exhibited thixotropic behaviour at pH 11 at solids concentrations of both 25% and 45%. This phenomenon has not yet been fully investigated.

ACKNOWLEDGEMENTS

The research described in this paper was funded by ICI (Aust.) Ltd. in the form of a Postgraduate Research Scholarship to Mrs. Janine Beazley.

REFERENCES

Dinsdale, A. and Moore, F., 1962, "Viscosity and it's Measurement," Chapman and Hall, London.

Fuerstenau, M.C., Miller, J.D., and Kuhn, M.C., 1985, Chemistry of Flotation, SME-AIME, New York.

Meagher, L. et al., 1988, "The effect of dissolved ions on the rheology of concentrated quartz suspensions," Xth International Congress on Rheology, Ausralian Society of Rheology,Sydney, Australia.

Osan, M.A., Horsley, R.A. and Barham, S.L., 1988, "The Rheological Characteristics of Several Western Australian Gold Ore Slurries and the Effectiveness of Viscosity Modifiers," Randol Perth International Gold Conference, Perth, Western Australia, Randol International Ltd., Golden, Colorado, U.S.A.

Snow, R.J., et al., 1987, Increasing Slurry Concentrations in the Carbon-in-Pulp Process," 12th Slurry Technology Association Conference, New Orleans, U.S.A.

Wilkinson, W.L., 1960, "Non-Newtonian Fluids," Pergamon Press, London.

Chapter 34

CIP CARBONS - SELECTION, TESTING AND PLANT MONITORING

Dr. Jim Avraamides

Mines Department of Western Australia
Chief, Mineral Processing Laboratory
Chemistry Centre, WA

The properties of activated carbon which are important for gold recovery applications include activity, capacity, wear resistance, particle size distribution and bulk density. There are experimental difficulties in attempting to simulate the plant environment in the laboratory in order to measure the above properties. However, standard tests can be set up to compare the characteristics of different carbons and to assist in monitoring the performance of carbon in CIP operations. These laboratory tests are critically evaluated and data are presented on the performance of carbon from a number of CIP plants.

REVIEW OF TEST METHODS

The rapid expansion of the gold industry has seen a corresponding increase in the use of activated carbon for gold (and silver) recovery from cyanide pulps and clear heap leach solutions. At present there are two types of activated carbon used widely for gold adsorption - namely coconut shell and coal or peat based extrudates. Activated carbons derived from coconut shell char are produced by several large and many more smaller manufacturers. The quality and the price are generally directly related. However, in recent years there has been an increased awareness of the necessity to purchase the most cost effective product. In response to this some simple tests were developed in the Mines Department to measure those carbon properties which the gold industry considered most important, ie kinetic activity and resistance to attrition.

There are many tests available for measuring the properties of activated carbon - a material that has a long history of applications in water treatment, gas adsorption, solvent recovery and food purification. Although the published, standard tests for activated carbon do have some relevance they are not specific for gold recovery carbons. Most activated carbon manufacturers have their own in-house methods for testing carbons and normally provide comprehensive data sheets for their customers. Unfortunately test methods vary between manufacturers so little benefit can be gained by comparing data sheets.

Most activated carbon data sheets include most of the following measurements:

a. Particle size range.

b. Surface area - by nitrogen absorption

c. Methylene blue, Iodine, or Carbon Tetrachloride (CTC) number.

d. Abrasion resistance or impact strength.

e. Kinetic activity.

f. Gold loading capacity.

An excellent discussion on the relationships (or lack of) among most of the above parameters was reported by Faulkner, Urbanic and Ruckel, 1987. It became apparent from the many requests from the Australian gold industry for carbon evaluation that the most sought after data were kinetic activity, particle size distribution, and wear characteristics. Over the last few years in conjunction with Murdoch University and Curtin University of Technology, and with the co-operation of the major distributors of activated carbon in Australia an extensive data base has been generated on the activity and wear properties of CIP carbons.

TEST METHODS

It should be emphasized from the start that the choice of test depends on whether the result is to be used for quality control in manufacture, plant design or comparison of carbons from different sources. Whereas measuring particle size distribution is a straight forward procedure using standard mesh sieves, other test methods such as kinetic activity (or gold loading rate) and abrasion or attrition resistance vary between laboratories. In particular, there seemed to be a tendency in all carbon manufacturer's data sheets for abrasion resistance values to lie above 99%!

It was decided at the beginning of this exercise to use two or three of the most successful (and highest priced) commercial carbons as reference materials. All measurements were then made using one or more of these as a control.

This was of particular importance for the activity test, the results of which were subject to changes in solution temperature, standard gold solution preparation and other experimental variables.

Activity Test

Several different methods for measuring the rate at which gold is adsorbed onto carbon from an alkaline cyanide solution have been proposed. There are differences in equipment (eg agitated baffled vessels, bottle rolls or flask shakers), solution composition, carbon concentration carbon particle size and carbon pre-treatment. For simplicity, speed and reliability the batch kinetic test developed and adapted by workers at Mintek in South Africa (Fleming and Nicol, 1984) and Murdoch University (Labrooy and Bax, 1985) was considered to be acceptable for the following reasons.

a) No special equipment was required apart from a bottle roll and a flame atomic absorption spectrophotometer (AAS).

b) Gold concentrations can be read directly on the AAS if in the range 0-10 ppm.

c) Time of the test is 3 hours and several samples are taken for gold analysis which smooths errors arising from sampling and measurement procedures.

d) Closely sized carbon samples are used in order to reduce the effects of external surface area differences between samples.

e) A control carbon is run with each batch of test carbons and the resulting activities are expressed either as a rate constant, k, or as a relative percentage activity.

The data obtained from the activity test only provide comparative activities. Results can be presented graphically as gold adsorption curves (Figure 1) or else used to calculate a rate constant, k, from the equation below which has been reported elsewhere in the literature (Labrooy and Bax, 1985).

$$\Delta[Au]_c^t = k[Au]_s^t \, t^n \text{ which can be rewritten as}$$

$$\log \Delta[Au]_c^t/[Au]_s^t = n \log t + \log k \quad \ldots(1)$$

where:

$\Delta[Au]_c^t$ = change in gold loading on the carbon since time, t = zero (ppm).

$[Au]_s^t$ = gold concentration in solution at time, t (ppm).

n = empirical constant dependent on carbon sample and having values in the range 0.6-1.5 and for fresh carbons around 1.0.

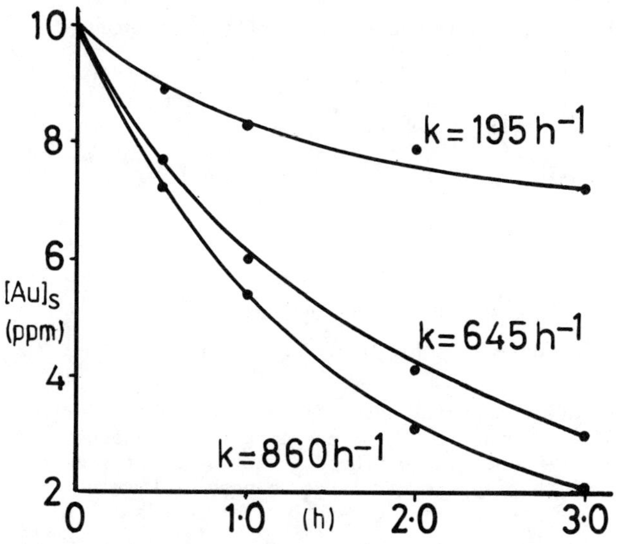

Figure 1. Activity data for coconut carbons

t = time of gold adsorption (h).

k = rate constant (h^{-1}, for n = 1)

Experimental conditions are as follows.

a) Weigh out 0.5 g screened carbon (dry weight basis)

b) Bottle roll at 100 rpm with 0.5 litres of a standard gold solution (10 ppm Au, 250 ppm NaCN, pH 10).

c) Take 5 mL samples at 0, 0.5, 1.0, 2.0, 3.0 hours and measure gold by flame AAS.

For coconut shell carbons the fraction 2.00-2.36 mm is normally taken. Extruded carbons are more regular in shape and size and are usually screened only at 1.00 mm to remove fines. High quality activated carbons have measured activity constants, k, in the range 700-1400 h^{-1} for temperatures between 15 and 35°C. The experimental variation in k values derived from the above test is about ±10% which is well within the scope of the method to rank carbon activities with confidence.

Relative activities are calculated from the gold adsorbed at time = 1 hour. This may be done graphically or can be calculated from the rate equation, which, at t = 1 hour simplifies to:

$$[Au]_s^{t=1} = 10{,}000/_{(k + 1000)} \quad \ldots (2)$$

Some values of k, solution gold concentrations and relative activities are given in Table 1. Equation 2 corrects for any variation in weight of carbon and initial concentration of gold. It is apparent from the data in Table 1 that there is a smaller incremental increase in relative activity than in k value. For the first two carbons in Table 1 the k values differ by a factor of about 7 whereas there is only a 4.5-fold increase in relative activity. Most clients prefer the

Table 1, Comparison between activity constants, k, and relative activities.

$k(h^{-1})$	$[Au]_s^{t=1}$ (ppm) Experimental	Calculated[1]	Relative Activity (%) (From Equn. 2)
114	9.1	9.0	17
817	5.7	5.5	76
1251	4.5	4.4	95
1465	4.1	4.1	100[2]

1) From k value using Equation 2.

2) Control sample arbitrarily set at 100%

relative activity data if they are seeking information on the merits of one carbon over another. At the upper end of the activity scale it can be seen that apparently large changes in k value translate into rather smaller increments in relative activity.

Assessment of fresh activated carbons is usually carried out using a distilled or deionized water solution of gold cyanide. Measurement of the activity of activated carbons from operating gold mines is preferably performed in process water. It has been established that carbons exposed to highly saline process waters exhibit diminished activities in low ionic strength test solutions. Consequently, requests are made to the client to provide sufficient process water if salinity levels are high (ie above 10,000 ppm). This is particularly significant in Western Australia where many CIP and heap leach operations are using ground waters containing up to 200,000 ppm total dissolved solids.

Wear Tests

Activated carbon is lost from CIP circuits through wet attrition in the abrasive agitated pulp environment of the adsorption circuit, during transport and screening operations and by oxidation or burning in the regeneration kiln. There will also be some losses through dry abrasion in the kiln. Several accelerated tests were devised or adapted from information available in the literature. All the tests involve subjecting the sized carbon sample to a wear regime and then measuring the proportion of undersize and oversize after screening at 1.00 mm. Carbons are always conditioned for a short time to remove imperfections and weak parts of the particles. This conditioning step often indicates how well the manufacturer has prepared the carbon for delivery. Some gold mining operaters condition all new carbon before it is sent to the adsorption circuit.

In the dry abrasion test a 50 g charge of dried sized carbon is pre-abraded for 10 minutes on a 1.00 mm standard brass sieve of 200 mm diameter using a 20 mm stainless steel ball and a "Ro-tap" sieve shaker. The +1.00 mm material (30 g) is then subjected to the same conditions for a further 60 minutes. The percentage ratio of oversize to total charge is the abrasion resistance [Equation (3)].

Table 2, Dry abrasion resistance for various activated carbons.

Sample	Particle Size (mm)	Dry abrasion resistance (percent)
A) Coconut shell	2.80-3.36	94.7
A) Coconut shell	2.00-2.36	93.7
A) Coconut shell	1.70-2.00	92.7
B) Coconut shell	2.00-2.36	85.0
C) Coconut shell	2.00-2.36	90.1
D) Extruded	+1.00 mm	96.5
E) Extruded	+1.00 mm	93.7

Abrasion Resistance (in percent)

$$= \frac{+1.00 \text{ mm oversize}}{\text{oversize + undersize}} \times 100 \qquad \ldots(3)$$

Abrasion resistances are related to particle size as shown by the data in Table 2.

Two wet attrition tests were devised - one using a water slurry of carbon and the second using an ore pulp. The equipment employed was a Wemco Attritioner comprising a baffled cylindrical vessel 160 mm diameter and 305 mm high and an electrically driven shaft at the bottom of which were attached 3 x 4 - bladed propellors, 34 mm apart, and covering an arc 100 mm in diameter, The carbon charge (30 g) is first conditioned at 1000 rpm for 5 minutes, screened at 1.00 mm and dried. The dry oversize (20 g) is returned to the baffled vessel and attritioned for a further 20 minutes. This procedure can be continued if necessary. Attrition resistance is calculated using Equation (3). Wet attrition tests using only water show an inverse relationship between the attrition and the particle size (Table 3).

Table 3, Wet attrition resistances for various activated carbons.

Sample	Particle size (mm)	Wet attrition resistance (percent)
A) Coconut shell	2.80-3.36	91.0
A) Coconut shell	2.00-2.36	92.5
A) Coconut shell	1.40-1.70	93.0
B) Coconut shell	2.00-2.36	93.0
C) Coconut shell	2.00-2.36	88.5
D) Extruded	+1.00 mm	91.5
E) Extruded	+1.00 mm	95.5

Some preliminary data are available for pulp attrition tests on a limited selection of carbons. Figure 2 shows the cumulative loss of carbon (-1.00 mm material) as the test is repeated several times on the oversize remaining from the previous test. Quite significant differences in behaviour were recorded for some commonly available activated carbons.

Wear cycle times for both the wet and dry tests were chosen to give at least a 5% weight loss as fines for the premium commercial carbons.

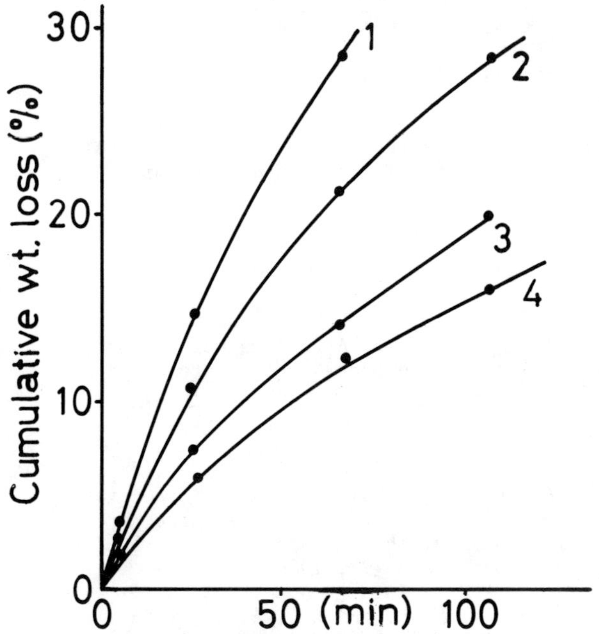

Figure 2. Attrition of carbon in 40% w/v pulp, carbon concentration 10 g/l.

This has made the tests less sensitive to operator and experimental error and gives a greater spread of results allowing for easier comparisons to be made between carbons.

Other Tests

Characterisation of carbons can also involve equilibrium loading tests, Methylene Blue,CTC or Iodine Numbers, bulk density and particle size analysis. Equilibrium loading or capacity is not considered as important as kinetic activity because in practice carbons are never loaded with gold to more than about 10 percent of their capacity. The combination of limited residence time and fouling by other pulp species limits gold loadings to 2000-10,000 g/t. Methylene Blue CTC,or Iodine Numbers are effectively measures of surface area which consequently may reflect activity. However, low activity brought about by fouling does not necessarily result in decreased values of Methylene Blue,CTC or Iodine Numbers. Surface areas based on nitrogen absorption are around 1000 m^2/g for good activated carbons but the way in which the surface is prepared is also important. Bulk density is another useful measure of carbon activity in that it indirectly relates to the extent of activation. The activation process involves treating the carbon with steam at temperatures above 800°C in order to create the pore structure and the right surface chemistry. Development of pores results in loss of carbon and hence lowers the bulk density to around 0.5 kg m^{-3}. Very low bulk densities are associated with very high activity but poor wear characteristics. Finally the particle size range of any batch of activated carbon should average around 2 mm for coconut shell type. Particles larger than 3 mm will have low activity and those below 1 mm will pass through the interstage screens and be lost into the tailings pulp.

It is important to use a combination of tests to properly assess a particular carbon. For example, a carbon submitted for testing showed a high activity, a reasonable dry abrasion resistance but a very low Carbon Tetrachloride Number. Examination of the wear test data revealed a very high pre-abrasion value. Discussion with the client confirmed that the steam activation had been carried out rapidly resulting in over-activation of the outer surface of the coconut shell char. This outer surface was of low bulk density and, consequently, was easily abraded exposing the harder core, which was largely un-activated.

PLANT MONITORING

The activated carbon used for gold recovery in a CIP plant is also an efficient adsorber for many other pulp constituents both organic and inorganic (Labrooy and Bax, 1985). Consequently the activity of carbon decreases with increased contact times in the pulp. The loaded carbon is then separated from the pulp, acid washed, eluted, regenerated and returned to the adsorption section. An activity profile of carbon through a plant is often a useful guide to overall plant performance when compared to the gold loading profile. Table 4 compares the behaviour of two plants treating oxide ore. Points to note are:

Table 4, Activity and gold loading profiles of CIP plants treating oxide ores.

Carbon sample	Plant A Activity (%)	Plant A [Au] (g/t)	Plant B Activity (%)	Plant B [Au] (g/t)
Fresh	100	-	100	-
Kiln Regen.	55	-	70	-
Adsorber 6	47	80	22	80
5	42	175	15	90
4	36	520	13	270
3	33	1500	12	535
2	33	2140	6	1460
1	33	4500	8	2330
Eluted	40	14	40	50

a) Better kiln regeneration in Plant B.

b) Maintenance of higher carbon activity through the adsorption section of Plant A.

c) Lower gold loadings on barren carbon from Plant A.

d) Different gold loading profiles. Plant B has some reserve capacity in the last adsorber (No. 6) to cope with any variations in gold grade or temporary shut down of the regeneration kiln.

e) Significant improvement in carbon activity following elution in the case of Plant B. This particular operation treats lateritic ore from a near surface deposit and the presence of naturally occurring vegetation products results in organic fouling of the carbon. These organics are probably removed by the strongly alkaline, hot strip solution.

It is often instructive to measure the levels of other species adsorbed onto carbon. Those most often present on carbons from Western Australian CIP plants include inorganics such as silver, nickel, copper, sulphate, carbonate, silica, iron, magnesium and calcium, and organics such as flotation reagents, humic acids and lubricating oils. Detection of organics by thermal analysis methods has proven to be quite successful, particularly for flotation reagents. A common contaminant on activated carbons is calcium carbonate and at levels above 5% w/w there is a profound deactivating effect. Acid washing of such carbons usually results in significant increases in activity.

CONCLUSIONS

The availability of appropriate procedures for measuring the properties of activated carbon is of great value to the metallurgist in making the initial selection of the most cost effective carbon and in the subsequent monitoring of the performance of the carbon in the plant. Those properties of carbon which are of immediate concern appear to be activity, wear resistance and particle size. An effective and simple activity test enables the plant operator to carry out measurements on-site which allows daily checking of the extent of carbon fouling and the effectiveness of the regeneration kiln. Although carbon losses by attrition can only be accurately measured through actual plant operating experience the availability of laboratory test methods is useful for ranking carbons in order of wear resistance. The final decision in carbon selection will also depend on other factors such as availability, consistency of product, moisture content, particle size distribution and last, but not least, price.

AKNOWLEDGEMENTS

The author wishes to thank Dr. S. R. Labrooy of Murdoch University for his many helpful comments and the technical staff of the Mineral Processing Laboratory for carrying out the experimental work. This paper is published by permission of the Director, Chemistry Centre WA.

REFERENCES

Faulkner, W. D., Urbanic, J. E. and Ruckel, R. W., 1987, "Activated Carbon For Precious Metals Recovery" Paper presented at 110th Annual Meeting SME and Metallurgical Society, Denver, Colorado, Feb 23-27, 1987.

Fleming, C. A. and Nicol, M. J., 1984, "The Adsorption of Gold Cyanide onto Activated Carbon. III. Factors Influencing the Rate of Loading and Equilibrium Capacity", J. S. African I.M.M., April pp 85-92, 1984.

Labrooy, S. R. and Bax, A. R., 1985. "Fouling of Activated Carbon by Organic Reagents", Proceedings of the 13th Australian Chemical Engineering Conference, Perth, Western Australia, pp 187-191, 1985.

Van Deventer, J. S. J., 1984, "Criteria for Selection of Activated Carbon Used in Carbon-in-Pulp Plants", Reagents in the Mineral Industry, (Editors Jones, M. T. and Oblatt, R.) I.M.M., London pp155-160, 1984.

Vennekens, M. J. A. and Van Vliet, B. M., 1985, "Aqueous Suspension Based Abrasion Resistance Assessment for Particulate Activated Carbons", Water S.A. 11(2), pp 49-56, April.

10

Processing III
Treatment of Refractory Ores

L.D. Hartzog
Coeur d'Alene Mines, Inc.
Coeur d'Alene, Idaho

J. Canterford
MinProc Technology Pty. Ltd.
East Melbourne, Vic., Australia

Chapter 35

REFRACTORY GOLD: THE ROLE OF PRESSURE OXIDATION

Roman M.G.S. Berezowsky
Principal Scientist

D. Robert Weir
Vice President, Technology

Sherritt Gordon Limited, Fort Saskatchewan, Alberta, Canada, T8L 2P2

Abstract. Pressure oxidation can virtually always render refractory auriferous sulphidic ores and concentrates amenable to cyanidation. Its effectiveness as a pretreatment of refractory gold feedstocks is reflected by the successful startup and operation, over the past four years, of four commercial plants. At least three more are scheduled to come on stream during the next three years. Over the past eight years, a variety of ores and concentrates have been subjected to integrated miniplant testing and process demonstration by Sherritt, and data derived from these studies have been used in the design and construction of several of the commercial facilities. Some factors affecting the selection of pressure oxidation in gold recovery processing are discussed.

INTRODUCTION

Gold ores are classified as refractory when a significant portion of the gold cannot be extracted efficiently by conventional means, even after fine grinding. A common cause of refractoriness is dissemination of fine grained or submicroscopic gold within sulphide minerals, usually arsenopyrite or pyrite. Such ores require a pretreatment to alter or destroy the sulphide matrix to render the gold amenable to recovery. Aqueous pressure oxidation can virtually always render gold bearing sulphidic materials amenable to cyanidation, and is one of the more recently developed techniques for the commercial pretreatment of refractory gold feedstocks (25).

Pressure oxidation is conducted at relatively high temperatures of 180° to 225°C, in either an acid or alkaline medium, and achieves essentially complete oxidation of the sulphides and liberation of the refractory gold within 1 to 3 h. Although the concept of pressure oxidation pretreatment is not new, dating back to the mid fifties, the renewed interest owing to high gold prices has gained considerable impetus with Homestake's decision to employ pressure oxidation at their McLaughlin property (13,17). Several reviews and publications have become available on pressure oxidation (6,7,12,24,32), and a recent study of various process options concluded that pressure oxidation is the prime choice for the treatment of refractory gold ores and concentrates (21).

During the past eight years, Sherritt has evaluated the applicability of pressure oxidation to a variety of refractory gold ores and concentrates (5,24-32). The studies have ranged from batch tests to completely integrated miniplant testing consisting of pressure oxidation, precious metals recovery, and waste control. Data derived from these studies have been used in the design and construction of Homestake's McLaughlin gold plant in California, which came on stream in 1985. The testwork also led to the design of the process flowsheets for the Sao Bento Mineracao Gold Complex in Brazil, which came on stream in late 1986 (9,10), the Olympias Gold Project in Greece, scheduled for startup in late 1990 (14), and for the Porgera Gold Project in Papua New Guinea (26,27), which is scheduled to come on stream in late 1992. All of the above are based on acid pressure oxidation technology. Two more commercial plants have started up recently, one at the Barrick Mercur Gold Mine in Utah, employing alkaline oxidation (3), and the other at the Getchell Mine in Nevada, using an acid medium (18).

OXIDATION CHEMISTRY

An account of the pressure oxidation chemistry has been given in previous publications (6,7,32). The major reactions in the elevated temperature oxidation of sulphides and arsenides in an acid medium involve total oxidation of the sulphidic sulphur to the sulphate form and of arsenic to the arsenate form.

$$4FeS_2 + 15\ O_2 + 2H_2O \longrightarrow 2Fe_2(SO_4)_3 + 2H_2SO_4$$

$$2FeAsS + 7\ O_2 + H_2SO_4 + 2H_2O \longrightarrow 2H_3AsO_4 + Fe_2(SO_4)_3$$

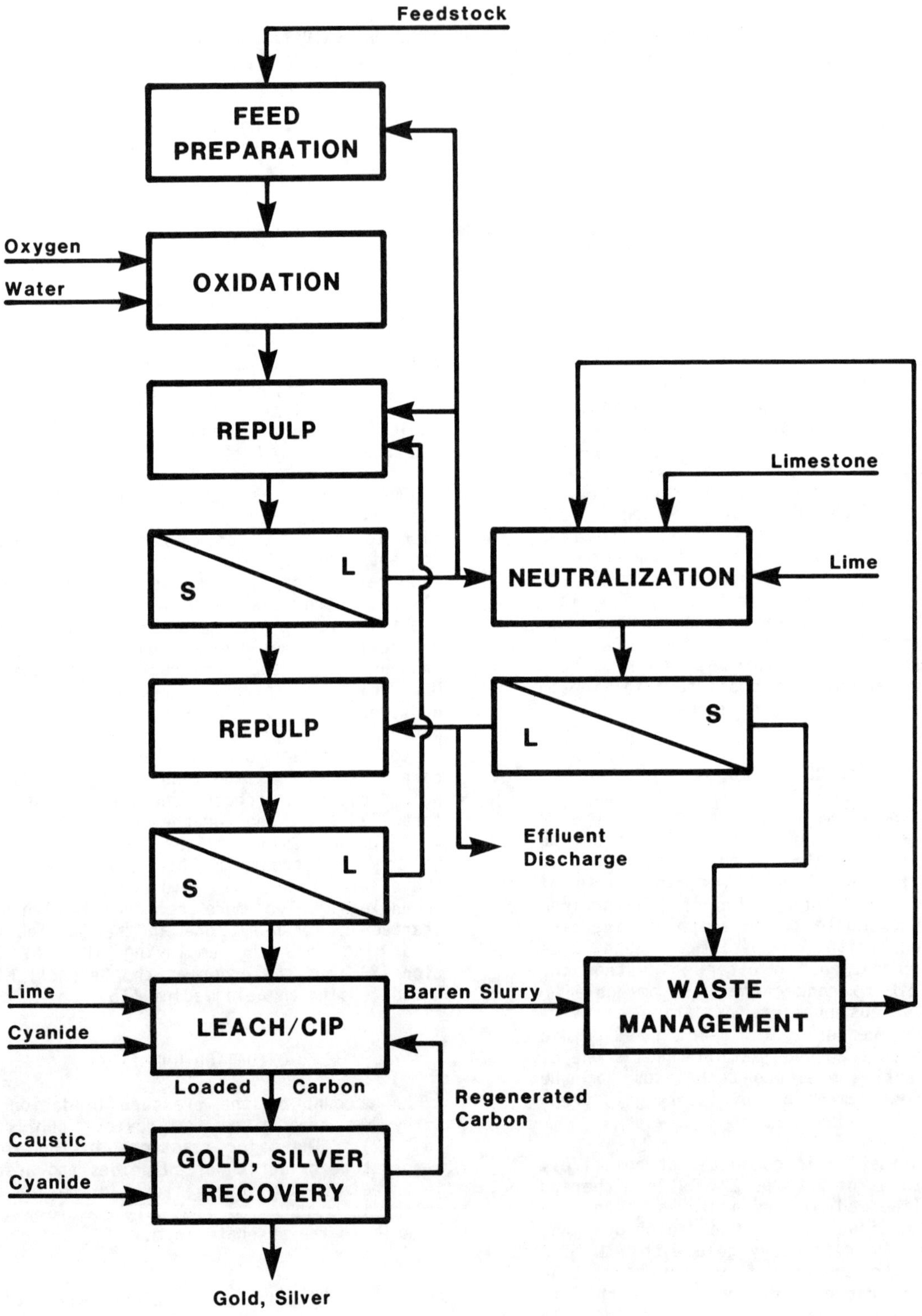

FIGURE 1. CONCEPTUAL FLOWSHEET FOR ACID PRESSURE OXIDATION PRETREATMENT OF REFRACTORY FEEDSTOCKS

$MS + 2\ O_2 \text{ ----> } MSO_4$
where M = Pb, Zn, Cu, Ni, Co, etc.

Consumption of some of the acid by gangue components, and the elevated temperature operation, promote appreciable hydrolysis and precipitation of the arsenate and ferric iron, as ferric arsenate, hematite, basic ferric sulphates, and jarosites. The major prerequisites for effective oxidation of the sulphides include sufficient acidity to maintain adequate iron levels in solution to promote the oxidation, and operation under conditions which minimize the incidence of elemental sulphur formation.

In the alkaline oxidation process, the sulphate released by the oxidation of the sulphides is neutralized, by added caustic reagents such as soda ash, or by the acid consuming gangue components such as calcite or dolomite. The discussions in this paper, however, will focus primarily on acid pressure oxidation.

FLOWSHEET DESCRIPTION

The choice of processing ores or concentrates is governed by a number of factors. For illustrative purposes, a conceptual flowsheet is given in Figure 1. Since many feedstocks may contain appreciable levels of carbonates, the feed is preconditioned with acid, preferably an acidic recycle stream from the washing of autoclave discharge slurry, to decompose the carbonates and remove as much carbon dioxide as possible prior to autoclaving. This serves to improve subsequent oxygen utilization. The pretreated material then is oxidized in the autoclave with oxygen, using the heat generated by the sulphide oxidation to sustain the reaction. Water may be added, as required, for temperature control. The autoclave discharge slurry is cooled by flashing, and is subjected to a liquid-solid separation to remove a portion of the acid, dissolved arsenic, and metal sulphates. Washing of the oxidized pulp prior to cyanidation is generally advisable. This serves to remove not only cyanicides, such as copper and zinc, released during the pressure oxidation, but also aluminum, iron and magnesium which would otherwise precipitate as slimy hydroxides, increasing the slurry viscosity, the possibility of gold losses through adsorption by the slimes during cyanidation, and fouling of activated carbon in the recovery circuit.

A portion of the acid liquor is recycled to the preconditioning stage while the remainder is treated in a neutralization circuit to neutralize the acid and precipitate the arsenic, metals and associated sulphate. Water is reclaimed for recycle to the wash circuit. In the pressure oxidation of flotation concentrate, the flotation tailings generally contain appreciable amounts of carbonates, and may often advantageously be added to the neutralization circuit to reduce limestone and lime requirements.

The washed, oxidized solids proceed to a gold recovery operation, for adjustment of the pH with lime and recovery of the gold, generally by cyanidation and precipitation, cyanidation and carbon-in-pulp (CIP) processing, or carbon-in-leach (CIL) processing. The sludge from the neutralization circuit, preferably dewatered by thickening, and the barren slurry from the gold recovery operation are combined for deposition in a tailings impoundment area. Runoff water from the tailings area normally is recycled.

APPLICABILITY OF PRESSURE OXIDATION

A favorable response of a refractory gold ore to pressure oxidation may not necessarily justify the pretreatment. The applicability of pressure oxidation depends on numerous factors including ore reserves, sulphur grade, quantity of refractory gold relative to the sulphide sulphur content, nature of the gold association with the sulphides, power costs, lime costs, etc. Variability of the ore grade and of the degree of refractoriness are also important considerations.

Treatment of Ore Versus Concentrate

Faced with a choice of whether to treat ore or concentrate by pressure oxidation, the major factors to be considered include response of the ore to flotation, the sulphur content of the ore, the content of carbonates and other acid consuming gangue components and, the nature of the autoclaved pulp. In either case, the use of indirect heat exchange for preheating the feed slurry should be avoided since the slurries are essentially gypsum saturated and prone to scaling. Judicious recovery of flash steam and direct heat exchange may enable essentially autogenous oxidation of ores containing ~3% S.

Recovery of acid containing solution from the autoclave discharge pulp in a liquid-solid separation stage prior to CCD washing may meet a portion of the acid requirements of the fresh ore. This will depend on the relative amounts of sulphur and acid consuming gangue components in the ore. Lower sulphur grades and high carbonate levels may justify the added cost of steam and sulphuric acid, depending on the grade of gold and the response of the ore to flotation. Poor response to flotation, very low sulphur grade, and relatively high carbonate levels (e.g. many of the gold ores in Nevada), may warrant consideration of pressure oxidation in an alkaline medium.

In most instances, some degree of upgrading of the sulphides by flotation is preferred. The autoclave size requirements are primarily dictated by the quantity of contained sulphur. Therefore, the processing of a concentrate rather than an ore may not necessarily have a major impact on the autoclave size. It will, however, reduce the size requirements of the other unit operations. The sulphur grade requirements for autogenous pressure oxidation are appreciably lower than those for autogenous roasting. Pressure oxidation of a lower grade concentrate than required for roasting offers the possibility of improved gold recovery, by virtue of reduced losses of gold to the flotation tailings. One of the main objectives of flotation thus becomes rejection of carbonates and acid consuming gangue to the flotation tailings, which in turn can be utilized in the neutralization circuit.

The pressure oxidation kinetics of sulphides are strongly dependent on particle size. For example, batch oxidation tests conducted on one of the Porgera concentrates showed that a reduction of particle size from 80% minus 110 µm, to 13 µm, decreased the retention time requirement for oxidation of 95% of the sulphur, from 92 min, to 23 min. Comparison of the relative capital and operating costs associated with concentrate regrinding versus reduced autoclave size and power requirements indicated an optimum particle size of about 80% minus 30 µm for this particular concentrate (26). Although a regrind of an ore will have a similar effect on the oxidation kinetics, the higher costs associated with regrinding of the ore and the possibility of deteriorated liquid-solid separation characteristics may offset some of the benefits derived in the pressure oxidation.

Another factor which may favor the processing of a concentrate rather than an ore is the variability of the ore sulphur content. Although pressure oxidation can accommodate a wide range of feed sulphur grades by means of autoclave feed pulp density adjustment, most efficient operation is achieved with a relatively constant sulphur grade. This is usually more readily achieved with concentrates than with ores.

It is evident, therefore, that the decision of whether to process ore or concentrate and the definition of optimal concentrate sulphur grade and particle size depend on a number of closely interrelated factors which must be thoroughly evaluated.

Pretreatment

Preacidification of the autoclave feed serves to enhance the subsequent oxygen utilization by removing carbonates prior to autoclaving. Another important role of the pretreatment is to ensure sufficient initial levels of acid and iron to promote a rapid initial rate of oxidation to achieve the required temperature profile in the autoclave. After pressure oxidation, there is considerable flexibility in the recovery of acid generated by the oxidation of the sulphides for meeting the requirements of the pretreatment. For most ores, a primary liquid-solid separation (oxidation discharge thickener) will be desirable prior to CCD washing to recover a relatively high strength acid stream for recycle, with higher carbonate ores requiring the addition of make up acid. In the case of concentrates, although the carbonate content is appreciably lower, some degree of preacidification almost always is warranted. The autoclaved pulp may proceed directly to a two stage CCD wash circuit, as shown in Figure 1, and a portion of the diluted acid waste liquor is recycled to the pretreatment stage.

Solids Recycle for Elemental Sulphur Control

The primary objective of the pressure oxidation is complete oxidation of the sulphidic sulphur to the sulphate form. This is achieved by operating at elevated temperatures, preferably higher than 180°C. Although under these conditions the pyritic sulphur is oxidized easily, many sulphides, such as arsenopyrite, pyrrhotite, and base metal sulphides tend to form elemental sulphur as an intermediate or primary oxidation product. Any elemental sulphur that is formed will selectively 'wet' the unreacted sulphide minerals, which can then agglomerate.

The high pulp density encountered in the oxidation of ores effectively suspends and disperses any elemental sulphur, such that with adequate retention time the sulphur will be oxidized completely. In the processing of concentrates, at low pulp densities, agglomeration due to the formation of elemental sulphur can lead to severe problems. By recycling a portion of the oxidized solids to build up the autoclave feed pulp density, the sulphide wetted particles can be effectively dispersed and excellent reactivity can be achieved. In addition, the recycle of oxidized solids in effect provides disproportionate solids retention time (29).

The process flowsheet allows for several choices for the recycle of solids. In the case of high carbonate concentrates, the preferred choice is a portion of the autoclave flash tank discharge slurry, which has a high acidity. Otherwise, a portion of the first wash thickener underflow is recycled. In either case, the slurry is recycled to the pretreatment stage to promote the evolution of carbon dioxide. Prior cooling of the recycle slurry may be advantageous in that it will provide an additional 'heat sink' for the heat of the oxidation reaction, allowing for a reduction in the cooling requirements in the autoclave.

Washing of Oxidized Pulp

The pressure oxidation can result in extensive extraction of base metals and appreciable dissolution of gangue components. The primary objective of the wash circuit is the removal of cyanicides and potential slime forming metals such as aluminum, iron and magnesium prior to gold recovery. High dilution of the pulp is employed to promote wash efficiency and flocculant utilization. This allows for the possibility of using a hydrocyclone for washing, and the recovery of coarse solids which may contain unreacted sulphides and possibly refractory gold encapsulated in gangue components. A small regrind circuit for these solids prior to recycling to the autoclave may reduce the regrind requirements for the feed concentrate and enhance the recovery of gold.

Environmental

A significant feature of the use of a process employing pressure oxidation for sulphidic feedstocks is that the handling and disposal of environmentally sensitive species is greatly facilitated. Noxious species are converted into relatively stable compounds in the pressure oxidation and neutralization operations, and are readily retained in an impoundment area.

Antimony. In the pressure oxidation, antimony compounds in the sulphidic feed are oxidized, hydrolyzed and essentially completely precipitated. Dissolution of antimony from the oxidized solids in the leach and CIP operations is virtually negligible.

Arsenic. Arsenic is initially dissolved, as arsenate, but is then precipitated as ferric arsenate or complex ferric arsenate-sulphate species. The extent of arsenic precipitation in the autoclave is dependent on a number of factors, including arsenic level in the feed, the relative amounts of iron and arsenic, solution acidity, and operating temperature, but generally between 85% and 95% of the arsenic is precipitated (8). Dissolution of the arsenic from the oxidized solids in subsequent gold recovery operations is virtually negligible. Treatment of the acidic water from the CCD wash in the neutralization circuit usually removes arsenic to <0.1 mg/L. The low levels of arsenic, compared to solubility levels of 2 to 3 mg/L As often quoted in the literature for ferric or calcium arsenates are attributed to a high Fe:As ratio (15) in the feed liquor and the large mass of precipitated hydroxide slimes which effectively adsorbs or chemisorbs the arsenic.

Base Metals. Base metals including cadmium, cobalt, copper, nickel and zinc are largely extracted during pressure oxidation and are removed from the oxidized solids in the wash circuit. Consequently, these metals are present in only trace amounts in the gold recovery circuit, resulting in lower cyanide consumption, cleaner leach solutions and lower contaminant levels on the carbon. The metals in the acid stream are precipitated to trace levels as their respective hydroxides or hydrated oxides in the neutralization circuit. Lead containing sulphides are converted to the relatively insoluble sulphate and jarosite species during the pressure oxidation, as is the bulk of the silver.

Mercury. During pressure oxidation, mercury containing species are apparently largely converted to mercuric sulphate which, under the prevailing conditions, is precipitated as a mercuric jarosite or is otherwise incorporated in the complex iron containing species. Although the pressure oxidation renders the bulk of the mercury refractory, some may still be amenable to extraction in the subsequent cyanidation, and provisions are required for control of mercury during the recovery and refining of the gold.

Sulphur. The autoclaving process oxidizes essentially all of the sulphur to the sulphate form. A portion reports to the solids as alunite, jarosites, basic ferric sulphates and calcium sulphates, while the remainder is present as acid and associated with soluble metals in the acid liquor. Treatment of the acidic solution in the neutralization circuit precipitates the sulphate as gypsum. The more labile sulphate containing species in the oxidized solids are also readily converted to gypsum during the pH adjustment prior to cyanidation. Ultimately, therefore, the sulphur in the feed stocks is discharged either as gypsum or a jarosite-gypsum mixture.

Cyanide. After precious metals recovery from the oxidized solids, the barren pulp will usually be dewatered to recover cyanide and reduce the cyanide content in the barrens. It can then be combined with dewatered sludge from the neutralization circuit for deposition in a tailings impoundment area. If it proves necessary to detoxify the cyanide containing barren slurry prior to impoundment, effective removal of cyanide can be achieved by treatment with the metals containing acid liquor from the wash circuit (27).

Potential for By-Product Recovery

During pressure oxidation, silver is initially also liberated. Under the prevailing conditions, however, which favour hydrolysis and precipitation of ferric iron, the silver becomes associated with refractory jarositic species, limiting subsequent extraction by cyanidation to less than 40%, and often to less than 10%. Effective enhancement of the recoverability of the silver can be achieved by atmospheric conditioning of the oxidized pulp for 1 to 2 h at 90°C with lime to decompose the jarositic species (11,27,32). The enhancement treatment may also result in a modest improvement in the extraction of the gold. Lime requirements are appreciably higher than for the normal ambient temperature neutralization and pH adjustment prior to cyanidation, so that the applicability of a silver enhancement treatment will depend on the grades and the relative improvements in silver and gold recoveries.

Base metals, which are effectively extracted in the pressure oxidation, may be recovered at an intermediate stage of the neutralization circuit. This offers the possibility of treatment of otherwise unattractive feeds such as gold or silver containing flotation middlings or low grade concentrates recoverable from tailings of flotation operations on cobalt, copper, nickel and zinc operations, in conjunction with the major gold feeds.

COMMERCIAL PLANTS

In the past four years, four commercial plants employing pressure oxidation for refractory gold pretreatment have come on stream. These include the McLaughlin and Getchell plants, both using acid oxidation of ore, the Sao Bento complex, using acid oxidation of concentrate, and the Barrick Mercur plant, using alkaline oxidation of ore. Three more are scheduled during the next three years. A brief account is provided here, primarily of the pressure oxidation process, at the four operating plants as well as the three projected to start up by 1993.

McLaughlin Gold Mine

The first commercial application of pressure oxidation for the pretreatment of refractory gold was at Homestake Mining Company's McLaughlin Gold Mine, where acid pressure oxidation is used to treat 2700 t/d sulphidic ore. Autoclave operations began in July 1985, and the pressure oxidation plant was fully operational in September, although design capacity was not attained until several months thereafter. Several accounts of the process selection and development, plant construction and operation are available (1,4,13,16,17,23).

The ore, ground to 80% minus 75 µm and thickened to between 40 and 45% solids at the mine, is pumped 7.5 km to the processing plant. Prior to autoclaving, the ore is reacted with recycled solution from the CCD wash circuit, containing acid produced by the oxidation of the sulphides in the autoclave, to decompose/remove the bulk of the carbonates. The acidulation is conducted in the "preoxidation" circuit, comprising several agitated stainless steel tanks in series. After thickening of the acidified pulp in a 16.8 m diameter stainless steel high rate Enviro-Clear unit, the solution is treated with lime to neutralize remaining acid and to precipitate metals. Reclaimed water is cooled and is recycled to the CCD wash circuit for the oxidized pulp. The underflow from the preoxidation thickener is sent to one of three autoclave trains.

The ore contains about 3% sulphide sulphur. A heat recovery system, augmented at times with steam injection into the autoclave, is required to maintain the desired temperature in the autoclave for efficient sulphur oxidation. In each autoclave train, the feed slurry is pumped by centrifugal pumps through two stages of direct contact splash-flash titanium heat exchangers with steam recovered from the autoclave discharge flash tanks. A Geho positive displacement diaphragm pump then pumps the slurry into the autoclave, via a third direct contact titanium heater utilizing autoclave vent gas. The slurry enters the autoclave at a temperature between 90 and 120°C, and at a pH of 1.8 to 1.9.

Each autoclave is a horizontal, four compartment, steel shell, lead and double brick lined pressure vessel with internal dimensions of 4.2 m diameter by 16.2 m long. Slurry in each compartment is agitated by 50 to 60 HP agitators with titanium shafts and ceramic blades. The temperature is staged from 160 to 180°C, utilizing the heat of reaction of the sulphides, augmented with steam injection when processing lower sulphur feeds. Oxygen, supplied from a 300 t/d oxygen plant, is sparged into the autoclave, to provide an oxygen overpressure of between 140 and 280 kPa. Slurry retention time in the autoclave is about 90 min.

The autoclave discharges to brick lined flash vessels in two stages, and the flashed steam is used to preheat fresh autoclave feed slurry. Flash tank discharge is washed and further cooled in a two-stage countercurrent decantation circuit, utilizing 16.8 m stainless Enviro-Clear units. The acidic waste water is recycled to the preoxidation section for acidifying fresh ore, while the washed pulp is neutralized to pH 10.8 with milk of lime prior to gold and silver recovery by cyanidation and carbon in pulp processing.

The commercial experience has confirmed the prerequisite of a high degree of sulphide oxidation to ensure consistently good gold recovery in cyanidation. Oxidation variables include temperature, pressure, oxygen flow, steam addition and pulp density. An emf of 450 mV in the discharge slurry has proven to be a minimum requirement for greater than 85% sulphur oxidation, and the oxidation variables are controlled so as to achieve this potential. In the four years of commercial operation, significant improvements in autoclave availability and throughput have been made at McLaughlin, and the pressure oxidation pretreatment has enabled well over 1 g/t incremental gold recovery over that attainable by direct cyanidation.

Sao Bento Mineracao

The Sao Bento Mineracao S.A. metallurgical process plant, situated in the Minas Gerais province of Brazil, was the second to utilize pressure oxidation technology for the recovery of refractory gold. The ore body comprises massive sulphide at an average in situ grade of approximately 4.5% As and 6.2% S. The sulphide minerals consist mainly of arsenopyrite, pyrrhotite and pyrite, with arsenopyrite the predominant mineral. The gold is largely associated with the sulphides as included (52%), interstitial (33%) and attached (13%) grains. Arsenopyrite is the major host for gold inclusions. Following evaluation of several alternatives for the destruction of the sulphide matrix, including roasting, bacterial oxidation, and pressure oxidation, the Sherritt process utilizing acid pressure oxidation of a flotation concentrate was selected for the commercial plant (9,10). The process consists of milling, flotation, pressure oxidation of the concentrate, carbon in leach, elution and electrowinning, and effluent disposal. The nominal capacity of the plant is 20 000 tonnes of ore per month.

Laboratory testwork on the process development by Sherritt consisted of batch oxidation tests on both ore and concentrate, followed by continuous testing of the various unit operations, on concentrate, and subsequently, a continuous fully integrated run in a miniplant. This work was conducted between November 1983 and February 1984 (5). A process design package, and detailed specifications for specialized equipment were then developed from the miniplant testwork data. Sherritt undertook the process design of the pressure oxidation, countercurrent decantation washing and neutralization sections, as well as the detailed design of the autoclaves, while General Union Mining Corporation provided the process design of the milling and gravity, flotation, carbon in leach and electrowinning and gold room. This enabled the engineering contractor to complete the design for a grass roots plant to treat nominally 240 t/d of concentrate. Process engineering design was initiated in October 1984, and gold was produced in January 1987.

Detailed accounts of the process selection, process development, design, commissioning and operation of the Sao Bento complex are provided elsewhere (9,10). A brief account is given here of the pressure oxidation, CCD and waste solution treatment (neutralization) circuits. Flotation concentrate, of a design composition of 9.9% As, 1.0% Ca, 6.0% CO_2, 34.0% Fe, 1.1% Mg, and 18.7% S, is reground to a nominal size of 90% minus 44 µm, thickened to 65% solids and stored in tanks ahead of the pressure oxidation section. From here, it is pumped to the feed preparation reaction tank where it is mixed with oxidized slurry recycled from the first wash thickener underflow tank, at a ratio of recycled solids:concentrate of 4:1. The primary purpose of the recycle is to maintain a high pulp density in the autoclave, to promote suspension and dispersion of elemental sulphur. The recycle also acts as a heat sink, thus assisting temperature control. The recycle slurry contains acid, which reacts with a portion of the carbonates, although siderite, the major carbonate, is only partly decomposed prior to autoclaving. The mixture then overflows to the oxidation feed storage tank which provides surge volume for the oxidation circuit and additional residence time for the reaction of acid and carbonates.

From the storage tanks, the slurry is pumped into each of two parallel autoclave trains. Each train consists of a positive displacement diaphragm autoclave feed pump, a horizontal 5 compartment, 6 agitator, 3.5 m diameter by 19.0 m long lead and brick lined autoclave, and a flash tank. Oxygen is fed continuously into each compartment, and the autoclaves operate at 190°C and at a pressure of 1600 kPa. Clarified river water is also pumped into the autoclaves, on an as required basis, for temperature control. Slurry from the last compartment of each autoclave is discharged through a choke to separate lead and brick lined flash tanks operating at atmospheric pressure. Autoclave vent gases are passed through a knock out pot prior to venting to the flash tanks. Steam from the flash tanks, after passing through a demister, is vented to the atmosphere.

Flash tank discharge is pumped to a repulp tank, where it is mixed with second wash thickener overflow solution before flowing to the first of two 25 m diameter wash thickeners. Overflow solution from the first thickener is stored in the tank ahead of neutralization. Approximately 80% of the thickener underflow is pumped to the recycle storage tank and the balance to the second wash thickener. Underflow from the second thickener is pumped to the carbon in leach section.

Two phase neutralization is used for the acidic first wash thickener overflow, including three stages of treatment with limestone, to pH 4.6, and two stages of liming, to pH 11. The sludge is dewatered in a thickener, and reclaimed water is cooled in a tower and recycled to the wash circuit. The thickened solids are discharged to the final tailings tank, where they are combined with CIL tailings and are advanced to the tailings dam.

Since reaching full capacity in April 1987, the plant has consistently performed at its design rate, and has demonstrated its robustness by treating greater sulphur grade variations and tonnages than originally envisioned (10).

Barrick Mercur Gold Mines

The third commercial plant to utilize pressure oxidation for the pretreatment of refractory gold, and the first to use alkaline pressure oxidation, came on stream in February 1988, at the Barrick Mercur Gold Mine in Mercur, Nevada. A single autoclave is used to oxidize 680 t/d of refractory carbonaceous sulphidic ore grading about 2 g/t Au and 1.7% sulphide sulphur.

Initial investigations into the treatment of the refractory ore had been undertaken during 1980 to 1982 by Getty Oil Company, a former owner of the property. The refractoriness of the gold was attributed to its inclusion in pyrite and marcasite, its adsorption onto organic carbon, and its complexing with hydrocarbons in the ore. The work led to the development of a patented aqueous pressure oxidation process in an alkaline environment (20,22,34). Subsequent evaluations of the ore body, however, revealed a lower than previously estimated proportion of the ore to be refractory, and in 1983, a 2730 t/d CIL mill commenced operation on oxidized ore, while refractory ore was stockpiled.

American Barrick Resources Corp. acquired the property in mid 1985 and expanded the original CIL mill to more than 4550 t/d. Consideration was given to separate treatment of the refractory ore. Additional testwork confirmed and refined the criteria developed in the previous testwork and confirmed that gold extraction could be improved from the variable range of 20 to 60% attainable by leaching of untreated ore, to between 90 and 91% after pressure oxidation. Following a favorable prefeasibility study, American Barrick decided, in October 1986, to proceed with construction of a commercial pressure oxidation circuit to treat refractory ore (3,19,33).

In the process, the refractory ore is crushed and ground to minus 75 µm in existing circuits, thickened in a new 19.8 m diameter high capacity thickener to 50% solids and pumped to a storage tank. The relatively low calorific value of the ore and the high temperature of 225°C required for effective pretreatment in the autoclave dictated the need for heat recovery from the discharge for feed preheating. The feed is preheated in three splash tower slurry heaters and a condenser in series, utilizing recycled steam from the autoclave discharge and autoclave vent gas. The temperature is successively raised from ambient, to 84°, 138°, 177°, and 183°C, respectively, with subsequent pressures of about 80, 380, 1100, and 3350 kPa, respectively. Pumping is accomplished by a series of ceramic lined centrifugal pumps and a Geho positive displacement diaphragm pump.

The slurry is then introduced into the autoclave. The autoclave is a horizontal, cylindrical four compartment, carbon steel vessel 3.66 m in diameter by 15.2 m long. The autoclave, as well as the splash and flash towers, are lined with 10 cm thick basic brick. Oxygen is sparged into each compartment, and single impeller mixers are used for agitation in each compartment. Projected oxygen requirement is about 24 t/d. The oxidation is conducted at 225°C and at a total pressure of about 3350 kPa. The heat of reaction of the sulphides and hydrocarbons is insufficient to maintain the temperature, and 4200 kPa steam is injected into the autoclave at a rate of approximately 110 kg/t of 1.7% sulphide sulphur ore. Slurry retention time in the autoclave is about 70 min.

Autoclave discharge slurry passes through three stages of pressure letdown (flash towers), which also reduce the temperature to about 96°C. Ceramic trimmed chokes are situated between each stage of letdown. Slurry from the final flash tower is cooled to about 27°C in a cooling tower or heat exchanger, before entering the CIL circuit.

Getchell Gold Mine

The Getchell Mine, owned by FirstMiss Gold Inc. and located near Winnemucca, Nevada, is the fourth commercial plant to employ pressure oxidation to liberate gold from sulphides. The refractory gold is primarily associated with pyrite. The ore also contains arsenic minerals, including realgar and orpiment, which apparently contain very little gold. Some gold is found with arsenic minerals associated with pyrite.

Although acid pressure oxidation testing of Getchell ore dates back as far as 1953, it was only in the past five years that interest in pressure oxidation for this ore was revived. Based on recent batch, semicontinuous, and continuous pilot plant testwork, the current flowsheet was developed (18). The flowsheet consists of blending run of mine ore from blend piles, SAG and ball mill grinding, acid pressure oxidation with heat recovery, neutralization, CIL, pressure elution, gold precipitation with zinc, mercury retorting and smelting. The nominal capacity of the pressure oxidation plant is 2730 t/d, and startup of the plant took place in early 1989.

Blended ore is expected to vary in composition, from 0.5 to 3.5% As, 2 to 4% sulphide sulphur, and 1.5 to 7.5% CO_2. The relatively high carbonate dictates acidification of the ore prior to autoclaving. Unlike the McLaughlin operation where acidic oxidized pulp is washed and the acid containing liquor is used to pretreat fresh ore, fresh acid alone is used at Getchell. It is unlikely that the acid produced in situ by the oxidation would meet more than a portion of fresh ore acid demand.

Blended ore, after grinding to 80% passing 74 µm, is thickened to between 45 and 50% solids in a 45.7 m thickener and is then reacted with sulphuric acid in three 6.1 m by 9.4 m stainless steel, stirred, conditioning tanks. Air is sparged into the slurry to promote CO_2 stripping and to provide an oxidizing environment. The slurry then proceeds to one of three autoclave trains.

Each slurry stream is preheated to about 150°C in two 1.5 m by 9.4 m ID acid brick lined carbon steel splash condensers and is then fed, by a positive displacement pump, into an autoclave. Each autoclave is a horizontal cylindrical multicompartment pressure vessel with shell dimensions of 3.35 m by 26.8 m and is lined with lead membrane and acid brick. Eight 75 HP agitators are used for slurry agitation and oxygen dispersion, and pure oxygen is injected below the impellers. The operating temperature in the autoclave is 210°C, and with feed containing 3% sulphide sulphur, the system is thermally autogenous. The temperature in the first reaction stage is controlled by adjusting the feed slurry temperature, or by injecting steam, while temperature control in subsequent stages is by water injection. A total pressure of about 3200 kPa, and an oxygen partial pressure of about 700 kPa are maintained by adjusting oxygen supply and autoclave vent flows.

The discharge slurry passes through two stages of pressure letdown through ceramic chokes. Each stage consists of a throttling choke and a cylindrical choke. Modulation of the throttling chokes controls the levels in the final autoclave compartment and in the first flash tank, respectively. The flashed slurry discharges into two brick lined flash tanks in series, the first 2.44 m by 6.1 m, and the second 3.66 m by 9.3 m. Single pass shell and tube heat exchangers, employing cooling water, bring the temperature down to about 50°C. The cooled acidic slurry is then neutralized directly, in a series of four 8.8 m by 9.4 m tanks, with limestone, then lime and, after dilution to 33% solids, proceeds to carbon in leach.

American Barrick's Goldstrike Mill

The next commercial application of pressure oxidation is scheduled to be at American Barrick's Goldstrike Mill in Nevada, where acid pressure oxidation will be used to treat refractory sulphide ore from the Betze Pit (2). The first of six 4.5 m diameter by 22.6 m long autoclaves is scheduled to come on stream in January 1990 to treat 1600 t/d of ore. By 1992, three autoclaves will be in operation, with the remaining three scheduled to come on stream by 1993, treating a total of 11 600 t/d of ore.

METBA's Olympias Gold Mine

The Sherritt Pressure Oxidation Process has been selected by Aegean Metallurgical Industries S.A. (METBA) for the pretreatment of refractory arsenical pyrite concentrate from the Olympias Mine in Northern Greece (14). A process similar to the one at Sao Bento will be used to treat 315 t/d of concentrate with an average composition of 25 g/t Au, 12% As, 40% Fe, and 41% S through two autoclave trains. Oxidation will be conducted at 190°C and 1800 kPa, and 98% pure oxygen will be supplied to the autoclaves from a 330 t/d oxygen plant. Three stages of CCD washing will be employed on autoclave discharge slurry, and a portion of the acidic oxidized slurry will be recycled. Extensive batch and continuous miniplant testing at Sherritt has demonstrated the ability of pressure oxidation to improve gold extractability from about 10% to beyond 97%. The plant is in the detailed design phase, with construction slated to start in July 1989, and plant start up scheduled for fall of 1990.

Porgera Gold

Sherritt's Acid Pressure Oxidation Process will also be employed at the Porgera Gold deposit in Papua New Guinea (26,27). Six autoclaves are scheduled to process 2700 t/d of pyritic concentrate, containing about 9% S, using a flowsheet similar to that developed for the Sao Bento and METBA feeds. This project is at the detailed design phase, and is scheduled to go on stream in late 1992.

CONCLUSIONS

The basic pressure oxidation process for pretreating refractory auriferous sulphidic feedstocks is proven; the challenge is to improve it, to reduce capital and operating costs, and to adapt it to a wider range of feed materials.

REFERENCES

1 Argall, G.O. Jr., "Perseverance and Winning Ways at McLaughlin Gold", Engineering and Mining Journal, October 1986, pp. 26-32.

2 Anon., "Barrick's Goldstrike Autoclave to be World's Largest", Kilborn Update, April 1989.

3 Anon., "Mercur Starts up its New Alkaline Pressure-Oxidation Autoclave Plant", Engineering and Mining Journal, June 1988, pp. 26-31.

4 Anon., "Homestake's McLaughlin Mine: One of the Largest U.S. Gold Mines Comes Onstream in 1985", Mining Engineering, January 1986, pp. 24-26.

5 Berezowsky, R.M.G.S., Haines, A.K., and Weir, D.R., "The Sao Bento Gold Project - Pressure Oxidation Process Development", Projects '88, 18th Annual Meeting of the Hydrometallurgical Section of the Canadian Institute of Mining and Metallurgy, Edmonton, Alberta, Canada, May 1988.

6 Berezowsky, R.M.G.S. and Weir, D.R., "Pressure Oxidation Pretreatment of Refractory Gold", Minerals and Metallurgical Processing, Vol. 1, May 1984, pp. 1-4.

7 Berezowsky, R.M.G.S. and Weir, D.R., "Pressure Oxidation for Treating Refractory Uranium and Gold Ores", Presented at the 22nd Annual Conference of Metallurgists of the CIM, Edmonton, Alberta, August 1983.

8 Collins, M.J., Berezowsky, R.M., and Weir, D.R., "The Behaviour and Control of Arsenic in the Pressure Oxidation of Uranium and Gold Feedstocks", Arsenic Metallurgy - Fundamentals and Applications, Reddy, R., Hendrix, J., and Queneau, P. eds., TMS-AIME, Warrendale, Pennsylvania, 1987, pp. 115-133.

9 Carvalho, T.M., et al., "Startup of the Sherritt Pressure Oxidation Process at Sao Bento", Perth International Gold Conference 1988, Perth, Australia, Oct. 28-Nov. 1, 1988, pp. 152-156.

10 da Silva, E.J., et al., "Process Selection, Design, Commissioning and Operation of the Sao Bento Mineracao Refractory Gold Ore Treatment Complex", Gold Forum on Technology and Practices - World Gold '89, Reno, Nevada, Nov. 5-8, 1989.

11 Genik-Sas-Berezowsky, R.M., and Weir, D.R., "Recovery of Silver from a Residue Essentially Free of Elemental Sulphur", U.S. Patent 4,632,701, December 30, 1986.

12 Gormely, L.S. and O'Kane, P.T., "Pressure Oxidation - a New Tool for Refractory Gold Ores and Concentrates", Annual Alberta-British Columbia Meeting of Canadian Mineral Processors, Vancouver, B.C., November 1984.

13 Guinevere, R., "McLaughlin Project: Process, Project and Construction Development", The First International Symposium on Precious Metals Recovery, Reno, Nevada, June 10-14, 1984.

14 Kontopoulos, A. and Stafanakis, M., "Process Selection for the Olympias Refractory Gold Concentrate", Precious Metals '89, Jha, M.C., and Hill, S.D., ed., The Minerals, Metals, and Materials Society, 1988.

15 Krause, E. and Ettel, V.A., "Ferric Arsenate Compounds - Are They Environmentally Safe?" Solubilities of basic ferric sulphates. 15th Annual Hydrometallurgical Meeting of the Canadian Institute of Mining and Metallurgy, Vancouver, B.C., August 1985.

16 Kunter, R.S. and Turney, J.R., "Gold Recovery Process", U.S. Patent 4,578,163, March 25, 1986.

17 Kunter, R.S., Turney, J.R., and Lear, R.D., "McLaughlin Development: The Project, The Problems, The Process", International Symposium on Precious Metals, sponsored by TMS-AIME and IPMI, Los Angeles, CA, February 1984.

18 Mason, P.G. and Nanna, R.F., "A New Beginning for the Getchell Mine", Precious Metals '89, Jha, M.C. and Hill, S.D., ed., The Minerals, Metals and Materials Society, 1988.

19 Mason, P.G., Skeldon, R.B., and Wicks, F.D., "Preparation for the Treatment of Refractory Ore at Barrick Mercur Gold Mine", 1st International Gold Conference, Vancouver, B.C., November 1987.

20 Mason, P.G., Wicks, F.D., and Gathje, J.C., "Process for the Recovery of Gold from Refractory Ores by Pressure Oxidation", U.S. Patent 4,552,589, November 12, 1985.

21 Mason, P.G. and O'Kane, P.T., Refractory Gold - Process Options. 24th Annual Conference of Metallurgists of the Canadian Institute of Mining and Metallurgy, Vancouver, B.C., August 1985.

22 Mason, P.G., et al., "Selection of the Process Flowsheet for the Mercur Gold Plant", Precious Metals: Mining, Extraction and Processing, Kudryk, V., Corrigan, D.A. and Liang, W.W., ed., TMS-AIME 1984, pp. 435-445.

23 Turney, J.R., Smith, R.J., and Janhunen, W.J. Jr., "The Application of Acid Pressure Oxidation to the McLaughlin Refractory Ore", Precious Metals '89, Jha, M.C. and Hill, S.D., ed., The Minerals, Metals and Materials Society, 1988, pp. 25-45.

24 Weir, D.R. and Berezowsky, R.M.G.S., "Aqueous Pressure Oxidation of Refractory Gold Feedstocks", International Symposium on Gold Metallurgy, 26th Annual Conference of Metallurgists of the Canadian Institute of Mining and Metallurgy, Winnipeg, Manitoba, 1987.

25 Weir, D.R. and Berezowsky, R.M.G.S., "Refractory Gold: The Role of Pressure Oxidation", Gold 100, Proceedings of the International Conference on Gold, Volume 2: Extractive Metallurgy of Gold, South African Institute of Mining and Metallurgy, Johannesburg, 1986, pp. 275-285.

26 Weir, D.R., King, J.A., and Robinson, P.C., "Preconcentration and Pressure Oxidation of Porgera Refractory Gold Ore", Minerals and Metallurgical Processing, Vol. 3, No. 4, November 1986, pp. 201-208.

27 Weir, D.R., Robinson, P.C., and King, J.A., "Precious Metals Recovery from Pressure Oxidized Porgera Concentrates", Minerals and Metallurgical Processing, Vol. 3, No. 4, November 1986, pp. 208-215.

28 Weir, D.R., "Process for the Pressure Oxidation Acid Leaching of Non-Ferrous Metal and Iron-Containing Sulphidic Material", U.S. Patent 4,606,763, August 19, 1986.

29 Weir, D.R., "Recovery of Gold from Refractory Auriferous Iron-Containing Sulphidic Material", U.S. Patent 4,605,439, August 12, 1986.

30 Weir, D.R. and Genik-Sas-Berezowsky, R.M., "Recovery of Gold from Refractory Auriferous Iron-Containing Sulphidic Ore", U.S. Patent 4,571,264, February 18, 1986.

31 Weir, D.R. and Genik-Sas-Berezowsky, R.M., "Recovery of Gold from Refractory Auriferous Iron-Containing Sulphidic Concentrates", U.S. Patent 4,571,263, February 18, 1986.

32 Weir, D.R. and Berezowsky, R.M.G.S., "Gold Extraction from Refractory Concentrates", 14th Annual Hydrometallurgical Meeting of CIM, Timmins, Ontario, October 1984.

33 Wicks, F.D., "The Barrick Mercur Gold Mine", Mining Magazine, November 1987, pp. 398-405.

34 Wicks, F.D., and Brewer, R.E., "A Practical Approach to Treatment of Low Grade Micron Size Carbonaceous Gold Ores", The First International Symposium on Precious Metals Recovery, Reno, Nevada, June 1984.

Chapter 36

THIOUREA LEACHING OF GOLD FROM AN OXIDISED GOLD-COPPER ORE

Cheryl M. McInnes,[1] Graham J. Sparrow,[2] and James T. Woodcock[3]

[1] Technical Assistant, CSIRO Division of Mineral Products, P.O. Box 124, Port Melbourne, Vic., 3207, Australia.
[2] Principal Research Scientist, CSIRO Division of Mineral Products, P.O. Box 124, Port Melbourne, Vic., 3207, Australia.
[3] Senior Principal Research Scientist, CSIRO Division of Mineral Products, P.O. Box 124, Port Melbourne, Vic., 3207, Australia.

ABSTRACT

Three composites, assaying about 3 g/t Au and 0.75% Cu, were investigated. Each contained oxide copper minerals, iron oxides, and silicates. One composite also contained copper and iron sulphides. Maximum gold extractions by thiourea leaching were obtained after copper removal by acid pre-leaching at pH 1.5 and redox potentials of 400 mV (Pt versus SCE). Best thiourea leaching conditions were pH 1.5, redox potentials over 200 mV, and 0.4% thiourea. This gave 90% gold extraction in 2 h with thiourea consumptions of 25 kg/t. Lower potentials gave poorer extractions. Ferrous iron additions before leaching halved thiourea consumptions. Longer leach times and the presence of sulphide minerals sometimes resulted in reprecipitation of gold.

INTRODUCTION

Cyanide is used extensively to dissolve gold from ores and concentrates. However the dissolution of gold with cyanide is not rapid, and leach times of 24 h or more are often required. Moreover the presence of cyanide-consuming minerals, particularly copper minerals, in an ore can result in high cyanide consumptions. For these reasons, and a growing environmental concern about the use of cyanide, alternative solvents have been sought for several years.

One alternative to cyanide is thiourea (Groenewald, 1976), which has been suggested by Hiskey (1984) and by Haque (1987), in particular, as being useful for treating copper-containing gold ores. In this paper the results of dissolution of gold with thiourea from an oxidised gold-copper ore are presented, and it is shown that while gold can be extracted, copper does interfere. An explanation for this interference is given, and the results are compared with those of cyanidation after acid pre-leaching.

This work represents part of the results of a collaborative research programme between the CSIRO Division of Mineral Products, Melbourne, and the Selwyn Joint Venture (Cyprus Gold Australia Corporation, Elders Resources Limited, and Arimco N.L.) to find the most suitable treatment route for recovering gold and copper from the Selwyn gold-copper deposit, near Cloncurry, Qld.

EXPERIMENTAL

Experimental work conducted included sample preparation, characterisation of the samples by assay and mineragraphic examination, acid pre-leaching to remove copper, gold leaching tests with thiourea and cyanide, and sampling and analysis of test products. Melbourne tap water, which is of high quality (<50 ppm total dissolved solids), was used in all leaching tests.

SAMPLE PREPARATION

The Selwyn deposit is composed of a primary sulphide zone with an oxide capping. Between the two is a transition zone of variable thickness (Morrison, 1986). The samples supplied were sawn drill core intercepts from the transition zone and the oxide zone. They were crushed to minus 10 mesh (1.651 mm aperture) and combined according to advice from Elders Resources Limited to make three composite samples.

One composite sample was prepared by combining drill core from the intermediate and oxide zones to give a sulphide-oxide sample (referred to as sample A). The other two (B and C), were prepared from the oxide zone only. Composites B and C were similar, but were from different, but comparable, drill cores. All the composites were riffled into 500 g lots for test work.

Before each test, the sample was wet ground at 60% solids for 20 min in a 200 mm diameter iron ball mill using a standard charge of steel balls. This gave products containing 84, 88, and 74% minus 74 μm (200 mesh) for composites A, B, and C respectively.

LEACHING CONDITIONS

Leaching tests included a pre-leaching step to dissolve copper, and then either thiourea leaching or cyanidation to dissolve gold.

Pre-leaching

For many tests the ground samples were given a preliminary leach with sulphuric acid under oxidising conditions to remove oxidised copper minerals before conducting the gold leach. This is termed a "pre-leach" in this paper.

Pre-leaching was conducted in open beakers at a liquid:solid (L:S) ratio of 3:1 by weight for 4-24 h at pH 1.5 or 2.5 and at a redox potential of 100-250 mV, or at 400 mV. In this paper all redox potentials (Eh) are given as the potential of a platinum electrode versus a saturated calomel electrode (SCE).

During pre-leaching the pH was controlled by a pH-stat consisting of a pH electrode (a combined glass and reference electrode) and a Radiometer auto-titrator system. Control was maintained by addition of 2.5 M sulphuric acid. Redox potential was controlled in some pre-leach tests by an Eh-stat consisting of a platinum/SCE pair and a Radiometer auto-titrator system. Control was maintained by addition of ferric sulphate or hydrogen peroxide solution. In other tests the redox potential was not controlled but was allowed to assume its natural level of 100 mV rising to 250 mV after 24 h.

After the completion of the pre-leach the slurry was filtered and the pregnant solution was assayed for copper. The filter cake was washed first with 500 mL of acid at the same pH as that of the pre-leach and then with 500 mL of distilled water.

Thiourea Leaching

Thiourea leaches were conducted in open beakers on ground ore, or on pre-leach residue, or, in a few tests, on unfiltered pre-leach slurry (i.e. with dissolved copper present) at 3:1 L:S ratio by weight for 6-24 h under controlled pH and redox potential and partial control of thiourea level. The pH was controlled in a similar manner to that noted for the pre-leach by addition of 2.5 M sulphuric acid, and the redox potential was controlled by addition of diluted 100 volume hydrogen peroxide (1:1 or 1:10 v/v) or ferric sulphate (400 g/L $Fe_2(SO_4)_3.9H_2O$) in 1 M sulphuric acid. In some tests sufficient ferrous ammonium sulphate, dissolved in water, was added to give 1500 or 18,000 ppm iron in solution. When ferric sulphate was used it was difficult to maintain the pH at the desired value because of the acid added with the oxidant. Similarly the redox potential usually decreased (became less positive) on addition of thiourea, and at times rose above the set point.

Thiourea levels were maintained at 0.4% except for two tests where 1.0% was used. Samples of solution were taken at appropriate intervals, the thiourea concentration determined, and then sufficient thiourea added as a slurry in water to return the concentration to the desired level. Solution samples were also taken at selected intervals for analysis for gold, copper, and iron.

After leaching, the slurry was filtered to recover pregnant solution for assay. The filter cake was washed with 1000 mL of distilled water, dried at 70°C, weighed, sampled, and assayed.

Cyanidation

Cyanidation tests were conducted using conventional bottle roll techniques. In most tests, an acid pre-leach at pH 1.2-2.5 and a redox potential of 400 mV for 18-24 h was used, but one test was conducted on an unleached sample of composite C.

The pre-leached residue or ore was pulped with water to give a L:S ratio of 3:1 by weight and sufficient lime was added to give 0.005-0.010% CaO in solution (pH 11-11.5). Sufficient cyanide was then added to give 0.083% NaCN in solution and the pulp agitated for 24 h. Periodic tests were made of cyanide and lime strengths and appropriate additions made. No samples were taken for gold assays during the tests. After agitation the pulp was filtered to recover pregnant solution for assay, and the residue was washed with distilled water, dried at 70°C, weighed, and assayed.

ANALYTICAL METHODS

Gold was determined on solution and solid samples by atomic absorption spectrometry (AAS) after aqua regia digestion, baking to remove nitric acid, re-dissolution in 3 M hydrochloric acid, and extraction into methylisobutyl ketone (MIBK).

Copper, iron, zinc, and lead assays were done by standard AAS techniques. Silica was determined by a gravimetric technique, and sulphur by Leco combustion with an infra-red spectrometric finish.

Thiourea concentrations were determined by titration with potassium iodate using VITEX indicator (an iodine indicator) after acidification with 1 M orthophosphoric acid. Formamidine disulphide concentrations in solution were not determined.

Sizing analyses were conducted with Tyler screens using a standard wet and dry technique based on a 74 µm aperture (200 mesh) screen.

MINERAGRAPHIC EXAMINATION

Polished sections of drill cores were examined under a Cambridge S100 scanning electron microscope (SEM) and a search for gold in the samples was made by the technique of Harrowfield, MacRae, and Simmonds (1988). Individual gold particles were analysed with a Cameca Camebax Micro scanning electron microprobe.

Minerals present in the feed and residue samples were identified by SEM examination and by x-ray diffraction techniques.

RESULTS

Results discussed here are concerned with characterisation of the composite samples by assays and mineral abundances, gold leaching tests with thiourea on the sulphide-oxide composite and the oxide composites, and comparative tests with cyanide.

HEAD SAMPLE

Assay data for the composite samples prepared from the drill cores are given in Table 1. A mineralogical examination by I.R. Pontifex and

Associates (unpublished report supplied to CSIRO by Elders Resources Limited) confirmed the results obtained in this work. This indicated that the composites were mainly composed of quartz and iron oxides (magnetite, martite, hematite, goethite, and limonite). The martite content of composite A was less than that of the other two composites. As well, there were small amounts of muscovite, biotite, chlorite, kaolin, calcite, and siderite present in all samples. Composite A contained a finely altered quartz-albite-chlorite-biotite schist incorporating iron oxides and clays that was absent in the oxide composites B and C.

Copper Minerals

The copper minerals identified in the oxide composites (B and C) were malachite and chrysocolla with traces of cuprite, chalcotrychite, and native copper. The sulphide-oxide composite (A) contained less malachite and chrysocolla, but contained chalcopyrite, bornite, chalcocite, and pyrite. These sulphides were absent in composites B and C.

Gold

Examination of polished drill core sections under the SEM indicated that the gold was present in an iron oxide phase (limonite or goethite) or in a siliceous phase (quartz or copper-containing quartz that is probably chrysocolla). Several grains were also observed associated with a copper sulphide. Most gold grains were 1-3 µm in diameter although grains up to 40 µm were found (C.M. MacRae and P.F. Simmonds, unpublished CSIRO Report MCC 834). The composition of the gold particles analysed was found to be different for the coarse and the fine particles.

A 30 µm particle in quartz, analysed at ten points across the particle, was very uniform in composition with 6.0% silver and 0.1% copper in the gold. By contrast, analyses of 1-3 µm particles in a chrysocolla phase indicated that they all contained 0.1% silver and 1.6% copper. It is possible that the variation in copper content of the gold affects its dissolution with thiourea as discussed later.

THIOUREA LEACHING OF SULPHIDE-OXIDE COMPOSITE

After a relatively unsuccessful preliminary leach with thiourea on the sulphide-oxide composite, as noted below, the effect of several variables on gold extraction and thiourea consumption was investigated. These were the effect of acid pre-leaching under various conditions to remove soluble copper, the effect of redox potential, and the effect of the type of oxidant. The copper and iron contents of the leach solutions were measured during these tests.

Table 1. Analytical data for the composite samples.

	Composite A	Composite B	Composite C
Au (g/t)	2.9	3.0	3.5
Ag (g/t)	3.4	2.8	3.8
Cu (%)	0.75	0.59	0.74
Fe (%)	26.5	25.9	35.4
Zn (%)	0.01	0.01	0.01
Pb (%)	0.04	0.01	0.01
SiO_2 (%)	40.7	32.8	36.6
S (%)	0.22	0.020	0.015

Preliminary Leaches

In the initial thiourea leaches at pH 2.5 with composite A, the redox potential was maintained at 150-180 mV for the first hour and then allowed to fall to its natural value of 115-125 mV. These conditions were chosen in order to minimise thiourea consumption. However, less than 20% of the gold dissolved in 6 h (Fig. 1) and only 53% in 24 h. Thiourea consumption was 10 kg/t after 6 h.

At pH 1.5, when the redox potential was maintained at 150-180 mV throughout the leach, 56 % of the gold was dissolved in 6 h as shown in Fig. 1. The thiourea consumption was 14 kg/t.

Effect of Acid Pre-leaching

Previous work in the Division (Bilston, LaBrooy, and Woodcock, 1984; Bilston, Millett, and Woodcock, 1987) indicated that gold could be dissolved at redox potentials of 180 mV, but poor results were obtained with the Selwyn sample. Hence an acid pre-leach was used in an attempt to dissolve copper minerals to make the gold more accessible to the thiourea solution and to reduce thiourea useage.

Fig. 1 shows the gold extraction by leaching with thiourea at pH 1.5-2.0 and 150-180 mV following no pre-leach, a pre-leach at 100-250 mV (the uncontrolled values), and a pre-leach at 400 mV. The results indicate that for thiourea leaching at pH 1.5, the pre-leach had little effect on the amount of gold dissolved since 53-57% of the gold was extracted in 6 h for each leach. However, it is evident that the rate at which the gold dissolved increased as the pre-leach conditions became more vigorous (i.e., the lower the pH and the higher the redox potential). After 24 h, only 74% of the gold had dissolved, and increasing the thiourea concentration from 0.4 to 1.0% did not increase the amount of gold extracted in 24 h at 180 mV, although the gold did dissolve faster initially.

Since not all the gold was dissolved by a thiourea leach at 180 mV, further leaches were carried out at 200-300 mV with iron(III) sulphate or hydrogen peroxide as the oxidants, even though a higher thiourea consumption was expected. In these leaches the redox potential was raised to at least 250 mV in 5-10 min and maintained at 250 mV for the first hour. Thereafter it was maintained at 250 mV for the remainder of the leach or allowed to fall to around 200 mV in an attempt to minimise the thiourea consumption. The use of the higher redox potential resulted in faster dissolution of the gold, and the amount of gold extracted increased from 74% to 96% (Fig. 2).

Effect of Type of Oxidant

Two oxidants (apart from oxygen in the air) were used - ferric sulphate and hydrogen peroxide.

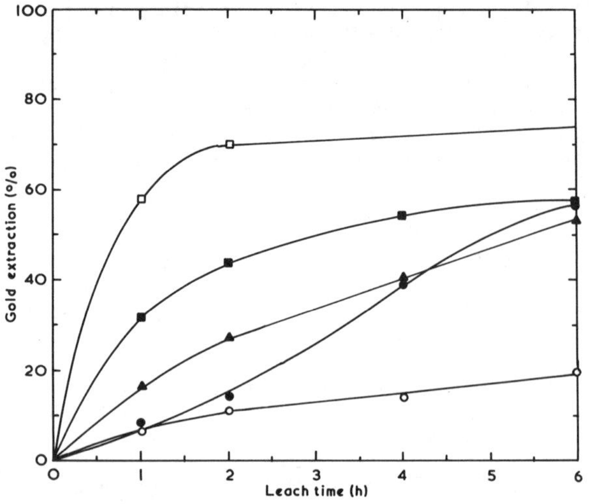

Fig. 1 - Effect of pre-leach conditions and thiourea leach conditions on dissolution of gold from composite A.
□ Pre-leach: pH 1.5, Eh 400 mV.
Thiourea leach (1%): pH 1.5, Eh 180 mV.
■ Pre-leach: pH 1.5, Eh 400 mV.
Thiourea leach (0.4%): pH 2.0, Eh 170 mV.
▲ Pre-leach: pH 2.5, Eh natural.
Thiourea leach (0.4%): pH 1.5, Eh 180 mV.
● Pre-leach: No pre-leach.
Thiourea leach (0.4%): pH 1.5, Eh 165 mV.
○ Pre-leach: No pre-leach.
Thiourea leach (0.4%): pH 2.5, Eh 150 mV.

Fig. 2 shows that when the redox potential was maintained above 200 mV during the leach with ferric sulphate, 95-97% of the gold was dissolved in 2-4 h. When hydrogen peroxide was the oxidant, a similar amount of gold was dissolved in 12-24 h.

The results in Fig. 2 (with hydrogen peroxide as the oxidant) also show that dissolution of gold was faster after the more vigorous pre-leach, as was found at 180 mV.

Details of representative leaches for the sulphide-oxide composite, at 180 mV and at over 200 mV, using iron(III) sulphate and hydrogen peroxide as the oxidants are given in Table 2.

Thiourea Consumption

While the higher redox potential increased the amount of gold extracted, it also significantly increased the thiourea consumption. Following a pre-leach at pH 1.5 and 400 mV, the thiourea consumption increased from less than 10 kg/t for the leach at 180 mV, to around 40 kg/t at 250 mV. Much of this increased thiourea consumption was the result of oxidation of thiourea in the first hour when the redox potential was maintained above 250 mV. For example, with iron(III) sulphate as the oxidant, 95% of the gold was extracted in 2 h and the thiourea consumption was 26 kg/t (Table 2).

The thiourea consumptions also depended on the pre-leach conditions. For thiourea leaches at 180 mV, thiourea consumption was highest after no pre-leach, and lowest after the most vigorous pre-leach (pH 1.5, 400 mV) as shown in Fig. 3. A similar

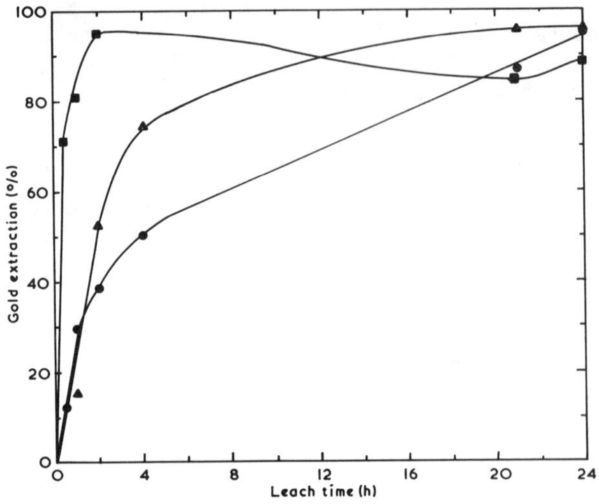

Fig. 2 - Effect of oxidant on dissolution of gold from composite A using 0.4% thiourea at pH 1.5 and Eh 200-300 mV after a pre-leach at pH 1.5.
■ Oxidant: Fe^{3+}. Pre-leach: Eh 400 mV.
Thiourea leach: Eh 210-265 mV.
▲ Oxidant: H_2O_2. Pre-leach: Eh 400 mV.
Thiourea leach: Eh 260-300 mV.
● Oxidant: H_2O_2. Pre-leach: Eh natural.
Thiourea leach: Eh 215-300 mV.

result was also observed for thiourea leaches above 200 mV.

Thiourea consumption, as shown later, is higher the greater the amount of soluble copper present. Solution assays for leaches at 180 mV showed that the copper content of the liquor during the thiourea leach was highest without a pre-leach (280 ppm Cu) and lowest after a pre-leach at pH 1.5 and 400 mV (8.9 ppm Cu).

THIOUREA LEACHING OF OXIDE COMPOSITES

It was shown above for the sulphide-oxide composite that for satisfactory gold extraction and minimum thiourea loss it was first necessary to pre-leach the sample to remove copper. To dissolve gold during thiourea leaching it was then necessary to operate at redox potentials above 200 mV. These conditions were therefore applied to the oxide composites, B and C. Topics investigated were the effect of the type of oxidant, reprecipitation of gold during leaching, and the effect of iron(II) additions. Thiourea consumptions were also noted.

Effect of Type of Oxidant

When iron(III) sulphate was the oxidant, 82% of the gold was dissolved from composite B in 2 h, while the corresponding value for composite C was 95%. With hydrogen peroxide as the oxidant, gold extraction was lower and more variable as shown in Table 3. The best result with hydrogen peroxide for composite B was obtained following a pre-leach at pH 1.5 and 400 mV which was not filtered, and the redox potential was maintained at 250 mV throughout the thiourea leach. Over 90% of the gold was dissolved in 2 h, but reprecipitation of the gold occurred at longer leach times (Table 3).

Reprecipitation of Gold

It can be seen from Fig. 4 that gold precipitated from solution after the maximum extraction on composites B and C. A similar result was also observed in tests with the sulphide-oxide composite (A). Addition of thiourea at 21 h did result in some of the gold redissolving as also shown on Fig. 4. The dissolution of gold from composite B was not improved by increasing the thiourea concentration to 1% after 2 h.

Effect of Ferrous Iron Additions

When the slurry was not filtered after a pre-leach in which hydrogen peroxide was used as the oxidant, the iron concentrations in solution were higher during the thiourea leach. The better gold dissolution under these conditions (Table 3) suggested that it may be necessary to not only maintain the redox potential at 250 mV during the leach, but to also maintain the iron concentration in solution above a particular level.

Groenewald (1976) has reported that iron(III) is the best oxidant for the dissolution of gold with thiourea. Hydrogen peroxide can oxidise iron(II) to iron(III), so that iron(III) could be the active oxidant for the dissolution of gold in the leaches. In the absence of added iron, the iron must be leached out of the sample. This is very slow with the oxide composites, but as noted previously, iron was dissolved from the sulphide-oxide composite during the thiourea leach.

To check whether the concentration of iron in the solution does affect the dissolution of gold, iron was added as iron(II) in the form of ferrous ammonium sulphate to the re-slurried pulp after a pre-leach at pH 1.5 and 400 mV, and hydrogen peroxide was added to raise the redox potential to the desired value. In this way it was expected that while the ratio of the concentration of iron(III) to iron(II) would be constant, since it is determined by the redox potential, the actual concentration of iron(III) in solution would be increased as the total iron concentration in solution increased. Figs. 5 and 6 show gold

Table 2. Leaching of sulphide-oxide composite A with 0.4% thiourea.

	Test A361		Test A362		Test A360	
Acid pre-leach						
Oxidant	Fe^{3+}		Fe^{3+}		H_2O_2	
pH	1.5		1.5		1.5	
Eh (mV)	400		400		400	
Time (h)	24		4		24	
Cu extr. (%)	65.5		57.4		63.4	
Thiourea leach						
Oxidant	Fe^{3+}		Fe^{3+}		H_2O_2	
pH	2.0		1.5		1.5	
Eh (mV)	160-185		210-265		260-400* 150-175§	
Time (h)	Au extr. (%)	Tu used (kg/t)	Au extr. (%)	Tu used (kg/t)	Au extr. (%)	Tu used (kg/t)
0.5	-	-	70.9	10.0	-	-
1	31.8	3.1	80.8	17.4	15.4	10.0
2	44.0	4.0	94.9	25.8	54.9	22.0
4	54.4	4.5	-	-	74.7	34.0
6	57.2	5.0	-	-	-	-
21	-	-	84.5	35.5	95.8	41.2
24	74.0	6.7	88.6	40.1	95.8	-
Cu extr. (%)	0.35		7.4		7.9	
Residue						
Au assay (g/t)	0.86		0.36		0.14	
Au loss (%)	26.0		11.4		4.2	
Total extraction						
Au (%)	74.0		88.6		95.8	
Cu (%)	65.9		64.8		71.3	
Calc. feed grade						
Au (g/t)	3.19		3.07		3.20	
Cu (%)	n.d.		n.d.		n.d.	

*During first 4 hours.
§After 4 hours.
n.d. not determined.
Copper extractions based on a head of 0.75% Cu.

Table 3. Leaching of oxide composite B with 0.4% thiourea.

	Test B352		Test B361		Test B363	
Acid pre-leach						
Oxidant	Fe^{3+}		H_2O_2		H_2O_2	
pH	1.5		1.5		1.5	
Eh (mV)	400		400		400	
Time (h)	24		18		21	
Cu extr. (%)	53.3		58.4		55.8*	
Thiourea leach						
Oxidant	Fe^{3+}		H_2O_2		H_2O_2	
pH	1.5		1.5		1.5	
Eh (mV)	195-250		200-250		250-290	
Time (h)	Au extr. (%)	Tu used (kg/t)	Au extr. (%)	Tu used (kg/t)	Au extr. (%)	Tu used (kg/t)
0.5	46.4	6.3	38.1	8.1	58.1	16.8
1	60.0	9.1	53.0	15.6	74.6	22.7
2	81.6	11.4	25.2	21.6	91.8	28.6
4	78.8	12.0	43.2	22.6	70.9	34.4
21	48.1	12.6	62.7	22.0	95.4	41.9
24	66.6	13.2	45.2	20.7	28.3	48.1
Cu extr. (%)	0.2		0.4		3.1*	
Residue						
Au assay (g/t)	1.10		1.48		1.86	
Au loss (%)	33.4		54.8		71.7	
Total extraction						
Au (%)	66.6		45.2		28.3	
Cu (%)	53.5		58.8		3.1*	
Calc. feed grade						
Au (g/t)	3.20		2.61		2.54	
Cu (%)	n.d.		n.d.		n.d.	

*The liquor was not filtered after the pre-leach.
n.d. not determined.
Copper extractions based on a head of 0.59% Cu.

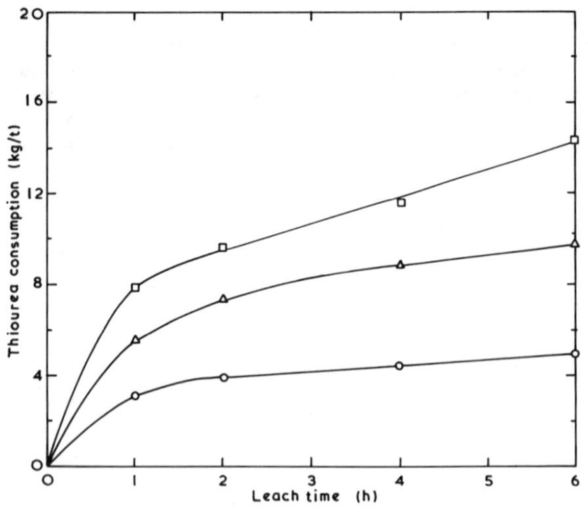

Fig. 3 - Effect of pre-leach conditions on thiourea consumption during leaching of composite A with 0.4% thiourea at pH 1.5 and Eh 180 mV.
□ No pre-leach.
△ Pre-leach: pH 2.5, Eh 100-250 mV.
○ Pre-leach: pH 1.5, Eh 400 mV.

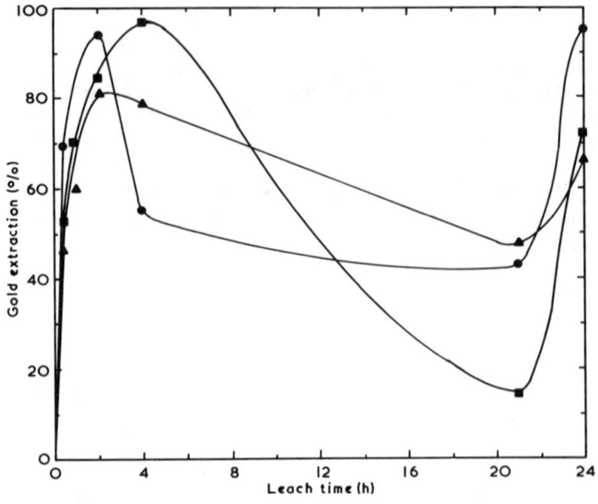

Fig. 4 - Reprecipitation of gold during leaching with 0.4% thiourea at pH 1.5 and Eh 200-250 mV after a pre-leach at pH 1.5 and 400 mV.
■ Composite A.
▲ Composite B.
● Composite C.

extractions for tests at total iron concentrations of 1500-2000 ppm and 17,500-20,000 ppm respectively.

With a total iron concentration of 1500-2000 ppm (Fig. 5) and a redox potential of 250 mV throughout the leach, 83% of the gold was dissolved in 4 h; there was 80% of the gold in solution at 21 h but between 21 and 24 h, gold precipitated and there was only 64% in solution at 24 h. With a similar total iron concentration and a lower redox potential of 200 mV, gold dissolution values increased with time, but after 24 h only 68% of the gold had dissolved (60% in 4 h).

With a total iron concentration of 17,500-20,000 ppm (Fig. 6), rapid dissolution of the gold was achieved at 250 mV with 95% of the gold being dissolved in 2 h and 90% of the gold was still in solution at 24 h. At a redox potential of 200 mV the gold extraction increased with time and at 24 h, 95% of the gold had dissolved. By contrast, with a redox potential of 175 mV, the gold dissolution values were similar to those at 200 mV for the first 2 h with 70% of the gold dissolved, but thereafter very little more gold was dissolved. This is similar to the amount of gold dissolved from the oxide-sulphide composite in 24 h at 185 mV (Table 2).

Thiourea Consumptions

The results presented in Figs. 5 and 6 show not only how the initial rate at which the gold dissolved increased with the iron concentration, and with the redox potential at which the leach was carried out, but also show how the thiourea consumptions changed with the leach conditions. It can be seen that that thiourea consumption was lower with the higher iron additions. Dissolution of 95% of the gold was achieved at 250 and 200 mV with thiourea consumptions of 8 and 11 kg/t respectively with the higher iron levels (Fig.6).

CYANIDATION

All three composites were cyanided for 24 h following an acid pre-leach at pH 1.5 and 400 mV. Extractions of 91, 94, and 90% of the gold were achieved for composites A, B, and C respectively. Cyanide consumptions were 6.7, 0.93, and 0.45 kg/t NaCN respectively. Lime consumptions were 0.9, 2.3, and 1.4 kg/t CaO respectively. During cyanidation of the sulphide-oxide composite (A), the cyanide level was increased to 0.2% NaCN to overcome the known detrimental effects in cyanidation of the sulphide minerals present.

Oxide composite C was also cyanided without a pre-leach. Gold extraction was similar to that following a pre-leach (93%), but the cyanide consumption was significantly higher at 7.9 kg/t NaCN. Lime consumption was 0.3 kg/t CaO.

DISCUSSION

The results for the three composite samples indicate that thiourea will dissolve gold from Selwyn gold-copper ore, but contrary to the suggestion of Haque (1987), copper minerals interfere with the extraction of gold by thiourea. To apply thiourea to such an ore it appears to be necessary to first pre-leach thiourea-soluble copper from the ore. Deschênes and Ghali (1988) found that a similar treatment was necessary before thiourea dissolution of gold from a chalcopyrite concentrate.

A summary of the optimum leaching conditions and possible reactions in the system is given here with reference to the ore mineralogy, particularly of the copper minerals and the gold. The problem of

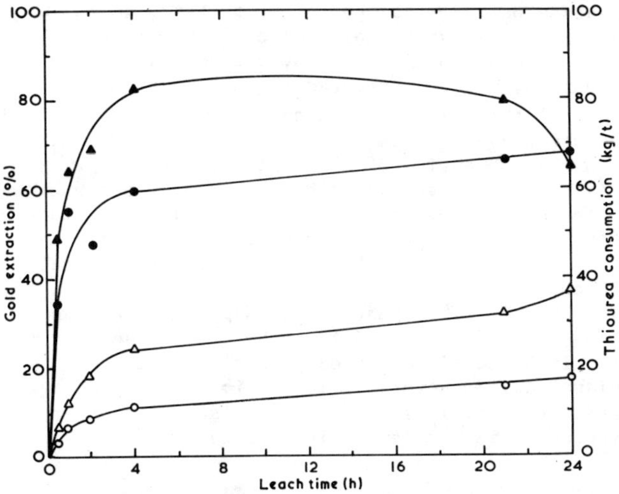

Fig. 5 - Effect of redox potential on gold dissolution and thiourea consumption during leaching of composite B with 0.4% thiourea and 1500 ppm ferrous iron at pH 1.5.
▲ Gold extraction at 250 mV.
● Gold extraction at 200 mV.
△ Thiourea consumption at 250 mV.
○ Thiourea consumption at 200 mV.

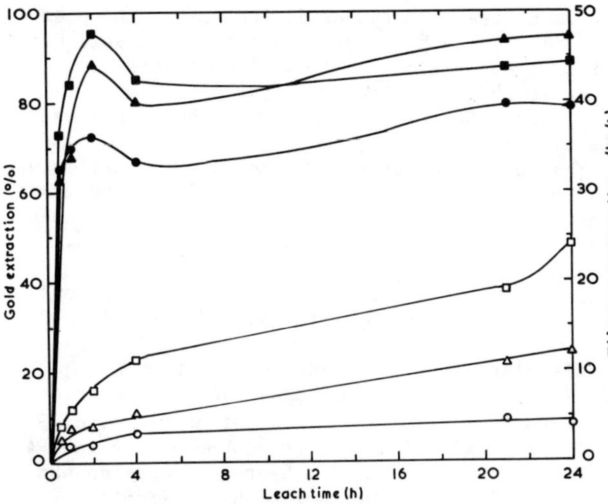

Fig. 6 - Effect of redox potential on gold dissolution and thiourea consumption during leaching of composite B with 0.4% thiourea and 18,000 ppm ferrous iron at pH 1.5.
■ Gold extraction at 250 mV.
▲ Gold extraction at 200 mV.
● Gold extraction at 170 mV.
□ Thiourea consumption at 250 mV.
△ Thiourea consumption at 200 mV.
○ Thiourea consumption at 170 mV.

reprecipitation of the gold is considered, and the effect of redox potential, and its control by ferrous iron (iron(II)) additions, on thiourea consumption and the extraction of gold is discussed.

LEACH REACTIONS AND CONDITIONS

Numerous reports on the dissolution of gold with thiourea have been summarised by Hiskey (1981, 1984), and Gabra (1984) has investigated the kinetics of gold dissolution from pyrite.

Gold dissolution with thiourea can be written as in Eqn (1).

$$Au^0 + 2(NH_2)_2CS + Fe^{3+} \rightarrow Au((NH_2)_2CS)_2^+ + Fe^{2+} \quad (1)$$

An oxidant such as iron(III), or hydrogen peroxide, is added to the ore or concentrate in the presence of thiourea at pH 1-3 to maintain the redox potential above the standard redox potential (134 mV with reference to a SCE; Skibsted and Bjerrum, 1977), in order to dissolve the gold.

At redox potentials above 200 mV, oxidation of thiourea to formamidine disulphide (Eqn 2) or cyanamide (Eqn 3) becomes important.

$$2(NH_2)_2CS + 2Fe^{3+} \rightarrow NH_2(NH)CSSC(NH)NH_2 + 2Fe^{2+} + 2H^+ \quad (2)$$

$$NH_2(NH)CSSC(NH)NH_2 \rightarrow (NH_2)_2CS + NH_2CN + S^0 \quad (3)$$

While formamidine disulphide has been shown to be an oxidant for gold (Groenewald, 1976), the formation of cyanamide results in a loss of thiourea from the system and so redox potentials above 200 mV should be avoided when possible.

The reaction between thiourea and Fe(III) has been studied by Zhu and Zhau (1988). While their methodolgy was not entirely clear, it was shown that besides the concentrations of the reactants, additions of cupric ions, sulphide minerals, and metallic iron affected the reaction rate, as well as the dissolution rate of gold wire in the solution.

The reaction of thiourea solutions with copper minerals does not seem to have been investigated. However, it is known that cupric ions are reduced by thiourea to form cuprous-thiourea complexes, such as $[Cu((NH_2)_2CS)_3]_2SO_4$ and $[Cu((NH_2)_2CS)_3]Cl$ (Nesbitt, Hendrix, and Nelson, 1985). Since all copper minerals are soluble in the acid oxidising conditions needed for gold dissolution in thiourea, the copper minerals can be expected to dissolve and the resultant cupric ions to react with thiourea. Alternatively, oxide and sulphide copper minerals could dissolve by direct reaction with thiourea. Any copper in solution is thought to be present as a cuprous species.

A white copper-thiourea compound precipitated from pregnant liquor filtrates containing over 200 ppm copper. Presumably during the leach this copper compound precipitates on the solids. A similar compound was observed to precipitate from solution by Xia and Wang (1988). Hiskey (1984) suggests that copper is precipitated from solution during the leach by reaction with sulphinic intermediates from the oxidation of the thiourea. An acid pre-leach was used to remove the acid soluble copper from the Selwyn composites to minimise the formation of copper-thiourea complexes. Between 58 and 76% of the copper was dissolved.

The best gold extraction from the composites was achieved after a pre-leach at pH 1.5 and 400 mV to remove copper. About 90-95% of the gold was then dissolved in 0.4% thiourea solution at pH 1.5. However, a redox potential of 200-250 mV was required resulting in high thiourea consumptions. At 180 mV only 75-80% of the gold dissolved in 24 h.

Dissolution of the gold was faster when iron(III) sulphate rather than hydrogen peroxide was used to control the redox potential. Over 90% of the gold was dissolved from Composite A in 2-4 h compared with 21-24 h when hydrogen peroxide was used (Fig. 2). However additions of iron(II) to give 18,000 ppm iron in solution were effective in minimising thiourea consumption while maintaining rapid gold dissolution.

ORE MINERALOGY

The presence of the copper and iron minerals, and the form of the gold in the composites, affected the composition of the leach liquors and the dissolution of the gold.

Oxide and Sulphide Minerals

Insufficient data were obtained to fully correlate ore mineralogy with various chemical reactions, including copper extraction during both pre-leaching and thiourea leaching. Copper extraction during pre-leaching depended on how vigorous the conditions were, but maximum extractions were about 65, 58, and 76% for composites A, B, and C respectively. The form of the copper remaining in the solids (assay 0.3% Cu) was not established, but it was not thiourea soluble as less than 1% additional copper was extracted during thiourea leaching. Part, at least, of the copper may be chrysocolla. It appeared that the sulphide copper minerals needed more vigorous pre-leach conditions to dissolve them.

Pyrite was thought to contribute to reprecipitation of gold as noted later, and possibly contributed iron to the leach solution. However, the iron oxides present (magnetite, martite, hematite, goethite, and limonite) probably had an influence on the amount of iron dissolved and control of the redox potential. Further work is needed to evaluate these effects.

Gold

Two forms of native gold were observed in the drill core sections from which the composites were prepared. Fine gold particles (1-3 μm) were found to contain 1.6% copper and 0.1% silver, while a 30 μm gold particle contained virtually no copper and 6% silver.

The leaching results in Tables 2 and 3, and elsewhere, show that the amount of gold dissolved depends on the redox potential of the leach. In order to achieve dissolution of over 90% of the gold a redox potential of over 200 mV was required. It is suggested that the more common form of the gold (the low-copper gold) is dissolved by thiourea at redox potentials around 175 mV while a higher potential of 250 mV is required to dissolve the high-copper gold. In previous work by Bilston, LaBrooy, and Woodcock (1984) with samples from Western Australia that did not contain appreciable amounts of copper, it was found that thiourea dissolved the gold at redox potentials of 180-200 mV. This tentatively suggests that the copper-containing gold is less reactive and requires the higher redox potential to be dissolved. However, it is possible that the higher redox potential is required to dissolve another (unidentified) gold-bearing mineral.

REPRECIPITATION OF GOLD

A problem in many of the leaches above 200 mV, particularly with the oxide composites, was reprecipitation of the gold from solution between 2 and 21 h (Fig. 4). A similar observation has been reported by Deschênes and Ghali (1988).

The precipitation of the gold during the leach is probably due to a number of factors, all of which may contribute to the re-precipitation of gold in this system. These factors are

1. reactions of the gold thiourea complex with decomposition products of thiourea (Hiskey, 1984),
2. reactions of the gold thiourea complex with H_2S generated from sulphide minerals,
3. dissociation of the thiourea complex and subsequent reduction of the gold back to the elemental state as a result of a reduction in thiourea concentration in solution with time,
4. cementation onto metallic iron from grinding media, and
5. adsorption onto minerals in the pulp.

With the oxide composite B, increasing the concentration of iron in solution reduced the amount of gold that precipitated during the leach and the thiourea consumptions as shown in Figs. 5 and 6. It is likely that the species in solution change as the iron concentration increases and that the thiourea is complexed, decreasing its rate of oxidation above 200 mV. In practice the precipitation of the gold may be avoided by using short leach times (less than 4 h).

REDOX POTENTIAL AND IRON CONCENTRATIONS

It has been proposed by Groenewald (1976), and confirmed in this work, that iron(III) is the best oxidant for the dissolution of gold with thiourea as shown in Eqn (1). With an increase in the total iron concentration in the leach liquor the effective concentration of iron(III) is increased at a particular redox potential (adjusted in this work with hydrogen peroxide) since the redox potential is controlled by the ratio of the concentration of iron(III) to iron(II) species.

The results for the Selwyn oxide composites (Figs. 5 and 6) indicate that the rate of dissolution of gold increases with the concentration of iron in the leach liquor, and that there appears to be a critical concentration of iron required in order to achieve 95% dissolution of the gold in 2 h.

The best way to achieve the desired iron concentrations is to add an iron(II) salt, not an iron(III) salt, since additions of iron(III) oxidise a significant proportion of the thiourea in reaching an equilibrium iron(III)/iron(II) ratio. Most of the iron in solution is present as iron(II) species, a point which many investigators fail to realise.

Chen (1982) and Xia and Wang (1988) used metallic iron to precipitate gold from thiourea leach pulp, and to reduce iron(III) to iron(II). This would also contribute iron(II) to the solution, as would some of the iron abraded from grinding media.

Without additions of iron, the iron contents of the thiourea leach solutions of the oxide composites were lower than those for the sulphide-oxide composite. This may reflect the presence of pyrite and chalcopyrite in the sulphide-oxide composite, as well as the iron oxide minerals that are also present in the oxide composites.

THIOUREA CONSUMPTION

High thiourea consumptions were obtained as a result of the requirement to carry out the thiourea leaches above 200 mV. With the sulphide-oxide composite, the thiourea consumption increased from less than 10 to 40 kg/t at 24 h on raising the redox potential from 180 mV to over 200 mV. However, at redox potentials over 200 mV the gold was dissolved in 2-4 h under the best conditions, and the thiourea consumptions were 25-30 kg/t. For the oxide composite C, it was only 10 kg/t. It should be pointed out that thiourea was added during the leach to maintain a constant level of 0.4%. There was not simply a single addition of thiourea at the commencement of the leach.

Thiourea consumptions were reduced when iron(II) was added to the leach as shown in Figs. 5 and 6. This reduction in thiourea consumption is likely to be associated with the formation of iron(III) and iron(II) complexes such as $[Fe(SO_4)((NH_2)_2CS)_2]^+$ and $[Fe((NH_2)_2CS)_2]SO_4$ respectively (Groenewald, 1976). Presumably when complexed with the iron the thiourea is oxidised less rapidly but is still available to complex with the gold.

The pre-leach conditions also affected the thiourea consumption as shown in Fig. 3. Thiourea consumptions were highest when there was no pre-leach, and when the slurry was not filtered after a pre-leach. After a pre-leach at pH 2.5 and 100-250 mV, the copper assay value of the solution was 790 ppm. This was decreased to 29 ppm on addition of thiourea for the gold leach, presumably due to the precipitation of a copper-thiourea complex. Thus copper in solution increases thiourea consumption by precipitation of a complex, and any copper in solution is probably present in the cuprous form since excess thiourea is present.

Also, with the sulphide-oxide composite, iron levels in solution were greatest after a pre-leach at pH 1.5 and 400 mV. The high iron level did not increase thiourea consumption and did not appear to be detrimental to dissolution of the gold.

Schulze (1984) has reported that thiourea consumption can be reduced by adding sulphur dioxide to the system to reduce formamidine disulphide back to thiourea. This was not attempted in this work. However, there are conflicting reports in the literature on the effectiveness of sulphur dioxide in reducing the thiourea consumption (Bilston, Millett, and Woodcock, 1987; Nomvalo, 1986; Yen and Wyslouzil, 1986; Deschênes and Ghali, 1988).

COMPARISON OF THIOUREA AND CYANIDE LEACHING

The application of thiourea leaching or cyanidation to the extraction of gold from Selwyn ore requires a relatively similar circuit in that both need a pre-leaching step to remove copper before gold extraction.

Leaching with 0.4% thiourea at pH 1.5 and a redox potential of 200-250 mV following an iron(II) addition can extract 90-95% of the gold in 2-4 h. A thiourea consumption of about 10 kg/t would be expected.

Cyanidation with 0.08% NaCN for 24 h can extract 90-95% of the gold. Cyanide consumption on oxide ore could be less than 1 kg/t, but would be much higher (7 kg/t) on sulphide-containing ore.

While not studied in this work, controlled potential sulphidization and flotation (Jones, Wong, and Woodcock, 1986), to produce a gold-containing copper concentrate for smelting, is also an option for treatment of the oxide and the sulphide-oxide ore of the Selwyn deposit.

CONCLUSIONS

Thiourea was shown to dissolve gold from the Selwyn oxidised gold-copper ore, but the copper interfered. Acid pre-leaching under oxidising conditions to remove copper followed by a short thiourea leach (4 h) at a redox potential above 200 mV gave the best results (i.e., high gold extraction and moderate thiourea consumption). Addition of iron(II) to the leach pulp reduced thiourea consumption and minimised reprecipitation of the gold during the leach.

ACKNOWLEDGEMENTS

The authors would like to thank the Selwyn Joint Venture (Cyprus Gold Australia Corporation, Elders Resources Limited, and Arimco N.L.) for its collaboration in this research project and for providing appropriate samples. Experimental assistance by Mr D.W. Bilston and Mr C.M. MacRae of the CSIRO Division of Mineral Products is gratefully acknowledged.

REFERENCES

Bilston, D.W., LaBrooy, S.R., and Woodcock, J.T., 1984. Gold and silver leaching from an oxidized gold ore with thiourea under controlled conditions, in Extractive Metallurgy Symposium, pp. 51-60 (Australasian Institute of Mining and Metallurgy: Melbourne).

Bilston, D.W., Millett, R.N., and Woodcock, J.T., 1987. The effect of manganese dioxide and sodium sulphite on thiourea leaching of a gold ore, in Research and Development in Extractive Metallurgy - 1987, pp. 19-28 (Australasian Institute of Mining and Metallurgy: Melbourne).

Chen Deng Wen, 1982. Studies and prospects of gold extraction from carbon bearing clayey gold ore by the thiourea process, in Preprints XIV International Mineral Processing Congress, Vol. 1, paper II-8, 11 pp. (Canadian Institute of Mining and Metallurgy : Montreal).

Deschênes, G., and Ghali, E., 1988. Leaching of gold from a chalcopyrite concentrate by thiourea, Hydrometall., 20: 179-202.

Gabra, G., 1984. A kinetic study of the leaching of gold from pyrite concentrate using acidified thiourea, in Precious Metals: Mining, Extraction, and Processing (Ed. V. Kudryk, D.A. Corrigan, and W.W. Liang), pp. 145-172 (Metallurgical Society of AIME: Warrendale).

Groenewald, T., 1976. The dissolution of gold in acidic solutions of thiourea, Hydrometall., 1: 277-290.

Haque, K.E., 1987. Gold leaching from refractory ores - literature survey, Mineral Process. Extr. Metall. Rev., 2: 235-253.

Harrowfield, I.R., MacRae, C.M., and Simmonds, P.F., 1988. The automated scanning electron microscope as a tool for gold microprospecting, in Microbeam Analysis 1988 (Ed. D.E. Newbury), pp. 481-482 (Microbeam Analysis Society and San Francisco Press: San Francisco).

Hiskey, J.B., 1981. Thiourea as a lixiviant for gold and silver, in Gold and Silver: Leaching, Recovery and Economics (Ed. W.J. Schlitt, W.C. Larson, and J.B. Hiskey), pp. 83-91 (American Institute of Mining and Metallurgical Engineers: New York).

Hiskey, J.B., 1984. Thiourea leaching of gold and silver - technology update and additional applications, Minerals Metall. Process., 1: 173-179.

Jones, M.H., Wong, K.Y., and Woodcock, J.T., 1986. Controlled-potential sulphidization and rougher-cleaner flotation of an oxide-sulphide copper ore, in Publications of the 13th CMMI Congress, Vol. 4, Metallurgy (Ed. L.E. Fielding and A.R. Gordon), pp. 33-40 (13th Congress of Mining and Metallurgical Institutions and The Australasian Institute of Mining and Metallurgy: Melbourne).

Morrison, I., 1986. The Starra Au-(Cu) project, N.W. Queensland, in Gold Exploration and Development North Queensland Conference, pp. 87-93 (Australasian Institute of Mining and Metallurgy: Melbourne).

Nesbitt, C.C., Hendrix, J.L., and Nelson, J.H., 1985. Use of thiourea for precipitation of heavy metals in metallurgical operation effluents, in Extraction Metallurgy '85 (Ed. M.J. Jones), pp. 355-374 (Institution of Mining and Metallurgy: London).

Nomvalo, Z.T., 1986. Thiourea leaching of Witwatersrand ore, in Gold 100, Vol. 2: Extractive Metallurgy of Gold (Ed. C.E. Fivaz), pp. 565-577 (South African Institute of Mining and Metallurgy: Johannesburg).

Schulze, R.G., 1984. New aspects in thiourea leaching of precious metals, J. Metals, 36(6): 62-65.

Skibsted, L.H., and Bjerrum, J., 1977. The standard electrode potentials of the gold aqua ions and the stabilities of gold(I) and gold(III) complexes in aqueous solution, J. Indian Chem. Soc., LIV: 102-108.

Xia Guangxiang and Wang Dan, 1988. A study of physical chemistry on gold extraction by thiourea process by IIL-technique, in Proceedings of the First International Conference on Hydrometallurgy (ICHM '88) (Ed. Zheng Yulian and Xu Jiazhong), pp. 521-525 (International Academic Publishers: Beijing).

Yen, W.T., and Wyslouzil, D.M., 1986. Pressure oxidation and thiourea extraction of refractory gold ore, in Gold 100, Vol. 2: Extractive Metallurgy of Gold (Ed. C.E. Fivaz), pp. 579-589 (South African Institute of Mining and Metallurgy: Johannesburg).

Zhu Tun and Zhau Ling, 1988. Some properties of acidic thiourea solution, in Proceedings of the First International Conference on Hydrometallurgy (ICHM '88) (Ed. Zheng Yulian and Xu Jiazhong), pp. 513-525 (International Academic Publishers: Beijing).

Chapter 37

THE PRESSURE AQUEOUS PRE-OXIDATION OF A REFRACTORY GOLD ORE FROM THE GOLDEN MILE, KALGOORLIE, WESTERN AUSTRALIA

Grazyna Dziurdzak*, James H. Kyle* and Robert C. Dunne#

*Western Australian School of Mines, Kalgoorlie, W.A.
#Newmont Australia Ltd., Belmont, Western Australia

ABSTRACT

The pressure pre-oxidation of refractory gold concentrate from the Golden Mile has been investigated and the conditions of temperature, oxygen pressure and percent sulphide breakdown for maximum gold extraction optimised. Both tellurides and sulphides were oxidised under optimal conditions and deposited in the residue as oxides along with haematite. Under non-optimal conditions, a reactive basic ferric sulphate was formed instead on the unreactive haematite. Maximum gold recovery after cyanidation (tail < 1 ppm) is well in excess of that produced by the present roasting process (tail 3-4 ppm).

INTRODUCTION

Gold ores are classified as refractory when a significant portion of the gold cannot be extracted by conventional cyanidation. The refractoriness may be of a physical or a chemical nature. The former type is usually due to sub-microscopic particles of gold being locked within mineral particles, for example in sulphides or silicates, whereas the latter is due to the unreactive nature of the gold compounds present in the ore (e.g. calaverite $AuTe_2$, petzite Ag_3AuTe_2, maldonite Au_2Bi, aurostibite $AuSb_2$).

If the gold is physically locked within sulphide minerals such as pyrite (FeS_2) or arsenopyrite (FeAsS), then partial or complete oxidation of the minerals is necessary to release the gold. This process will also lead to the decomposition of refractory gold compounds such as those listed above.

In Australia, roasting is currently the most cost effective means of oxidising refractory pyrite and arsenopyrite concentrates to produce a product amenable to cyanidation. On the Golden Mile, roasting of pyrite concentrates followed by cyanidation produces a residue still containing 3.5 ppm gold. A recent study by Corrans and Hayes (1989) has indicated that, due to problems with the roasting process, this gold is still locked in partially oxidised pyrite and is difficult to recover.

Another major drawback of roasting is that it produces large quantities of sulphur dioxide gas which is released into the atmosphere. In the semi-arid regions of Western Australia, where refractory gold ores are currently being mined, this is currently acceptable practice although some limitations are put on emissions. In many other regions of Australia, this is not acceptable and the sulphur dioxide would need to be collected, or an alternative technology used.

One such alternative technology that has previosly been used successfully in the United States, and more recently also in Brazil, is that of pressure aqueous pre-oxidation of the sulphide minerals. This process utilises air or oxygen gas at high pressures and temperatures to oxidise an aqueous slurry of the ore or concentrate to produce haematite, iron sulphates and sulphuric acid.

The main objective of the present study was to find the optimum conditions for the aqueous presssure pre-oxidation of a refractory gold ore concentrate from the Golden Mile, containing both sulphides (mainly pyrite) and tellurides, such that the maximum subsequent recovery of gold was achieved by conventional cyanidation.

Chemistry of the Aqueous Oxidation of Pyrite

The reactions occurring during the aqeuous oxidation of pyrite are not simple and are affected by temperature, pH, and the solubility of the products. In the course of the oxidation the following reactions may take place (Beigler & Swift, 1978; Hiskey, 1981; Chander & Briceno, 1987):

At low pH:

$$2FeS_2 + O_2 + 2H_2SO_4 \text{ ---> } 2FeSO_4 + 2H_2O + 4S \quad ..(1)$$

At higher pH:

$$2FeS_2 + 7O_2 + 2H_2O \text{ ---> } 2FeSO_4 + 2H_2SO_4 \quad ....(2)$$

$$FeS_2 + H_2SO_4 \text{ ---> } FeSO_4 + H_2S + S \quad ....(3)$$

$$4FeSO_4 + O_2 + 2H_2SO_4 \text{ ---> } 2Fe_2(SO_4)_3 + 2H_2O \quad ..(4)$$

The hydrogen sulphide produced can be oxidised either by oxygen or by ferric sulphate:

$$2H_2S + O_2 \text{ ---> } 2H_2O + 2S \quad ....(5)$$

$$H_2S + 2O_2 \text{ ---> } H_2SO_4 \quad ....(6)$$

$$H_2S + Fe_2(SO_4)_3 \text{ ---> } 2FeSO_4 + H_2SO_4 + S \quad ..(7)$$

$$H_2S + 4Fe_2(SO_4)_3 + 4H_2O \text{ ---> } 8FeSO_4 + 5H_2SO_4 \quad ....(8)$$

Ferric sulphate may also contribute to the oxidation of pyrite:

$$FeS_2 + 4Fe_2(SO_4)_3 + 4H_2O \text{ ---> } 9FeSO_4 + 4H_2SO_4 + S \quad ..(9)$$

$$FeS_2 + 7Fe_2(SO_4)_3 + 8H_2O \text{ ---> } 15FeSO4 + 8H_2SO_4 \quad ....(10)$$

Elemental sulphur may also be further oxidised to sulphate by oxygen or by ferric sulphate:

$$2S + 3O_2 + 2H_2O \text{ ---> } 2H_2SO_4 \quad ....(11)$$

$$S + 3Fe_2(SO_4)_3 + 4H_2O \text{ ---> } 6FeSO_4 + 4H_2SO_4 \quad ...(12)$$

The oxidation of ferrous to ferric ion (reaction 4) is very slow under normal conditions but is greatly accelerated in autoclaves at elevated temperature and pressure. The rate of this reaction can be further enhanced by the use of catalysts such as cupric ion or activated carbon (Posner, 1953).

The oxidation of sulphur is strongly temperature dependent and very slow below 120°C, the melting point of monoclinic sulphur. Above 160°C no sulphur is formed at any pH (Umetsu et al., 1977).

Depending on the temperature and the pH, soluble ferric ion is removed from the solution either as goethite, a basic ferric sulphate, a jarosite or as haematite. At low temperatures goethite is formed whereas above 110°C, the following reactions may take place:

At low acidity:

$$Fe_2(SO_4)_3 + 3H_2O \text{ ---> } Fe_2O_3 + 3H_2SO_4 \quad ......(13)$$

At moderate acidity:

$$3Fe_2(SO_4)_3 + 14H_2O \text{ ---> } 2(H_3O)Fe_3(OH)_6(SO_4)_2 + 5H_2SO_4 \quad .(14)$$

$$2(H_3O)Fe_3(OH)_6(SO_4)_2 + 3H_2SO_4 \text{ ---> } FeOHSO_4 + Fe_2(SO_4)_3 + 10H_2O \quad .(15)$$

$$2FeOHSO_4 + H_2O \text{ ---> } Fe_2O_3 + 2H_2SO_4 \quad ....(16)$$

At high acidity:

$$Fe_2(SO_4)_3 + 2H_2O \text{ ---> } 2FeOHSO_4 + H_2SO_4 \quad .....(17)$$

The percentage of sulphur in the precipitate depends on temperature, the concentration of free sulphuric acid in solution and the concentration of precipitating ions such as Zn^{2+} and/or Cu^{2+} (Chander & Briceno, 1987).

The boundaries between low, moderate and high acidity are temperature dependent. In the present study, at 200°C, the basic ferric sulphate was formed at concentrations of sulphuric acid greater than 68 g/L.

Chemistry of Gold Tellurides

Geochemistry. Gold combines with Tellurium to form a number of hypogene tellurides and telluride-sulphides of which the most common are listed below:

Sylvanite	$AuAgTe_2$
Calaverite	$AuTe_2$
Krennerite	$AuAgTe_4$
Montbroyite	Au_2Te_3
Petzite	$AuAg_3Te_2$

The behaviour of tellurium during oxidation of gold tellurides is varied. It would seem from previous investigations that, when the oxidation potential is relatively low, the products are mainly tellurous acid and soluble tellurites.

$$AuTe_2 + 2O_2 + 2H_2O \text{ ---> } Au + 2H_2TeO_3 \quad .(18)$$

At higher oxidation potentials, telluric acid and soluble tellurates are produced.

$$AuTe_2 + 3O_2 + 6H_2O \text{ ---> } Au + 2H_6TeO_6 \quad (19)$$

Hydrolysis of solutions containing tellurites appears to precipitate TeO_2 close to the primary tellurides. Soluble tellurites and tellurates can be precipitated by a number of cations including ferric ion to produce insoluble minerals such as emmonsite ($Fe_2Te_3O_9.2H_2O$). Adsorption of the anions by hydrous ferric oxide and coprecipitation and/or adsorption by jarosite may also take place (Boyle, 1979).

Chemistry of Extraction. There is considerable doubt as to the actual refractoriness of gold tellurides. Almost invariably gold tellurides are found in conjunction with metallic sulphides, and it is thought that processing difficulties may result principally from gold locked in pyrite rather than from the lack of reactivity of the tellurides. However, recent evidence suggests that the lack of reactivity is due to the formation of a passivating film of tellurous acid on the surface of the mineral (Jayasekera et al., 1988). In highly oxidising media such as hypochlorite this surface film does not form and leaching of the mineral is fast (Jayasekera, personal communication).

Tellurides subjected to cyanidation react as follows:

$$4AuTe_2 + 8CN^- + 9O_2 + 12OH^- \text{---> } 4Au(CN)_2^- + 8TeO_3^{2-} + 6H_2O \quad (20)$$

or if oxygen is insufficient:

$$4AuTe_2 + 8CN^- + 3O_2 + 12OH^- \text{ ---> } 4Au(CN)_2^- + 4Te^{2+} + 4TeO_3^{2-} + 3H_2O \quad (21)$$

During roasting, gold tellurides are decomposed liberating gold while tellurium volatilises.

EXPERIMENTAL

All pressure oxidation testwork was performed in an electrically-heated 3.8 L pure titanium autoclave using pure oxygen gas. Temperature was controlled to within 1^oC using a thermocouple and temperature bath. Pressure was controlled manually and measured with a pressure gauge. Gases were slowly bled from the reactor during each test to ensure the reactor overpressure was due solely to oxygen and water vapour. The autoclave was provided with a flat-blade turbine impeller (run at 1500 rpm), means for venting the gases, introducing high pressure gases, and sampling the slurry.

A standard bottle roller was used for all cyanidation testwork.

All reagents used were of laboratory reagent grade and fresh tap water was used to prepare the slurries.

Procedure

A measured amount of flotation concentrate and water were placed in the autoclave. The autoclave was sealed, the temperature set to target and the heater and agitator started. The slurry was heated under nitrogen. When the target temperature was achieved, compressed oxygen was introduced with the venting valve slightly open to rid the reactor of nitrogen. The pressure was then adjusted to the desired value and the timeclock started. The reaction was allowed to proceed for a set time. At the end of a run the system was quickly cooled to 50^oC and the pressure decreased to atmospheric.

The parameters investigated in these tests were temperature ($170\text{-}200^oC$), oxygen partial pressure (340- 1730 kPa), percentage solids in the slurry (10-20%) and autoclave retention time (0.25-4.0 hrs). The total autoclave pressure ranged from 1070 to 3130 kPa.

The slurry was then discharged through the blowpipe, filtered, washed and dried. The solution was analysed for acidity, iron, sulphate and in some cases tellurium. The solid residue was subjected to a standard 24 hour bottle cyanidation test at pH > 11.5 with 0.5% NaCN.

RESULTS AND DISCUSSION

The material used in this study was flotation concentrate from the North Kalgurli Mines Croesus Plant. It comprised a mixture of sulphide minerals, mostly pyrite, together with non-sulphide gangue, mainly quartz, sericite and calcite. Small grains of native gold and telluride minerals (mainly petzite and coloradoite HgTe) are present as discrete particles, inclusions in sulphide or in non-sulphide gangue. An analysis of some of the major elements in the flotation concentrate is shown below:

Fe	30.8%	S	32.5%	SiO_2	17.2%
Au	79 ppm	Te	100 ppm	CaO	2.4%
Ag	31 ppm	Hg	39 ppm	Cu	631 ppm
Co	126 ppm	Ni	259 ppm	Zn	185 ppm

Size analysis indicated that 67% of the material was minus 75 um.

Gold Liberation

The results of the pressure oxidation tests, showing the gold recoveries achieved at maximum sulphur breakdown are listed in Table 1.

Standard cyanidation tests run prior to pressure oxidation indicated that the flotation concentrate contained 25 ppm of refractory gold. Of this, 9 ppm could be released by selective oxidation of the tellurides. The remaining 16 ppm was therefore locked within the sulphide minerals as elemental or combined gold.

The relationship between residual gold values and sulphide breakdown is linear and independent of the temperature and pressure at which the pressure oxidation is performed. Therefore to totally release this remaining gold, it is necessary to completely destroy the pyrite matrix (Figure 1).

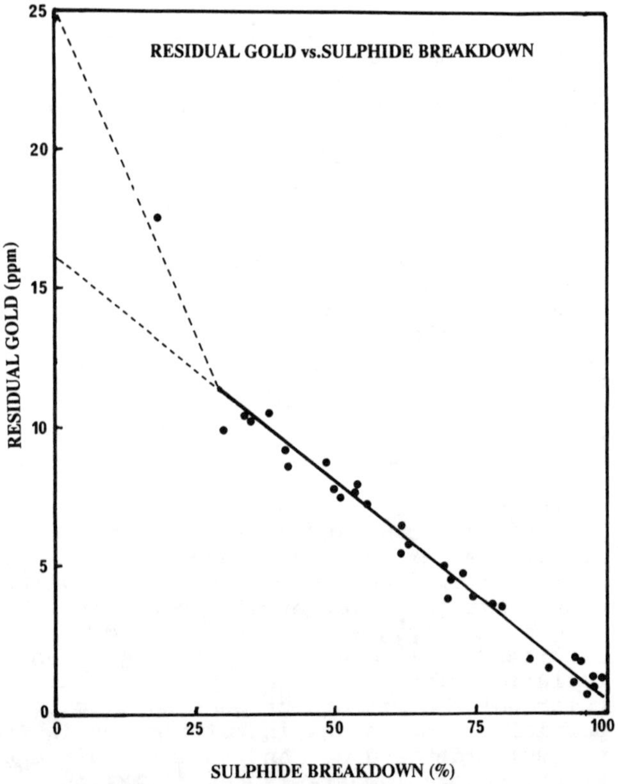

FIGURE 1. The Relationship between Residual Non-leachable Gold in the Concentrate and Sulphide Breakdown.

Mineralogy of Leach Residue

A mineralogical examination of the leach residue showed that the sulphides had been replaced by oxides, mainly haematite. The resulting structure was very porous thus allowing cyanide and dissolved oxygen to gain access to the previously locked gold.

Table No. 1

RESULTS OF PRESSURE OXIDATION TESTS

Temp. (°C)	Oxygen Pressure (kPa)	Reaction Time (Hrs.)	Sulphide Breakdown (%)	Refractory Gold Recovery(%)	Total Gold Recovery(%)
170	520	4	71.0	49.6	84.0
	920	4	97.5	84.8	95.2
	1330	3	96.0	77.2	92.8
	1730	2.5	94.2	84.8	95.2
180	510	4	96.3	69.6	90.4
	910	3	98.2	69.6	90.4
	1320	2	95.4	90.0	96.8
	1720	1.5	94.1	82.4	94.4
190	460	3	97.4	88.8	96.4
	660	2	93.5	74.8	92.0
	1060	1.5	96.8	84.8	95.2
	1470	1.0	94.8	84.8	95.2
200	340	3.0	97.2	91.2	97.2
	750	1.5	96.3	88.0	96.2
	950	1.25	98.3	92.4	97.6
	1360	0.75	98.5	88.8	96.4

Formation of $FeOHSO_4$

The flotation concentrate used contained a high percentage of sulphur (32.5%) which produces considerable quantities of free sulphuric acid during the oxidation process. This, in turn, encourages the formation of a basic iron sulphate, termed here $FeOHSO_4$, whose exact composition and crystalline structure is at present unknown. The formation of this precipitate is undesirable for two reasons. Firstly, the leachate slurry is very difficult to filter, and secondly, the basic iron sulphate decomposes to haematite at higher pH releasing sulphuric acid and causing excessive consumption of caustic soda before and during cyanidation.

The optimum solids concentration to avoid basic iron sulphate formation was found to be 10% w/w. However, even at this low pulp density, $FeOHSO_4$ is formed under conditions of high temperature and low oxygen pressure. At 190 and 200°C, it was formed at pressures below 1000 kPa whereas at lower temperatures, it was not formed at any pressure. This implies that, at low slurry densities, basic iron sulphate may be formed by a kinetic effect rather than it being a thermodynamically stable product. However, without a knowledge of the kinetics of the various reactions that may occur in this complex system, this cannot be stated with any certainty.

Kinetics of Sulphide Decomposition

The kinetics of the sulphide decomposition was investigated at a slurry concentration of 10% solids. The range of pressures and temperatures studied were 340 - 1710 kPa and 170 - 200°C respectively. The results obtained are shown in Figures 2 - 5.

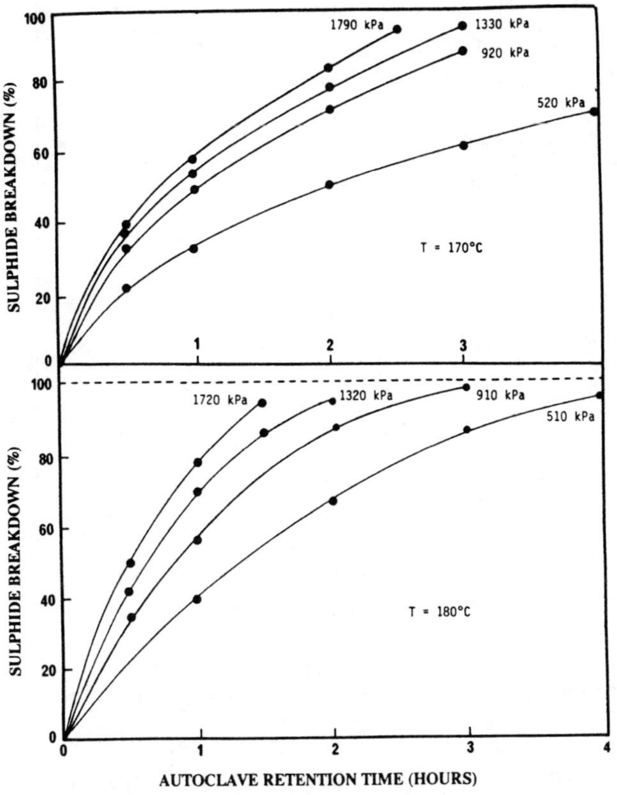

FIGURES 2 & 3. The Percentage Sulphide Breakdown in the Concentrate vs. Autoclave Retention Time at 170 & 180°C and a Range of Pressures.

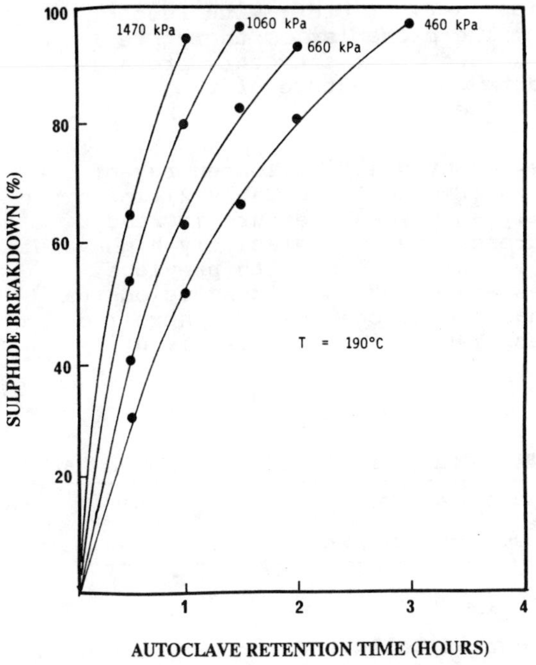

FIGURE 4. The Percentage Sulphide Breakdown in the Concentrate vs. Autoclave Retention Time at 190°C and a Range of Pressures.

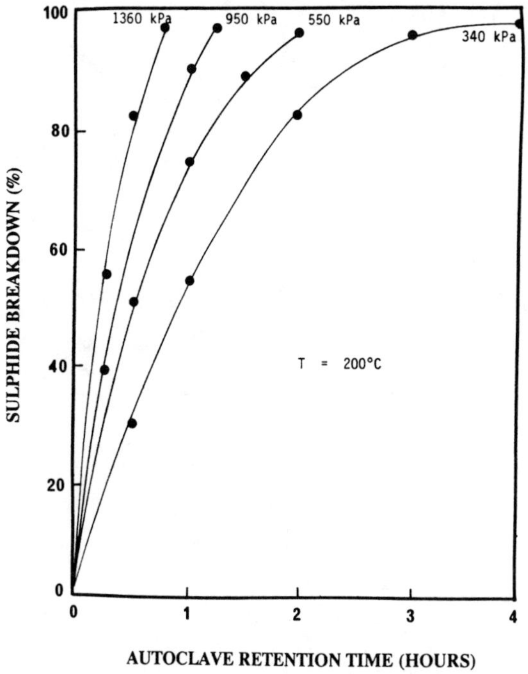

FIGURE 5. The Percentage Sulphide Breakdown in the Concentrate vs. Autoclave Retention Time at 200°C and a Range of Pressures.

In Figure 6, the data are plotted according to the surface reaction model:

$$1 - (1 - \alpha)^{1/3} = k_o t \qquad (22)$$

where α is the fraction of sulphur breakdown, k_o is the rate constant and t is the time in hours. The straight line fit passing through the origin indicates that the rate is controlled by the chemical reaction at the surface of the pyrite. The curved plot obtained at 170°C and 520 kPa is indicative of a change of mechanism occurring, probably from chemical to mass transport control at lower temperatures and pressures.

A plot of the rate constant k_o as a function of oxygen pressure is shown in Figure 7. The straight line fit in the temperature range 170-190°C indicates first order dependence of the rate on oxygen pressure (PO_2), whilst the non-zero intercept (k_1) indicates a fast reaction prior to the oxidation of pyrite, probably the oxidation of the more reactive sulphides in the concentrate (e.g. pyrrhotite).

$$k_o = k_1 + k_2.P(O_2) \qquad (23)$$

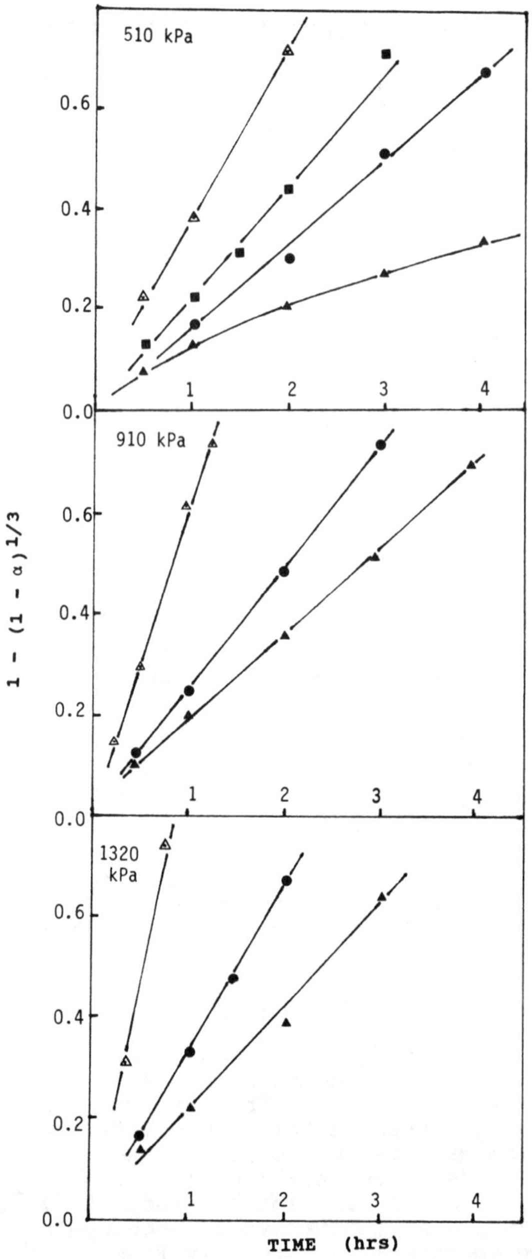

FIGURE 6. Surface Reaction Model Plots for the Oxidation of Concentrate at a Range of Pressures. The Temperatures are 170°C(▲), 180°C(●), 190°C(□) and 200°C(△).

At 200°C, the dependenceof the rate on oxygen pressure becomes second order (see insert in Figure 7):

$$k_o = k_3 + k_4.P(O_2)^2 \text{ ....(24)}$$

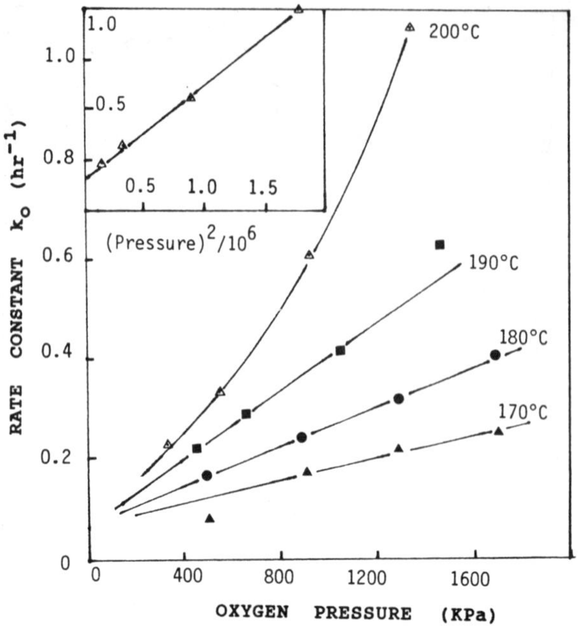

FIGURE 7. Pressure Dependence of the Rate Constant k_o at a Range of Temperatures.

In contrast to these results, previous studies on the rate of oxidation of pyrite under similar conditions (Warren, 1956) indicated a half-order oxygen dependence. However, McKay and Halpern (1958), showed that at lower pressures (100 - 400 kPa) and temperatures (100 - 130°C), the reaction is first order with respect to oxygen. The mechanism is therefore complex, and to some extent, probably dependent on the nature of the concentrate.

The measured activation energy of the reaction (Figure 8 and Table 2) increased slightly as oxygen pressure increased. This trend, and the relatively high values obtained, are in line with previous studies and support the conclusion that the reaction is chemically controlled (Warren, 1956; King & Lewis, 1980).

Table No. 2

MEASURED ACTIVATION ENERGIES FOR PYRITE OXIDATION

Oxygen Pressure (kPa)	500	920	1330
Activation Energy (kJ/mol)	62.6	73.4	85.7

Behaviour of Tellurium

The chemical analysis of the leachate solutions did not indicate the presence of any tellurium. It was, however, present in the solid residue.

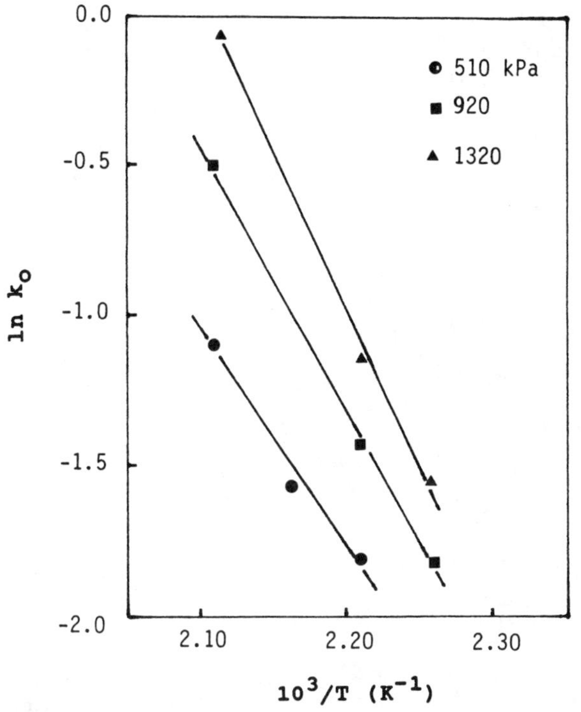

FIGURE 8. Activation Energy Plots for the Oxidation of Concentrate at a Range of Pressures.

The Eh of all leach solutions after pressure oxidation was between +450 and +550 mV and the pH below 1. According to the pH - potential diagram for the tellurium-oxygen system, tellurous acid should be formed under these conditions. Presumably this has co-precipitated with or been adsorbed onto the insoluble iron compounds in the leachate residue.

CONCLUSION

Aqueous pressure pre-oxidation has proven to be effective in treating a pyrite-telluride refractory gold ore from the Kalgoorlie Golden Mile. Optimum conditions for sulphide breakdown so as to produce a residue comprising mainly haematite were: a slurry concentration of 10% solids, a temperature of 170-180°C, and an oxygen pressure of 900-1700 kPa, depending on the reaction time required. Under these conditions , in excess of 95% of the refractory gold could be recovered compared with about 85% presently recovered by a roasting process.

The rate of the process was found to be first order in oxygen pressure, with a measured activation energy ranging from 62.6 kJ/mol at 500kPa to 85.7 kJ/mol at 1330 kPa.

REFERENCES

Beigler, T., and Swift, D.A., 1978, "Anodic Behaviour of Pyrite in Aqueous Solutions", Electrochimica Acta, Vol. 24, 415-420.

Boyle,R.W.,1979, "The Geochemistry of Gold and its Deposit", Bulletin No. 280, Canadian Geological Survey (584p).

Chander, S., and Briceno. A., 1987, "Kinetics of Pyrite Oxidation", Minerals and Metallurgical Processing, August, 171-176.

Corrans, I.J., and Hayes, J., 1989, "The Recovery of Gold fron Leached Calcine Residues," MERIWA Report No. 46, Minerals & Energy Research Institute of Western Australia, Perth, Western Australia.

Hiskey, J.B., and Schlitt, W.J.. 1981,"Aqueous Oxidation of Pyrite", in Interfacing Technologies in Solution Mining, Proceedings of the 2nd SME-SPE International Mining Symposium, Colorado, Nov.,

Jayasekera, S., Ritchie, I.N., and Avraamides, J.,1988, "Electrochemical Behaviour of Gold Telluride in Perchloric Acid Solutions," 7th Australian Electrochemistry Conference, Sydney, February, Royal Australian Chemical Institute, Melbourne, Australia.

King, W.E. & Lewis, J.A., 1980, "Simultaneous Effects of Oxygen and Ferric Ion on Pyrite Oxidation in an Aqueous Slurry," Ind. Eng. Chem. Process Des. Dev., Vol. 19, 719-722.

McKay, D.R. & Halpern, J., 1958, "A Kinetic Study of the Oxidation of Pyrite in Aqueous Suspension," Metallurgical Transactions, Vol. 212, 301-309.

Posner, A.M., 1953, "Kinetics of the Charcoal-Catalysed Autoxidation of Ferrous Ion in Dilute Hydrochloric Acid Solutions," Transactions Faraday Society, Vol. 49, 389-395.

Umetsu, Y., Tozawa, K., and Sasaki, K., 1977, "The hydrolysis of Ferric Sulphate Solutions at Elevated Temperatures", Metallurgical Society of C.I.M. Annual Volume, 111-117.

Warren, I.H., 1956, "The Generation of Sulphuric Acid from Pyrite by Pressure Leaching," Australian Journal of Applied Science, Vol. 7, 346-358.

ACKNOWLEDGEMENTS
Funding from North Kalgurli Mines and WASM Education Fund.
CSIRO Div. Mineral Products (Melb.) for use of autoclave.

Chapter 38

PROCESS SELECTION, DESIGN, COMMISSIONING AND OPERATION OF THE SAO BENTO MINERACAO REFRACTORY GOLD ORE TREATMENT COMPLEX.

by

Mr Edward J. da Silva.
General Mining, Metals and Minerals Limited
Metallurgical Manager

Dr Alan K. Haines
General Mining, Metals and Minerals Limited
Senior Manager: Metallurgical Engineering

Mr Tarcisio M. Carvalho
Sao Bento Mineracao
General Manager

Mr Marcos P. de Melo
Sao Bento Mineracao
Metallurgical Manager

Mr Barry N. Doyle
Sherritt Gordon Mines Limited
Manager, External Technology

ABSTRACT

A metallurgical complex to recover gold from highly refractory arsenical ore was recently commissioned by Sao Bento Mineraco. The plant, with a nominal capacity of 20 000 tonnes per month, is situated in the Minas Gerais province of Brazil . The process consists of run-of-mine milling, flotation, pressure oxidation of the concentrate, carbon-in-leach, elution and electrowinning, and effluent disposal.

The paper provides an account of the investigations that resulted in the decision to install the pressure oxidation circuit, process and engineering design and plant start up.

A comparison is made between the results achieved during the first two years of production and the design parameters.

INTRODUCTION

The Sao Bento Mineracao Metallurgical process plant was the first in Brazil to incorporate fully autogenous run-of-mine milling and only the second worldwide to utilize pressure oxidation technology for the recovery of gold from a refractory ore body.

The plant, with a nominal capacity of 20 000 tonnes per month, is situated in the state of Minas Gerais, approximately 100 kilometres north east of Belo Horizonte, the third largest city in Brazil. It is located between the towns of Barao de Cocais and the historic Santa Barbara.

The earliest indications of exploration and mining date back to 1860 when an auriferous outcrop was discovered at Pinta Bem, one of the highest points of the region. Mining activity continued until 1905, whereafter operations were abandoned due to the depletion of the oxidised material and the low recoveries obtained from the unoxidised sulphidic ore.

In 1923 an attempt to reopen the mine was made. However, operations were once again abandoned in 1925 due to the poor recoveries obtained. To this day a roaster stack and chimney exist as a monument to this abortive attempt.

In 1944 a program of drilling revealed the presence of economic ore reserves. However investors were loath to come forward with the necessary finance due to the risk of uneconomic recoveries. The mine was mothballed again and transferred to Sao Bento Mineracao.

In 1978 the first contacts with Sao Bento Mineracao were made by Gencor. Based on the results of a preliminary study in 1979 it was decided to commence with detailed metallurgical investigations. These, which were concluded at the beginning of 1984, indicated the possibility of economic recovery of the gold.

It was decided to prove the viability of the chosen extraction route on a pilot scale. At the same time a feasibility study was conducted. Based on the pilot plant and feasibility study results the project was approved in August 1984.

MINERALOGY

The ore body comprises massive sulphide at an average in-situ grade of approximately 4,5 % arsenic and 6,2 % sulphur.

The sulphide minerals consist mainly of arsenopyrite, pyrrhotite and pyrite, with arsenopyrite the dominant mineral. The gangue minerals comprise mainly quartz, siderite and ankerite.

The gold is largely associated with the sulphides as included, attached and interstitial grains in approximately the following percentages:

inclusions in sulphides	52%
interstitial to sulphides	33%
attached to sulphides	13%
free gold in silicate/carbonate matrix	2%

Gold occurring as inclusions has a strong tendency to be in the arsenopyrite. Gold interstitial and attached to the sulphides shows a slight preference to arsenopyrite and appears to have an affinity for the pyrrhotite. The gold appears to have the least affinity to the pyrite. The occurrence of the gold as described clearly shows the ore to be refractory in nature and as such requires complex treatment.

PROCESS SELECTION

Processes for the treatment of sulphidic refractory gold ores generally comprise three basic steps, namely concentration, sulphide destruction and gold extraction. The heart of the process is the sulphide destruction step, with the other two dependant somewhat on the basic route chosen. Three sulphide destruction possibilities existed. These were:

- roasting
- chemical oxidation at an elevated pressure
- bacterial oxidation

The roasting route is the oldest and was at the time the only one in commercial operation. Two types of roasters are mainly used; Edwards roasters as used at Fairview Mine in South Africa, and Fluosolids roasters. The Edwards roasters were rejected due to their low unit capacity and Gencor's own experience of the enormous difficulty in removing arsenic and SO_2 from the gas emissions. Only Fluosolids roasters were evaluated.

The pressure leaching route had been known for some thirty years but had at that stage not been commercially utilized for gold bearing ores. Nonetheless Homestake Mining was to use the process for its McLaughlin deposit in California which came on stream in 1985

The bacterial route was in its infancy and totally unproven. For this reason it was not included in the evaluation of this particular project even though Gencor is actively involved in its development.

The process selected for the Sao Bento mine was the pressure oxidation route. This decision was based on extensive testwork, Gencor's own experience with roasting of arsenopyritic concentrates, worldwide visits to operating plants and discussions with engineering contractors specializing in the field, taking into account the following factors:

- gold recovery
- operability and technical risk
- market for byproducts
- environmental impact
- plant hygiene
- financial aspects

Gold Recovery

Extensive pressure leaching tests on Sao Bento concentrates were carried out by Sherritt Gordon and yielded a consistent extraction in the range 95-97 %. Further to the tests on Sao Bento material, Sherritt had conducted numerous tests on concentrates from various deposits throughout the world. These are depicted in Table 1 and as can be observed recoveries are consistently good.

	ANALYSIS %		GOLD	CYANIDATION EXTRACTION Au %	
ORIGIN	As	S	g/t	DIRECT	OXIDIZED
ONTARIO	5,7	18,7	232	74,4	99,4
PAPUA NEW GUINEA	0,9	16,3	77	22,3	98,1
NWT, CANADA	8,9	21,8	74	18,4	98,1
AUSTRALIA	9,8	16,2	72	38,1	98,3
SOUTH AMERICA	19,2	28,0	51	21,0	98,6
AFRICA	5,0	15,7	49	42,6	92,4
PAPUA NEW GUINEA	0,4	28,2	39	55,2	98,2
GREECE	11,6	41,6	32	4,7	98,6
PAPUA NEW GUINEA	0,3	22,4	30	42,5	95,9
IDAHO	4,6	16,5	24	7,3	98,2
PAPUA NEW GUINEA	0,2	32,4	23	53,9	99,0
PAPUA NEW GUINEA	0,4	38,5	14	39,6	98,7
DOMINICAN REPUBLIC	0,1	25,9	9	24,4	96,6
GREECE	1,3	48,2	3	5,3	89,5

Table 1. Cyanidation Recovery of Refractory Gold

Roasting testwork was undertaken by Gencor Laboratories, and two of the world's leading authorities on roasting. The roasting tests yielded variable gold extractions which ranged from 85 % to 93 %, with an average of 90 % for similar levels of arsenic and sulphur removal.

Experience with commercial Fluosolids roasters was that the recovery in practice would be at least 2 % down on the above figures. Roasting was thus expected to yield a gold recovery of no better than 88 %.

Operability and Technical Risk

The higher recovery of gold was one of the main advantages of pressure oxidation. Coupled to this is the fact that it is capable of handling ROM material thus eliminating the losses across the concentration step. However, abrasion in the key areas of slurry pumping, autoclave internals and pressure let-down was an important risk factor and this could only have been enhanced with the treatment of ROM material. It was advisable to treat a flotation concentrate with its much lower abrasive gangue content, particularly if gold recovery across the flotation step was high.

A further inherent advantage of incorporating flotation was that it provided a surge thereby minimising fluctuations in grade of sulphur and arsenic which markedly affect operability.

The Fluosolids roasting process required a concentrate feed in order to maintain autothermal conditions. In the case of Sao Bento, the concentrate would contain approximately 20% sulphur and 15-18% arsenic. Such a level of arsenic had never been treated before and doubt existed on the part of the roaster manufacturers that such a plant could be successfully operated. At that time the highest level of arsenic treated successfully by Fluosolids roasting was 8%.

The roasting process comprised two stages. The first stage was operated under reducing conditions to promote the sublimation of the arsenic and the second under oxidising conditions to oxidize the sulphur, with both steps necessary for gold liberation. A major uncertainty was the removal of these components from the gas stream. Brazilian legislation does not allow discharge of arsenic or sulphur to the atmosphere in the quantities that would have been produced by Sao Bento. Recovery of these was therefore essential. Gold production would thus have been dependant not only on successful operation of these recovery units, but also on the attainment of a dependable market for these by-products.

To Gencor's knowledge, there were no plants producing sulphuric acid from pyritic concentrates with such high levels of arsenic. Investigations showed that the only successful method of removing arsenic from roast gas streams at such high levels of arsenic was by means of a bag house. This method, however, reduces the arsenic in the gas to levels of only 5-10 mg/NM^3, which is not low enough to feed to a conventional sulphuric acid plant, which has a requirement of less than 1,2 mg/NM^3.

Market for By-Products

The selected pressure oxidation route would produce no by-products, although the potential to produce elemental sulphur did exist.

The roasting route would have produced of necessity sulphuric acid and arsenic trioxide. The market in Brazil for sulphuric acid could easily have absorbed the production from Sao Bento. This market was however dependant on the phosphoric acid demand, which was well known for its cyclic nature and there could conceivably have been a time when the acid would have been difficult to dispose of. This would have placed the mine at risk.

The biggest marketing problem would have been disposal of arsenic trioxide. The mine would have produced 12 000 tons per annum at a milling rate of 25 000 tonnes per month. This was equal to 25% of the then world consumption, and its placement would have had a serious impact on price.

The purity of the arsenic trioxide produced by two-stage Fluosolids roasting is low at 90-95 %. This product is not directly marketable to consumers and the only outlets would probably have been the two major refineries, Boliden in Sweden and Koppers in the U.S.A.

The long term market for arsenic trioxide is uncertain. It is highly toxic and carcinogenic and consequently products produced from it are under constant threat of substitution.

Thus the market could shrink whereas potential production is on the increase due to the fact that exploration for gold is moving rapidly with many of the prospects being refractory in nature and containing arsenic.

Environmental Impact

The pressure oxidation process would produce a liquid effluent only, whereas roasting would result in both gaseous and liquid effluents, the latter being scrubber discharge.

The maximum discharge level for sulphur into the atmosphere in Brazil is 80 mg/NM^3. The permissible level for arsenic was not specified at that time, but Sao Bento would have been required to meet the level specified in the country of origin of the technology. In the case of Boliden technology, the Swedish specification of 0,05 mg/NM^3 was applicable. The levels being achieved on the two Canadian operations were in the order of 5-10 mg/NM^3 at the time.

The maximum level for arsenic in liquid discharges to so-called secondary rivers in Brazil is 0,2 mg/ℓ. This is achievable by lime/ ferric sulphate precipitation techniques, as had been demonstrated in the pressure oxidation testwork. Simple liming would not have produced an acceptable effluent.

Plant Hygiene

The experience of Gencor is that special plant hygiene precautions are necessary in the handling of dry arsenic trioxide. Small quantities on the skin lead to extreme irritation and disorders curable only by the use of cortisone creams, and periods away from the arsenic environment.

Arsenic trioxide vapours, particularly in the presence of acid vapours lead to a similar phenomenon and particular care has to be taken in controlling the atmosphere around tanks containing hot sulphuric acid solutions with high concentrations of arsenic. It was thus Gencor's opinion that baghouse, arsenic containerisation and scrubber areas of a roaster/acid plant would constitute a major health problem.

The pressure oxidation route had a major advantage in that the process is all aqueous albeit producing arsenic bearing acid solutions. These solutions are diluted and the arsenic appears to be complexed with iron. No plant hygiene problems were expected and this was verified by the fact that no problems were experienced during the pilot plant campaign at Sherritt Gordon.

In both cases there is the ever present danger of arsene should nascent hydrogen be produced by accidental contact of zinc or aluminium with arsenic bearing acid solutions. Special measures need to be taken when handling these solutions.

Economic Aspects

The working costs of a plant using pressure oxidation were estimated to be some 13 % higher than those of the roasting route. However, gold recovery was expected to be 7 % higher, and even when taking into account the loss of acid revenue, the net revenues over the life of the mine for both routes about break even. No value was attached to the arsenic because of the extreme uncertainty of its market.

The capital cost of the pressure oxidation route including the oxygen plant was about 20-25 % higher than the roasting route. Nevertheless, when the various technical factors were taken into account, particularly the serious environmental and plant hygiene problems associated with roasting of concentrates at these very high levels of arsenic, the pressure oxidation route with its inherent cleanliness was preferred.

PROCESS DESCRIPTION

Milling and Gravity Separation.

Ore with a top size of 250 mm is trucked from the mine to a receiving bin at the plant, from where it is fed by a conveyor belt into a silo. Ore is fed from the silo to the Run-of-Mine mill by a horizontal feed conveyor. The mill operates under fully autogenous conditions with automatic feed and moisture control. The ROM mill product discharges onto a scalping screen, operating in a closed circuit with it, where plus 19 mm material is separated and returned to the mill. The screen underflow is pumped to a secondary ball mill, operating in closed circuit with a hydrocyclone. The secondary mill discharges over two plane tables, where the coarse free gold is recovered. This gold concentrate is further cleaned by two stages of shaking tables. The product has a size analysis of approximately 70% minus 74 microns.

Flotation and Regrind.

The mill product is pumped to the flotation section, where the emphasis is on maximum recovery i.e. essentially bulk flotation. A rougher/scavenger circuit is utilized with the rougher concentrate being subjected to regrinding to produce a product in the size range of 90% minus 44 microns. The concentrate is thickened to 65% solids and stored in tanks ahead of the pressure oxidation section.

Pressure Oxidation and CCD Washing

Concentrate from the concentrate storage tanks is pumped to the feed preparation tank, where it is mixed with recycled first wash thickener underflow pulp. This recycle acts both as a heat sink, thus assisting temperature control, and as a dispersant for elemental sulphur that may form in the autoclave. In addition the acid component reacts with carbonates in the concentrate to release carbon dioxide gas thereby increasing the oxygen utilization in the autoclave. To this feed preparation tank a foam suppressor, calcium lignosulphonate, is added. The mixture then overflows to the feed reaction tank where excess carbon dioxide is allowed to dissipate ahead of feeding the autoclave.

Feed to the autoclaves is by means of positive displacement diaphragm pumps. The autoclaves operate at a pressure of 1 600 kPa and a temperature of 190°C. Oxygen at a purity of 95,0 % is fed continuously to each of the five compartments. Quench water for temperature control is fed on an as required basis. Feeding of slurry to the first compartment is continuous with the solids being maintained in suspension by mechanical agitation. The slurry discharges from the last compartment to a flash tank, where it attains atmospheric conditions. It is then pumped to a repulp tank ahead of the C.C.D. circuit.

The autoclave discharge slurry is mixed with second wash thickener overflow solution in the repulp tank before flowing to the first wash thickener. Overflow solution from this thickener is stored in a tank ahead of neutralization.

Thickener underflow is pumped to a splitter which delivers approximately 80% to the recycle storage tank and the balance to the second wash thickener. Underflow from the second thickener is pumped to the Carbon-in-Leach Section.

Neutralization

A two step neutralization process is employed. In the first step the acid solution is contacted with limestone slurry in three stages, and attains a pH of 4,8 at the final stage. The slurry then flows through two liming stages at the end of which a pH of 11 is attained. The neutralized slurry containing calcium sulphate, ferric arsenate and other precipitated metallic ions is thickened prior to discharge to the final tailings tank, from where it gravitates to the tailings dam.

Carbon-In-Leach

The second wash thickener underflow is neutralized with lime ahead of two leaching stages. The slurry then gravitates through eight stages of adsorption with a counter current carbon flow. Cyanide concentration is maintained at 0,15% NaCN in the final tank.

The slurry from the final stage gravitates to a fine carbon safety screen. The fine carbon (+0,3 mm) is collected in rocker bins and stored ahead of further processing, while the slurry is pumped to the tailings thickener for cyanide solution recovery. The thickened residue is then pumped to the final tailings tank where it combines with the other plant residues before gravitating to the tailings dam.

Loaded carbon is pumped from the first stage to the loaded carbon screen, where it is washed and stored ahead of acid treatment.

Elution and Regeneration

The loaded carbon is acid treated with hydrochloric acid prior to elution. The acid treatment is done at a temperature of 90°C over a period of approximately 4 hours.

Elution is carried out over a period of 16 hours at a temperature of 125°C and with a 1,5% sodium cyanide and 1,0% sodium hydroxide electrolyte solution. Heating is done in two plate exchangers using column discharge eluant and steam, while final cooling to below boiling point is done in a third exchanger using cooled water. The cooled eluant then flows to an electrowinning cell where the gold is plated out onto steel wool cathodes. The spent electrolyte then returns to the electrolyte tank and is recycled back to the elution column.

At the end of the elution cycle the gold loaded cathodes are placed in a calcine furnace. The eluted carbon is educted from the elution column to the regeneration section.

Regeneration of the carbon is carried out at a temperature of 750°C in an externally heated rotary kiln. On discharging from the kiln the carbon is quenched in water and educted to the sizing screen. Carbon passing through the screen gravitates to the fine carbon tank while the coarse carbon returns to the adsorption circuit.

Refining and Smelting.

Cathodes are calcined for 24 hours at 700°C.

At the end of this period the calcine is treated chemically to obtain refined gold.

The process produces gold of 99,95 % and silver of 99,5 % purity in a single stage. The refined gold and silver are then dried, melted in an induction furnace and poured into bars.

Overall recovery of the gold in the refining stage is in excess of 99,9 %. All residues from the refinery are returned to the leach circuit.

Residue Disposal

All tailings are discharged into a cross-valley type dam using a beaching technique. All water is returned to the neutralization plant for treatment prior to the excess being disposed of.

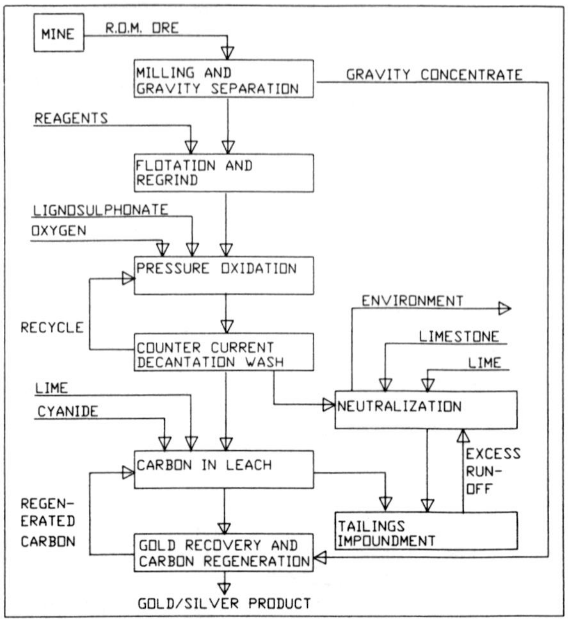

Figure 1. Process Flowsheet.

PROCESS DESIGN, ENGINEERING AND CONSTRUCTION

Equipment process specifications were completed in South Africa, and Canada in the case of the pressure oxidation circuit, prior to the project transferring to Brazil. This enabled conceptual layout design to reach an advanced stage resulting in much time saving.

Davy McKee, Stockton,in conjunction with its Brazilian subsidiary, Sertep, was appointed the project management contractor, Paulo Abib, the detailed design engineers for the metallurgical plant and Sertep the detailed design engineers for the balance of the surface installations.

Expatriate co-ordination engineers were supplied by Davy McKee with Gencor sending over an expatriate team, consisting of a project manager, cost engineer, design engineer and project metallurgist.

The detailed design and engineering were done in Sao Paulo, the industrial heart of South America. This work was completed with the use of 328 000 manhours and the production of 2 500 drawings. In addition, a further 265 000 manhours were used in project management, procurement, planning, cost control and financial services. To complete the picture 141 000 manhours were used in site supervision, measurement and control.

The earthworks commenced on 23 February 1985 and lasted for nine months. The tailings dam earthworks commenced on 5 July 1985 and were completed on 15 September 1986. In this operation 690 000 m^3 of earth were used in the construction of the wall. In addition 1900 m^3 of concrete were poured into a permeable fault, along the dam basin, to minimise seepage.

Civil construction commenced on 10 August 1985, almost one year to the day on which the project was formally approved, and continued for 13 months. It was completed on 20 September 1986.

The MEIP (mechanical, electrical, instrumentation and piping) installations commenced on 24 February 1986 and were completed at the end of December 1986.

COMMISSIONING.

The overall commissioning of the metallurgical plant was sub-divided as follows:

- pre-commissioning.
- mechanical and electrical commissioning.
- cold commissioning (non process materials).
- metallurgical commissioning (process materials).

Pre-Commissioning

This comprised the following:

- checking of equipment against specification.
- checking of installed equipment with respect to access, ease of operation and safety.
- the monitoring and checking of pipeline systems against specification and the checking of valve access and safety.
- the thorough testing of the control system by use of a dynamic computer simulator.

Mechanical and Electrical Commissioning.

This was carried out by the MEIP (mechanical,electrical, instrumentation and piping) contractor, under the direct supervision of commissioning engineers from Davy McKee.

Cold Commissioning

Cold commissioning of the plant was, in general, done by the metallurgical operations team. In the pressure oxidation section, however, it was done under the technical supervision of Sherritt Gordon personnel.

Close planning and co-ordination were necessary to ensure that the mechanical and electrical commissioning resulted in the programmed release of equipment, so as not to delay cold commissioning (hence metallurgical commissioning). In order to achieve this the following were done.

- the plant was grouped into logical systems. The systems were then listed in bar chart form to ensure programmed commissioning.
- all the equipment in each system was detailed. This resulted in a minimum delay as a result of efforts being concentrated in the required area at the correct time.

The actual cold commissioning involved the following:

- the thorough flushing of all pipelines with water(in the case of process fluids) or air (in the case of compressed and instrument air). Special measures were required in the case of the oxygen lines. The cleaning of these were entrusted to the oxygen supplier.
- operation of the operational system using water. An exception was the pressure oxidation autoclaves, where an acid/ferric solution was used due to the internal refractory lining.

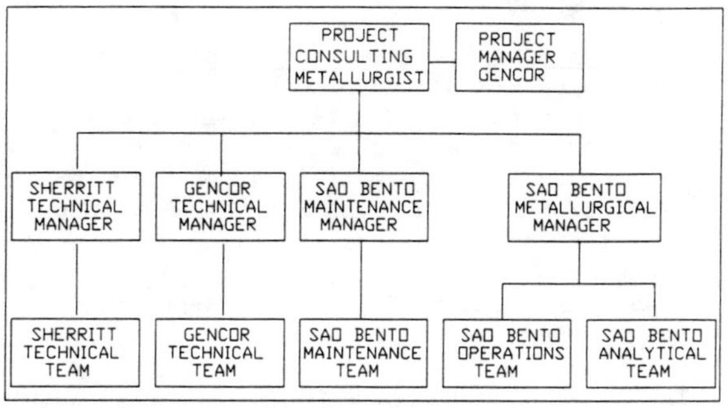

Figure 2. Commissioning Management Structure.

Metallurgical Commissioning.

The planned start date for metallurgical commissioning was 10 November 1986, with a build up period of six months to reach full production.

The actual commencement was as follows:

Utilities	:	25 September	1986
Milling	:	20 October	1986
Flotation and Regrind	:	20 November	1986
Pressure Oxidation	:	06 December	1986
Carbon-In-Leach	:	29 December	1986
Lime Preparation Plant	:	24 December	1986
Limestone Preparation	:	15 January	1987

The first elution was carried out on 5 January 1987 and the first gold pour on 7 January 1987.

Figure 3. Shows the commissioning tonnages.

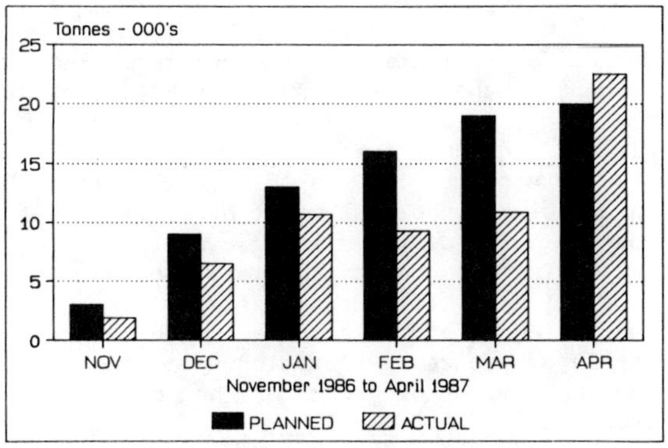

Figure 3. Actual Versus Planned Tonnage.

The Gencor technical team was in place by the end of the first week of October 1986 and the complete Sherritt technical team by the end of October 1986. (The first member had arrived at the end of September 1986 to supervise the pre-commissioning of the autoclaves).

The analytical laboratory commenced operations in February 1986. A certain amount of confidence thus existed by the time metallurgical analyses were required. However the services of an experienced Gencor analyst were required to train the local team in the managerial aspects of a gold mine analytical laboratory.

Commissioning Management Philosophy. The commissioning was planned according to the following philosophy. The technical teams would maintain around the clock coverage of the operations in their respective areas. They would be responsible for the minute to minute technical supervision and control of the plant. They would also ensure sampling, collection of data, make on the spot calculations and recommend process changes where necessary.

The plant operations team would be responsible for the start up and operation of the equipment, under the technical supervision of the technical teams. Changes to plant operational procedures were not to be done directly by the technical teams, but through the area foreman, or in the case of night shift, the night shift supervisor.

The operating responsibility of the section would belong to the foreman.

The maintenance team would be the regular plant maintenance crew, under the direction of the plant engineer. However, around the clock coverage would be necessary during this phase. This would include the services of an instrument technician. In addition a special commissioning modifications team consisting of pipe mechanics and boilermakers would be formed to assist the operations team in the plant commissioning.

The laboratory team would comprise its regular personnel. However, should the need arise, arrangements were to be made for analyses to be done around the clock.

Overall co-ordination of the commissioning was to be done by the Gencor Consulting Metallurgist.

Planned Sequence of Material Flow

- the ore storage silo was to be filled with ore in anticipation of the start of milling.
- commissioning of the milling circuit (excluding the gravity section) and the production of ground ore to allow commissioning of the autoclaves. This involved by-passing flotation until the required amount of ground ore had been produced.
- during production of the ground ore, the establishment of a ferric sulphate containing sulphuric acid solution recycle in the pressure oxidation wash circuit.
- when sufficient ground ore had been produced, the flotation by-pass was to be discontinued and commissioning of flotation and the production of concentrate would then commence.
- the acidifying of the ground ore and commissioning of the first autoclave to commence and be continued until the oxidized slurry recycle inventory had been attained.
- the commencement of concentrate feed to the autoclave on attaining the recycle inventory.
- on attainment of steady operation, the commissioning of the second autoclave to commence.
- the advancement of the acidic solution to the neutralization on attainment of the operating level in the first wash thickener.
- the advancement of oxidized solids to the carbon-in-leach section on attainment of inventory in the wash circuit.
- neutralization of the acidic solution in the neutralization section with air-separated hydrated lime, pending completion of the lime and limestone preparation facilities.
- neutralization of the oxidized slurry in the carbon-in-leach section with air-separated hydrated lime, pending completion and commissioning of the lime preparation plant.
- the filling of the carbon-in-leach tanks with neutralized slurry and leaching of the gold.
- when all the tanks had been filled with pulp the advancement of activated carbon.
- on attainment of the carbon inventory in the adsorption tanks, pulp feed to the circuit to commence.
- the first batch of loaded carbon to be removed and acid treated.
- the first elution and electrowinning.
- the subsequent smelting of the first gold.
- when the general plant operations had stabilized the commissioning of the gravity concentration section would commence.
- once the plant had been commissioned and had attained steady state operation, chemical refining of the gold and silver to commence.

METALLURGICAL COMMISSIONING PROBLEMS

Only the major problems will be described.

Milling Section.

The-Run-of Mine mill feed hopper was subjected to continual chokes, resulting in high downtime.

The feed hopper was designed as a rectangular type box with its outlet, situated approximately 500 mm above the floor, internally overlapping the mill inlet trunnion. The principle of operation was that a "dead" box would be formed beneath the outlet, with feed into the mill being effected by the ore pressure and the exceeding of the angle of repose. Problems were experienced with continual chokes resulting in excessive downtime. In addition, the large volume of water added to assist the feeding resulted in inefficient grinding in the mill. In South Africa the problem is greatly alleviated by the presence of the cyclone underflow material due to the closed circuit operation, which is not the case at Sao Bento. The problem was solved by the elimination of the "dead" box and the modification of the hopper so that the ore is deflected directly into the mill inlet trunnion.

Pressure Oxidation

Slurry Discharge Valves. These were the first items to present problems. Two manually operated plug type isolating valves were installed on each autoclave discharge line. These lines are 125 mm in diameter and fabricated in Inconel 625. The valves are in Inconel 625 with teflon internals.

During autoclave operations, closure of the valve resulted in the teflon seat being ripped by the plug and subsequently being lodged in the let-down choke. The immediate result of this was an increase in the pulp level in the autoclave. Subsequent failure to dislodge the teflon from the choke required the depressurization of the autoclave via the emergency vent. This resulted in the discharge of the solids to the atmosphere and their subsequent settlement over a large part of the plant, causing serious housekeeping and maintenance problems.

The abovementioned occurred with all the installed valves resulting in no usable valves on site after less than two month's operation.

The reasons for the teflon being ripped from the valve body were two-fold. Firstly, poor quality control in the manufacture of components, and secondly the incorrect grade of teflon for the temperature duty. During autoclave operation the temperature is in the range 185°C to 190°C and, with the block valves in the open position, the teflon seat "creeps". Operation of the valves under these conditions resulted in the plug ripping the teflon. On restart of the autoclave the teflon lodged in the discharge choke, forcing the depressurization through the emergency vent.

The problem was aggrevated by the following:

- non availability of spare valves.
- non availability of teflon or a substitute.
- non standard diameter thus special manufacture.
- lack of interest by the only manufacturer in Brazil (who subsequently went out of business 4 months after autoclave start up).

As a consequence it was necessary to operate the autoclaves without isolating valves. This resulted in both production losses and increased costs since the autoclaves could not be placed "on hold" necessitating extended pre-heat and pressurization periods on start up.

The plug isolating valves have since been replaced with imported Herbar Angle Valves resulting in rectification of the problem.

Vent Control Valves. The autoclave venting system consists of a vent line to knock-out pot, with a nuclear level detector, an on-line oxygen analyzer and a vent flow control valve. Solids in the vent gases are precipitated in the knock-out pot and discharge via the activation of an automatic valve by the nuclear level detector.

During commissioning, these valves, comprising Alloy 20 body with Alloy 20/Steelite internals, were severely damaged by solids as a result of the nuclear level detectors having been mounted at the incorrect height on the knock-out pots. The valves were returned to the supplier for repairs. In place of the valve an orifice plate was installed. This thus resulted in no control over the oxygen partial pressure hence oxygen utilization, which in turn resulted in the inefficient oxidation of the sulphides.

These valves have since been replaced with Herbar Angle Valves, with acceptable results.

Autoclave Agitator Seals. An excessive amount of autoclave downtime was recorded due to failure of the agitator seals. At one stage a complete stoppage of operations was threatened due to the non availability of further spares. This was averted by a concerted effort by the maintenance team. The underlying problem was eventually found to be the incorrect assembly of the seal. (The seals were assembled by the manufacturer prior to delivery). Once the defects had been corrected no further abnormal problems were experienced.

Gaskets. The gaskets specified for the autoclaves, the ancillary equipment and the piping were of the compressed asbestos sheet type. Theoretically, this material should have withstood the operational conditions. However it was found to be totally unsuitable for the duty. The first gasket failed on the slurry discharge line after 100 hours of autoclave operation. Subsequently a further 290 hours of operations were recorded before the gaskets on the slurry discharge and vent lines failed in quick succession. As a result of the non availability of a better material the gaskets were replaced with the same compressed asbestos sheeting type. During the following 20 days a total of 30% downtime was recorded as a result of gasket failures. In the meantime, teflon (PTFE) gaskets were installed on a trial basis on certain flanges. No problems were experienced with these and it was decided to replace all the gaskets in the pressure lines with teflon.

Performance of these gaskets was way above expectation and as a result downtime due to gasket failure has virtually been eliminated.

Neutralization

Failure of the mechanical agitators was the major problem experienced in this section. In fact this was a major problem throughout the plant and at one stage threatened to bring production to a halt. The problems ranged from loose agitator paddles and excessive oil leaks to gearbox failure as a result of a multitude of lubrication problems. In the short term, production was maintained only by the use of compressed air for agitation purposes. To ensure continuity of production the agitators were returned on a programmed basis to the supplier for definitive repairs and modifications to the gearbox and lubrication system.

OPERATOR TRAINING

The highly successful commissioning of the plant within the projected time period is a tribute to the heavy emphasis placed on the training of the operations personnel. This training could be categorized into three phases:

- training in South Africa at Gencor plants.
- pre-commissioning training at the plant.
- on-the-job training during commissioning.

Training in South Africa.

This training applied to the management, down to foremen level. It included intensive technical and managerial training at three operations with similar unit processes or equipment to that of Sao Bento. These were Fairview mine (flotation of sulphide minerals and gravity separation of free gold), St Helena mine (run-of-mine milling, carbon in pulp, elution, electrowinning/smelting) and Impala Refineries (pressure leaching).

Pre-Commissioning Training.

This was applicable to personnel down to control room operator level and comprised the following:

- the writing of preliminary operating manuals by the metallurgical foremen, making use of flowsheets, pipe and valve diagrams, process and instrumentation diagrams, process specifications and supplier catalogues.
- a limited amount of training on the simulator. It was possible to train only to the shift supervisor level due to time restraints.
- training of the operators in the theoretical and practical aspects of the plant operations by use of the preliminary operating manuals, and actual training in the plant during the erection phase. The training day was split into two. The morning session was set aside for theory while the afternoon session involved the accompanying of the erection and checking of the equipment and piping systems, to bring irregularities to the attention of the contractor as early as possible. In this manner the supervisor became aquainted with his area of operation well in advance of commissioning and, in addition, provided an extremely important pre-commissioning service.
- a series of lectures and discussion sessions given by senior Sherritt Gordon personnel to the plant management and pressure oxidation supervisory staff on the various control parameters and their inter-relationships.

The general training of the supervisors and the control room operators was implemented and co-ordinated by the metallurgical trainer with technical assistance by the metallurgical foremen and the plant management.

On-The-Job Training

This training commenced during the mechanical and electrical commissioning stage, during which the operation of equipment was the responsibility of the control room operator.

During commissioning, on-the-job training was entrusted mainly to the expatriate technical commissioning teams. This was particularly difficult at the attendant level, due to the background of this type of person. They were all recruited, to coin a phrase, "off the streets", with the majority having been previously employed as agricultural labourers. However, their willingness to learn, which compensated partly for their lack of experience, resulted in them attaining the required level of competence.

The training of the shift supervisors, although not without its problems, was not too difficult. All were technically qualified, with metallurgical operational experience ranging from three to twelve years.

The foremen, while not requiring much training in the operational aspects, required guidance in the managerial aspects of the job, especially with regards to planning and organization.

The overriding problem was that of language. It was particularly difficult in the pressure oxidation section where, because of the exacting conditions, there was little room for error and instructions had to be clear. To alleviate the problem, an interface shift team, between the Sherritt Gordon technical team and the shift operations team, was formed. This interface team was made up of the plant metallurgist, the metallurgical trainer, the pressure oxidation foreman and a member of the Gencor technical team, as they were the only available people able to communicate in both languages. This caused the disruption of many other important aspects of the plant start-up.

Notwithstanding all the problems, the training went surprisingly well, which is to the credit of the Gencor and Sherritt Gordon technical teams.

Table 3 outlines the periods during which the technical teams provided shift coverage:

SECTION	SHIFT COVERAGE		TECHNICAL TEAM
	START	END	
MILLING FLOTATION	20 OCT. 1986	12 DEC. 1986	GENCOR
PRESSURE OXIDATION, CCD WASH, NEUTRALIZATION	13 NOV. 1986	24 JAN. 1987	SHERRITT GORDON
CARBON-IN-LEACH ELUTION, SMELTING	12 DEC. 1986	16 JAN. 1987	GENCOR

Table 3. Technical Teams Shift Coverage.

OPERATIONAL RESULTS

The design parameters are compared with the actual results during the first two years of commercial operation. The actual results reflected are on a monthly average basis.

Milling

		DESIGN	ACTUAL
Milling Rate	t/m	20000	25010
Product	% -74 microns	72,0	67,0
Availability	% of Total	85,0	84,0

Table 4. Milling Results

Although the actual availability is lower than the design figure, this was due mainly to problems in other sections of the plant resulting in mill shutdown. Actual time lost due to milling section breakdowns was less than 5,0%.

Flotation

		DESIGN	ACTUAL
Gold Recovery	%	95	97,4
Sulphur Recovery	%	97	98,4
Concentrate	% -44 microns	95	88

Table 5. Flotation Results.

Notwithstanding the coarse analysis of the flotation feed (67% minus 74 microns compared to the design of 72%), the recovery across the flotation exceeded the design figure.

Pressure Oxidation

		DESIGN	ACTUAL
Feed Rate	t/m	6000	6120
Sulphide Conversion	%	98,5	94,0
Availability	% of total	85	87,2

Table 6. Pressure Oxidation Results

The lower sulphide conversion is due to many reasons, the following being most significant.

- problematic oxygen sparging system.
- insufficient oxygen dispersion by the axial flow agitators.
- problems with the autoclave venting valves.
- excessive carbon dioxide in autoclave.

The original autoclave venting valves were replaced with angle valves resulting in improved control. With regard to the carbon dioxide, modifications to the flotation process have reduced the carbonates to a more acceptable level. Optimization of the flotation circuit is carried out continually to give the required gold recovery, sulphur grade and carbonate content.

The agitators and oxygen sparging system are the subject of present investigations and it is hoped to resolve the problem shortly by re-design and the installation of radial flow agitators.

Neutralization

		DESIGN	ACTUAL
Effluent Solution	As mg/ℓ	0,2	0,24

Table 7. Neutralization Results

Problems were experienced in this section when the ferric/ferrous ratio was low due to incomplete oxidation.

An anticipated problem in this section was that of scale build-up in the tanks and on the agitators. Although a certain amount of scaling was evident, this was relatively minor in nature and presented no problems. The reason for the low scaling is thought to be due to the use of a two stage neutralization technique whereby a gradual increase in pH is induced, thereby resulting in formation of coarse precipitates.

Gold Recovery

	DESIGN	ACTUAL
Adsorption Gold Extraction %	93	92,2
Elution Efficiency %	99	99,4
Regeneration Efficiency %	95	93,5
Overall Gold Recovery %	91,0	90,0

Table 8. Gold Recovery Results.

The lower gold recovery is due to two factors.

Firstly the lower sulphide conversion in the autoclaves, and secondly the excessive generation of fine carbon, which reports to the residue. The fine carbon problem was alleviated by importing high quality carbon from France. This has reduced the carbon loss from over 300 g/t to less than 100 g/t. In addition the fine carbon gold loss, as a percentage of the overall adsorption residue, has decreased from 0,4 g/t to 0,02 g/t.

The effect of residual sulphides on the gold extractability is shown in Figure 4.

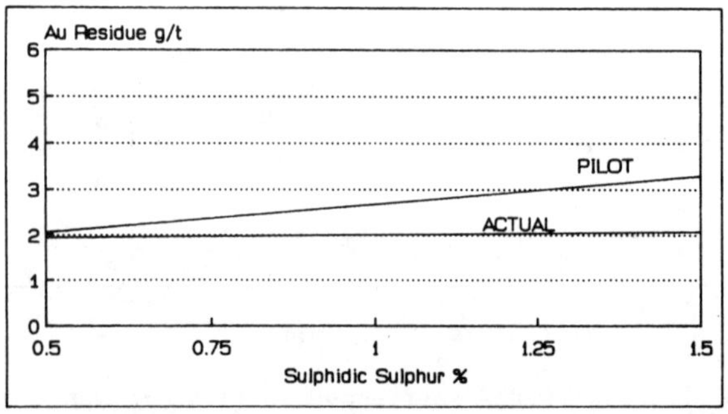

Figure 4. Gold Extractability.

Figure 4 shows the relationship between the residual sulphide and gold residue. A comparison is provided of the data from the commercial process with that of the integrated pilot plant runs conducted by Sherritt. It has been found that in the commercial operation a high degree of gold liberation is attainable without as extensive a degree of sulphur oxidation as was required in the pilot operation. This may be attributed to a lower degree of association of refractory gold with pyrite in the commercial plant feed and is consistent with the behaviour of several other arsenopyrite/pyrite concentrates tested by Sherritt, where the refractory gold was largely associated with arsenopyrite.

Reagent Consumptions (based on concentrate tonnage).

		DESIGN	ACTUAL
Limestone	kg/t	450	395
Lime	kg/t	140	159
Sodium Cyanide (100%)	kg/t	2,3	1,9
Lignosulphonate	kg/t	0,8	0,77
Flocculant	kg/t	1,0	1,0
Activated Carbon	kg/t	0,10	0,36

Table 9. Reagent Consumption.

CONCLUSION

After some two years of full production the process has proven to be technically and economically viable. It has demonstrated its robustness by treating greater sulphur grade variations and tonnages than originally envisaged. The oxidation plant itself, comprises sophisticated engineering and demands a high level of skills in both its operation and maintenance. The close process control and exacting conditions dictate the use of exotic materials. A carefully planned maintenance program, as well as precision workmanship is an absolute necessity to avoid excessive downtime.

ACKNOWLEDGEMENTS

The authors wish to thank General Mining Union Corporation (Gencor) and Sao Bento Mineracao for permission to publish this paper.

REFERENCES

Connell, Larry, 1981 "Roasting Process at Giant Yellow Knife (Arsenic?) Mine". Trans. CIMM, 20th Annual Conference of Metallurgists.

Roberts, J. Scott, 1976 "Controlling Roaster Off Gases At Campbell Red Lake Mines", Canadian Mining Journal, July, pages 54 - 56.

Chapter 39

CHANGING TRENDS IN GOLD ORE TREATMENT IN WESTERN AUSTRALIA. THE PROBLEM OF REFRACTORY ORES

Dr. Jim Avraamides

Mines Department of Western Australia
Chief, Mineral Processing Laboratory
Chemistry Centre, WA

Historically, much of Western Australia's gold production has come from the refractory sulphide-telluride ores located in Kalgoorlie's Golden Mile. Most of the recent gold boom has been associated with low grade, open pit free-milling oxide deposits. Primary sulphides are commonly encountered at depth and in many cases are highly refractory. Different approaches to the problem of recovering gold from these ores have been adopted subject to local conditions both technical and environmental. The least preferred option is the preparation of a gold-rich flotation concentrate which is shipped to a smelter. Roasting of flotation concentrates can be practised in certain remote locations where sulphur emissions are tolerated. Fine grinding of flotation concentrates followed by direct cyanidation is another popular approach although this often leads to fouling of carbon by flotation reagents in the adsorption section and high soluble gold losses.

HISTORY OF GOLD METALLURGY IN WESTERN AUSTRALIA (BLASKETT, 1952)

Coolgardie was the site of the first major gold find in 1892 followed by the discovery of Kalgoorlie and the "Golden Mile" in 1893. Both goldfields started as largely alluvial deposits but production moved underground within a few years. Along with Colorado's Cripple Creek, Ontario's Kirkland Lake and Fiji's Vatukoula deposits the Golden Mile is notable for its refractory tellurides. The Kalgoorlie refractory ores are sulphide-tellurides in which gold is associated with pyrite and arsenopyrite, and also occurs as calaverite ($AuTe_2$), sylvanite ($AuAgTe_4$) and krennerite (Au_4AgTe_{10}). By the turn of the century there were 12 mills treating oxidised ore by conventional cyanidation - zinc precipitation and 11 mills treating sulphide-tellurides. The latter ores were usually roasted following dry grinding, and the calcine pan amalgamated to recover coarse free gold. The amalgamation residue was split into sands and slimes fractions and cyanided in the conventional manner. Overall gold recovery was reported to be 92-95%.

Some attempts were made to treat the sulphide-tellurides in an operation comprising stamp battery crushing, amalgamation, concentration of the tellurides using strakes and finally roasting and amalgamation of the calcine. Because the concentration step was rather ineffective only about 70% of the tellurides reported to the roaster. Strake tailings were cyanided. Direct leaching of the tellurides was also carried out using bromocyanide to enhance dissolution rates.

Flotation as a means of producing a concentrate was tested on the Golden Mile in 1924 but was not used commercially until 1930 at the Lake View and Star mine. The economic advantages of flotation followed by roasting were soon apparent. A comparison of the three treatment options practised in Kalgoorlie in 1933 is shown in Table 1.

Table 1. Relative Treatment Costs in Kalgoorlie in 1933.

Process	Gold Recovery(%)	Relative Milling Cost
Float/Roast/Cyanide	87 - 90	1.00
Roast/Cyanide	93 - 95	2.16
Bromo-cyanidation	92	1.84

Apart from some technological improvements the basic float/roast/cyanide method remains in widespread use in Kalgoorlie. Static roasters have been replaced by fluidized bed designs and zinc precipitation has given way to carbon-in-pulp gold recovery.

Elsewhere in Western Australia are to be found other refractory gold ores, although not in the massive quantities which have made the Golden Mile the State's greatest single gold producing area to date. Arsenopyrite/pyrite ore is presently being treated at Mt. Windarra near Laverton, and arsenic and antimony ores occur at Nullagine and Wiluna. The Wiluna antimony/arsenic ore was at one time treated by flotation, smelting and refining. Flotation concentrates, arsenic residues and arsenical flue dusts were treated in a blast furnace producing bullion (5-15 oz/ton Au, 9.8% Sb, 0.8% As and 0.6% Cu), speiss (18.7% As, 8.9% Sb, 4.3% S, 57.8% FeO, 0.6% Pb, 1.1% Cu and 1.6 oz/ton Au) and slag. Gold was recovered from

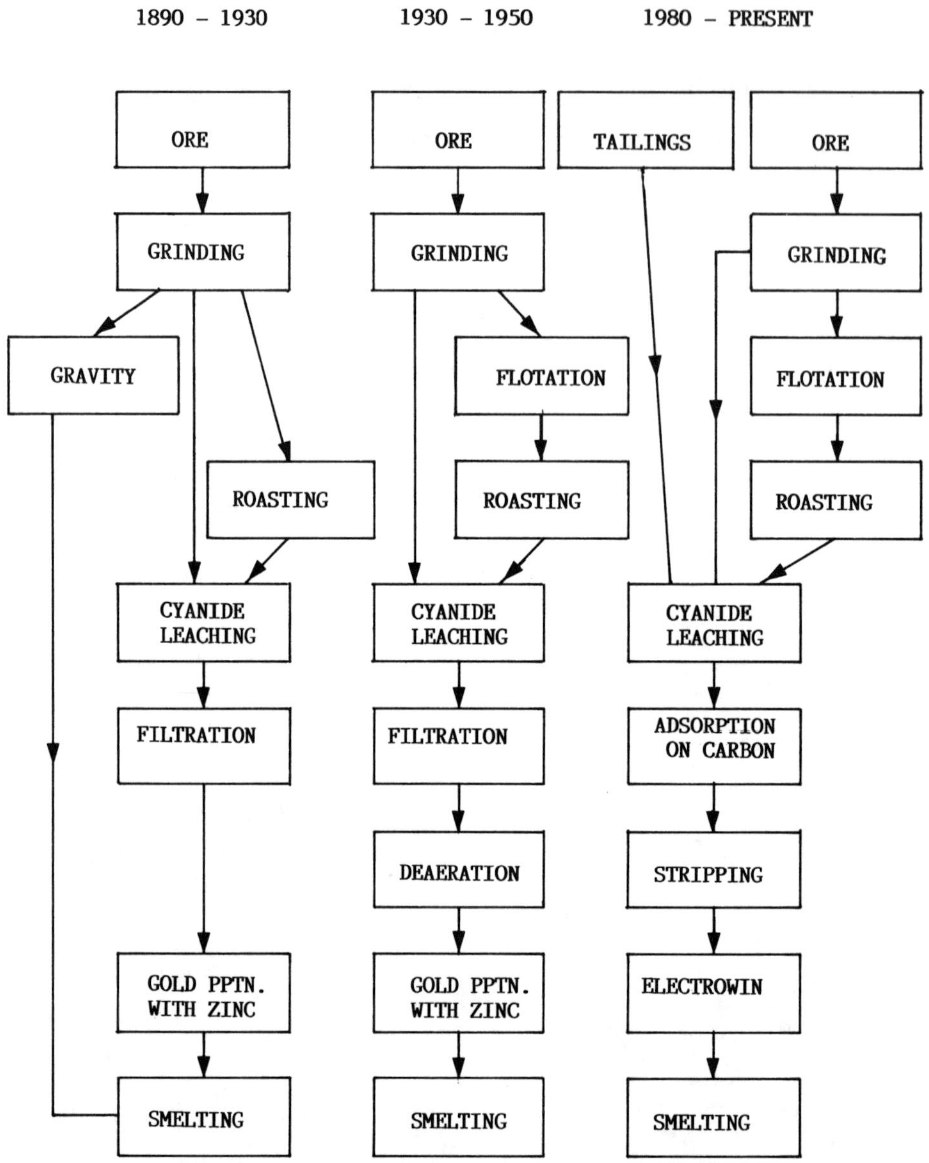

FIGURE 1

GENERALISED SIMPLIFIED FLOWSHEETS OF GOLD RECOVERY PROCESSES ILLUSTRATING METALLURGICAL PRACTICES SINCE 1890

the bullion by adding zinc and heating to 500°C. The Au-Zn-Sb alloy which separated out was sent for refining which comprised removal of antimony and zinc and cupellation to produce a bullion assaying 850 parts Au, 100 parts Ag and 50 parts Cu and base metals. Smelting was carried out from 1935 to 1939. By this time better roasting techniques had improved the cyanidation recovery by around 3%. This made the more conventional processing route of float/roast/cyanide more attractive than smelting. Figure 1 gives a generalized comparison of ore treatment methods.

ROASTING PRACTICE (VON BERNEWITZ, 1940)

The efficacy of roasting as a pre-treatment for Kalgoorlie refractory ores was demonstrated by the presence at one time of a total of 87 furnaces treating 73,000 tons of ore and 7,300 tons of concentrate each month. Two types of Edwards furnaces were employed. The first was the Simplex type which had one firebox and a single row of 15 rabbles. Throughput was around 20 tons per day of 3-5% sulphides or 7 tons of 20% sulphide gold concentrates. In 1912, when roasting had peaked in popularity in Western Australia there were a total of 36 Edwards Simplex furnaces in operation.

Although the Simplex furnace is compact, simple and cheap to install, it is less economical in fuel than the Duplex furnace. The Duplex furnaces had 3 fireboxes and up to 30 pairs of rabbles. Retention time of feed was 6 hours and up to 100 tons per day of 5% sulphide ore could be roasted or 30 tons of 35% sulphide concentrate. The advantage of this furnace is that the sulphides were rabbled more efficiently and over a longer period of time than in other roasters, resulting in more complete removal of sulphur and arsenic.

The third type of furnace used widely in Western Australia was the Merton. This had three hearths. The ore/concentrate was introduced into the upper section and was moved by rabbles from end to end eventually passing downwards to the finishing hearth and to discharge. Capacity was 30 tons per day of 3% sulphur ore which gave a calcine of about 0.01% sulphide sulphur. Most Mertons were used for whole ore roasting - of a total of 27 in operation at one time only 3 were treating concentrates.

By the end of the 1930's the trend in the industry was to roast only concentrates using Edwards Duplex furnaces. At the Paringa and Croesus plants on the Golden Mile, flotation concentrates of the sulphide-telluride ore were roasted in a 60 rabble furnace to a maximum temperature of 650-700°C. Throughput of 36% sulphide was 1 ton per hour and residence time 4 hours. Calcine containing 90 g/ton gold, 100 g/ton silver and 0.36% sulphide sulphur was cyanide leached at 55% solids for 45 hours and then sent to the Merrill-Crowe precipitation plant. The calcine residue contained 5-6 g/ton gold.

At Wiluna in the 1930's an oil fired rotary kiln was used to roast a 12% sulphide concentrate to 950°C to produce a clinker suitable for feeding to a blast furnace. The rotary kiln's capacity was 40 tons/day and it was capable of removing 95% of arsenic, 90% of sulphur and 70% of antimony.

By the early 1950's the development of fluidized bed roasters was being noted by the industry in Kalgoorlie and the new technology has now replaced static roasters. All of the four roasters in current use in Western Australia are Dorr-Oliver fluo-solids types incorporating a cyclone for fines collection, gas cooler and baghouse/electrostatic precipitator. One of the roasters is a two-stage device located at the Lancefield mine. It is treating an arsenical gold concentrate of the following approximate composition:

Iron	27%	Gold	41 g/tonne
Sulphur	25%	Silica	25%
Arsenic	2%	Antimony	200 ppm

A two-stage roast was chosen in order to minimise arsenic fixation as iron arsenate and maximise the production of arsenious oxide. The first stage operates at 550°C with less than stoichiometric air for sulphur oxidation. Arsenious oxide, As_2O_3, is produced which is volatile and is eventually trapped in the bag-house as a solid. The overflow from the first stage roast is then fully oxidized at 650°C in the second stage and the haematitic calcine sent to cyanidation and carbon-in-pulp (CIP) gold recovery. Roaster feed rate is 70 tonnes/day producing 0.81 tonnes of As_2O_3. Gold recovery is 93-94%. The roaster was installed in 1982 and recently has been replaced by a single stage roaster. Gold recoveries are approximately the same probably due to the decreasing arsenic levels in the ore, and there is the added advantage of the greater operational simplicity of the single stage roast. (Quadrio and Kellett, 1984)

In response to environmental concerns over sulphur dioxide emissions from the three roasters on the Golden Mile near Kalgoorlie, a new satellite roasting facility has been constructed some 20 km to the north. Once again a fluidized bed roaster has been chosen but with a larger cyclone, the oversize from which is returned to the roaster. The reason for choosing this Lurgi recirculation bed roaster is to improve the extent of roasting of finer particles of sulphides which might otherwise have limited residence time in the fluidized bed. Improved roasting should increase gold recovery and decrease cyanide consumption. Plans are already in hand to install a second identical unit at the same site within the next two years.

OTHER GOLD ORE TREATMENT METHODS

Most of Western Australia's current gold production comes from free milling oxide or sulphide ores using CIP or CIL. Some heap leaching is carried out and this practice is increasing, particularly as low grade ore stock piles grow. Large heap leaching operations ($> 10^6$ tonnes) are in production at Kambalda, Telfer and Whim Creek. Many more smaller heap and vat leach operations are treating ore and tailings.

There are a number of other options to roasting for handling refractory or problem ores. At the Telfer gold mine operated by Newmont Australia copper in the oxide and sulphide ore has resulted in significant alterations to ore treatment. The presence of copper as chalcocite and chalcopyrite results in fouling of carbon and excessive cyanide consumption. Oxide ore containing cuprite is currently stock piled if copper content is above 0.5%. Sulphide and transition ore are floated and the concentrate shipped to Japan to a copper smelter where credit is also received for the gold content. Flotation tailings are high in copper (up to 2%) and are not being processed. Several new technologies are being investigated including bacterial leaching, a cyanide regeneration process to recover complexed cyanides, and the addition of ammonia to minimize cyanide consumption by copper during leaching.

Flotation is also used at the Harbour Lights gold mine in Leonora to produce two concentrates. An initial bulk column flotation is followed by a differential flotation to separate arsenopyrite and pyrite. Arsenopyrite concentrates assaying 70 to 100 g/tonne gold are being shipped to a smelter while the pyrite concentrate is cyanide leached and put through the CIP circuit. Residual gold in the leached tails can be up to 20 g/tonne and options for its treatment include further upgrading to remove talc by flotation in order to increase sulphide sulphur content to a sufficient level for custom roasting.

At the Paddington gold mine near Kalgoorlie the open pit oxide contains around 5% sulphides as pyrite/arsenopyrite. All the ore is column floated, the bulk concentrate fine ground and then recombined with oxide float tails and sent to the CIL section. Current throughput is around 10^6 tonnes/year.

Until recently an antimonial gold flotation concentrate from the Blue Spec/Golden Spec mine near Nullagine was treated by cyanide pressure leaching and gold recovered by CIP.

CONCLUSION

The Western Australian gold industry has introduced a wide range of technologies to extract gold from the many and varied deposits being worked at the present time. Where roasting is permissible this is the favoured pre-treatment method for sulphide/telluride or arsenopyrite concentrates. Regrinding is an option if liberation can be achieved and excessive levels of cyanicides are absent. Arsenopyritic and copper rich sulphide ores are being exported as concentrates to smelters. Other emerging technologies such as the Cutech process, Sirosmelt and bacterial leaching are being considered for certain highly refractory ores. Choice of treatment method is often dictated by location, environmental considerations and ore reserve limitations.

ACKNOWLEDGEMENTS

The author wishes to acknowledge the assistance rendered to him by the Western Australian gold mining industry in the form of technical data. This paper is published by permission of the Director, Chemistry Centre WA.

REFERENCES

Blaskett, K. S., 1952, "Ore Treatment at Western Australia Gold Mines", Tait Publishing Company Pty Ltd, Melbourne, Australia.

Quadrio, A. J. and Kellett, K. R., 1984, "Recovery of Gold From Lancefield Arsenical Sulphide Ore". Proceedings AusIMM Conference: Gold-Mining, Metallurgy and Geology, Kalgoorlie, October.

von Bernewitz, M. W., 1940, "Roasting Gold-Silver Sulphide Ores and Concentrates" Bulletin Vol. R3-B1 Denver Equipment Co (Canada) Ltd. Reprinted from Canadian Mining Journal.

11

Processing IV General Processing

R.R. Beebe
Homestake Mining Co.
San Francisco, California

J.T. Woodcock
G.J. Sparrow
CSIRO Division of Mineral Products
Port Melbourne, Vic., Australia

Chapter 40

A DEDICATED RESIN FOR GOLD-THE STIMULUS NEEDED FOR UNIVERSAL ACCEPTANCE OF A RESIN-IN-PULP PROCESS

Brian R Green, Kenneth G Ashurst and Tracy E Chantson

Council for Mineral Technology
Private Bag X3015
Randburg, 2125, South Africa.

Abstract. Work that has culminated in the manufacture and testing of a gold-selective resin on a pilot-plant scale is described. Following a brief discussion about the nature of commercially available resins which identifies them as poor candidates for a gold extraction application, the improved properties of the new type of resin are illustrated using results from comparative loading tests. Further results from continuous countercurrent extractions using synthetic leach liquors and elution work using thiourea as the eluant demonstrate the superiority of the new resin over commercial resins and hence, its potential for use in a resin-in-pulp process for gold.

INTRODUCTION

Although activated carbon is widely used in the Western World for the extraction of gold from cyanide liquors, resin seems to be the preferred reagent in the Soviet Union (Anon, 1978). Work carried out some time ago at Mintek on the evaluation of resins for this purpose (Fleming, C.A., and Cromberge, G., 1984) highlighted chemical advantages of resin over carbon under plant conditions prevailing at certain mines. More recent work in the Hydrometallurgy Division dispelled fears about problems associated with the screening of resins and their physical stability. Also, a great improvement in elution methods (Fleming, C.A., 1985) and the successful implementation of resin-in-pulp (RIP) at the Golden Jubilee Mine (Fleming, C.A., 1989) have made the use of strong-base resins more attractive. In all these investigations commercial resins, originally designed for other applications, were used.

The advantages noted for ion-exchange resins over activated carbons were that they are not poisoned as readily by organic matter or fine solids and that their loading rates are higher. Negative aspects of existing ion-exchangers are poor selectivity, complex and generally slow elutions, and relatively high costs.

In South Africa increasing quantities of gold solution are being produced which contain poisons. Recirculated water tends to accumulate organic compounds such as flotation reagents which contaminate the pulps. Organic matter is encountered in shallow low-grade alluvial deposits and in old dump material, whereas underground sludges contain fine solids. Notwithstanding some of the disadvantages of resins they sometimes offer the only economic method of recovery of gold values. Golden Jubilee (Fleming, C.A., 1989) is an example.

Work on the development of resins for gold was undertaken at Mintek for reasons emanating from the above discussion. Weak-base resins were originally the focus of attention (Green, B.R., and Potgieter, A.H., 1984; Schwellnus, A.H., and Green, B.R., 1988) because of the expected ease of elution with aqueous NaOH. Subsequently, work with strong-base resins commenced after results of preliminary experiments indicated that substantial improvements could be made by simple modifications to the resin structure.

For the development of more suitable resins, account was taken of knowledge offered by the literature and results of gold extraction testwork carried out ten years ago at Mintek on a range of commercial resins which gave an idea of how various types of strong-base resins performed. It was believed that certain structural parameters of the resins would have great influence on resin performance. Therefore a programme for the systematic investigation of these parameters was drawn up. As a result, a resin structure that shows great promise has been developed over the last two years. An example of this resin type was recently prepared on a large scale (300 kg) in cooperation with NCP* and pilot-plant trials have begun at the Kloof gold mine.

This paper describes screening tests which are presently used to evaluate resins developed for the extraction of gold and laboratory trials of a prototype of the new strong-base resin which was subsequently prepared for the pilot plant.

RESULTS AND DISCUSSION

A decision to conduct intensive investigations aimed at the improvement of ion-exchange resins for use in the resin-in-pulp process in the gold mining industry and to assess resins from manufacturers created an urgent requirement for the development of standard tests for evaluation of the resins.

---

*NCP is a division of Sentrachem Ltd, South Africa.

Specifications were an initial screening test of about 24 hours duration followed by a more rigorous test for promising resins that could be completed in a few days. These tests should indicate relative loading kinetics, selectivity, and capacity. Resins that exhibited the required standards would proceed to a laboratory scale countercurrent trial that would simulate plant practice and would require two to three weeks for completion.

Amberlite IRA 400 has in the past been evaluated and found to exhibit gold loading characteristics typical of most commercial strong-base resins. Hence this resin was chosen as the standard for comparison with experimental resins.

Five experimental strong-base resins from different manufacturers and several from our own laboratory were subjected to the sequence of tests. The results presented here are those for the most promising strong-base resin evaluated so far.

Batch Screening Test

In the rapid and effective batch test (standard conditions given in Table 1), a Mintek strong-base resin showed a high capacity for gold (20 000 ppm at 24 hours) and good selectivity for gold over the base metals, especially copper, cobalt, and iron (Figure 1). In contrast, IRA-400 revealed a moderate capacity for gold (15 600 ppm) and relatively poor selectivity over the base metals, especially nickel.

The batch test proved to be very effective for initial screening of the resins. However, there were two conditions that were different from the countercurrent process used in plant practice. These were contact of the resin with concentrations of metals in the solution that fell significantly as the test proceeded and the artificially high concentrations of metals in the liquor that were chosen to facilitate rapid analysis. These drawbacks were rectified in a more rigorous column test to which promising resins were subjected.

Column Tests

The passage of the synthetic leach liquor through the resin in a column (Table 2) ensured contact with constant concentrations of metals, similar to conditions in the first stage of the countercurrent circuit used in plant practice. Reproducible hydrodynamic conditions were attained by pumping the leach liquor down the column at a constant flow rate.

The column test was carried out on the same two resins (Figure 2). The test indicates greater selectivity for gold over zinc and nickel for the Mintek resin compared with the batch test (Figure 1). However, the batch test correctly indicated that the resin has a high gold capacity and good selectivity for gold over the other metals. The column test with IRA-400 confirmed indications from the batch test that moderate gold loading and poor selectivity could be expected. Standard conditions that were adopted are summarized in Table 2.

TABLE 1

Batch screening test - standard conditions

Resin mass, 0.3 g (referred to equilibration in pH 7 solution)
Solution composition: 10 ppm each of Au, Ag, Zn, Ni, Co, Fe(II), Fe(III), Cu as the cyanide complexes; pH 10.5
Solution volume: 800 ml
Sampling times (hours): 0.5, 1, 2, 4, 6, 24
Sample volume: 7 ml (in duplicate)

TABLE 2

Standard conditions for the column test

Resin mass, 0.3 g (as Table 1)
Solution composition: (ppm) Au(5), Ag(0.5), Zn(2), Ni(5), Co(1), FeII(5), FeIII(5), Cu(10) as the cyanide complexes; pH 10.5
Solution flow rate: 2.5 ml/min
Periods of flow (hours): 2, 7, 24, 48, 72

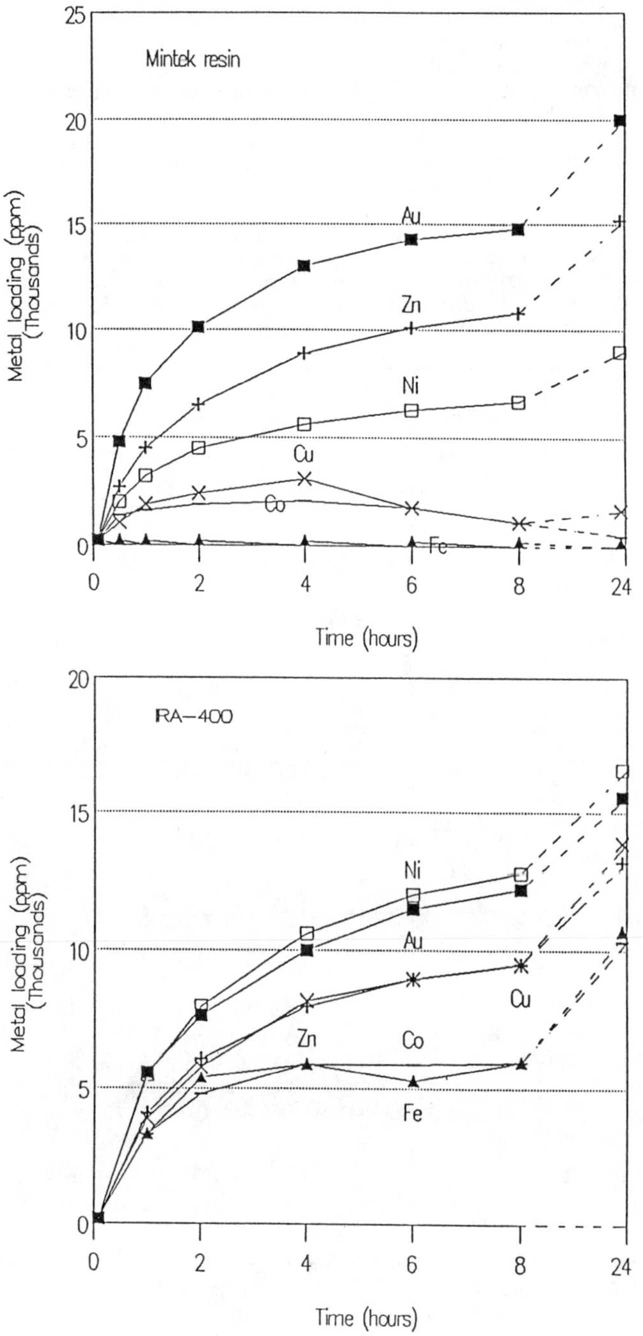

Figure 1: Batch tests for the Mintek strong-base resin and IRA-400

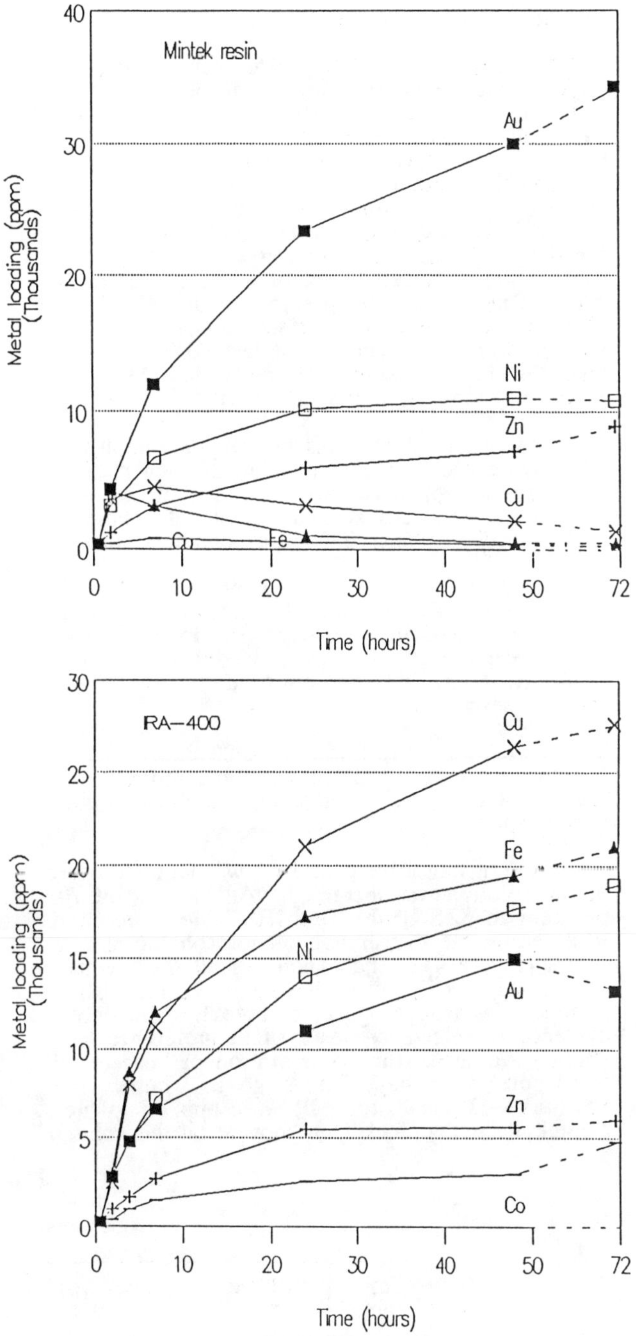

Figure 2: Column tests for the Mintek strong-base resin and IRA-400

Resins that were assessed as outstanding from the results of the column test were subjected to a countercurrent ion-exchange trial.

Countercurrent ion-exchange test

The countercurrent apparatus comprised six stages. Each stage consisted of a column containing a specified volume of resin. The top of each column was connected to the lower inlet of the next with provision for sampling the solution between the columns. The test solution was pumped upwards through the stages from one to six from a large reservoir at a constant flowrate for a specific length of time. At this point the first column was removed for elution and the other columns moved up one position so that the direction of resin movement was opposite to that of the solution flow. The vacant last position was taken by the first column after elution. This procedure constituted the first cycle.

Analysis for the metals in solution samples taken from each stage at the end of each cycle indicated the approach to the steady state condition. At steady state, a sample of resin from each stage was analysed for the content of metals.

The two resins investigated by the batch and column tests, IRA 400 and the Mintek resin, were subjected to the countercurrent trial in order to provide experience with resins possessing moderate properties (IRA 400) and outstanding properties (Mintek resin).

A preliminary test with IRA400 under the conditions 1A (Table 3) resulted in significant loading of gold and other metals on the first stage only (Figure 3).

Further stages were utilized when the flowrate of the solution was increased. An increase of 79 per cent to 52.5 BV/h (Test 1B, Table 3) resulted in a reasonable distribution of gold on the six stages (Figure 4). The loading of all the metals on all the stages (Figure 4) showed increasing loading in the series from stage 6 to stage 1. The only evidence of selectivity was a slight preference for copper over iron (the concentration of both metals is 10 ppm) from stage 3 to stage 1. These conditions (Table 3, test 1B) were applied to the Mintek resin that had been showed by the column test to be much more selective for gold (Figure 2).

Application of the conditions that were optimum for IRA-400 to the Mintek resin (Table 3, 2A) resulted in gold adsorption on the first two stages only (Figure 5). This feature reflects the much higher capacity for gold of this resin compared with IRA-400. In addition, the loading of all the metals on the six stages reveals a higher selectivity for gold (liquor concentration 5 ppm) over some of the base metals, especially copper and iron (10 ppm).

An increase of the flowrate of the liquor by 42.9 per cent to 75.0 BV/h (Table 3, 2B) extended the adsorption of gold to three stages with higher loading (Figure 6) and improved the selectivity of gold on the first stage, especially over nickel.

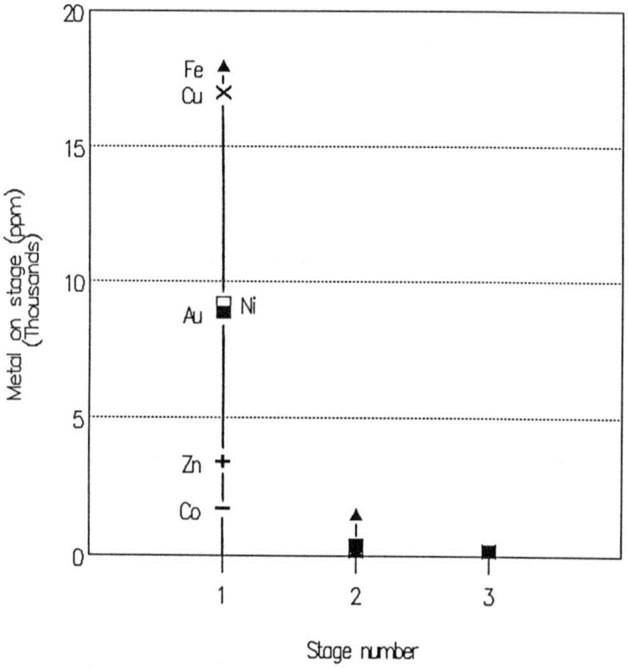

Figure 3: **Metals loaded on IRA-400 at steady state for countercurrent test 1A (Table 3)**

TABLE 3

Conditions for countercurrent tests*

Test no	Resin	Bed vol. ml	Flowrate soln.BV/h	Ratio soln./ resin flowrates	Cycles to steady state	Mass of BV, g
1A	IRA-400	8.0	29.3	703	12	2.86
1B	IRA-400	8.0	52.5	1259	12	2.86
2A	Mintek	8.0	52.5	1259	7	2.50
2B	Mintek	8.0	75.0	1799	9	2.50
2C	Mintek	6.0	100	2400	13	1.90

*Flowrate of resin, 0.0417 BV/h; Cycle period, 24 hours

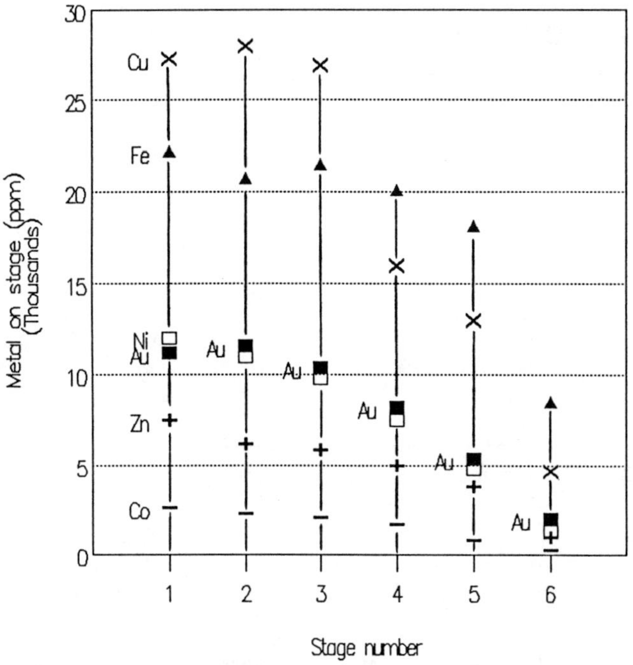

Figure 4: **Metals loaded on IRA-400 at steady state for countercurrent test 1B (Table 3)**

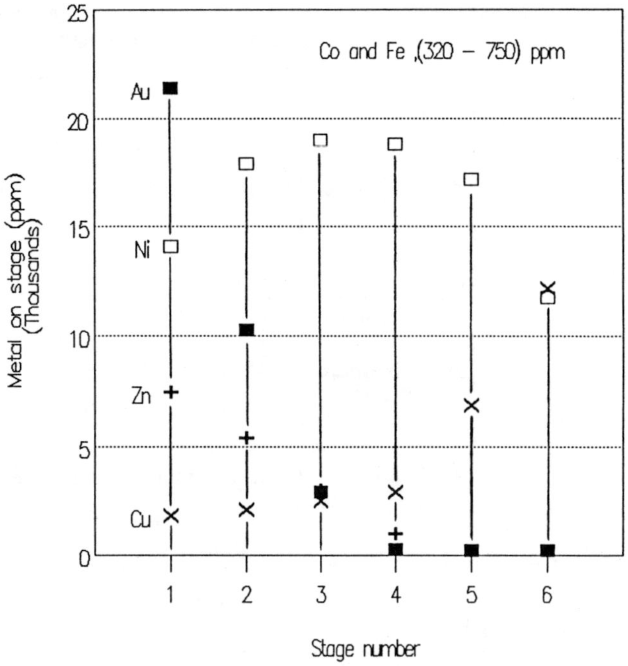

Figure 6: **Metals loaded on the Mintek resin at steady state for countercurrent test 2B (Table 3)**

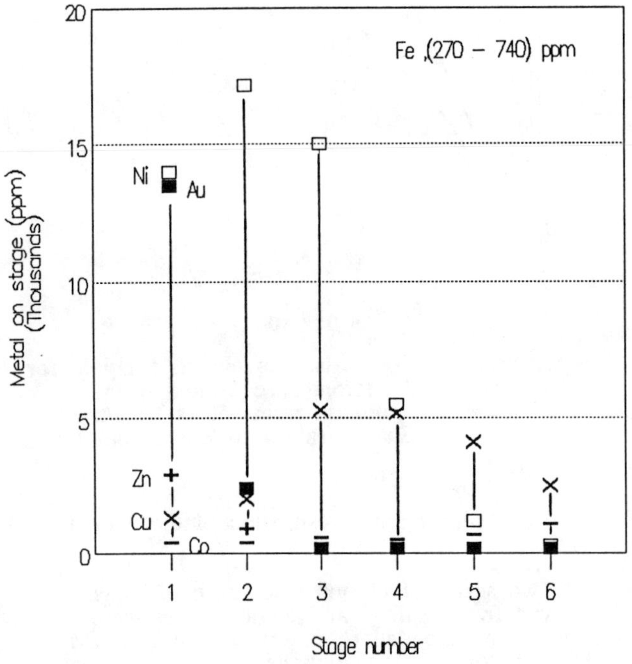

Figure 5: **Metals loaded on Mintek resin at steady state for countercurrent test 2A (Table 3)**

Further increase of the flowrate of the liquor was undesirable because of the risk of leaks at the joints of the apparatus. Instead, a reduction of the bed volume of resin from 8 ml to 6ml in all the stages increased the flowrate of the liquor from 75 to 100 BV/h (Table 3, 2C). A satisfactory profile of the gold adsorption on the six stages was achieved (Figure 7) with further improvements in gold capacity and selectivity. The characteristics for each stage (Table 4) show even adsorption of gold on the first four stages. The fact that the loading of gold and base metals across the stages is so dependent upon the operating conditions suggests that the design of a RIP plant will require careful consideration in order to achieve optimum results.

The loss of 0.08 ppm of gold from the last stage (Table 4) amounts to 1.15 mg of gold per cycle, or 2.08 per cent of the total gold passed. It is clear from Figure 7 that the loss could be virtually reduced to zero without a significant effect on the total gold loading on stage 1 by slight adjustment of the parameters.

Elution

It was considered important to investigate methods of elution after having established good loading characteristics for a strong-base resin. The results showed (Figure 8) that the Mintek resin could be stripped by the known methods for strong-base resins. Rates of elution were slow with zinc-cyanide and thiocyanate, but rapid with thiourea eluant (1M in 0.1 M H_2SO_4) at 60°C.

TABLE 4
Characteristics* for each stage of test 2C (Mintek resin)

	Stage					
	1	2	3	4	5	6
Liquor concn. leaving stage (ppm)	3.08	2.30	1.43	0.90	0.37	0.08
Total Au on resin (ppm)	26500	21100	15700	10700	5630	2100
Au adsorbed per cycle	5400	5400	5000	5070	3530	2100

*14.4 l of leach liquor pumped per cycle of 24 hours; Au concentration in leach liquor 3.85 ppm.

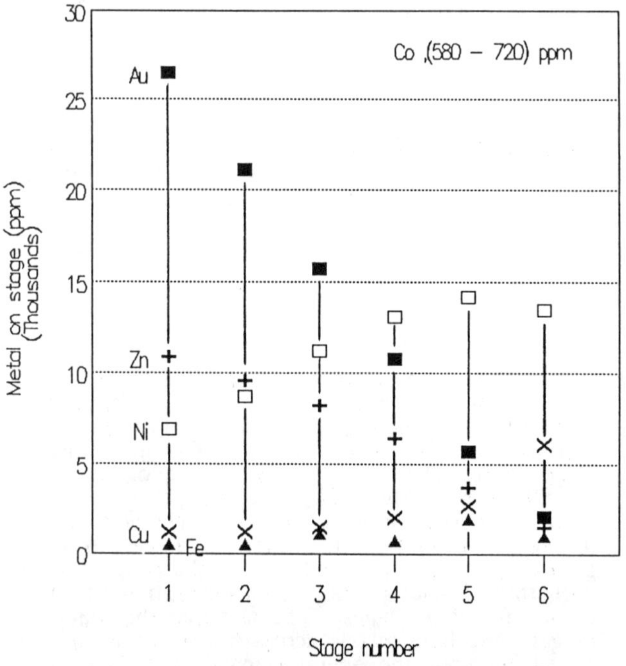

Figure 7: **Metals loaded on the Mintek resin at steady state for the counter-current test 2C (Table 3)**

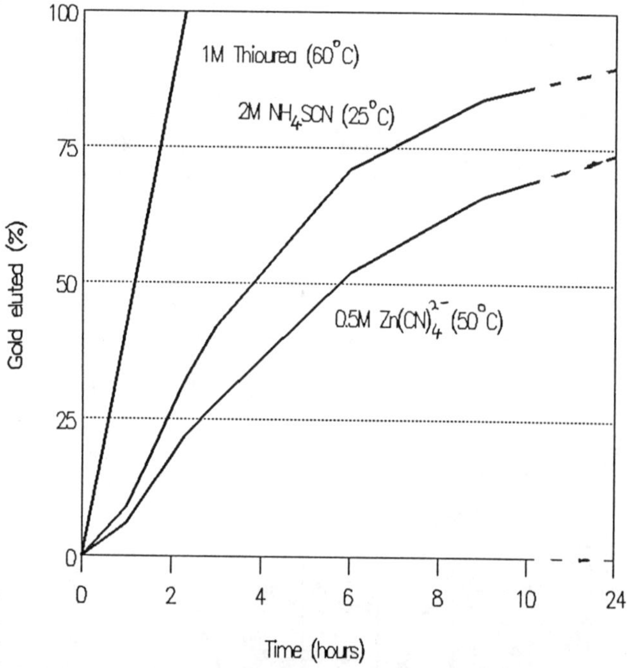

Figure 8: **Comparison of eluants for gold for the strong-base Mintek resin (Gold on resin ~50 000 mg/kg. Elution rate 1 to 2 bed volumes/h).**

A thiourea elution is impractical where substantial amounts of cobalt and iron are loaded on a resin, as would normally be the case with conventional strong-base resins (eg IRA 400). The low pH value employed during elution causes precipitation of cobalt and iron as cobalticyanide and Prussian blue-type complexes respectively, and hence the resin becomes poisoned. The Mintek resin is selective against these metals which, under usual solution conditions, will have loaded sufficiently poorly to make a thiourea elution feasible. Work described below was done in order to assess this method of elution and to pave the way for a pilot-plant investigation of the Mintek resin.

Two series of elution tests, each of which involved five loading* and elution cycles were undertaken. Loaded resin was washed with 1M H_2SO_4 before elution, and the eluted resin was

*The concentrations of metals (particularly the base metals) in the loading solution at equilibrium were typical of what could be expected in a leach liquor to ensure that the metal concentrations on the resin would be realistic. Gold was loaded to a level of about 20 000 ppm.

TABLE 5
Cyclic loading and elution tests with new eluant

Cycle	Metal extracted from solution* %							Metal in acid wash,# % of metal loaded				Metal in Thiourea eluate#, % of metal loaded				Metal on resin after elution, ppm						
	Au	Ag	Cu	Fe	Co	Ni	Zn	Au	Ag	Ni	Zn	Au	Ag	Ni	Zn	Au	Ag	Cu	Fe	Co	Ni	Zn
1	>89	33	0	0	7	51	78	0	0	22	100	100	99	78	0	83	48	<200	<200	<200	<200	<200
2	86	35	0	0	13	43	78	0	0	28	100	100	-	72	0							
3	86	33	0	0	0	35	75	0	0	-	100	100	-	-	0							
4	87	29	0	0	0	32	71	0	0	-	100	-	-	-	-							
5	89	29	0	0	0	31	69	0	0	46	100	100	100	54	0	20	100	<200	277	633	<200	<200

*3 ml of resin were loaded from 2.5 l of solution containing Au (10 ppm), Zn (10 ppm), Ni (10 ppm), Co (2 ppm), Cu (10 ppm), Fe (10 ppm) and Ag (5 ppm).
#Cu, Fe and Co are not applicable

TABLE 6
Cyclic loading and elution tests with recycled eluant

Cycle	Metal extracted from solution, %*					Metal in acid wash,+ % of metal loaded				Metal in Thiourea eluate#, % of metal loaded			Metal in Thiourea# eluate, ppm			Metal on resin after elution, ppm						
	Au	Ag	Co	Ni	Zn	Au	Ag	Ni	Zn	Au	Ag	Zn	Cu	Ni	Fe	Au	Ag	Cu	Fe	Co	Ni	Zn
1	88	59	0	52	92	0	0	42	100	100	99	0	67	72	210	83	48	<200	<200	<200	<200	<200
2	81	68	0	52	91	0	0	16	100	100	-	0	90	298	337							
3	78	66	0	41	87	0	0	15	100	100	-	0	53	187	205							
4	84	50	0	30	76	0	0	79	100	100	-	0	67	205	295							
5	78	48	0	36	82	0	0	33	100	100	100	0	81	258	337	13	4	11	320	670	36	345

*Same solution used in previous series. Extraction results for Cu and Fe were high (~ 50 %) do not conform with results from elution or with previous results. They are certainly incorrect.

+Cu, Fe and Co are not applicable.

#Co not applicable. Base metal concentrations complicated by recycling of eluant.

washed with water before being reloaded. In the first series fresh eluant was used for each elution. Results are given in Table 5 and conclusions are summarized below.

i. Gold extraction appeared to drop slightly after the first cycle, and then stabilized.

ii. Selectivity against copper, iron and cobalt was confirmed.

iii. Base metal loading (Co, Ni, Zn) appeared to decrease with successive cycles.

iv. All of the zinc and some of the nickel were removed during the acid wash.

v. Gold, silver, and the rest of the nickel were efficiently eluted with thiourea.

vi. There was a very slight build-up of iron and cobalt on the resin. It is possible that as the affinity for base metals decreases with successive cycles (see iii) the build-up of cobalt and iron might tail off. This would need further investigation.

In the second series of loading and elution cycles (Table 6); the eluate was passed through a bed of steel wool to remove the gold and then recycled. The results showed similar trends to those observed for the first series. A concern about the effect of base-metal build-up in the eluant (originating from the loaded resin and the steel wool) seems groundless. Concentration levels of copper, nickel and iron stabilized at about 80, 250 and 300 ppm

TABLE 7

Effect of thiourea eluant* on resin capacity

	Metal loaded on resin, ppm						
	Au	Ag	Zn	Ni	Co	Fe	Cu
Before	19200	8500	15200	10200	1100	0	1400
After	16300	5100	10400	4802	1000	2600	2600

*Thiourea (1M in 0.1 M H_2SO_4), 60°, 15 days.

respectively.* Whether this was due to co-precipitation of the metals by S^{2-} together with gypsum during pH adjustment (1 to 2) with $Ca(OH)_2$, or because of a 15 per cent bleed (sample removed for analysis) with each cycle, is not known.

The elution step probably presents the most hostile environment for chemical stability which the resin will encounter in the RIP process envisaged for the Mintek resin. A sample of the resin was therefore treated with the eluant for 15 days at 60°C - the time that it would be exposed to this environment during one year of use. The metallurgical performance was tested by a batch loading test. The results summarized in Table 7 are not too discouraging. Gold capacity decreased by about 15 per cent. Microanalysis of the resin showed that there was little change in the composition of the resin indicating perhaps, that the lost capacity might have been caused by poisoning rather than by chemical degradation. This is an area which would require further work.

CONCLUSIONS

The increasing interest in resins for recovering gold from certain pulps has resulted in attempts to develop more suitable strong-base resins. A series of standard screening tests was developed because of the need to evaluate and compare resins emanating from the laboratories of resin manufacturers and of Mintek.

No more than two resins have been identified so far which show advantages over existing strong-base resins. Only the one dealt with in this paper exhibits selectivity against cobalt and iron, thus allowing the rapid thiourea elution to be used. Small changes in resin performance observed during cyclic loading and elution tests, presently ascribed to the treatment of resin with thiourea, have not yet been characterized but would need to be monitored during pilot-plant testwork.

The next stage of evaluation for this and other promising resins which might appear on the scene would be to determine suitability for the processing of various gold-containing solutions or pulps. This work has begun in South Africa with the identification of gold-bearing materials which are difficult to process by conventional methods.

*The $CaSO_4$ precipitate which should have been white, was a grey colour. H_2S was detected during the titration of resins which had been in contact with thiourea.

ACKNOWLEDGEMENT

The work was done in collaboration with the Chamber of Mines as part of its research programme and is published by permission of the Council for Mineral Technology. The contribution to this work by I. Tyc and G. Steinbach is greatly appreciated.

REFERENCES

Anon., 1978, "Gold from Russia's Muruntau deposit", *Coal Gold Base Miner.*, vol. 26, no. 7.

Fleming, C.A., 1985, "Novel process for the recovery of gold cyanide from strong-base resins," *Extraction Metallurgy '85*. London, The Institution of Mining and Metallurgy, pp.757-787.

Fleming, C.A., 1989, "Recovery of gold by resin-in-pulp at the Golden Jubilee Mine," *Precious Metals '89* Jha, M.C., and Hill, S.D. (eds.) The Minerals, Metals and Materials Society, Warrendale, Pennsylvania pp. 105-119.

Fleming, C.A., and Cromberge G., 1984, "Pilot-plant studies on the extraction of gold from cyanide solution by anion-exchange resins," *J.S. Afr. Inst. Min. Metall.*, vol. 84, no. 11 pp. 369-378.

Green, B.R., and Potgieter, A.H., 1984, "Unconventional weak-base anion-exchange resins useful for the extraction of metals, especially gold". *Ion-exchange Technology*. Naden D, and Streat M, (eds.). Chichester, Ellis Horwood, pp.626-636.

Schwellnus, A.H., and Green, B.R., 1988, "Structural factors influencing the extraction of gold cyanide by weak-base resins", *Ion-exchange for Industry*, Streat M. (ed). Ellis Horwood Ltd. pp. 207-218.

Chapter 41

RECOVERY OF FINE GOLD FROM A PLACER ORE BY COAL GOLD AGGLOMERATION

S R Bellamy, C I House and C J Veal

BP Research Centre, Sunbury-on-Thames, Middlesex

INTRODUCTION

Coal gold agglomeration (CGA) is a new concept in gold processing. It involves the recovery of oleophilic (oil-loving) gold grains by causing them to adhere selectively to the surfaces of coal oil agglomerates. Under optimum conditions very high recoveries of gold can be achieved. The agglomerates are recycled to build up gold loading to typically 1000 - 5000 g/tonne. Bullion can be produced by burning the loaded agglomerates with subsequent recovery of gold from the highly loaded ash either by dissolution/electrowinning or smelting.

A wide-ranging programme of laboratory testwork has been performed to investigate the response of different ore types to the process. This has shown CGA to be most applicable to ores in which the gold is liberated or easily liberatable, relatively coarse (ca 500 μm) and fine (<5 μm) gold grains being recoverable with equal ease.

The best opportunities for CGA seem to be:

- placers containing fine gold not recoverable by gravity concentration
- low grade gravity concentrates to avoid multiple magnetic/gravity separation stages and amalgamation
- free milling ores where the gold can be liberated at a reasonable ore particle size eg >d_{80} 50 μm
- tailings past or present from gravity concentration plants.

CGA testwork ranging in scale from laboratory to pilot plant on tailings produced by gravity concentration in the 19th century Australian gold rush has already been reported (Bonney, C F, 1988; House, C I, Townsend, I G and Veal, C J, 1988). An economic appraisal for a Canadian free milling gold ore has also been published (Buckley, S A, House, C I and Townsend, I G, 1989).

This paper describes results of a detailed laboratory programme to assess CGA as a means of recovering gold from a beach sand deposit in New Zealand.

The area from which the samples were taken consists of quarternary interglacial marine, lagoonal, fluvial and eolian deposits. The heavy minerals present are predominantly ilmenite and garnet and small concentrations of gold. In certain locations the sands and gravels are consolidated by a limonitic cement. In the 19th Century some parts of the deposit were mined and the ore processed using stamp milling and gravity concentration. Gold recoveries were low, probably below 50%, due primarily to the fine size of the gold grains, the flaky nature of the gold and also imperfect liberation by the stamp mills.

Our laboratory testwork was divided into three parts:

i) Preliminary studies on an initial sample to assess the feasibility of treating the ore by CGA

ii) Extended studies on further samples from various locations in the deposit to confirm predictions from (i)

iii) Conventional gold recovery techniques.

EXPERIMENTAL PROCEDURES

All CGA tests were performed by agitating a slurry of ore in water in a 0.6 dm³ pyrex beaker fitted with four vertical baffles. When required, potassium amyl xanthate collector was added and the slurry conditioned for 5 min. Coal (ex BP Coal Australia's Tahmoor Colliery) previously ground to a d_{80} 50 μm was added followed by a commercial grade of gas oil (density 840 kg/m³). The agglomeration time was 30 min. The agglomerates were separated from the tailings by screening, both were then dried and analysed for gold by fire assay. This test procedure provides a reasonably accurate estimate of gold recoveries that might be expected from our 1 tph pilot plant.

Ore samples were ground either dry in an agate ring disc mill or wet in a steel batch tumbling ball mill with steel grinding media. Particle size distribution of mill products was measured using a laser diffraction size analyser.

Attrition scrubbing was performed using a Denver D12 agitator fitted with dual opposed-pitch, axial-flow impellers. Slurry pulp density was 70% solids and agitation was performed at 1500 rpm (equivalent to a tip speed of 5.3 ms^{-1}) for 30 min. Commercial attrition scrubbers operate under similar conditions of pulp density and tip speed.

Gravity concentration was performed using a Mozley laboratory separation table fitted with a flat deck. Gold recovery by cyanidation was investigated in bottle roll tests, 65 hours duration at pH 11.0 - 11.5 adjusted with lime and with 3.2 kg of sodium cyanide added per tonne of ore. At the end of each test the pulp solids were filtered and washed. The combined washings were analysed for gold by atomic absorption spectroscopy, and the dried solids by fire assay.

Flotation tests were carried out in a Denver D12 flotation cell. The pulp was conditioned with 220 g/t potassium amyl xanthate collector, 30 - 40 g/t Dowfroth 250 frother and agitated at 1200 rpm. A succession of concentrate fractions was taken, up to a total flotation time of 30 min.

MINERALOGICAL CHARACTERISATIONS

The samples from the different sites varied considerably in appearance. In particular the sample colour varied considerably due to differing levels of iron oxides and ilmenite.

As-received, Sample 1 was mainly composed of fine light sand but also contained some darker aggregates with a stratified colour variation. This lamination, characteristic of marine deposition and consolidation is due to light coloured quartz/feldspar and dark ilmenite/garnet minerals. Quartz was the most abundant mineral with significant amounts of muscovite, mica and feldspar. The majority of the coarser material consisted of aggregates of smaller particles held together by brown coloured cement. The average particle shape observed was angular and not well rounded as is common in most beach sand ores.

Samples 5 and 6 contained coarse siliceous gravel and large pebbles up to 10 mm in size, whereas Sample 4 contained none.

Microscopic examination of all of the samples indicated that the modal particle size of the sand was the same in each sample being 200 - 300 μm.

The chemical composition of Samples 1, 3 and 5, as determined by XRF, are given in Table 1. The most significant differences are in the Ti and Fe contents, reflecting differences in proportions of ilmenite and iron oxide slimes. There is also a correlation between Au (determined by fire assay) and Ti which would be expected since the beach sand has already been subjected to a degree of natural gravity concentration.

Table 1

CHEMICAL ANALYSIS BY X-RAY FLUORESCENCE

Element	Sample 1 %	Sample 3 %	Sample 5 %
Na	0.9	1.5	2.3
Mg	1.3	1.1	1.3
Al	5.5	6.3	6.7
Si	31.0	27.0	35.0
K	1.9	1.4	3.2
Ca	1.1	1.7	0.8
Ti	2.1	2.7	0.2
Mn	0.5	0.9	ND
Fe	6.4	8.9	2.6
Zn	0.1	D	D
Pb	1.1	ND	ND
Au (ppm)	0.78	0.77	0.21

D - Detected
ND - Not detected

GOLD DISTRIBUTIONS

The samples were all wet screened at 150 μm. The undersize and oversize fractions were dried, weighed and assayed for gold. The results are given in Table 2.

Table 2

GOLD DISTRIBUTION IN ORE SAMPLES

Sample No	No Pretreatment		Scrubbing Pretreatment	
	Mass % -150 μm	Gold % -150 μm	Mass % -150 μm	Gold % -150 μm
1	14	62	17	59
2	8	74	13	99
3	27	63	29	94
4	13	44	25	99
5	5	78	ND	ND
6	36	97	39	95

ND - Not determined

The data for all samples are similar in that there is a preponderance of % gold over % mass in the -150 μm size fraction. However the absolute values of per cent mass and gold vary considerably.

The effect of attrition scrubbing on gold and mass in the -150 μm size fraction is also shown in Table 2. For Samples 2, 3, 4 and 6 the vast majority of the cemented aggregates were broken up and more than 94% of the gold released into the final size fractions. The relatively small increase in per cent mass after scrubbing confirms the natural grain size of the sand at >150 μm. Sample 1 appears atypical perhaps because the cemented aggregates are more strongly compacted in this part of the deposit or perhaps the gold is genuinely more coarse. Several other samples, not reported here, have been assessed and confirm the atypical behaviour of Sample 1.

Thus preconcentration of the ore by attrition scrubbing and size splitting at ca 150 μm appears to be a technically viable method for processing the deposit and worthy of further testwork. This route would have a major impact on the economics of the project.

CGA TESTWORK

The Effect of Grinding

In CGA the prime factor which affects the recovery of gold grains is their state of liberation. In the preliminary work, which was carried out, on Sample 1, the main variable investigated was grind size.

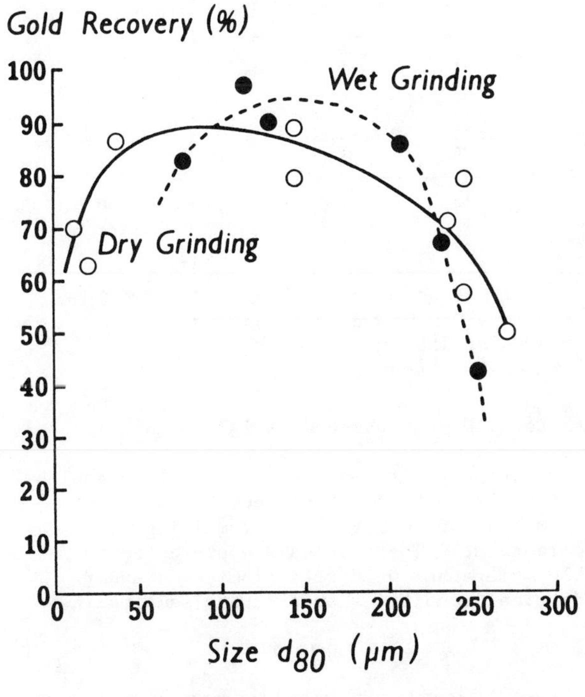

Fig 1. Effect of Grinding on Gold Recovery by CGA

The results from dry grinding the ore in a ring disc mill are shown in Figure 1. Decreasing d_{80} from 250 to about 180 μm increased the gold recovery from 40 to 90%. However, with a longer grinding time (d_{80} 20 μm) the recovery fell to about 60 per cent. The initial increase in the gold recovery is attributed to the breaking up of aggregates which released the adherent or cemented fine gold. A decreasing number of strongly bonded aggregates in the finer ground samples was observed under an optical microscope. Under optimum grinding conditions (d_{80} 130 μm) the majority of the aggregates had been broken up and further grinding resulted in general particle size reduction. The decrease in recovery with fine-grinding may be due to the reduction in hydrophobicity of the gold caused by embedding of gangue particles or the deformation of the soft metallic flakes.

The effect of wet ball milling was very similar to that produced by dry grinding (Figure 1). Optimum gold recovery (>85%) occurred for d_{80} sizes of between 200 and 90 μm.

Thus, only a mild treatment is required for gold (up to 90%) to be liberated from gangue particles. A grinding pretreatment could therefore simply consist of a rotary scrubber or pebble mill followed by attrition scrubbing to disperse the aggregates. Size reduction of the primary sand particles should not be necessary.

Results with Scrubbed Material

Six small-scale CGA batch tests were performed on each sample that had previously been attrition scrubbed and screened at -150 μm (Table 3). The operating parameters investigated were:

- Agitation speed during ore/agglomerate contacting — 750 and 1500 rpm
- Collector addition — 0, 200 and 500 g/t
- Agitation speed during collector conditioning — 750 and 1500 rpm

Operating conditions which gave tailings grades less than or equal to 0.04 g/t were identified for four of the samples (Nos 2, 3, 5 and 6). The single exception was Sample 4 where tailings grades did not fall below 0.13 g/t. Possible reasons for this apparent anomaly are (i) the attrition scrubbing conditions, which were not optimised, were not sufficiently severe to break down all the cemented aggregates or (ii) the attrition scrubbing failed to remove iron oxide coating of the liberated gold grains. Sample 4 contained a smaller proportion of coarse (+0.5 mm) material than the other samples.

In many of the CGA tests gold recoveries well above 90% were achieved. Gold recovery is a function of the tailings grade and the head grades of the test sub-samples (calculated from grades of agglomerates and tailings). The calculated head grades varied considerably within a test sequence for a given sample, owing to the small size of sub-sample available for each test. In addition the calculated head grade varied throughout the five samples indicating considerable variation across the deposit.

The results are in good agreement with those with Sample 1 for which gold recoveries of 90% on a head grade of ca 0.8 g/t were achieved. The light grind on Sample 1 and attrition scrub on Samples 2, 3, 5 and 6 appear to have had the same effect.

Table 3

CGA RESULTS ON SCRUBBED ORE (-150 μm)

Sample No	Test No	Impeller Speed rpm	KAX Addition kg/t Ore	Conditioning Speed rpm	Calc Head g/t	Tails Grade g/t	Gold Recovery %
2	A	750	0.2	750	1.48	0.90	41.0
	B	750	0.5	750	1.76	0.66	63.7
	C	750	0.2	1500	0.74	0.09	89.0
	D	1500	-	-	1.31	1.39	2.9
	E	1500	0.2	750	1.82	0.04	98.0
	F	1500	0.5	750	1.05	0.04	96.5
3	A	750	0.2	750	1.84	0.05	97.3
	B	750	0.5	750	2.39	0.05	97.9
	C	750	0.2	1500	2.33	0.22	90.6
	D	1500	-	-	1.93	0.19	90.2
	E	1500	0.2	750	2.03	0.09	95.6
	F	1500	0.5	750	2.23	0.05	97.8
4	A	750	0.2	750	0.61	0.18	70.4
	B	750	0.5	750	0.52	0.25	52.2
	C	750	0.2	1500	0.53	0.13	75.6
	D	1500	-	-	0.16	0.15	3.8
	E	1500	0.2	750	0.53	0.41	22.2
	F	1500	0.5	750	0.52	0.29	44.0
5*	A	750	0.2	750	0.48	0.21	56.6
	B	750	0.5	750	0.47	0.11	76.6
	C	750	0.2	1500	0.16	0.06	62.0
	D	1500	-	-	0.50	0.29	41.8
	E	1500	0.2	750	0.28	0.04	85.6
	F	1500	0.5	750	0.20	0.04	79.7
6	A	750	0.2	750	0.54	0.05	90.7
	B	750	0.5	750	0.66	<0.03	>95.4
	C	750	0.2	1500	0.73	0.12	83.5
	D	1500	-	-	0.79	<0.03	>96.2
	E	1500	0.2	750	0.63	<0.03	>95.3
	F	1500	0.5	750	0.68	<0.03	>95.6

*Sample 5 scrubbed but not screened at 150 μm since insufficient undersize for all testwork

The ore samples showed considerable variation in their response to CGA even after attrition scrubbing. In the case of Samples 3 and 6 excellent recoveries were achieved in all tests independent of the operating parameters within the ranges investigated. In contrast, Samples 2 and 5 gave no significant gold recovery in the absence of collector even at high agitation speed. The degree of agitation was however, influential in increasing gold recovery in the presence of collector either during conditioning or during ore/agglomerate contacting.

Overall the results suggest that the gold surface properties vary considerably in different ore samples probably due to the presence of iron oxide staining and increased surface hydrophilicity. Some of the variations in surface properties might have been due to differences in the effectiveness of the scrubbing technique used.

It was possible to overcome variations in surface properties in 4 of the 5 samples examined by adding collector to increase surface oleophilicity and using relatively high agitation rates which increase the energy of interparticle collisions and chances of successful adhesion of the gold grains to the coal oil agglomerates.

Results with Unscrubbed Material

Some experiments were carried out on Samples 2, 3 and 5 that had previously been screened at 1.7 mm but had no scrubbing or milling pretreatment. The results are presented in Table 4 together with data from tests under indentical conditions on scrubbed samples.

Table 4

EFFECT OF ATTRITION SCRUBBING ORE -1.7 mm ORE

Sample No		Impeller Speed rpm	KAX Addition kg/t Ore	Calc Head g/t	Tails Grade g/t	Gold Recovery %
2	Scrubbed	1500	0.2	1.82	0.04	98.0
	Unscrubbed	1500	0.2	0.27	<0.03	>89.1
3	Scrubbed	1500	0.2	2.03	0.09	95.6
	Unscrubbed	1500	0.2	1.02	0.30	71.0
5	Scrubbed	1500	0.2	0.28	0.04	85.6
	Unscrubbed	1500	0.2	0.61	0.22	63.7

In the case of Samples 3 and 5 gold recoveries were better on scrubbed materials, reflecting greater liberation of the gold grains. The scrubbing made little difference with Sample 2.

With such a heterogeneous ore with widely differing degrees of consolidation the scrubbing stage would have to be incorporated into the final flowsheet as an insurance. The improvement in recovery potentially achievable would almost certainly justify any increased cost.

COMPARISON OF CGA WITH CONVENTIONAL GOLD RECOVERY TECHNIQUES

Gravity Concentration

Gravity concentration of gold from a -150 μm fraction screened from as-received Sample 1 was reasonably effective (Figure 2) with 71.4% recovery into 20 per cent mass, to produce a concentrate grade of 7.2 g Au $tonne^{-1}$. A final concentrate had a grade of 30 g Au $tonne^{-1}$ but recovery was only 10%.

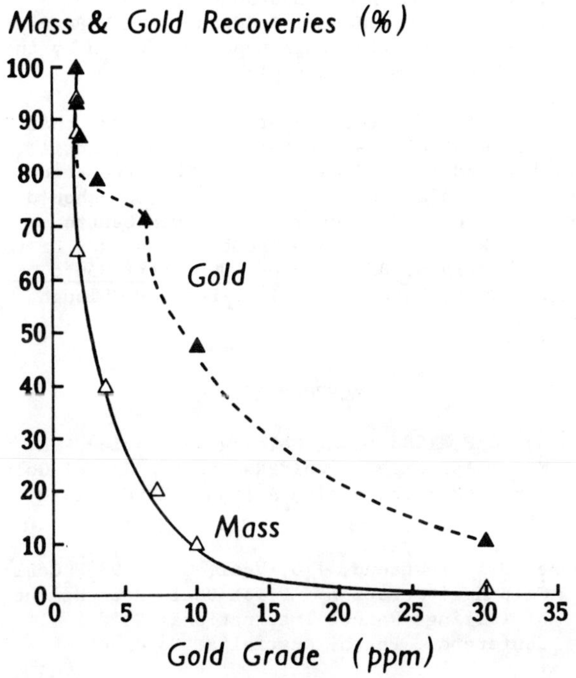

Fig 2. Gravity Concentration of -150 μm Size Fraction

The extreme flakiness of the gold grains was also thought to adversely affect the performance of gravity concentration.

Assuming the shape factor f from Gy's formula (Otley, D H, 1966) to be:

$$f = \frac{\text{mass of particle}}{(\text{particle diameter})^3 \times \text{density}}$$

then the f value for observed grains was only 0.01 compared with a value of 0.2 more usual for flaky alluvial gold.

It is unlikely that better results could be achieved on industrial equipment as the Mozley table is more suited to the recovery of fine particles than, for example, spirals or shaking tables. The overall process gold recovery would be expected to be less than 70%.

Gravity concentration does not compare well with CGA where recoveries in excess of 90% should be achievable. Also with recycle agglomerate grades of 2000 to 5000 g/t will be obtained.

Cyanidation

Samples 2 - 6 (-1.7 mm, unscrubbed) were cyanided to give data on the percentage of leachable gold and also lime and cyanide consumptions. The results are presented in Table 5.

In all cases very little gold remained in the residue after leaching. Gold recoveries were close to or above 90% except for Sample 4 (66.7%) which was artificially low due to the low calculated head grade of the test sample. Lime requirements were high (7.1 - 13.1 kg/t) due to the acidic nature of the ore. Cyanide consumptions varied from 0.24 to 1.10 kg/t in part due to varying amounts of iron dissolution.

Gold recoveries by cyanidation and CGA compare well for Samples 2, 3, 5 and 6. Comparison of relative performance on Sample 4 is complicated by the low head grade of the cyanidation test sample.

Table 5

CYANIDATION RESULTS ON SAMPLES 2 - 6

Sample No	Natural pH	NaCH Consumed kg/t	Lime Requirement kg/t	Gold in Solution g/t	Gold in Residue g/t	Calculated Head g/t	Gold Extracted %
2	6.8	0.31	7.5	0.24	0.03	0.27	88.9
3	6.2	0.88	8.2	1.17	<0.03	1.19	>97.4
4	6.3	1.00	10.8	0.07	<0.03	0.09	>66.7
5	5.4	1.10	13.1	0.43	0.04	0.43	91.5
6	7.3	0.24	7.1	0.57	<0.03	0.59	>94.9

Flotation

Flotation tests were performed with Samples 3 and 4, attrition scrubbed but not screened at 150 μm. The results are shown in Table 6. In both tests the pulp frothed well before adding any reagents. After addition of frother (Dowfroth 250) the froth was around 10 mm deep. In 15 - 20 min the froth became thin and Dowfroth 250 (0.01 kg/t) ore was added.

Table 6

RESULTS OF FLOTATION TESTWORK

Sample No	Flotation Time min	Cumulative Weight %	Cumulative Gold Grade g/t	Recovery %
	5	0.89	77.82	>95.0
	10	1.33	52.48	>95.5
(Scrubbed -1.7 mm)	20	2.39	29.31	>95.9
3	30	3.00	23.35	>96.0
	Tails grade	-	<0.03	-
	Calc head	-	0.73	-
	5	0.85	8.24	>66.7
	10	1.40	5.25	>69.9
(Scrubbed -1.7 mm)	20	2.62	2.86	>71.6
4	30	3.48	2.18	>72.4
	Tails grade	-	<0.03	-
	Calc head	-	0.10	-

The gold in both Samples 3 and 4 was sufficiently hydrophobic for flotation because after 30 min the tailings grades had fallen below 0.03 g/t equivalent to recoveries of >67% and >96% dependent on calculated head grade.

The final concentrate grades were 23.35 g/t for Sample 3 and 2.18 g/t for Sample 4. If the flotation feed had been preconcentrated with removal of the +150 μm material then the float concentrates would have been higher in grade but probably not much higher since it is the fines which tend preferentially to become entrained in the froth. Therefore the flotation concentrates would need a considerable amount of subsequent processing to produce bullion.

In the case of Sample 3 CGA and flotation compare well in performance. With Sample 4 flotation gave lower tailings grades than CGA due either to the lower flotation gold head grade or to the surface properties of the gold being sufficiently hydrophobic for flotation but not for agglomeration. As noted earlier further better understanding of Sample 4 is thought likely to lead to improved CGA recoveries.

CONCLUSIONS

The performance of CGA on samples from a beach sand deposit in New Zealand was extremely encouraging. A processing flowsheet involving attrition scrubbing, size classification and CGA on the undersize is thought to be the best route for treating this ore. Low tailings grades and high gold recoveries were achieved and almost certainly can be improved still further with more optimisation. From previous experience the CGA agglomerate grades would be 2000 - 5000 g/t or higher and after combustion the ash would contain 2 to 5% or more gold.

CGA performance was superior to gravity concentration, the fine, flaky nature of the gold leading to relatively low gravity recoveries (ca 70%) into concentrates containing only ca 7 g/t. Although giving gold recoveries similar to CGA flotation also produced low grade concentrates which would require extensive further treatment to produce bullion. CGA performance was also similar to cyanidation but would be expected to be considerably cheaper to operate and would have the environmental advantage of not entailing the large scale use of cyanide.

ACKNOWLEDGEMENTS

The authors wish to acknowledge contributions to the work from Dr H A Bustamante, Mrs E A Crathorne, Mr A R Goddard, Mr K D Pearse. Permission to publish this paper was given by the British Petroleum Company plc.

Note: Readers of this paper will be aware that BP has sold most of its mineral operations, until 1989 handled by BP Minerals, to RTZ Corporation plc. Commercial enquiries regarding CGA should be addressed to Dr C J Veal, BP Research Centre, Chertsey Road, Sunbury-on-Thames, Middlesex or to Dr D J Westlake, BP Ventures, Worcester House, Langley Quay, Waterside Drive, Langley, Slough, SL3 6EY.

REFERENCES

Bonney, C F, 1988, Coal Gold Agglomeration - A Novel Approach to Gold Recovery. Randol Gold Forum 88, Scottsdale, Arizona, 23rd - 24th January.

House, C I, Townsend, I G, Veal, C J, 1988, Coal Gold Agglomeration - Pilot Scale Retreatment of Tailings Randol International Gold Conference, Perth, Australia, November.

House, C I, Townsend, I G, Veal, C J, 1988, Coal Gold Agglomeration, International Mining September.

Buckley, S A, House, C I, Townsend, I G, 1989. A Technico-Economic Assessment of the Coal Gold Agglomeration Gold Recovery Process, The Canadian Mineral Processors Conference, Ottawa, January.

Otley, D H 1966, Gy's Sampling Slide Rule, World Mining August 1966, 40 - 44.

Chapter 42

INCO SO_2-AIR CYANIDE REMOVAL PROCESS UPDATE

E.A. Devuyst*, B.R. Conard*, G. Robbins** and R. Vergunst*

*J. Roy Gordon Research Laboratory, Inco Limited, Mississauga, Ontario
** Inco Tech, Vancouver, B.C.

Abstract. By the middle of 1989 Inco will have applied its patented cyanide removal technology to seventeen commercial operations.

Removal of cyanide from CIL slurry tailings is applied at Equity Silver and Inco's new 1,200 mtpd mill at Inco/Golden Knight. Detoxification of CIP tailings is carried out by Colosseum California (3,600 mtpd), Canamax's Ketza River (360 mtpd) and Westmin's Silbak Premier (2000 tpd) operations. Repulped tailings from Merrill-Crowe flowsheets are treated at Total Erikson's Mount Skukum (300 mtpd) and Skyline's Johnny Mountain (200 mtpd) mills and barren bleed at Inco/Queenston's McBean mill (500 mtpd). Lynngold's MacLellan mill (1800 mtpd) and ERG (40,000 mtpd) treat pond supernatant water. Plating shop cyanide effluents are decyanized at Kuntz Electroplating and two other plating shops in Ontario.

Details regarding some of the above operations are given. Operating and capital cost estimates for various treatment options are discussed.

INTRODUCTION

The Inco process for continuous cyanide removal from industrial wastewaters was patented by G.J. Borbely, and co-workers (1984). The process requires SO_2, air, lime and a copper catalyst. Both free and complexed cyanide are oxidized to cyanate in the pH range 8 to 10. Cyanide bound to iron is removed by precipitation as a ferrocyanide insoluble salt and other base metals are removed as hydroxides.

The process was piloted in 1982 on barren bleed solution at Campbell Red Lake Mines Ltd., Ontario (Devuyst, 1982). Since 1982, the Inco SO_2-air cyanide removal technology has been applied in seventeen commercial installations. The process has significant advantages over other commercial processes in the completeness of cyanide removal and lower reagent requirements and costs.

PROCESS APPLICABILITY

A wide variety of Canadian and U.S. gold and/or silver mill effluents, as well as effluents such as flotation tailings, metal finishing waste solutions and coke oven and blast furnace scrubber waters have been successfully treated in the laboratory.

Laboratory experiments at Inco are generally carried out in one or two reaction vessels placed in series. Sulfur dioxide is added pre-mixed in air (or separately as a sulfite solution). The tests are carried out at room temperature, maintaining the desired pH using a lime-water suspension (4 wt.% lime). When needed, copper is added as a cupric sulfate solution to the feed.

To date, Inco has successfully tested over fifty different effluents without failure, but treatment conditions and control strategy vary considerably. Some typical effluent treatment results obtained in the laboratory are given in Table 1 for operations which were eventually implemented in the plant.

Gold mill effluents generally contain, in addition to free cyanide, variable amounts of thiocyanate, and copper, nickel, iron and zinc cyano-complexes. The results in Table 1 show that CN_T was consistently removed from barren solutions, CIP or CIL tailings and pond waters containing 13 to 1124 mg/L down to less than 1 mg/L. Oxidation of SCN was generally limited to 10-20%. Metal values such as Cu, Fe, Ni and Zn are removed to <1 mg/L with occasional higher residual concentrations of copper. Reagent requirements vary depending on the type of feed, but are usually in the range 3 to 5 g SO_2/g CN_T. Copper, as Cu^{2+}, additions depend on the iron content of the feed and are in the range 0 to 0.5 g/g CN_T.

COMMERCIAL INSTALLATIONS

By the middle of 1989 Inco will have applied its patented cyanide removal technology to seventeen commercial operations. Typical installations are described in this paper. A summary of operating results is given in Table 2.

Equity Silver

Equity Silver Mines, Houston, British Columbia has operated a CIL circuit to scavenge gold and

TABLE 1 Laboratory Cyanide Removal Results from Continuous Treatment

Company	Type of Effluent	Stream	Total Ret'n (min)	pH	Assays (mg/L) CN_T	Cu	Fe	Reagents ($g/g\ CN_T$) SO_2	Cu^{2+}
Campbell	Barren	Feed	60	11.1	940	39	118		
Red Lake	Solution	Eff		10.0	0.7	0.6	0.2	3.0	0.15
Equity	CIP Tailings	Feed	30	10.0	195	42	25		
	(30% Solids)	Eff		8.0	0.4	1.1	0.2	3.8	0.25
Mount	Tailings	Feed	40	9.8	215	3.0	40		
Skukum	(45% Solids)	Eff		8.0	0.3	0.2	0.1	2.1	0.50
Lynngold	Pit Water	Feed	20	8.7	13	1.5	0.1		
		Eff		9.9	0.6	0.4	0.1	4.8	0.40
Colosseum	CIP Tailings	Feed	80	10.5	243	84	85		
	(53% Solids)	Eff		8.0	0.3	0.9	0.2	4.3	0
Kuntz	Electroplating	Feed	100		100	54	4.5		
	Rinse Water	Eff		8.5	2.1	3.2	0.6	3.6	0.50
McBean	Barren Solution	Feed	90	11.7	404	35	26		
		Eff		9.0	0.3	0.5	<0.2	3.5	0
Skyline	Repulped Tailings	Feed	120	10.4	1124	455	73		
		Eff		8.0	1.5	2.7	0.1	4.0	0
Inco/Golden	CIL Tailings	Feed	60	10.5	200	15	2.4		
Knight		Eff		8.0	0.2	0.9	0.2	4.6	0.2
ERG	Pond Water	Feed	40	9.0	155	88	6.5	6.5	0
		Eff		7.7	0.1	0.3	<0.1		
Westmin	CIP Tailings	Feed	120	10.7	278	24	13	5.3	0.07
		Eff		8.0	0.3	0.8	0.1		

TABLE 2 Commercial Operations - Cyanide Removal Results

Company	Stream	pH	Monthly Average Assays (mg/L)			Reagent Consumptions ($g/g\ CN_T$)		
Equity	CIL Tails	11.0	100	35	2			
	Effluent	8.0	1-5	2-5	0.2	5.9	0.27	0
Mt. Skukum	Repulp	11.0	100	5.0	15			
	Effluent	8.2	0.9	1.0	0.2	4.0	0.25	0
McBean	Barren	11.5	370	30	20			
	Effluent	9.0	0.2	0.7	<0.2	4.0	0	4.0
Lynngold	Pond	8.7	100	10	2.0			
	Effluent	9.5	0.6	0.1	0.1	6.0	0.10	8.0
Colosseum	CIP Tails	10.6	375	129	2.2			
	Effluent	8.7	0.4	1.5	0.2	5.6	0.11	2.9
Ketza River	CIP Tails	9.8	150	8.0	<0.1			
	Effluent	8.4	5.0	15	<0.1	6.0	0.30	0
Skyline	Repulp	10.5	450	300	10			
	Effluent	8.1	<1.0	2.0	0.3	6.0	0	-
Kuntz	Rinse	9.5	150	90	2.8			
	Effluent	8.5	0.2	1.2	<0.2	8.0	-	-

silver from flotation tailings since 1985. The plant can handle up to 10,000 mtpd of tailings at 40 wt.% solids, containing typically 100 mg/L CN_T. Inco's cyanide destruction process is carried out in a single 560 m^3 reactor equipped with a 250 hp radial impeller using liquid SO_2, lime and $CuSO_4$ solution as the reagents. Reagent additions are adjusted from a control panel and the pH is maintained at 7.8-8.2. The product of cyanide destruction is typically 1-5 mg/L CN_T, and reagent consumptions are 5.4 g SO_2/g CN_T and 0.27 g Cu^{2+}/g CN_T.

Mount Skukum

The Mount Skukum 300 tpd gold cyanidation mill in the Yukon was commissioned in April 1986. Gold recovery is by standard Merrill-Crowe zinc cementation and the washed leach solids contain sufficient barren solution to provide the required bleed. Cyanide is destroyed in flotation cells adapted to the SO_2-air process. Because of the remote location and small scale, $Na_2S_2O_5$ is used as SO_2 reagent. A preferred pH of 8.0 is maintained by acidifying repulp water to counteract excess base present in the tailings. Cyanide is removed from 100 mg/L CN_T to below 1 mg/L. Residual Cu and Fe are also below 1 mg/L after treatment. Reagent consumptions are typically 3.5-4.0 g SO_2/g CN_T and 0.25 Cu^{2+}/g CN_T. Most of the added copper is for the removal of iron cyanide as $Cu_2Fe(CN)_6$.

McBean

The Inco process was implemented on barren bleed at Inco/Queenston's 500 tpd McBean mine, Ontario, in June 1984. The treatment is carried out in a single stage equipped with flotation cell agitators, using $Na_2S_2O_5$ as the SO_2 reagent. Cyanide is removed from 370 mg/L to 0.2 mg/L CN_T using 4 to 5 g SO_2/g CN_T. Very little or no Cu^{2+} is required.

Lynngold Operation

The Lynngold, Lynn Lake, Manitoba, cyanide destruction circuit for treatment of pond supernatant water was put on stream in June, 1987. The operation comprises two 16 m^3 stages equipped with flotation mechanisms for air addition. Aqueous $Na_2S_2O_5$ and $CuSO_4$ solutions are added to the first stage to oxidize the cyanide. No air or $Na_2S_2O_5$ is added in the second stage where base metals are precipitated by pH adjustment. Cyanide is currently removed from 100 mg/L to 0.6 mg/L CN_T and Cu and Ni are removed to 0.1 and 0.4 mg/L, respectively. Reagent consumptions are 6 g SO_2/g CN_T and 0.1 g Cu^{2+}/g CN_T.

Colosseum California

Colosseum was commissioned in October 1987 and treats CIP tailings from its 3600 mtpd operation at 45 wt% solids. Cyanide is removed from 375 mg/L CN_T down to < 1 mg/L CN_T in a 400 m^3 single reactor equipped with a 250 HP dual hydrofoil impeller. Reagent consumptions are typically 5-6 g/g CN_T liquid SO_2 and 0.11 g/g CN_T Cu^{2+}. A pH of 8 to 9 is maintained with dry lime.

Ketza River

Canamax's Ketza River 360 mtpd mill in the Yukon was put into operation in February, 1988. Cyanide removal on CIP tailings at 40 wt.% solids is carried out in a single stage reactor equipped with three 25 hp flotation devices for agitation and aeration. Average cyanide removal from 150 mg/L to 5 mg/L is currently achieved. Treated effluent is stored in a settling pond from which supernatant will be treated for As and residual CN removal prior to discharge. Reagent consumptions are 6 g SO_2/g CN_T and 0.3 g Cu^{2+}/g CN_T.

Skyline

Cyanide destruction on repulped tailings from a Merrill-Crowe operation was started in July 1988 at Skyline's 200 tpd Johnny Mountain mill in British Columbia. The slurry is treated with a 20% $Na_2S_2O_5$ solution at pH 8 controlled with milk of lime in a single stage 60 m^3 reactor equipped with a 50 hp radial impeller. Cyanide is removed from 450 to below 1 mg/L. The feed contains high copper and therefore no Cu addition is required. About 6 g SO_2/g CN_T is used in the process.

Inco/Golden Knight

Inco's Golden Knight cyanide destruction circuit was commissioned in February, 1989. CIL tailings at 45 wt.% solids is treated in a single stage reactor equipped with four 20 hp Dorr Oliver flotation devices. Liquid SO_2 is used to remove cyanide from 100-200 mg/L to 1-10 mg/L, using 4 to 5 g SO_2/g CN_T. The pH is controlled at 8.0-8.5 with lime. Copper addition is typically 0.1 g/g CN_T.

Kuntz Electroplating Inc.

At Kuntz Electroplating Inc., Kitchener, Ontario, 600 gallon tanks are used to treat plating waste solutions. $Na_2S_2O_5$ is used as SO_2 reagent and sodium hydroxide for pH control. Cyanide is removed from 150 mg/L to 0.2 mg/L CN_T. Copper, nickel, iron and zinc are all below 0.2 mg/L in the treated wastes.

CAPITAL AND OPERATING COSTS

CIP Tailings

An example of approximate capital and operating costs for a 1000 mtpd CIP operation treating a 45 wt.% solids pulp containing 270 mg/L CN_T, 270 mg/L SCN^- and 110 mg/L $S_2O_3^{=}$, as well as soluble metals is given in Table 3. The example is for a NaCN consumption of 2 lbs/mt.

The capital cost estimate does not include foundations and installation costs. Reagent costs in Cdn $/mt of 250 for liquid SO_2, 100 for lime (as $Ca(OH)_2$ and, 1000 for $CuSO_4 \cdot 5H_2O$ were used to estimate the operating costs.

Pond Water

An example of approximate capital and operating

TABLE 3 CIP Tailings (1,000 mtpd)

Capital Items		Canadian $
Tank (100 m^3)		35,000
Mixer-Motor-Bridge (75 hp)		43,000
Air Blower (1000 cfm, 10 psig)		21,000
New SO_2 Tank		86,000
Tank	42,000	
Padding System	18,000	
Valves-Piping	26,000	
Copper Sulfate System		19,000
Lime System		39,000
Instrumentation (2 probes, controller)		5,000
	TOTAL	248,000

Operating Costs		Canadian $/mt
Reagents		
SO_2	(5 g/g CN_T)	0.42
$CuSO_4 \cdot 5H_2O$	(0.16 g/g CN_T)	0.05
Lime	(7 g/g CN_T)	0.23
Royalty	(33 cents/kg SO_2)	0.55
Manpower	(20% Manshift)	0.10
Electrical power	(5 cents/kw-hr)	0.16
Maintenance	(5% of capital)	0.10
	TOTAL	1.61

TABLE 4 Pond Water (1,000 mt/d)

Capital Items	Canadian $
Tank (30 m^3)	25,000
Mixer-Motor-Bridge	25,000
Air Blower (300 cfm, 5 psig)	15,000
$Na_2S_2O_5$ System	10,000
Copper Sulfate System	10,000
Lime System	20,000
Instrumentation (on tank)	5,000
TOTAL	110,000

Operating Costs		Canadian $/mt 35% Natural Degradation	Canadian $/mt 85% Natural Degradation
Reagents			
$Na_2S_2O_5$	(6 g/g CN_T)	1.00	0.20
$CuSO_4 \cdot 5H_2O$	(0.16 g/g CNT)	0.05	0.01
Lime	(6 g/g CN_T)	0.14	0.03
Royalty	(33 cents/kg SO_2)	0.20	0.04
Manpower	(10% Manshift)	0.05	0.05
Electrical power	(5 cents/kw-hr)	0.05	0.05
Maintenance	(5% of Capital)	0.05	0.05
	TOTAL	1.54	0.43

costs for a 1000 mt/d operation using 2 lbs NaCN/mt assuming a pond supernatant treatment rate of 100 m^3/h containing 100 mg/L CN_T and 50 mg/L SCN (35% natural degradation assumed) and 20 mg/L CN_T and 50 mg/L SCN (85% natural degradation), respectively, is given in Table 4.

The capital cost estimate does not include foundation and installation costs. Reagent costs in Cdn $/mt of 700 for $Na_2S_2O_5$• 100 for lime (as $Ca(OH)_2$) and, 1000 for $CuSO_4 \cdot 5H_2O$ were used to estimate the operating costs in each case.

ACKNOWLEDGEMENTS

The authors would like to thank B. Tandi who has played a major role in the laboratory testwork and P. Iamarino for analytical services; their assistance and that of several Inco operations personnel during plant startups is much appreciated.

REFERENCES

Borbely G. J., Devuyst, E. A., Ettel, V. A., Mosoiu, M. A., and Schitka, K. J., "Cyanide Removal from Aqueous Streams," Can. Patent 1165474, April 10, 1984. Covered by South African Patent 82/0372, UK Patent 2091713B, US Patent 4537686, Aug. 27, 1985 and Patent Applications pending in other countries.

Devuyst, E. A., Conard, B. R. and Ettel, V. A., "Pilot Plant Operation of the Inco SO_2/Air Cyanide Removal Process," Canadian Mining Journal, pp 27-30.

Chapter 43

TRANSPORTATION OF LIQUID SODIUM CYANIDE - RESPONSE STRATEGIES FOR SPILLS

Mr. W. P. Staunton, Ms S. Formby, Mr. R. S. Schulz and Dr. J. Avraamides

Mines Department, Western Australia
Chemistry Centre (WA)

Transportation of sodium cyanide from factory to minesite often attracts adverse comment from both the community and government agencies. In Western Australia the transportation of 30% w/w aqueous sodium cyanide solution raised fears in the community concerning the safety of such material in the event of spillage.

The Chemistry Centre of Western Australia initiated a study of the effects of such a spill, to determine appropriate short and long term response strategies. Factors which were considered included the rate of hydrogen cyanide gas release under various climatic and soil conditions and the effect of alternative treatments in reducing such emissions. The data collected have been used to plan emergency response procedures.

INTRODUCTION

Following a five year boom in the gold industry, gold production in Australia reached a record level in 1988, expected to be in the region of 138-140 tonnes (4.4-4.5 million ounces). Approximately 75% of this production comes from Western Australia.

In November 1988, Australian Gold Reagents (AGR) began producing sodium cyanide at a plant located in the Kwinana industrial area, 20 km south of Perth, Western Australia. The plant will produce 50,000 tonnes per annum (tpa) of a 30% w/w aqueous solution of sodium cyanide at a pH of 13 (equivalent to 15,000 tpa of 100% NaCN). Plans are already in hand to double production.

The solution is transported to the mine sites in 18,000 litre isotainers, specifically designed to resist mechanical damage. The transportation of such large quantities of cyanide in the metropolitan area and through the catchment area for much of the state's water supply gave rise to public debate on the safety of such operations. Concern by railway workers handling shipments resulted in a brief strike.

The Chemistry Centre (WA) undertook a preliminary investigation to determine both the likely levels of hydrogen cyanide gas which might be expected in the atmosphere near a spill, and possible methods of treating a spill to reduce any such emissions and detoxify contaminated soil.

Review of Treatment Options

Alkaline Chlorination. Oxidation of cyanide by chlorine or, more commonly hypochlorite is widely used in the plating industry and is a highly developed technique (Zabban & Helwich, 1980). More recently it has been used to treat mine effluents (Eccles, 1977). The reaction proceeds via a toxic intermediate, cyanogen chloride, CNCl, as shown in the equations below;

$$NaCN + NaOCl \rightleftharpoons CNCl + 2NaOH \qquad ...(1)$$

$$CNCl + 2NaOH \rightleftharpoons NaCNO + NaCl + H_2O \qquad ...(2)$$

Reaction (1) occurs almost instantaneously, however, reaction (2) requires 10-15 minutes at a pH of 10 (Goldstein, 1976). If sufficient hypochlorite is available the cyanate may be further oxidized via the following reaction;

$$2NaCNO + 3NaOCl + H_2O \rightleftharpoons N_2 + 2CO_2 + 3NaCl + 2NaOH \qquad ..(3)$$

pH should be maintained at a high level to assist in rapid conversion of toxic and volatile CNCl to less harmful products.

Brady (1975) recommended the use of calcium hypochlorite for the treatment of spills from cyanide plating baths. His report stated that hypochlorite was found to be the only treatment applicable to a spill. However no critical evaluation of alternatives techniques was provided.

The minimum quantity of sodium hypochlorite solution (10% w/v Cl_2) required to convert 1 litre of the sodium cyanide solution to cyanate (1000 x less toxic) is 4.3 litres. To convert the cyanide to nitrogen and carbon dioxide requires 10.8 litres.

Hydrogen Peroxide Hydrogen peroxide has been acknowledged as a possible reagent for cyanide treatment for many years and recently Ok Tedi Mining Ltd installed a plant using Degussa's Hydrogen Peroxide Process (Griffiths et al, 1987). As in alkaline chlorination the cyanide is oxidized to cyanate;

$$NaCN + H_2O_2 \rightleftharpoons NaCNO + H_2O \quad ...(4)$$

The minimum quantity of 30% w/w hydrogen peroxide required to convert 1 litre of the sodium cyanide solution to cyanate is 0.7 litres.

Precipitation of Insoluble Iron Cyanides. A different approach to those already described is to immobilise the cyanide in a stable compound. One possible scheme involves the addition of excess iron (II) sulphate leading to the formation of the insoluble precipitate "Prussian Blue" (ferric hexacyanoferrate (II), $Fe_4[Fe(CN)_6]_3$) via the following reactions;

$$FeSO_4 + 6NaCN \rightleftharpoons Na_4[Fe(CN)_6] + Na_2SO_4 \quad ...(5)$$

$$2FeSO_4 + 2Fe(HSO_4)_2 + 3Na_4[Fe(CN)_6] + O_2 \rightleftharpoons Fe_4[Fe(CN)_6]_3 + 6Na_2SO_4 + 2H_2O \quad ...(6)$$

This method has been investigated as a treatment for gold mill effluent (Randol International, 1987; Pettet and Ware, 1955; Schiller, 1983

The minimum quantity of iron(II) sulphate heptahydrate, $FeSO_4.7H_2O$, required to convert 1 litre of the sodium cyanide solution to insoluble Prussian Blue is 660 grams. However, even partial complexation of the cyanide to soluble ferrocyanide (eqn 5) will considerably reduce the toxicity of the effluent and the evolution of hydrogen cyanide gas.

A method using mixtures of iron(II) sulphate and an alkali or alkaline earth metal carbonate to treat spills has been patented (Seidenberger, 1977).

Maintenance of pH. A short term treatment strategy may be to use alkali or alkali salts to maintain the pH of the cyanide environment at a sufficiently high level. At pH 13 only 0.02% of cyanide in solution is present as molecular hydrogen cyanide. If the pH is allowed to drop to 9.3 the proportion present as HCN rises to 50%.

EXPERIMENTAL

Analytical Procedures

All analytical methods were based on the appropriate APHA (1985) procedures.

Free cyanide was determined either by titration with silver nitrate or by spectrophotometric determination of the pyridine-barbituric acid cyano complex at 578 μm.

Total cyanide was determined by digestive acid distillation of the sample. The liberated hydrogen cyanide was collected in an alkaline solution and determined spectrophotometrically as described for the free cyanide.

Free ammonia was determined by spectrophotometric determination at 630 μm using the direct indophenol method.

Hypochlorite was determined by thiosulphate titration in the presence of excess acid, iodide and starch indicator.

Chemicals

All chemicals used for the air scrubbing and subsequent analyses were of AR quality. Hydrogen peroxide was added as a 30% w/w solution. Hypochlorite was added as a 125 g/l solution of sodium hypochlorite. The sodium carbonate, calcium hydroxide and iron(II) sulphate heptahydrate ($FeSO_4.7H_2O$) applied to the soil during the tests were of technical grade quality. The sodium cyanide solution used was supplied by the manufacturer.

Solution Tests

The effectiveness of the sodium hypochlorite, hydrogen peroxide and iron(II) sulphate was initially assessed by testing each of the reagents using 20 ml of the cyanide solution. Any gases liberated were collected by scrubbing through 10% hydrochloric acid solution followed by sodium hydroxide solution (10 g/l). The acid solution was analysed for ammonia and the caustic solution for free cyanide.

Soil Tests

The apparatus used to determine the effect of spillage onto soil is shown in Figure 1. This consisted of a large perspex container with a soil surface area of 0.09 m^2 (30 cm x 30 cm). Air was sucked through the head space above the soil surface, first passing through a diffuser immersed in water, which acted as a humidifier. Prior to passing the pump the air stream was scrubbed by bubbling through a solution of 10% hydrochloric acid followed by a solution of 10 g/L sodium hydroxide. The sodium hydroxide scrubber removed HCN from the air stream while the hydrochloric acid removed any ammonia present. As no ammonia was detected, the acid was removed from the circuit after a short time and an additional caustic scrubber substituted.

After packing the containers a small quantity of water was used to wet the surface of the soil. Approximately 500 ml of the sodium cyanide solution was dispensed as evenly as possible over the soil surface. Penetration into the soil did not appear to be more than a few centimetres.

Soil

The soil used was a coastal sand obtained in the Perth metropolitan region. This soil is a highly leached quartz sand containing a small amount of organic material. Natural pH was about 6-7 with the sand having little buffering capacity. The soil was prepared by removing any leaves, branches etc and then air drying. As the material has very little structure, packing of the containers consisted simply of padding down the soil with a steel weight.

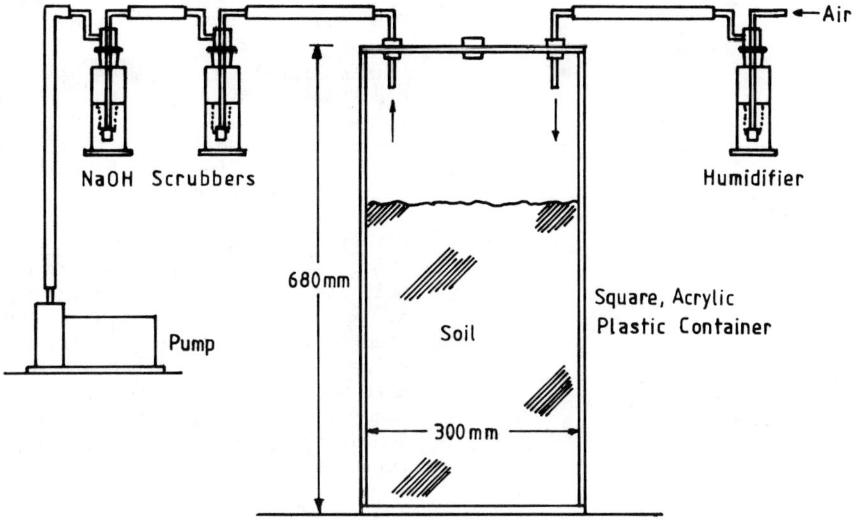

Figure 1. Soil Test Apparatus

RESULTS AND DISCUSSION

Solution Tests

Sodium Hypochlorite. Test results are summarised in Table 1.

Table 1. Solution Tests - Alkaline Chlorination

Ratio OCl:CN	HCN evolved, µg	NH_3 evolved, µg	CN left in residue,%*	OCl left, %*
3:1	<0.01	21	<0.001	26.5
1:1	50000		0.08	1.0

* as percentage of original CN or OCl present at start

In both cases the solution reached boiling point (>95°C) and the pH dropped to approximately 9.

Ferrous Sulphate Tests. Test results are summarised in Table 2.

Table 2. Solution Tests - Ferrous Sulphate

Ratio Fe:CN	$FeSO_4$: Na_2CO_3	HCN evolved, µg	Free CN remaining in residue as % of initial
3:1	75:25	14	1.2
3:1	100:0	140	9

Final pH was approximately 9 for tests without carbonate addition and 10.5 for those tests using 75:25 ratio of iron(II) sulphate to sodium carbonate. The distinctive "Prussian Blue" precipitate was observed amongst the brown iron hydroxides. The drop in pH is commensurate with the precipitation of iron hydroxides.

Hydrogen Peroxide Tests. At a 1:1 H_2O_2:CN molar ratio addition of the hydrogen peroxide was followed after about 4 seconds by a sudden violent reaction with the temperature rising to 90°C and the pH dropping to 9.5 prior to solidification of the residue. Quantitative analysis of evolved gases indicated that 480 µg of HCN had been liberated with 600 µg of NH_3.

Dilution of either the hydrogen peroxide or the sodium cyanide solution reduced the violent nature of the reaction, however the solution still reached boiling point within a minute.

Soil Tests

For each soil test the cumulative hydrogen cyanide evolved over time was determined. Figure 2 illustrates a typical set of data for the first 100 hours following commencement of the test. It can be seen that the rate of evolution is fairly constant. In fact evolution rates were shown to remain unchanged for up to 1000 hours.

Effect of Air Flow Rate. Air flow rate through the container was varied to determine its effect on gas emission rates. A plot of the HCN(g) emission rates versus air flow rate is given in Figure 3. It can be seen from the test data that evolution rates generally increase with increasing air flow through the tank, apparently reaching a plateau at 3-5 l/min air flow. The reasons for this trend would include the reduced emission rate due to near equilibrium concentrations of HCN(g) at the surface under low flow conditions and the possible acidification of the solution by CO_2(g) adsorption

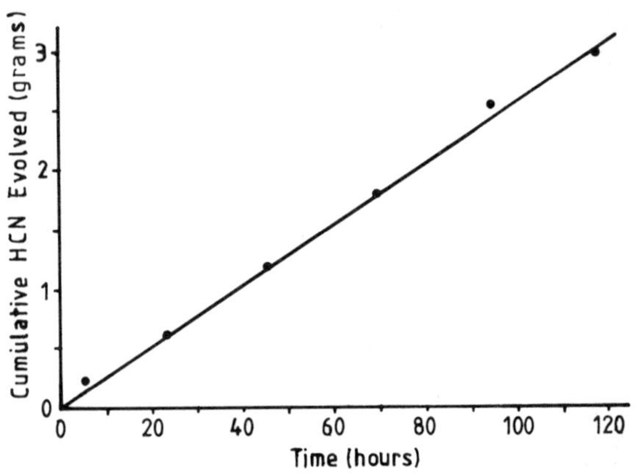

Figure 2. HCN(g) Evolution From Soil

from the atmosphere, presumably enhanced at higher flow rates.

Effect of Rainfall. The effect of water addition on emission rates was measured by adding the equivalent of 25 mm of rain to the surface. There was a decrease in rate, with a new HCN(g) evolution rate of approximately 20-40% of the original value eventually being established.

Effect of Temperature. All the tests with one exception were run at ambient temperature (20-25°C). To determine the effect of elevated temperature, one container was modified to allow the placement of electrical heating tape several centimetres below the soil surface. Sufficient current was then passed through the tape to maintain a temperature of 40°C in the first two centimetres of soil. The hydrogen cyanide evolution rate was 2.3-7.6 times higher than the rate for equivalent tests at room temperature. This was also the only test where the emission rate was seen to decrease markedly over time without any treatment.

Effect of Soil Cover. The effect of soil cover on HCN volatilisation was measured by placing a total of 100 mm of soil over the original surface. This cover had virtually no effect on emission rate. This limited effect may be partly due to the porous nature of the sand being used in these tests.

Effect of Treatment with Ferrous Sulphate. The effect of covering the surface with ferrous sulphate was tested by placing the equivalent of 8.5 kg/m^2 of iron(II) sulphate over the contaminated surface. The results are shown in Figure 4. As can be seen from the graph no change in emission rate was noted even 300 hours after the addition. Liberal applications of water (3 l/m^2) produced an immediate diminution in rate with "Prussian Blue" formation being observed.

Effect of Treatment with Ferous Sulphate/Sodium Carbonate. The effect of adding a mixture of ferrous sulphate and sodium carbonate to the soil surface was tested by applying an 85:15 mixture of iron(II) sulphate:sodium carbonate at a rate of 10 kg/m^2. Addition of the mixture caused an initial rapid drop in emission rate (by 2/3rd), falling to virtually no emission 150 hours after treatment.

Effect of Treatment with Slaked Lime. The effect of an application of slaked lime at a rate of 5.5 kg/m^2 was to produce an immediate cessation of HCN(g) evolution, suggesting that $Ca(OH)_2$ may be a most effective short term treatment option.

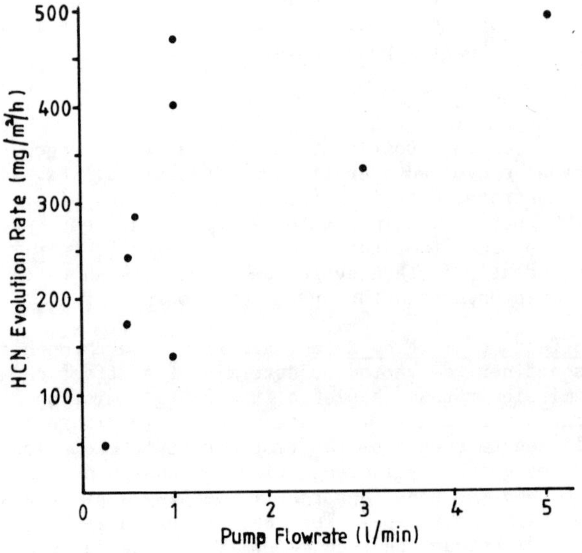

Figure 3. Variation of HCN(g) Evolution Rate with Air Flow Rate

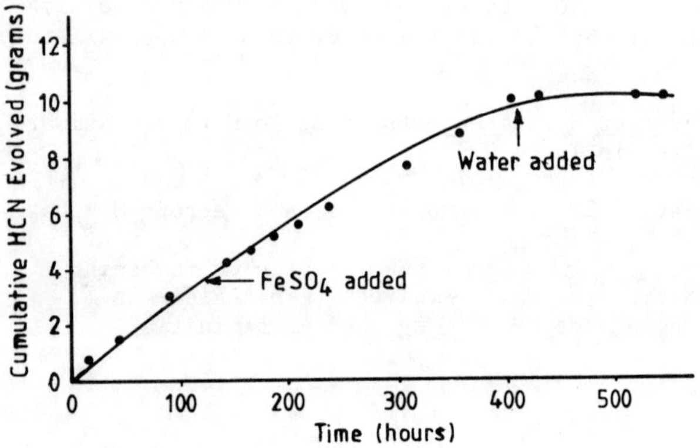

Figure 4. Effect of $FeSO_4/Na_2CO_3$ Treatment on HCN(g) Evolution

Hydrogen Cyanide Evolution from Liquid Surfaces.

The rate of evolution of hydrogen cyanide from liquid surfaces, such as puddles, was determined to be approximately 15% of the rate for the equivalent test on soil. This reduced figure could be due to either pH changes in the soil or a larger effective surface area due to the porous nature of the sand or a combination of both these factors.

Hydrogen Cyanide Evolution from a Hot Surface

The effect of spillage of cyanide solution onto a hot surface such as tarmac or concrete was determined by heating a piece of road bitumen on a hot plate, the whole apparatus being operated within one of the containers. The bitumen was heated to 70-80°C to mimic conditions which could be encountered in the Eastern Goldfields of Western Australia. Two tests were made, with results listed in Table 3.

Table 3. HCN Evolution from Hot Surfaces

Area of Spill, cm^2	Volume NaCN Soln mL	Evolution Rate HCN(g) $mg/m^2/h$ 10 min	1 hour
32	2	875	719
94	3	249	91

Due to the small areas involved these tests were not as reliable as the soil work. However, they do indicate that emissions do not exceed the values recorded for the soil evolution rates.

Air Dispersion Calculations

The levels of HCN(g) which could be expected in the atmosphere near a spill were determined using two computer models, WHAZAN (Technica International, Consulting Scientists and Engineers, London) and MAXMOD (Environmental Protection Authority of Western Autralia). For this exercise the highest rate of emission recorded during these tests (1070 mg $HCN/m^2/h$, for the test at 40°C) was used to provide a conservative estimate of concentration levels. The first set of conditions examined was a standard "worst case" described below:

Emission Rate:	1070 mg $HCN/m^2/h$
Wind Velocity:	1.5 m/s
Spill Area:	100 m^2 (5m x 20m)
Temperature:	20°C
Mixing:	1:1 with air with measurements made at ground level

Results are listed in Table 4.

Under conditions of no wind (0.5 m/s), an extreme case which might be encountered on still, cold nights, the following values were obtained;

Table 4. HCN Atmospheric Concentrations, 1.5 m/s Wind Velocity

Distance from Spill, metres	[HCN], ppm (Downwind)
15	13
25	5
55	1

Table 5. HCN Atmospheric Concentrations, 0.5 m/s Wind Velocity

Distance from Spill, metres	[HCN], ppm (Downwind)
55	68
75	37
150	10
300	2.6
500	1

These conditions would be very infrequent and are not taken as the basis for normal responses.

CONCLUSIONS

These investigations have provided information which should assist with the planning of controlled field trials. The results indicate;

1. Where sodium cyanide solution is spilt into neutral soils, such as the sand used in these tests, evolution rates of hydrogen cyanide gas should not pose an immediate danger to persons more than 50 metres from a spill.

2. Emissions from liquid surfaces will be lower than for soils and the comments in 1 apply.

3. In situ detoxification of soils using oxidants such as sodium hypochlorite or hydrogen peroxide produce vigorous reactions and should only be used by trained personnel under controlled conditions. These chemicals are inappropriate for emergency response use except for small spills.

4. Of the solid materials tested on contaminated soils, slaked lime and ferrous sulphate/sodium carbonate proved the most effective in limiting hydrogen cyanide gas emissions. Ferrous sulphate on its own was only effective after being washed into the soil with water.

5. Spilling of the sodium cyanide solution onto a hot surface produces gas evolution rates similar to that for soils.

ACKNOWLEDGEMENTS

The authors would like to thank Australian Gold Reagents for their cooperation during this work. This paper is published by permission of the Director, Chemistry Centre(WA).

REFERENCES

American Public Health Association, 1985, Standard Methods for the Examination of Water and Wastewater, 16th Edition.

Brady, R., 1975, "Contingency Plans for Control and Treatment of Accidental Spills of Toxic Chemicals from Electroplating Processes", National Technical Information Service, U.S. Department of Commerce.

Eccles, A. G., 1977, "Pollution Control at Western Mines' Myra Falls Operation", CIM Bulletin, pp. 141-147.

Goldstein, M., 1976, "Economics of Treating Cyanide Wastes", Pollution Engineering, pp. 36-38.

Griffiths, A., Knorre, H., Gos, S. & Higgins, R., 1987, "The Detoxification of Gold-Mill Tailings with Hydrogen Peroxide", J. S. Afr. Inst. Min. Metall., Vol. 87, pp. 279-283.

Pettet, A. & Ware, G., 1955, "Disposal of Cyanide Wastes", Chem. and Ind., pp. 1232-1238.

Randol International ltd, 1987, "Gold and Silver Recovery Innovations. Phase III." Vol. 9, pp 5361-5363.

Schiller, J., 1983, "Removal of Cyanide and Metals from Mineral Processing Waste Waters", U.S. Bureau of Mines, RI8836.

Seidenberger, J., 1977, "Spill Control Composition and Use Thereof", U.S. Patent 4,011,098.

Zabban, W. & Helwich, R., 1980, "Cyanide Waste Treatment Technology - The Old, The New and The Practical", Plating and Surface Finishing, pp. 56-59.

Chapter 44

RECOVERY OF GOLD FROM COPPER-BEARING ORES

D.M. Muir, S.R. La Brooy and C. Cao

(A.M.I.R.A. Gold Technology Group
Murdoch University,
Perth, Western Australia, WA 6150)

Abstract: Many gold ores have copper mineralisation which presents problems in a carbon-in-pulp operation. Various types of copper-gold ores and ways of tackling the problem are discussed. In some cases copper and gold can be upgraded to a concentrate by flotation; otherwise, copper or gold can be selectively leached. When both copper and gold cyanide are present in solution, it is possible to minimise copper loading onto carbon or to selectively strip the copper from the carbon. Alternative leach solutions like thiourea and bromine are discussed and results obtained with the ammonia/cyanide system are presented in the light of our present understanding of the chemistry involved.

INTRODUCTION

In a survey of 115 various primary gold deposits Schwartz (1944) found that, besides the common associations with pyrite and arsenopyrite, native gold was intergrown with chalcopyrite ($CuFeS_2$) in over 20% of the ores he examined. Significant associations were also found with tetrahedrite ($(Cu,Fe)Sb_4S_{13}$) and tennantite ($(Cu,Fe)As_4S_{13}$), as well as bornite (Cu_5FeS_4) and chalcocite (Cu_2S) in certain ores. Recent studies by Gasparrini (1983) however, show that, unlike pyritic or arsenopyritic gold ores, the gold particles in many chalcopyrite ores are often discrete allowing a significant portion of the gold to be liberated from the copper. This is, of course, an advantage when trying to selectively float or leach copper from gold but a disadvantage when trying to produce a copper-gold concentrate by flotation. With weathered or surface oxidised or transition ores, the copper mineralisation is commonly malachite ($Cu(OH)_2CuCO_3$) or cuprite (Cu_2O) arising from the oxidation of copper sulfides. In this form it is more difficult to float and separate the copper as a concentrate.

In Australia, examples of copper-gold ores which are being processed include primary supergene ores, transition mixed oxide/sulfide ores, surface oxidised ores and tailings dumps (Table 1). However, only certain fractions of such ores are currently amenable to processing economically and much of the ore is being stockpiled for future processing once suitable techniques are proven (Table 1). Over the years, four processing options have been identified and tested. These are:- float and produce a high grade copper-gold concentrate for smelting; or separate a low copper fraction ($<$ 0.5% Cu) and process by conventional cyanidation and C.I.P.; or selectively pre-leach high copper fractions ($>$ 1% Cu) followed by cyanidation of gold; or finally selectively leach gold using mixed CN^- reagents or alternative reagents. Clearly this last option is the most attractive to process a wide range of ores if suitable reagents can be found, but this approach also represents the greatest challenge in research and development.

This paper reviews the principles and problems associated with these options and considers the current state of the technology in the search for alternative or selective reagents. It is shown that, in principle, the ammonia-cyanide system has attractive features which could be exploited. Preliminary testwork data on ammonia/cyanide leaching of high copper sulfide ore is presented.

FLOTATION OF HIGH GRADE COPPER GOLD CONCENTRATES

With reasonably high grades of primary copper ores, the most economic approach to recovering gold is to float and smelt a copper concentrate, and extract the gold from the anode slimes of the copper electrorefining circuit. This is practiced in Australia by Newmont Ltd at their Telfer Gold Mine, where a concentrate assaying 25.7% Cu and 109 gt^{-1} Au is produced and shipped to Japan. Also by BHP Gold Ltd at their Brown's Creek Mine, where a concentrate assaying 22.0% Cu and 300 gt^{-1} Au is sent to an Australian copper smelter (Table 1). Unfortunately it may take months before the gold value is realised and significant amounts of gold may be left in low grade copper tailings and stockpiled for future possible processing.

With oxidised or transition ores it is more difficult to float and obtain an acceptable copper concentrate and various sulfidization pretreatment methods have been tried to improve flotation recoveries, particularly of mixed oxide/sulfide ores. Sulfidization of surface oxidised copper

Table 1: Some Australian Copper-Gold Operations

	Mine		Grade %Cu	Grade gt^{-1} Au	Mineralogy or ore type	Process
A.	Telfer (W.A.)	Ore(1)	4.2	19	Cc, Py, Q, clay	CPS Flotation
		Cons	25.7	109	Cc, Py	Sold to Smelter
		Tails	0.8	2	Q, clay Cc	Stockpiled
		Ore(2)	<0.5	2-50	low Cu oxidised	CIL/CIP
		Ore(3)	1.5-3	2-4	high Cu oxidised	Stockpiled
B.	Paris (W.A.)	Tailings Dump	1.2	6	high Cu oxidised	NH_3-CN^- vat leach/CIC
C.	Browns Creek (NSW)	Ore	0.3	5	Cp, B, Cc, Wo	Flotation
		Cons	22.0	300		Sold to smelter
		Tails	0.04	0.4		Discarded
D.	Red Dome (Qld)	Ore(1)	0.4	1.6	low Cu/oxide	Heap leach/CIC
		Ore(2)	0.9	1.9	high Cu/oxide	Stockpiled
		Ore(3)	0.1	2.4	low Cu/sulfide	Stockpiled
		Ore(4)	1.0	2.7	high Cu/sulfide	Stockpiled
E.	Chunderloo Prospect (W.A.)	Drillcore	6.2	20-50	transition	Flotation
		Cons	17	41	Py, Cc, (Cp), (L)	?
		Tails	6	12		?

Key: Cc = chalcocite; Py = pyrite; Cp = chalcopyrite; Q = quartz; Wo = wollastonite; L = limonite; B = bornite.

ores is usually carried out by pre-treating the ore with a solution of sodium sulfide, but reproducibly high recoveries of copper by subsequent flotation with xanthates is difficult to achieve in practice. Too little sulfide in solution gives poor recoveries because of inadequate sulfidization, but too much S^{2-} also results in poor recoveries because it causes depression of sulfide minerals in flotation and reacts with xanthates. Furthermore S^{2-} is readily oxidised by air to sulfur or polythionates. Detailed studies by Jones and Woodcock (1978 a,b) showed that optimum results could only be achieved by controlling both Na_2S and xanthate additions. Best results were obtained by using an ion-selective electrode and maintaining the solution potential at values around E_s = -450 mV in four successive cycles of S^{2-} addition, followed by xanthate addition and flotation, over a 23 minute period. In the first three stages, the added xanthate was rapidly consumed, but in the fourth stage residual xanthate was present in solution indicating sufficient for adequate flotation. Thus careful plant control and monitoring are necessary, particularly if the ore grade and mineralogy fluctuates.

Following optimisation of the process at the Telfer Gold Mine over the past year (Lumley, 1988) controlled potential sulfidization (CPS) of 47-75 μ sized ore gave improved copper and gold recovery. CPS followed by flotation carried gold previously associated with cuprite and quartz (due to weak sulfidization of quartz) and left tailings containing 0.8% Cu and 2 gt^{-1} Au (Table 1).

This pre-treatment technique, however, appears best suited to partly oxidised sulfide ores rather than simple oxidised ores. Recent tests with Na_2S pre-treatment on a low grade oxide ore (0.68% Cu) in China (containing mainly malachite with some chrysocolla, bornite, chalcocite, and chalcopyrite), gave only 45-70% recovery of copper and a concentrate containing < 10% Cu (Chen, 1983). Further research has found that sulfidization could be improved by heating the ore slurry with elemental sulfur in an autoclave at 180°C at pH 6-10. This procedure hydrothermally decomposed sulfur to produce S^{2-} in situ (equation 1) (Deng et al., 1984) and enabled 88% of the copper to be recovered as a 20% concentrate.

$$4S + 6OH^- \rightarrow 2S^{2-} + S_2O_3^{2-} + 3H_2O \qquad ...(1)$$

A less expensive option which is currently being evaluated (Deng and Chen, 1986) is to mix the ore with a kerosene solution containing elemental S and Aliquat 336 at 80°C for 2-3 hours. Under alkaline conditions and in the presence of quaternary ammonium ions as catalyst, sulfur again reacts according to equation 1 and hydrothermally sulfidizes copper oxides. Copper recoveries of 82%, yielding a 13% Cu concentrate, have been achieved in this manner.

Overall there appears to be scope for improved flotation and copper recoveries, but inevitably a significant fraction of the gold will remain in the flotation tailings, according to the mineralogy of the ore, making this approach second best to approaches dealing with whole ore.

CYANIDATION AND RECOVERY OF GOLD BY CARBON.

The problem with direct cyanidation of copper-gold ores is the reactivity of most copper minerals towards cyanide (Halferdahl, 1929). Indeed numerous processes have been proposed or tested for the recovery of copper from either high grade concentrates or flotation tailings, and a much greater understanding of copper cyanide chemistry has emerged (Shantz and Reich, 1978).

The reactivity of copper minerals in cyanide solution follows the order (Hedley and Kentro, 1945) azurite ≥ malachite ≈ copper ≈ chalcocite ≥ cuprite > bornite > enargite > tetrahedrite > chrysocolla > chalcopyrite. Typically at ambient temperatures in 0.1% NaCN, about 90% of cuprite, 20% of tetrahedrite and 5% chalcopyrite are leached over a 24 hour period (Leaver and Woolf, 1931) to give an equilibrium mixture of $Cu(CN)_2^-$, $Cu(CN)_3^{2-}$, and $Cu(CN)_4^{3-}$ species in solution according to the free CN^- concentration and molar ratio of CN^-:Cu. Any Cu(II) which is added or is dissolved from the ore, reacts with CN^- to form Cu(I) complexes and cyanogen (equation 2) and leads to higher lime and cyanide consumption (Stamboliadis et al., 1976).

$$Cu^{2+} + 3CN^- \rightarrow Cu(CN)_2^- + \tfrac{1}{2}(CN)_2 \qquad ...(2)$$

Thus according to this stoichiometry, oxidised copper ores which are highly reactive to CN^- will consume approximately 1.5 kg CN^- per kg Cu in the ore to produce mainly $Cu(CN)_3^{2-}$ species. Whilst ores containing only chalcopyrite or chrysocolla minerals will consume considerably less CN^-.

It is generally recognised that the leaching of gold does not take place until the molar ratio of CN^-:Cu in solution exceeds 3:1 (Hedley and Kentro, 1945; Shantz and Reich, 1978). This is because of the strong complexation with Cu(I) which leaves very little free CN^- in solution. The tetra-cyano complex, however, is quite weak ($K_{4,3} = 0.02$) and solutions of $Cu(CN)_4^{3-}$ will dissolve gold via dissociation of its fourth CN^- until the ratio of CN^-:Cu falls to 3:1. The rate of gold dissolution, however, is only about half that of the rate with NaCN. In practice, since it is rare to find pure mineralisation, an economic cut-off grade for direct cyanidation of copper-gold ores, based upon CN^- consumption, is around 0.5% Cu for oxide ores and 1% Cu for sulfide ores. However, this still leaves the problem of selectively recovering the gold from solutions which are rich in copper relative to gold, by carbon adsorption.

Processing of copper-gold cyanide leach liquors.

At the Red Dome heap leach operation (Table 1), pregnant liquors sent to carbon adsorption columns typically contain 1 ppm Au, 1500 ppm Cu and 500 ppm free CN^- at pH 11.5 (Ryan, 1988), whilst loaded carbons assay about 5,000 gt^{-1} Au and 6,000 gt^{-1} Cu. Thus carbon is relatively selective for gold over copper and relatively little copper loads onto carbon provided high levels of free CN^- are maintained in solution. The reason is that carbon strongly adsorbs monovalent cyanide complexes like $Au(CN)_2^-$ and $Ag(CN)_2^-$ but has only a weak affinity for multi-charged complexes like $Fe(CN)_6^{3-}$, $Ni(CN)_4^{2-}$, and $Cu(CN)_3^{2-}$ (Tsuchida et al, 1984). As recently demonstrated (Ibrado and Fuerstenau, 1988) there is a direct correlation between the loading of copper on carbon and the speciation of copper(I) cyanide in solution. High copper loadings on carbon occur when copper is present as $Cu(CN)_2^-$, (when there is little free CN^- in solution), but low copper loadings are obtained when the copper is in solution as $Cu(CN)_3^{2-}$. This observation also provides a means of selectively stripping copper from loaded carbon simply by washing with cold 5% NaCN solution in a pre-elution step to desorb copper as $Cu(CN)_3^{2-}$ and $Cu(CN)_4^{3-}$ without desorbing gold.

Thus provided copper levels in solution are not excessive and sufficient free CN^- can be maintained, activated carbon can be used successfully. The problem, then, is to minimise CN^- consumption and preferably recover and recycle cyanide, as well as recover copper. Generally this involves clarification and acidification with recovery of HCN by caustic scrubbing, but losses of CN^- due to oxidation to carbonate are, of course, irreversible. It is worth noting, therefore, that recent studies by Muir et al., (1988) show that dissolved copper, even at concentrations of 250 mgl^{-1}, is an excellent catalyst for the oxidation of CN^- by air, especially in the presence of activated carbon. Thus, in practice, it may be difficult to maintain high levels of free CN^- in a C.I.L. operation with copper ores unless staged cyanide additions are constantly made.

Processing copper gold cyanide eluant liquors.

For clarified carbon eluates rich in copper, gold is selectively electrowon from copper provided the molar ratio of CN^-:Cu exceeds 3:1. Below this ratio, copper starts to co-precipitate with gold to produce impure cathodes (Avraamides, 1989). An interesting alternative option, which avoids conventional gold electrowinning , is to acidify and recover a mixed CuCN.AuCN precipitate. This can be directly smelted to obtain high grade gold bullion with copper reporting to the slag (Sorensen, 1988). A quantitative recovery of gold is achieved at 50°C and pH 1-2, provided the molar ratio of Cu:Au in solution exceeds about 2:1, and HCN is recovered in a caustic scrubber.

Recovery of copper and cyanide from solution.

Various methods have been proposed for the recovery of copper and cyanide, and sometimes gold, from copper-rich clarified barren liquors including electrolysis, sulfide precipitation and acidification.

In early processes, a solution of either CaS or Na_2S was added to precipitate Cu_2S and release free CN^- (Rose et al., 1967). But high sulfide consumptions, presumably through oxidation with air, and difficulty in filtering or processing the slimy precipitate, presented problems.

Although direct electrolysis is used by the electroplating industry, it is not particularly suited to electrowinning of dilute copper cyanide liquors. Current efficiencies are generally low, and cyanide consumption due to oxidation reactions at the anode is over 1.5 kg CN/kg Cu (Shantz and Reich, 1978). In a patent taken out by Lower (1965) on recovering copper from leach solutions, high cyanide recovery was only achieved if less than about 50% of the copper in solution was electrolyzed.

Acidification with H_2SO_4 to pH 3 results in the precipitation of CuCN or double cyanide compounds (including AuCN.CuCN) and the volatilization of HCN. This was practised at the New Occidental Gold Mine in NSW, Australia earlier this century, but in the past few years this technique of acidification-volatilisation-reneutralisation (AVR) has been re-examined and developed by CANMET and Golconda Engineering and Mining Services Ltd mainly as a method of recovering cyanide (Kitney, 1988; McNamara, 1985). Little attention has been paid to processing the precipitate, until recently when AMDEL announced the development of the "Cutech Process" to recover both Au and Cu from cyanide leach liquors and regenerate the cyanide (Anon, 1988). Although no precise details of the process have been published to judge its range of application and economic viability, it appears to closely follow the AVR technique, but specifically recovers a gold sludge and metallic copper from the reprocessing of the initial precipitate. Like all these recovery methods, one of the economic drawbacks is the need to produce a clarified liquor from the spent pulp liquors before processing. Hence the overall recovery of CN^- and copper from any technique is not high and this limits the grade of copper ores that can be directly processed.

PRETREATING COPPER ORES PRIOR TO CYANIDATION.

The removal of copper by pre-leaching the ore with either dilute acid, iron(III) or ammonia solutions prior to cyanidation for gold is an obvious approach to decrease cyanide consumption and downstream C.I.P. processing problems. Most studies, however, have focussed on the hydrometallurgical processing of high grade copper concentrates, particularly chalcopyrite, as an alternative to smelting (Dutrizac and MacDonald, 1974). But considering the low reactivity of tetrahedrite, chrysocolla, and chalcopyrite towards CN^- it would appear that in this context it is only necessary to pre-leach reactive copper minerals if copper values are small relative to gold. The choice of an acid or alkaline pre-leach process will usually be determined by the ore mineralogy and reactivity of the gangue. But, of course, the solid-liquid separation and washing characteristics of the ore are also important, and if further cyanidation is to be carried out, ammonia pre-leaching has an intuitive advantage.

Most oxidised copper minerals, except chrysocolla, readily dissolve in dilute H_2SO_4 or $(NH_4)_2CO_3/NH_3$ solutions but the overall copper extractions from ammonia leaching are generally only 60-70%, unless the ore is first reduced in a kiln with coal or fuel oil at 700°C (Koch and Pietsch, 1974). Reduction converts oxide copper to metallic copper and alters the gangue mineralogy and refractory copper minerals to achieve both greater copper extraction (>95%) and lower adsorption of dissolved copper species. The Segregation Process (Wright, 1973) is a refinement involving reduction roasting with coal and NaCl in which copper powder segregates from the gangue and can be separated by flotation. It was operated commercially on low grade refractory copper oxide ores in Africa between 1930-1970 and has since been demonstrated to be applicable to copper sulfide or mixed oxide/sulfide concentrates following an initial dead-roast to remove sulfur (Marcuson, 1980). The advantage is that high grade metallic copper concentrates can be obtained from a range of ore types. This copper can be readily leached (Parker et al., 1981; Lu and Grimsey , 1983) or shipped, and any remaining copper is often locked up as refractory copper ferrites resulting from the initial dead-roast. A further advantage is that silver also segregates with copper together with up to 50% of the gold. Whether gold truly segregates, or is simply collected by flotation of the copper,is not clear.

The direct leaching of sulfide ores requires an oxidant such as Fe(III) or O_2, and the various copper sulfides exhibit complex leaching kinetics according to their mineralogy (Dutrizac and MacDonald, 1974). Furthermore, acid solutions produce elemental sulfur, whilst ammoniacal solutions produce a range of unsaturated sulfur species, including $S_2O_3^{2-}$, $S_4O_6^{2-}$, all of which could interfere with subsequent cyanidation by either reacting with CN^- or forming gold sulfide passive films. Generally, even the most reactive sulfides like Cu_2S and Cu_5FeS_4 require elevated temperatures (>60°C) to effect complete reactions with Fe(III) or NH_3/O_2 in reasonable times. Percolation leaching of these minerals at ambient temperatures is reported to take months to achieve >90% Cu extraction even with bacterial assistance, whilst the reactions with chalcopyrite are at least an order of magnitude slower (Dutrizac and MacDonald, 1974).

Overall, it would appear that the opportunities for economically pre-treating low grade copper ores, particularly sulfides, are limited unless integrated with a cheap acid or ammonia plant or efficient reagent recovery/recycle from the spent liquors. Ideally, copper should be recovered from dilute leach streams using solvent extraction/electrowinning, rather than by cementation, to obtain the highest by-product value. In many cases, long term weathering or bacterial leaching of heaps over several years would seem to be the most cost effective approach if copper values are low.

SELECTIVE LEACHING OF GOLD FROM COPPER.

The main object in any selective leaching approach is either to choose a reagent which is selective for gold over copper minerals or to choose reagent mixtures and conditions which render the copper minerals insoluble. Unfortunately, due to the fundamental similarities between the chemistry of copper and gold, all of the commonly accepted alternative reagents for gold, such as Cl_2, Br_2, $S_2O_3^{2-}$, SCN^- and $(NH_2)_2CS$ (Avraamides, 1982), are also reactive towards most copper minerals. Furthermore, copper(II) is known to catalyse the oxidation and decomposition of hypochlorite to chlorate (Warren and Mounsey, 1983), thiosulfate to tetrathionate (Zipperian et al., 1988; Tozawa et al, 1981; Byerly et al., 1973), thiourea to formamidine disulfide, cyanamide and sulfur (Deschênes and Ghali, 1988), as well as cyanide to cyanate (Muir et al, 1988). Reagents must, therefore, leach gold much faster than the copper mineral, or create passive films on the copper mineral such as oxides or carbonates, or complex with copper(II) to reduce its activity towards gold reagents.

Leaching with Br_2 by the K-Process.

Whilst Br_2 itself is non-selective towards gold and copper, the K-Process, as recently developed and patented by Kaljas Pty Ltd, is claimed to be selective for leaching gold from metallic copper and oxidised copper ores (Sehic, 1988). The process itself involves using a proprietary combination of bromine reagents at a neutral pH around 7 which rapidly dissolves gold yet passivates the leaching of copper metal and inhibits the dissolution of copper oxides such as azurite, malachite, and chrysocolla. Pilot tests have demonstrated that the process is suited to treating Cu/Au electronic scrap, coarse gold gravity concentrates, and oxidised copper-gold ores, but its application for sulfide copper-gold ores remains to be proven, unless the sulfide is pre-oxidised prior to leaching. Presumably bromine reacts with sulfides in the same way as chlorine to produce acid and sulfate (equation 3) with high attendant reagent consumptions. So methods of decreasing its reactivity would need to be developed.

$$CuFeS_2 + {}^{17}/_2Cl_2 + 8H_2O \rightarrow$$

$$Cu^{2+} + Fe^{3+} + 2SO_4^{2-} + 16H^+ + 17Cl^- \quad ...(3)$$

Results so far published (Sehic, 1988) indicate that gold is selectively leached in < 3 hours from various oxidised copper ores (> 1% Cu) containing from 1 to 80 gt^{-1} Au. Relatively low bromine consumptions are claimed relative to cyanide consumption, but no details or economics have been published. Current research aimed at recovering gold from bromine-C.I.P. pulps is well advanced (Sehic, 1989) which should assist the economic viability. The interesting aspect of this work is that gold(III) bromide is rapidly reduced to metallic gold on activated carbon surfaces, so that techniques of avoiding gold attrition from the carbon had to be developed.

Leaching with thiourea.

The use of thiourea to extract gold from sulfide concentrates has been investigated by Bilston et al. (1983), Gabra (1984) and Deschênes and Ghali (1988). The leaching of gold with thiourea normally requires addition of an oxidant, like Fe^{3+}, as shown in equation (4) since the E^0 for this reaction is ≈ 380 mV. The leach rate depends on pH, thiourea, and oxidant concentrations.

$$Au^0 + 2CS(NH_2)_2 + Fe^{3+} =$$

$$Au[CS(NH_2)]_2^+ + Fe^{2+} \quad ...(4)$$

With copper minerals, however, a considerable amount of oxidant can be provided by the soluble copper, as shown in equation (5) (Krzewska and Podsiadly, 1980) because thiourea also complexes strongly with copper(I).

$$2Cu^{2+} + 2(n+1)\ CS(NH_2)_2 =$$

$$2Cu[CS(NH_2)_2]_n^+ + [NH_3(NH)CSSC(NH)NH_3]^{2+} \quad ...(5)$$

where $n = 1,2,3,4$

Nevertheless, some improvement in ultimate gold extraction or the leaching kinetics was found on addition of ferric ions or hydrogen peroxide (Deschênes and Ghali, 1988). Because traces of oxidised copper minerals and copper sulfides apart from chalcopyrite readily react with the acidic thiourea leach solution, an acid pretreatment is required to remove as much acid soluble copper as possible. This limits the thiourea reaction with copper(II) ions that leads to excessive thiourea consumption, as shown in equation (5). Bilston et al. (1983) found that copper concentrations of 100-200 mgl^{-1} in the leach liquor did not appear to cause much of a problem. However, once the copper concentration exceeded 500-700 mg l^{-1}, reaction (5) could cause very serious loss of thiourea, leading to virtually complete precipitation of thiourea from a 1.2% solution, as the gelatinous copper(I) thiourea complex. Deschênes and Ghali (1988) also observed copper precipitation after solution copper concentrations had reached a maximum around 300 mgl^{-1}.

Where excessive amounts of copper go into solution, not only will the copper(I) thiourea complex be precipitated, but more thiourea will be oxidised to the less stable dimer formamidine disulfide. This is a standard problem with gold extraction via thiourea, as gold dissolution and thiourea oxidation both occur in the same potential range. This can, therefore, contribute to thiourea consumption via its decomposition through a sulfinic compound as shown in equation (6):

$$2CS(NH_2)_2 \longleftrightarrow NH_3^+(NH)CSSC(NH)NH_3^+ \qquad E^0 = 420\text{ mV}$$

$$\rightarrow CS(NH_2)_2 + \text{sulfinic compound} \rightarrow H_2NCN + S^0 \quad ...(6)$$

While an acid prewash was found to reduce the problem, some copper compounds are unattacked at ambient temperatures at pH 1, yet will subsequently react with acid thiourea solution at pH 1.5. Selective reduction of formamidine disulfide back to thiourea could worsen the copper problem. Fortunately, chalcopyrite can be leached with only minor thiourea loss (Bilston et al., 1983; Deschênes and Ghali, 1988), but the presence of more than minor bornite or oxide or carbonate copper, virtually rules out the use of acid thiourea leaching.

With pure chalcopyrite concentrate, Deschênes and Ghali (1988) found that the use of 12.5 kgt^{-1} SO_2 addition in conjunction with acid pretreatment was able to reduce the thiourea consumption from 14 kg t^{-1} to 4.6 kg t^{-1} for 92% gold extraction . Sulfur dioxide or sulfite addition is thought to reduce formamidine disulfide back to thiourea (Belgian Patent, 1977, and Schulze, 1984).

Leaching with CN^-/NH_3 and $CN^-/CO_3{}^{2-}$ mixtures.

Introduction. The leaching of oxidised copper-gold ores with cyanide mixed with ammonia has been practised on several occasions with variable success since it was first patented by Hunt (1901). It has been further studied by Jarman and Brereton (1905), Putman (1950), and by Churchill and Laxen (1966), who showed that certain mixtures of NH_3/CN^- dissolve less copper than either NH_3 or CN^- overall. On tests with an oxidised ore containing 1.5% Cu, using solutions containing 2.5 kgt^{-1} ore KCN and 1.1 kg/t^{-1} ore NH_3, Jarman and Brereton found that the bulk of gold extracted in the first few hours, with slow further extraction over 48 hours, whilst the copper concentration in solution rose to a mximum in 3 hours, then fell steadily. Typically 60-70% Au and < 5% Cu was extracted over a 24-48 hour period. Whilst the amount of ammonia required varied with the copper grade, the optimum molar ratio of $NH_3:CN^-$ was found to be about 2:1. In more recent tests on the Paris high-copper tailings dump containing 1.2% and 6 gt^{-1} Au (Table 1), Ruane (1988) found that the addition of 5 kg t^{-1} NH_3 reduced the overall cyanide consumption from 20 kgt^{-1} to 1.55 kgt^{-1} NaCN, whilst extracting about 70% Au and 35% Cu.

Table 2: Some relevant equilibria in the $Cu-NH_3-CN^--H_2O$ system

Cu^{2+}	+ $4CN^-$	⇌	$[Cu(CN)_4{}^{2-}]$	weak. cf.	$CuCl_4{}^{2-}$
Cu^{2+}	+ $4NH_3$	⇌	$Cu(NH_3)_4{}^{2+}$	log β_4	= 13.0
Cu^+	+ $2NH_3$	⇌	$Cu(NH_3)_2{}^+$	log β_2	= 10.6
Cu^+	+ $2CN^-$	⇌	$Cu(CN)_2{}^-$	log β_2	= 21.6
CN^-	+ H^+	⇌	HCN	pK_a	= 9.3
NH_3	+ H^+	⇌	$NH_4{}^+$	pK_a	= 9.5
Cu^{2+}	+ H_2O	⇌	$Cu(OH)_2$ + $2H^+$	pK_s	= 19.2

Chemistry of the $Cu-NH_3-CN^--H_2O$ system. The chemistry of the $Cu-NH_3-CN^--H_2O$ system is complicated by a number of factors. As some of the relevant equilibria listed in Table 2 show, there is not only competition of Cu^{2+} for NH_3 and CN^- but there are pH effects due to protonation of CN^- and NH_3 around pH 9.5 or precipitation of $Cu(OH)_2$. Furthermore Cu^{2+} is known to oxidise CN^- to cyanogen (equation 2) as well as catalyse the air oxidation of CN^- to carbonate (Muir et al., 1988). Because of the reaction between Cu^{2+} and CN^-, it can be predicted from analogous behaviour with Cl^- that any intermediate tetracyano-copper(II) complex will be weak, with ammonia readily displacing CN^- in the inner-solvation sphere. By contrast, the dicyano copper(I) complex is strong (log β_2 = 21.6) and ammonia will have no effect since it forms a much weaker diammino copper(I) complex (Table 2). Nevertheless, the complex readily oxidises with air to form insoluble malachite.

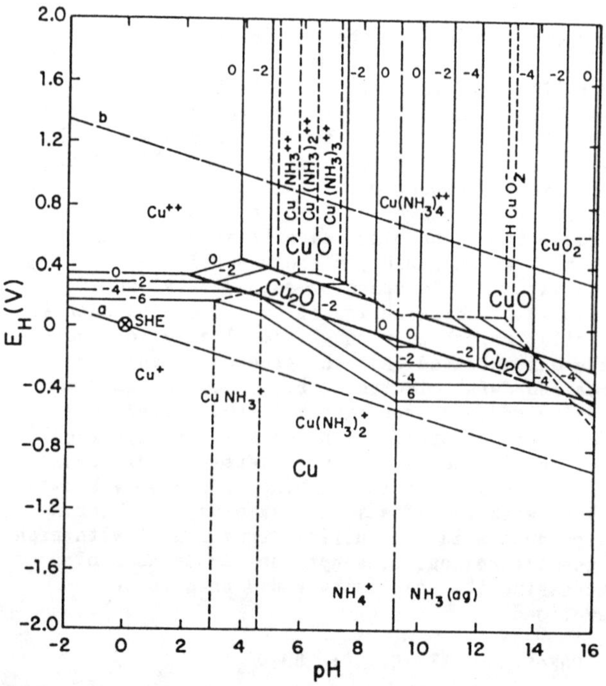

Figure 1. Potential - pH diagram for the copper-ammonia-water system at 25°C for unit activities of $NH_{3(aq)}$ and $NH_4{}^+$ (after Duby, 1977).

The Eh-pH diagram for the $Cu-NH_3-H_2O$ system (Figure 1) shows that there is only a narrow range of pH between pH 8-12 where $Cu(NH_3)_4{}^{2+}$ species are stable. Under more acidic or alkaline conditions CuO or $Cu(OH)_2$ precipitates, but in carbonate media at pH < 10, malachite or azurite will precipitate instead. Of course, the exact pH region over which $Cu(NH_3)_4^{2+}$ is stable is a function of total ammonia (NH_3 + NH_4^+), copper concentration (Luo, 1987), and temperature as shown by Figures 2 and 3 (Tozawa et al., 1976). Thus in a typical leach where the copper concentration in solution increases, and the free ammonia concentration decreases through complexation or volatility, it is possible to move out of the region of $Cu(NH_3)_4^{2+}$ stability and precipitate CuO. Figure 2 shows that a 1 M Cu(II) solution with a total NH_3:Cu molar ratio of 2:1, will precipitate CuO at pH 11, and that with < 0.5 M total NH_3 it is not possible to solubilise more than 0.1 M $Cu(NH_3)_4^{2+}$ at any pH. Figure 3 shows that as the temperature is raised, the region of stability of $Cu(NH_3)_4^{2+}$ shifts to lower pH and higher total NH_3.

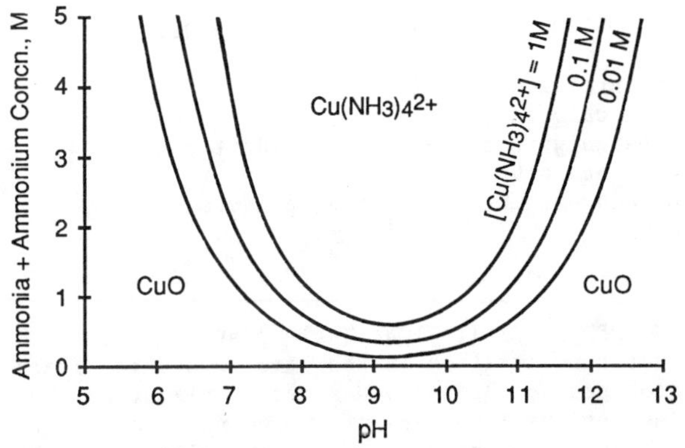

Figure 2. $Cu(NH_3)_4^{2+}$ stable region at 25°C (adapted from Tozawa, 1976).

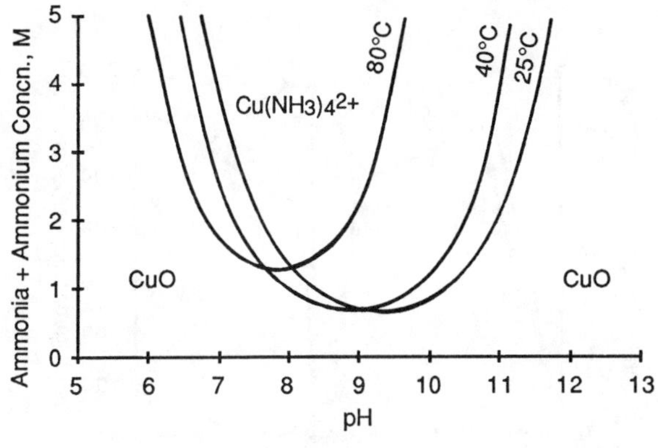

Figure 3. Change in $Cu(NH_3)_4^{2+}$ stable region with temperature (adapted from Tozawa, 1976).

Mechanism of Gold Leaching in $Cu-NH_3-CN^--H_2O$ System. From early testwork and more fundamental studies by Duke and Courtney (1952), it is now generally accepted that the role of ammonia is to lower the activity of Cu(II) and considerably retard the degradation of cyanide by Cu^{2+} (equation 2). The kinetics of the reaction between Cu(II) and CN^- is consistent with the decomposition of complexes of the type $[Cu(NH_3)_2(CN)_4]^{2-}$ to $Cu(CN)_3^{2-}$. As demonstrated by Putnam (1950), and later by Churchill and Laxen (1966) using the corresponding $Cu^{2+}-CN^--CO_3^{2-}-HCO_3^-$ system, a stabilised Cu(II) cyanide complex leaches gold at optimum rates when the molar ratio of CN^- to Cu^{2+} is around 2:1. This indicates that the reaction may proceed according to equation 7.

$$Cu(NH_3)_4(CN)_2 + Au \rightarrow Au(CN)_2^- + Cu(NH_3)_2^+ + 2NH_3 \quad ...(7)$$

The rate of leaching of gold with both the ammonia and carbonate-cyanide complex is slower than with free cyanide, but it is significant that leaching occurs even in the absence of air and is much faster at elevated temperatures (Churchill and Laxen, 1966). Putnam's work indicates that the complex is quite stable over a period of 30 hours in pure solution, but slow decomposition occurs at high pH (>12) with addition of excess lime, rendering the solution inactive in a few hours. Its stability in pulp media is not known. However there is no doubt that by leaching in the presence of air, Cu(II) will also catalyse the oxidation of CN^- to CO_3^{2-} (Muir et al., 1988) and lead to the precipitation of malachite. This oxidative degradation of CN^- may be more important in practical leach systems.

Leaching Tests with High-Copper Sulfide Ore: In order to gain some experience with the behaviour of the cyanide/ammonia leach system with sulfide rather than oxide ore, investigations were carried out on drill-core samples from the Chunderloo prospect in W. Australia. As shown in Table 1, the major mineral components of the material were pyrite and chalcocite (Cu_2S)/digenite (Cu_9S_5) with minor chalcopyrite ($CuFeS_2$) and limonite. The copper grade of the material was around 6% with gold head grades varying from 20-50 gt^{-1}.

Leach tests were generally run for 24 hr with 3:1 liquid:solid ratio on 250 gram batches of ore ground to 90% minus 63µ. Ammonia was added to the leach tests in the form of ammonium chloride.

Table 3 shows that in test A, provided the cyanide demand of 87 kg of sodium cyanide per tonne of ore was met, 88% gold and 52% copper extraction were obtained. Hence the gold was not refractory. The cyanide leach test B was preceded by a 24 hr preleach with 1 M sulfuric acid at ambient temperature. This did reduce the cyanide consumption to 55 kgt^{-1}, with a similar total copper extraction in the acid preleach and alkaline cyanide leaches. However gold extraction in test B was only 33%.

Table 3: Summary of Cyanide/Ammonia Leach Test Results with Chunderloo Prospect Sulfide Ore

	Reagent Additions, kg/t			$NH_3:CN^-$	% Au Extn	mg/l Cu in leach		pH Range
	NH_4Cl	NaCN	$Ca(OH)_2$	Ratio	at 24 hr	Max.	at 24 hr	
A	0	87	8	0	88	14000	14000	12.0-12.4
B	†0	55	15	0	33	6600	6600	11.9-12.5
C	80	34	45	2:1	60	970	430	9.4-10.8
D	40	16	42	2:1	55	1400	60	11.4-11.7
E	40	16	40	2:1	42	720	100	9.5-10.8
F	60	16	63	3:1	71	980	280	9.0-11.0
G	107	16	89	5:1	77	690	4	11.1-13.2
H	280	24	199	10:1	64	6500	6500	9.6-10.6
I	20	8	22	2:1	32	600	100	10.6-11.7
J	30	8	36	3:1	43	500	0	11.3-11.6
K	30	8	40	3:1	48	190	0	10.7-11.7
L	80	8	53	8:1	1	2300	1200	9.4-10.5
M	*43	12	40	3:1	72	520	140	10.1-11.5
N	#20	#8+4+4	20	#2:1-0.5:1	53	§200	60	9.6-10.8

Notes:

* staged cyanide addition - initial 8 kgt^{-1} then further 4 kg t^{-1} at 8 hr.

† 1 M H_2SO_4 preleach for 24 hr at ambient temperature.

\# Ammonia preleach for 24 hr with 60 kgt^{-1} ammonium chloride and 42 kgt^{-1} lime at pH 10.7. Leach with staged cyanide additions at 0, 6, and 8 hr.

§ copper concentration in preleach liquor 1200 mg/l.

While it is realised that the cyanide consumption is totally uneconomic, a possible 80% reduction in cyanide consumption using ammonia, as experienced at the Paris dump leach operation, may make the treatment of the Chunderloo prospect more attractive, especially if lower copper grades can be separated.

AMMONIA TO CYANIDE RATIO: The work of Jarman and Brereton (1905) has shown that cyanide consumption with oxidised copper containing ores can be reduced by use of the ammonia/cyanide system. In their work with 1.5-3% soluble copper in the ore, the optimum gold extractions and minimum copper extractions were obtained at an ammonia to cyanide ratio around 2:1. However there would appear to be some basis to the use of higher ammonia to cyanide ratios at higher copper contents (e.g. 4%).

Ammonia/cyanide leach tests reported in Table 3 investigated ammonia to cyanide ratios from 2:1 to 10:1. Gold extractions were not as good as with the straight cyanide leach in test A, with the best results around 70-77%. High ammonia to cyanide ratios (tests H & L) resulted in higher copper extractions,but lower gold extractions unless the cyanide addition was increased. Tests D to G with initial cyanide additions of 16 kg of sodium cyanide per tonne show that ammonia to cyanide ratios of 3:1 to 5:1 behave similarly. Tests I to L with lower cyanide additions of 8 kgt^{-1}, show insufficient cyanide was present in these cases, and consequently gold extraction suffered.

COPPER PRECIPITATION: Table 3 shows that the copper concentration in the ammonia/cyanide leach liquor passed through a maximum in almost all cases. At the lower ammonia to cyanide ratios it was possible to have no copper in solution at the end of a 24 hr leach. Possible mechanisms for this copper reaction have been discussed above. While the reaction was beneficial in limiting cyanide consumption, if it is allowed to go to completion the precipitation of all the gold in the solution was also observed, as shown in Figure 4 for the extended results from test J. In the initial phase of the leach a maximum copper concentration was

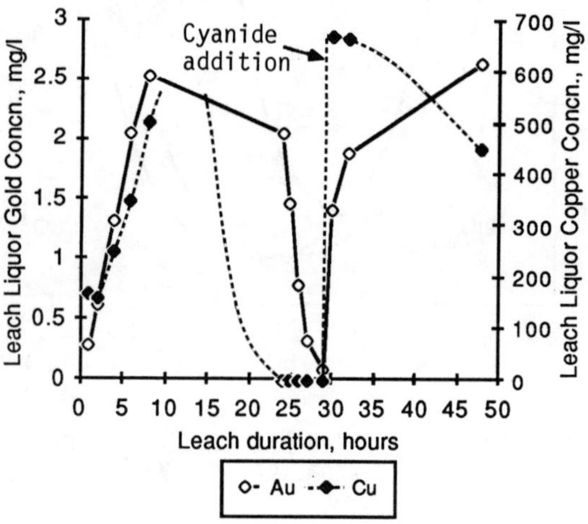

Figure 4.Gold and copper concentrations for test J

reached by 8 hr, with the copper concentration falling to zero by 24 hr. However the gold concentration in the leach liquor did not fall to zero until 29 hr. Addition of a further 4 kgt^{-1} sodium cyanide at this point produced an immediate jump in the copper concentration in the leach liquor to around 670 mg/l before copper levels again began to decline. On the other hand, the gold concentration jumped back to its original value and continued to increase to give 56% gold extraction after 48 hr for a cyanide addition of 12 kgt^{-1}. It would appear that either gold precipitates with copper as an AuCN.CuCN type compound, or is adsorbed as $Au(CN)_2^-$ onto copper oxide precipitates. Either way, gold and copper both readily redissolve upon addition of more CN^-, and further gold leaching is achieved.

STAGED CYANIDE ADDITION: A further means of reducing the cyanide requirement is staged cyanide addition as in test M. Gold and copper concentrations in the leach liquor are shown in Figure 5. It can be seen that the gold concentration in the leach liquor continues to increase until 24 hr, even though the copper concentration in solution has again declined to 140 mg/l. However, addition of a further 4 kgt^{-1} sodium cyanide at 24 hr had little effect on gold extraction, while increasing copper extraction. These results emphasise that the presence of free CN^- is necessary to prevent gold precipitation.

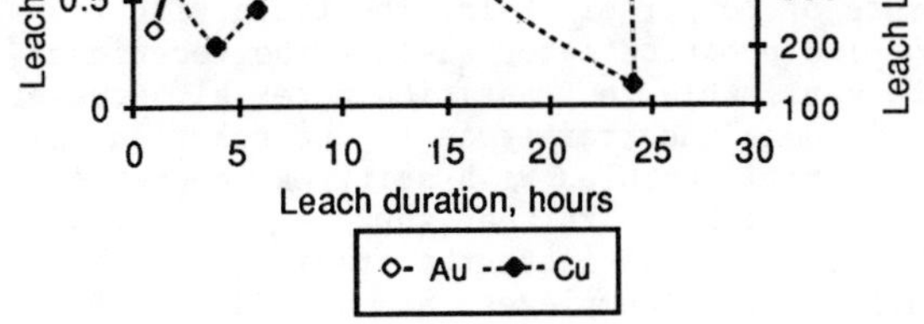

Figure 5.Gold and copper concentrations for test M

FREE CYANIDE CONCENTRATION: In conventional cyanide leaching the free cyanide concentration can be readily determined and more reagent added as required in order to maintain an acceptable rate of leaching. Unfortunately in the ammonia/cyanide system this is not always feasible, as the ammonia interferes with both the cyanide ion selective electrode and with silver precipitation, according to the NH_3 concentration in solution. For example, no end point can be detected by the cyanide electrode in a silver nitrate titration when 2% ammonia is present at pH 10.

In the presence of copper there is also the problem that copper(II) will oxidize free CN^-. The titration curves in Figure 6 illustrate the accepted fact that in the absence of ammonia, no appreciable free cyanide is present in solution until the $Cu^{2+}:CN^-$ ratio exceeds 1:3, due to the rapid and near complete oxidation of CN^- according to equation 2. However in the presence of ammonia (Figure 7) rather more free cyanide is present for the same $Cu^{2+}:CN^-$ ratio. As the $Cu^{2+}:NH_3$ ratio increases from 1:0 to 1:4 and 1:6 in the presence of 2 or 3 mole ratio CN^-, the amount of titratable free cyanide also increases. This supports the use of ammonia to decrease the cyanide demand of the ore, and the view that a stable copper(II) ammonia cyanide complex is formed.

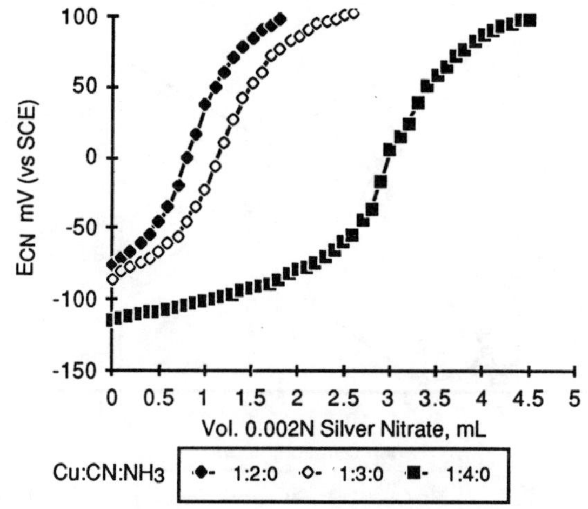

Figure 6. Effect of cyanide to copper ratio on the concentration of free cyanide measured by silver nitrate titration (in absence of ammonia).

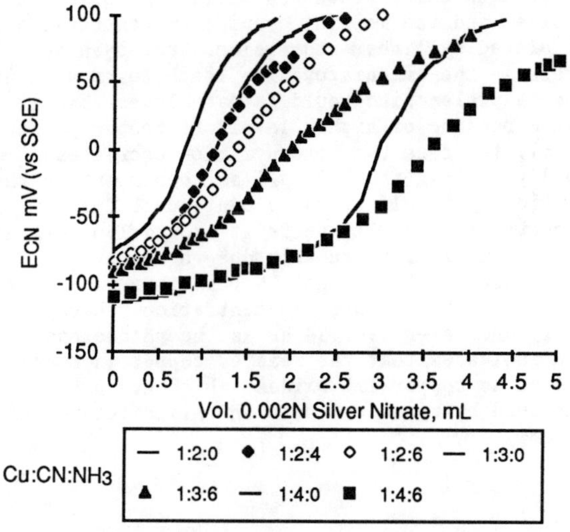

Figure 7. Effect of ammonia on the free cyanide concentration at various cyanide to copper ratios at pH 10.5-11.0.

An estimate of free cyanide levels in actual leach tests was obtained in test N, where the ore was initially pre-leached with 60 kgt^{-1} ammonium chloride and 42 kgt^{-1} of lime at pH 10.7, followed by staged cyanide addition. The ammonia to cyanide ratios varied from 2:1 to 0.5:1, as the initial 8 kgt^{-1} sodium cyanide addition was supplemented by further 4 kgt^{-1} sodium cyanide additions at 6 and 8 hr leach times. Gold extraction, as well as copper and free cyanide concentration profiles for this leach are shown in Figure 8. Free cyanide levels (expressed as NaCN) were estimated by cyanide ion selective electrode measurements on diluted samples of leach liquor. These were each

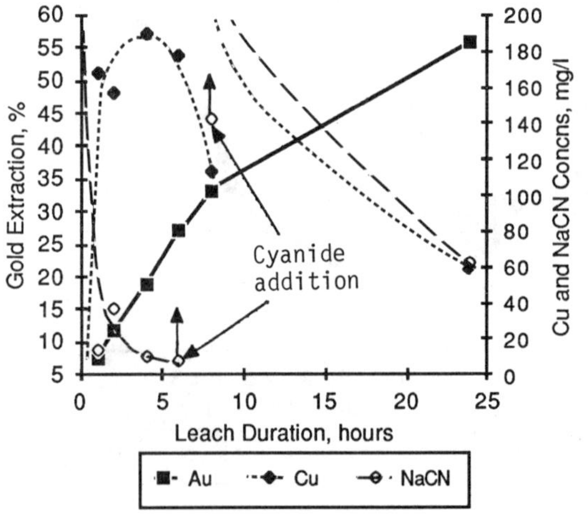

Figure 8. Gold extraction, copper and free cyanide concentration profiles for test N.

spiked with three known concentrations of sodium cyanide, and the initial cyanide concentrations calculated by linear regression. Free cyanide levels in the ammonia/cyanide leach tests without ammonia preleaching would be even lower than in test N because of higher levels of copper present. Clearly the free CN^- concentration decreases rapidly to less than 20 mg/l as copper enters the solution, yet gold leaching continues. Copper concentration quickly peaks at around 180 mg/l as CN^- is consumed, presumably due to oxide precipitation. Additional CN^-, in this case, has little effect on copper concentration, which slowly decreases over 24 hr as the pH decreases. This indicates that the reacted copper is mostly present as copper(II) oxides, which are not dissolved by cyanide itself in this dilute ammonia system.

CONCLUSIONS

The most economic route for treatment of copper sulfide containing gold ores is by flotation to produce an acceptable copper concentrate for gold and copper recovery by smelting. For transition ores or ores where conventional flotation does not work, controlled potential sulfidization pretreatment may be able to produce an acceptable flotation concentrate.

Direct cyanidation and selective gold adsorption on carbon is feasible for low copper ores (<0.5% Cu) and may be economic for higher grades of chalcopyrite containing ores, which are least reactive towards cyanide.

The economic application of sulfuric acid pretreatment for oxide copper removal is likely to be limited, because of acid consuming gangue minerals and neutralisation costs preceding cyanidation. Bacterial dump leaching of the sulfide copper is also likely to be uneconomic because of slow leaching and acid drainage, unless the copper can be recovered by solvent extraction and the acid recycled to the dump. Ammonia pretreatment for copper removal, while avoiding neutralisation costs, suffers from higher reagent costs for the ammonia, as well as low reactivity towards most copper sulfide minerals.

Selective leaching of gold by thiourea is limited to chalcopyrite concentrates because of high reactivity of the thiourea towards other copper minerals. There are also problems with thiourea stability in the presence of Cu(II) and with its adsorption on pulp minerals.

Bromine based leach systems seem to be limited to oxidised copper minerals because of the reactivity of the bromine with sulfides.

The ammonia/cyanide leach tests reported here confirm the observations of earlier workers that this system can reduce cyanide consumption in the presence of copper minerals. The tests with the Chunderloo prospect material show the technique may be applicable to transition material. However there is some uncertainty as to the reaction of the copper(II) complex with sulfides, competing with gold leaching. Further work is required on oxide ores to investigate the stability of the $Cu(NH_3)_4CN_2$ type complexes, and on the reactivity of the copper(II) complex towards simple sulfide minerals.

ACKNOWLEDGMENTS

The authors wish to thank the Australian Mineral Industries Research Association for a grant to allow testwork to be carried out and for permission to publish results from ammonia/cyanide leach tests on Chunderloo material. Also Endeavour Resources for providing the test sample and the Institute of Chemical Metallurgy, Beijing for granting leave to Cao Changlin to undertake research training in the Mineral Chemistry Research Group at Murdoch University.

REFERENCES

Anon. (1988) "Cutech cyanide process recovers both gold and copper". E & M.J. (February) p 45. See also Mining Mag. (January) pp 60-61.

Avraamides, J. (1982) "Prospects for alternative leaching systems for gold: a review" Proc. Conf. on C.I. P. Technology for the Extraction of Gold (Aus.I.M.M., Melbourne) pp 369-391.

Avraamides, J. (1989) W. Australian Government Chemical Laboratory, unpublished results.

Belgian Patent (1977) 847,441.

Bilston, D.W., La Brooy, S.R. and Woodcock, J.T., (1983) "Use of thiourea leaching for gold and silver recovery" Status report on AMIRA Project 80/P127. Australian Mineral Industries Research Association: Melbourne.

Byerly, J.J., Fouda, S. and Rempel, G.L. (1973) "Kinetics and mechanism of the oxidation of thiosulfate ions by copper(II) ions in aqueous ammonia solution" J. Chem. Soc., Dalton Trans. pp 889-893.

Chen, J. (1983) "Hydrometallurgy in China" in "Hydrometallurgy: Research, Development, and Plant Practice" (Eds K. Osseo-Asare and J.D. Miller) A.I.M.E. Warrendale pp 65-86.

Churchill, M. and Laxen, P.A. (1966) "The dissolution of gold in cyanide solutions containing copper in a carbonate/bicarbonate medium, preliminary results" N.I.M. Report No. 2, Mintek, Johyannesburg.

Deng, T., and Chen, J. (1986) Personal Communication.

Deng, T. Ke J., and Chen, J. (1984) "Kinetics of disproportionation of elemental sulfur in aqueous solutions" J. Chem. Industry and Engineering (China) 4 pp 328-334.

Deschênes, G. and Ghali, E. (1988) "Leaching of gold from a chalcopyrite concentrate by thiourea" Hydrometallurgy 20 (2) pp 179-202.

Duby, P. (1977) "Thermodynamic properties of aqueous inorganic copper systems" INCRA Monograph IV on the Metallurgy of Copper. International Copper Research Assoc. Inc., New York.

Duke, F.R. and Courtney, W.G. (1952) "Complexes in oxidation-reduction reactions. The copper(II)-cyanide reaction" J. Phys. Chem. 56 pp 19-21.

Dutrizac, J.E. and MacDonald, R.J.C. (1974) "Ferric ion as a leaching medium" Minerals Sci. Engng. 6 (2) pp 59-100.

Gabra, G., (1984) "Leaching of gold from pyrite and chalcopyrite concentrate using acidified thiourea", 23rd Annual Conference Meeting, C.I.M.M., Quèbec, Canada.

Gasparrini, C. (1983) "The mineralogy of gold and its significance in metal extraction" C.I.M. Bulletin (March) 76 (851) pp 144-153.

Halferdahl, A.C. (1929) "The problem of treating cupriferrous precious-metal ores by the cyanide process" Eng. Min. J. 128 pp 350-357.

Hedley, N. and Kentro, D.M. (1945) "Copper cyanogen complexes in cyanidation" Trans. Can. Inst. Min. Metall. 48 pp 237-251.

Hunt, B. (1901) U.S. Patent No. 689,190 (Dec 17).

Ibrado, A.S. and Fuerstenau, D.W. (1988) "The adsorption of silver cyanide on activated carbon". Paper presented at A.I.M.E. Annual Meeting, Phoenix. S.M.E. Preprint No. 88-176.

Jarman, A. and Brereton, E. Le Gay (1905) "Laboratory experiments on the use of ammonia and its compounds in cyaniding cupriferous ores and tailings" Trans. Inst. Min. and Metall. XIV pp 289-334.

Jones, M.H. and Woodcock, J.T. (1978 (a)) "Optimization and control of laboratory sulphidization of oxidised copper ores with an ion selective electrode" Proc. Australas. Inst. Min. Metall. 266 pp 11-19.

Jones, M.H. and Woodcock J.T. (1978 (b)) "Evaluation of ion-selective electrode for control of sodium sulphide additions during laboratory flotation of oxidised ores" Trans. Inst. Min. Metall. (Section C) 87 pp C99-105.

Kitney, M.J. (1988) "Cyanide regeneration from gold tailings; Golconda's Beaconsfield experience" Proc. Randol Perth Internat. Gold Conf. (Randol Internat. Golden, Co.) pp 89-95.

Koch, W.E. and Pietsch, H.B. (1974) "Treatment of oxide copper ores by reduction and ammoniacal leaching" Trans. Inst. Min. Metall. (Sec. C) pp C14-17.

Krzewska, S. and Podsiadly, H. (1980) "Studies on the reaction of copper(II) with thiourea - I" J. Inorg. Nucl. Chem. 42 pp 83-86.

Leaver, E.S. and Woolf, J.A. (1931) "Copper and zinc in cyanidation, sulfide-acid precipitation" U.S. Bureau Mines Tech. Paper No. 494.

Lower, G.W. (1965) "Electrolytic recovery of copper from copper cyanide leaching solutions" U.S. Patent 3,463,710.

Lu Z-Y and Grimsey, E.J. (1983) "Copper separation from iron and nickel-iron sulfide concentrates by a double-roast acetonitrile leaching process" Proc. Australas. Inst. Min. Metall. 288 (December) pp 19-27.

Lumley, B (1988) Senior Project Metallurgist, Telfer, Personal Communication.

Luo, R. (1987) "Overall equilibrium diagrams for hydrometallurgical systems: copper-ammonia-water system" Hydrometallurgy 17 (2) pp 177-200.

Marcuson, S.W. (1980) "Application of the segregation process to roasted copper concentrates" Mineral Sci. Engng 12 pp 21-36.

McNamara, V.M. (1985) "CANMET Acidification/Volatilization/Reneutralization Process for Cyanide Recovery and Pollution Control" CANMET (Ottawa) Report No. MRP/MSL 85-44 (IR).

Muir, D.M., Aziz, M. and Hoecker, W. (1988) "Cyanide losses under C.I.P. conditions and effect of carbon on cyanide oxidation" Proc. 1st Int. Conf. on Hydrometallurgy (I.C.H.M. '88) Beijing (Eds Zheng Y., and Xu., J.) Int. Academic Publishers, pp 461-465.

Parker, A.J., Muir, D.M., Smart, Y.C. and Avraamides, J. (1981) "An application of aceto-nitrile leaching and disporportionation: Refining segregated copper from roasted concentrates and ores" Hydrometallurgy 7 pp 213-233.

Putnam, G.L. (1950) "Copper-ammonia process for cyanidation of complex ores" Chem. Engng and Min. Review (June 10) pp 347-348.

Rose, D.H., Lessels, V., and Buckwalter, D.J. (1967) "Pilot plant testing of cyanide leaching of copper from White Pines tailings" Trans. A.I.M.E. 238 pp 221-224. See also Mining Engng. (August) pp 60-63.

Ruane, M. (1988) Glencoe Resources Pty Ltd. Personal communication.

Ryan, M. (1988) "Operation of the Red Dome Heap Leach" Proc. Conf. on Economics and Practice of Heap Leaching in Gold Mining, Cairns (Aus.I.M.M., Melbourne) pp 87-93.

Schulze, R.G., (1984) "New aspects in thiourea leaching of precious metals" J. Metals pp 62-65.

Schwartz, G.M. (1944) "The host minerals of native gold" Econ. Geol. 39 pp 371-411.

Sehic, O.A. (1988) "An update on the K-Process" Proc. Randol Perth Internat. Gold Conf. (Randol Internat. Golden, Co) pp 184-187.

Sehic, O.A. (1989) Personal communication.

Shantz, R. and Reich, J. (1978) "A review of copper-cyanide metallurgy" Hydrometallurgy 3 pp 99-109.

Sorensen, P.F. (1988) "Gold recovery from C.I.P. eluates by precipitation with mineral acid. Parts I-III" Hydrometallurgy 21 (2) pp 235-254.

Stamboliadis, E., McHardy, J. and Salman, T (1976) "The effect of copper ions on the cyanidation of a gold ore" C.I.M. Bulletin (March) 69 pp 128-130.

Tozawa K., Inui, Y. and Umetsu, Y (1981) "Dissolution of gold in ammoniacal thiosulfate solution" Paper presented to A.I.M.E. Annual Conference, T.M.S. Paper Selection, A81-25.

Tozawa, K. Umetsu, Y. and Sato, K. (1976) "On chemistry of ammonia leaching of copper concentrates" in "Extractive Metallurgy of Copper" (Eds J.C. Yannopoulos and J.C. Agarwal) A.I.M.E., New York, Chapter 36.

Tsuchida, N., Ruane, M., and Muir, D.M. (1984) "Studies on the mechanism of gold adsorption on carbon". Proc. Int. Conf. on Min. Sci. and Technol. "Mintek 50" (Ed. L.F. Haughton) Council for Mineral Technology, Johannesburg pp 647-656.

Warren, I.H. and Mounsey, D.M. (1983) "Factors influencing the selective leaching of molybdenum with sodium hypochlorite from copper/molybdenum sulphide minerals" Hydrometallurgy 10 pp 343-357.

Wright, J.K. (1973) "The Segregation Process" Minerals Sci. Engng 5 (2) pp 119-134.

Zipperian, D., Raghavan, S. and Wilson, J.P. (1988) "Gold and silver extraction by ammoniacal thiosulfate leaching from a rhyolite ore: Hydrometallurgy 19 (3) pp 361-375.

Chapter 45

CARBAD GOLD RECOVERY

MARK CADZOW AND RODNEY LAMB

CARBAD Pty Ltd
Sydney, Australia

ABSTRACT

CARBAD is a recently developed Australian process for the recovery of gold and platinum group metals (PGM). The process relies on the selective agglomeration phenomena of the gold and PGM to capture the mineral in a carbon-hydrocarbon media. The precious minerals are then extracted from the loaded media by a solvent gold room process producing a relatively pure concentrate. The media is then reformed and recycled. The process development from batch test work to a 5 t/h commercial scale CARBAD plant is discussed. The operation and results of processing a variety of ore types at several alluvial and hardrock gold mines is examined. The operating and capital costs for the CARBAD process shows that it is significantly less than that the conventional CIP/CIL.

The environmental safety of the process is a major feature where precious metal deposits are located within urban environments or within wilderness areas.

INTRODUCTION

The CARBAD process was conceived to exploit the natural oleophilicity/hydrophobicity of gold, and new innovative carbon-hydrocarbon pelletizing techniques, to produce a novel method for gold benefication.

Oil pelletization or agglomeration of carbonaceous materials such as coal has been demonstrated on various scales since the early 1900's, and has proved to be an effective process for recovery of fine oleophilic particles (Mehrotra, 1983). Coal - oil agglomeration has been shown to be extremely selective, rejecting hydrophilic mineral matter and producing, low ash, high yield coal products (Swanson, 1977).

The oleophilic nature of gold has been exploited in the past by such techniques as the grease pan and grease table. The CARBAD process uses this principle in a high shear environment which maximises the probability of contact and gold recovery.

PROCESS DESCRIPTION

The CARBAD process relies on preferential surface wetting of oleophilic/hydrophobic particles (gold and PGM) by a hydrocarbon (media oil) to capture these particles into a carbon-hydrocarbon pellet (CARBAD media). The gold (for example) is wetted and enveloped by the media oil and becomes an integral part of the media structure, as indicated in figure 1.

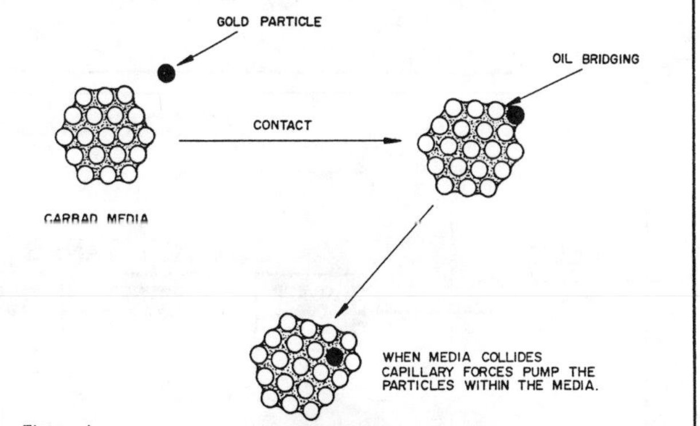

Figure 1. CARBAD MECHANISM

Other oleophilic minerals may be recovered, but high selectivity can be achieved by conventional reagent control if necessary. As with all surface wetting recovery techniques, some degree of liberation is required for gold recovery to occur. Complete liberation however is not essential, as recovered gold particles have shown to be up to 80% gangue, refer to figure 2.

The CARBAD process flowsheet (figure 3) describes gold bearing ore being contacted with media in a series of three continuous high shear contact tanks. The media is contained within each tank by screens, until it is sufficiently loaded with gold. The degree of loading is theoretically limitless, as new oil film sites are created as oleophilic particles are attached, but economic limitations restrict a normal loading to between 10 to 20 kg/tonne of media. A small amount of media oil is continuously fed into each tank to maintain the hydrocarbon balance in the media.

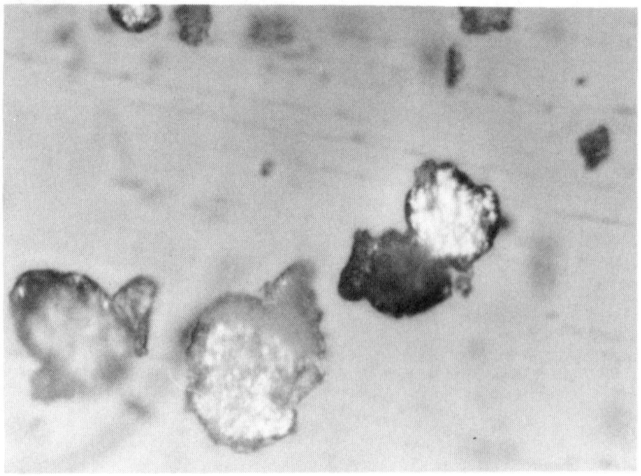

Figure 2. GOLD PARTICLES RECOVERED BY CARBAD

Excess and leached oil is recovered from the tailings and returned to the feed. Under normal operation the media remains discrete pellets neither growing nor disintegrating, with an equilibrium size of the media of approximately 3 mm diameter. Figure 4 shows the media after 3 days of operation in the 5 t/h CARBAD plant.

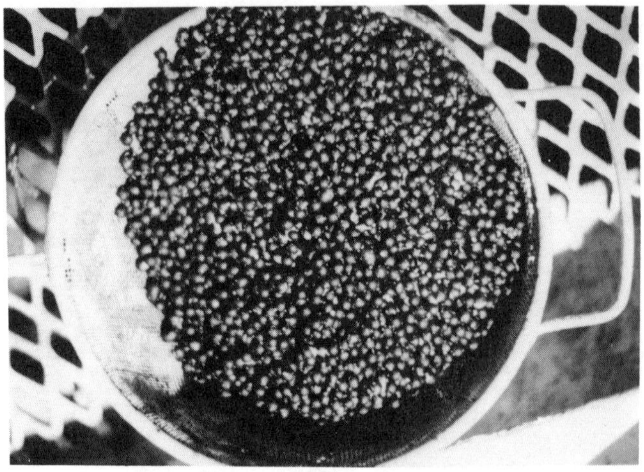

Figure 4. MEDIA AFTER 72 HOURS OPERATION

Loaded media is recovered, by isolating the particular tank and pumping the contents over a screen. The media is transferred to an agitated tank containing dense solvent (intermediate to the density of the carbon and gold) resulting in the dissolution of the oil bridges, and disintegration of the CARBAD media to discrete particles.

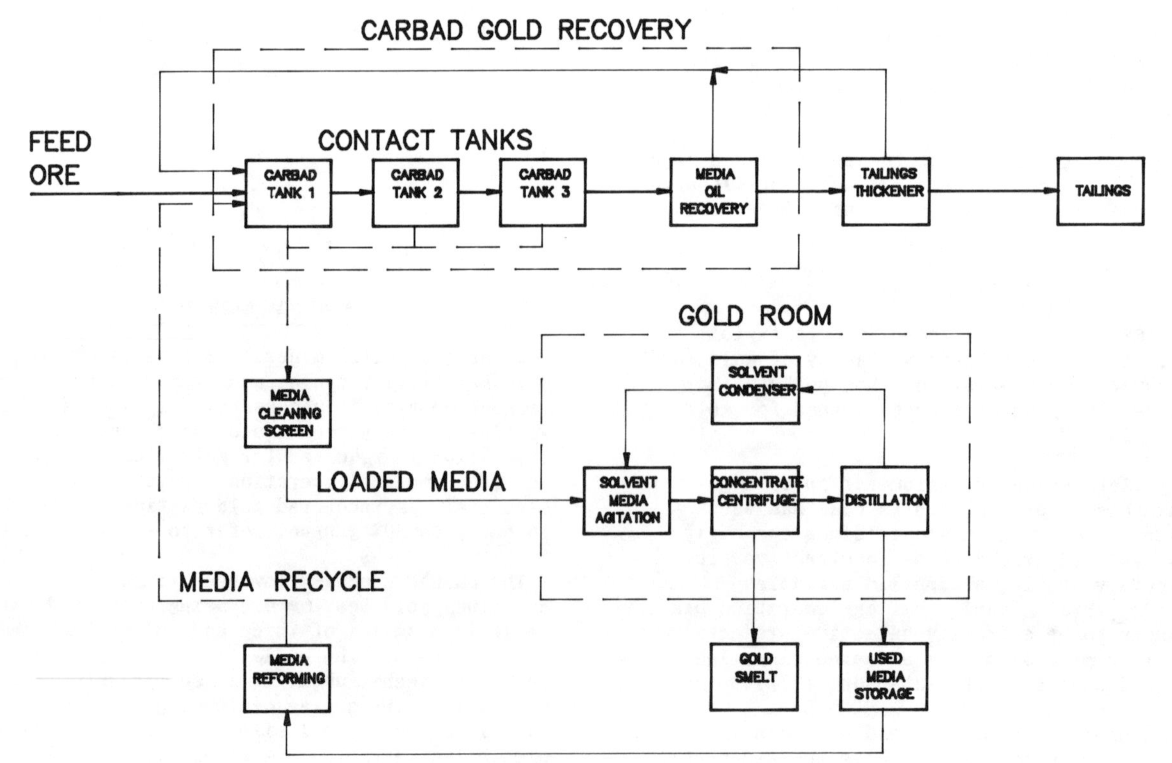

Figure 3.CARBAD PROCESS FLOWSHEET

The slurry is pumped through a centrifuge and the resulting concentrate, may be direct smelted or may require specific flux smelting depending on the quantity and type of oleophilic impurities recovered. The solvent is distilled for re-use and the media is reformed to provide a continuous recycle of all materials.

An alternative to the solvent gold room and recycle is burning of the media.

This is recommended for small scale operations where the economics of the gold room are not justified.

CARBAD APPLICATIONS

The CARBAD process relies on at least partial gold liberation for effective gold recovery. Based on this criteria, it is applicable to both hard rock and alluvial ores, as a primary recovery method, for gravity and flotation concentrate upgrade, and for tailings retreatment. Where gold surfaces are liberated (such as free-milling ores), the gold recoveries experienced have been high.

The CARBAD process has the ability to recover particles from sub-micron in size, up to 400 um, making the process useful for ores with widely varying gold particle size.

A list of some typical gold recoveries achieved on hard rock and alluvial ores is provided in table 1. Typically in a hard rock application the gold recovery is primary influenced by fineness of grind (liberation). A grind size verses gold recovery graph is shown in figure 5. This is for a copper enriched supergene ore, in North Queensland and shows the gold recovery increasing as a function of liberation. In alluvial deposits CARBAD can recover gold below the lower limit of gravity processes, less than 50 um.

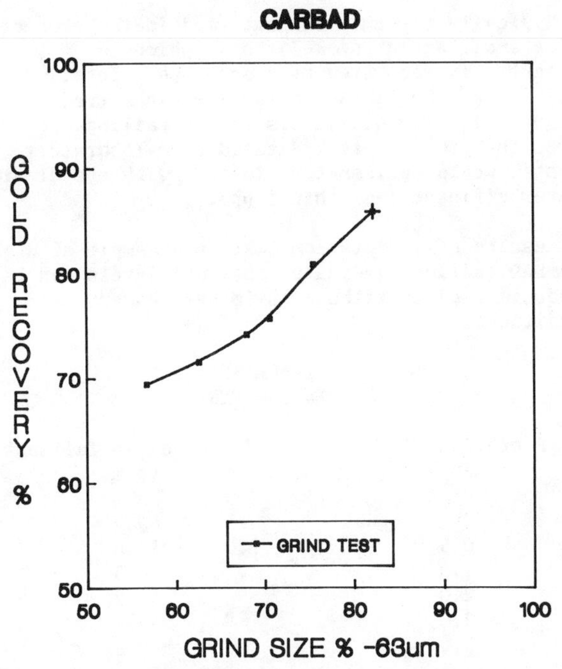

Figure 5 Grind Size Vs Gold Recovery, North Queensland Supergene Ore

Table No. 1
CARBAD RECOVERIES ON VARIOUS ORE TYPES

Ore	% -75 microns	Head Grade (g/t)	% Gold Recovery
1. Hard Rock Ores	80	18.5	99.2
Arsenopyrite-pyrite	80	6.11	93.3
-Quartz	90	1.23	94.3
Limonite/pyrite	70	3.43	94.8
- Quartz	85	2.4	95.5
Pyrite - Quartz	70	1.48	79.7
Chalcopyrite/pyrite	94	14.1	88.6
- Quartz	92	12.1	84.6
Pyrite/minor chalco	67	2.1	93.8
- Quartz	73	2.1	95.7
Sulphides/Quartz	65	2.28	92.8
	75	2.72	95.6
2. Alluvial Ores	(-250 m)	0.43	91.8
Alluvial Clay			
Alluvial Wash (Qld)	(-250 m)	0.39	98.8
		3.4	99.2
Alluvial (N.T.)		1.67	98.8
Beach Sand (N.Z.)		0.24	98.0

	% -75 microns	Head Grade Pt	Pd	% Recovery Pt	Pd
3. PGM Ores					
Beach Sand (N.Z.)		6.04	0.09	97.6	94.4
		45.4	0.43	99.8	99.8
Sulphide Ore	70	0.16	0.55	45.6	68.1

Gold ores which are difficult for cyanide processes because of high copper or arsenic contents are often amenable to the CARBAD process. For ores which contain hydrophobic sulphides, pH modification and depressants may be required. The CARBAD process has proved to be extremely efficient on a range of ores, however where the gold is extremely fine (<10 um) or in solid solution, recoveries have suffered.

Figure 6. 5-10 t/h CARBAD plant (lower right) showing the size comparison to a 10 t/h CIP installation.

PROCESS KINETICS

Gold recovery is extremely rapid in the CARBAD process, due to the intense agitation provided in the contact tanks. An indication of the kinetics is given in table 2, which shows % recovery of the total recoverable gold at various times for three different tank capacities.

Residence Time (mins)	% of total Gold Recovery 1 l.	10 l.	1000 l. (1m³)
1	84	78	
2	89	83	89
5	94	92	95
10	96	94	97
20	100	100	100

This table also indicates that the rapid kinetics at the laboratory scale can be maintained at the semi-commercial stage (1 m^3 tanks). The kinetics are also independent of head grade.

Rapid kinetics enables very small processing plants to handle large tonnages. Figure 5 shows the comparison between a 5 t/h CARBAD plant in the foreground and a 10 t/h CIP plant.

The 5 t/h CARBAD plant (3 x 1 m^3 tanks @ 40% solids) is based on a residence time of 20 minutes. Due to the rapid kinetics the capacity of this plant could be doubled, by reducing the residence time to 10 minutes, for only a small loss in gold recovery.

Economics will dictate the recovery/throughput relationship, but it is possible to operate a fixed size plant at a wide variety of capacities without significantly affecting gold recovery.

CARBAD DEVELOPMENT

The simple processing parameters and uncomplicated flowsheet enabled the process to be developed rapidly from the one litre laboratory batch cell to a 5 t/h semi-commercial plant. This plant has been operated for over 2000 hours at both alluvial and hard rock applications.

The simplicity of the process enabled it to be operated by inexperienced operators, providing 24 hour operation during this development.

The alluvial ores treated fluctuated in grade between 0.56 g/t and 0.105 g/t, but recoveries of up to 90% were recorded, which was marginally below the recovery predicted by the laboratory tests (99.3% on a head grade of 0.68 g/t).

Some plant modifications, enabled good correlations between the laboratory and the semi-commercial plant. On a hard rock ore laboratory batch testing yielded a gold recovery of 86%, and for the same sample the CARBAD plant recorded a recovery of 85%.

After establishing recoveries and scale up parameters at the semi-commercial scale, a 25 kg/hr transportable pilot plant was constructed, to enable minesite feasibility studies and to evaluate difficult ores.

CARBAD TAILINGS EFFLUENT, ENVIRONMENT

The CARBAD process has a significant advantage over most other processing techniques in that no toxic chemicals are used, making the process evironmentally benign.

Typically the only reagent used in the process is a small amount of media oil, which is essentially recovered by a scavenger flotation cell after the CARBAD contacting tanks (refer figure 3). The oil levels in the tailings emitted from the process, as indicated from laboratory tests, would be less than 10 ppm, with the cleared water effluent less than 1 ppm.

Results of a flotation test on a sample of CARBAD tailings indicates that oil levels can be reduced rapidly without any other reagent additions.

TABLE 3
FLOTATION TEST

Flotation Time (mins)	Oil Level in Tailings (ppm)
0	65.4
0.5	19.6
1.0	16.3
2.0	13.2
3.0	10.7
5.0	8.8
Final Tails (liquid)	<1.0

CAPITAL AND OPERATING COSTS

As a result of the extremely rapid kinetics, simple processing mechanism and gold room which recycles the media and solvent, the capital and operating costs are significantly lower than for CIP/CIL processes. Typical budget estimates of plant capital costs are:

1 t/h concentrate upgrade plant without solvent gold room	$US100,000
10 t/h gold recovery plant and gold room	$US300,000
50 t/h gold recovery plant and gold room	$US550,000

For higher capacities a maximum module size of 50 t/h would be recommended.

Operating costs will vary significantly with ore type and locations, as power costs and reagent usage will fluctuate. Estimates based on treatment of a simple free milling quartz reef ore, which requires no pH modification or depressants are provided in Table 4. Labour costs have not been included due to the large variation in $/tonne based on rates and plant throughput. The man power would be the order of that required to operate a CIP/CIL plant.

Table No. 4
OPERATING COST ESTIMATES

	$US/tonne
Reagent Costs	0.42
Media Consumption	0.11
Power @ 4 c/kW/h	0.25
	0.78

CONCLUSION

The CARBAD process has proven to be a significant development in gold and PGM process technology offering distinct advantages over current processes.

CARBAD'S rapid kinetics and high efficiency together with simplicity of design enables the construction of low cost, high throughput plant, with low reagent and operating costs.

The process has been demonstrated to be easily scaled up and operated with a minimum of experience.

The CARBAD process has been proven to recover gold from a wide range of hard rock and alluvial ore types, either as a primary process or as a concentrate upgrade stage.

REFERENCES

Mehrotra, V.P., et al., 1983, "Agglomeration Techniques for Processing of Fine Coals", Int. J. Min. Pro, 11, p 175.

Swanson, A.R. et al., 1977, "Some Fundamental Aspects of the Selective Agglomeration of Fine Coal" Proc. 2nd Ind. Symposium on Agglomeration, Atlanta p 939.

Index